THE INTEGRATED RADIOGRAPHY WORKBOOK

First Edition

ROBERT DeANGELIS
& MARY FRAN LLOYD

DEDICATION
TO ALL OF OUR STUDENTS PAST, PRESENT AND FUTURE

ABOUT THE AUTHORS

Robert DeAngelis is currently the program Director of Rutland Medical Center's School of Radiologic Technology in Rutland Vermont.

His previous publications include the Radiologic Science Workbook by Praeger Scientific, The Radiography Workbooks I & II and Simulated Registry Exam Series by the Health and Allied Science Publishers.

Mr. DeAngelis received an AAS degree from Westchester Community College, Valhalla, New York, and a BS degree from Downstate Medical Center, Brooklyn, New York.

Mary Fran Lloyd received her diploma from Rutland Regional Medical Center's School of Radiologic Technology in 1965 and has served as the clinical instructor for the same school for the past four years.

INTRODUCTION

The Integrated Radiography workbook is a comprehensive text containing thousands of questions and explanations to assist radiographic students in preparing for their academic course and registry examination.

The text was developed from the fusion of what would have been the fifth editions of the Radiography Workbooks I & II.

The Extensive modifications to the content and format will again make this the only text available which accurately conforms to the most recent content specification of the registry exam.

The Section I, II, III, V, VI, VII
The Section & artwork compliments of Paula Hughes.
Section 4 compliments of Nuclear associates.

ISBN • 0-943589-19-3

Revised 2000

First Edition ©1996

INDEX

Section I:

Basic Mathematical Principles

Section II

Radiation Protection and Biology

Section III

Equipment Maintenance And Operations

Section IV

Image Production and Evaluation

Section V

Anatomy & Physiology

Section VI

Radiographic Procedures

Section VII

Patient Care and Management

Section I
Basic Mathematical Principles

Millisecond Conversion

Exercise No. 1

1.-20. State the decimal equivalent in seconds for the following millisecond values.

Example: 100 ms = .10 seconds

Solution: The decimal is moved three places to the left.

1. 800 ms = ______ s
2. 130 ms = ______ s
3. 80 ms = ______ s
4. 16 ms = ______ s
5. 5 ms = ______ s
6. 750 ms = ______ s
7. 200 ms = ______ s
8. 88 ms = ______ s
9. 24 ms = ______ s
10. 3 ms = ______ s
11. 540 ms = ______ s
12. 125 ms = ______ s
13. 18 ms = ______ s
14. 30 ms = ______ s
15. 7 ms = ______ s
16. 350 ms = ______ s
17. 175 ms = ______ s
18. 35 ms = ______ s
19. 45 ms = ______ s
20. 1 ms = ______ s

21.-40. State the millisecond equivalent for the following decimal values.

Example: .020 seconds = 20 ms

Solution: The decimal is moved three places to the right.

21. .07 s = ______ ms
22. .001 s = ______ ms
23. .80 s = ______ ms
24. .03 s = ______ ms
25. .004 s = ______ ms
26. .65 s = ______ ms
27. .31 s = ______ ms
28. .095 s = ______ ms
29. .02 s = ______ ms
30. .06 s = ______ ms
31. .005 s = ______ ms
32. .09 s = ______ ms
33. .90 s = ______ ms
34. .17 s = ______ ms
35. .5 s = ______ ms
36. .61 s = ______ ms
37. .16 s = ______ ms
38. .345 s = ______ ms
39. .040 s = ______ ms
40. .021 s = ______ ms

41.-50. State the decimal equivalent for the following fractional values.

Example: 1/2 = .5 seconds

Solution: 1 ÷ 2

41. $\frac{1}{10}$ second = ______ seconds
42. $\frac{1}{20}$ second = ______ seconds
43. $\frac{1}{40}$ second = ______ seconds
44. $\frac{1}{60}$ second = ______ seconds
45. $\frac{1}{15}$ second = ______ seconds
46. $\frac{1}{120}$ second = ______ seconds
47. $\frac{1}{240}$ second = ______ seconds
48. $\frac{1}{480}$ second = ______ seconds
49. $\frac{3}{10}$ second = ______ seconds
50. $\frac{3}{4}$ second = ______ seconds

Millisecond Conversion

Exercise No. 2

1.-20. State the decimal equivalent in seconds for the following millisecond values.

Example: 300 ms = .30 seconds

Solution: Decimal place moved three places to the left.

1. 600 ms = ______ s
2. 230 ms = ______ s
3. 180 ms = ______ s
4. 36 ms = ______ s
5. 7 ms = ______ s
6. 950 ms = ______ s
7. 150 ms = ______ s
8. 84 ms = ______ s
9. 26 ms = ______ s
10. 3 ms = ______ s
11. 340 ms = ______ s
12. 225 ms = ______ s
13. 28 ms = ______ s
14. 40 ms = ______ s
15. 8 ms = ______ s
16. 130 ms = ______ s
17. 275 ms = ______ s
18. 75 ms = ______ s
19. 35 ms = ______ s
20. 5 ms = ______ s

21.-40. State the millisecond equivalent for the followind deicmal values.

Example: .400 seconds = 400ms

Solution: Decimal place moved three places to the right.

21. .09 s = ______ ms
22. .003 s = ______ ms
23. .50 s = ______ ms
24. .08 s = ______ ms
25. .007 s = ______ ms
26. .25 s = ______ ms
27. .53 s = ______ ms
28. .075 s = ______ ms
29. .04 s = ______ ms
30. .02 s = ______ ms
31. .001 s = ______ ms
32. .06 s = ______ ms
33. .10 s = ______ ms
34. .27 s = ______ ms
35. .6 s = ______ ms
36. .11 s = ______ ms
37. .12 s = ______ ms
38. .315 s = ______ ms
39. .048 s = ______ ms
40. .031 s = ______ ms

41.-50. State the decimal equivalent for the following fractional values.

Example: $\frac{1}{5}$ = .2 seconds

Solution: 1 ÷ 5

41. $\frac{1}{30}$ second = ______ seconds
42. $\frac{1}{50}$ second = ______ seconds
43. $\frac{1}{2}$ second = ______ seconds
44. $\frac{3}{10}$ second = ______ seconds
45. $\frac{2}{5}$ second = ______ seconds
46. $\frac{3}{50}$ second = ______ seconds
47. $\frac{1}{250}$ second = ______ seconds
48. $\frac{1}{4}$ second = ______ seconds
49. $\frac{3}{8}$ second = ______ seconds
50. $\frac{2}{3}$ second = ______ seconds

Decimal Multiplication

Exercise No. 1

1.-20. In the following, state the decimal that is two times the stated value.

Example: .05 x 2 = .10

1. .07 ______
2. .15 ______
3. .012 ______
4. .004 ______
5. .05 ______
6. .25 ______
7. .75 ______
8. .009 ______
9. .60 ______
10. .032 ______
11. .125 ______
12. .005 ______
13. .03 ______
14. .23 ______
15. .08 ______
16. .33 ______
17. .45 ______
18. .062 ______
19. .035 ______
20. .002 ______

21.-40. In the following, state the decimal that is four times the stated value.

Example: .05 x 4 = .20

21. .06 ______
22. .15 ______
23. .013 ______
24. .002 ______
25. .035 ______
26. .33 ______
27. .125 ______
28. .005 ______
29. .40 ______
30. .022 ______
31. .140 ______
32. .001 ______
33. .075 ______
34. .62 ______
35. .85 ______
36. .092 ______
37. .008 ______
38. .104 ______
39. .017 ______
40. .46 ______

41.-50. In the following, state the squared value for the stated value.

Example: $3^2 = 9$ (3 x 3 = 9)

41. 2 = ______
42. 36 = ______
43. 150 = ______
44. 72 = ______
45. 180 = ______
46. 40 = ______
47. 30 = ______
48. 200 = ______
49. 6 = ______
50. 100 = ______

Exercise No. 2

1.-20. In the following, state the decimal that is two times the stated value.

Example: .05 x 2 = .10

1. .09 ______
2. .16 ______
3. .012 ______
4. .007 ______
5. .07 ______
6. .35 ______
7. .85 ______
8. .006 ______
9. .80 ______
10. .052 ______
11. .125 ______
12. .003 ______
13. .06 ______
14. .44 ______
15. .038 ______
16. .66 ______
17. .48 ______
18. .032 ______
19. .075 ______
20. .002 ______

21.-40. In the following, state the decimal that is four times the stated value.

Example: .05 x 4 = .20

21. .04 ______
22. .10 ______
23. .023 ______
24. .004 ______
25. .025 ______
26. .16 ______
27. .125 ______
28. .003 ______
29. .80 ______
30. .033 ______
31. .240 ______
32. .007 ______
33. .095 ______
34. .82 ______
35. .65 ______
36. .072 ______
37. .006 ______
38. .204 ______
39. .037 ______
40. .03 ______

41.-50. In the following, state the squared value for the stated value.

Example: $4^2 = 16$ (4 x 4 = 16)

41. 4 = ______
42. 32 = ______
43. 140 = ______
44. 92 = ______
45. 180 = ______
46. 60 = ______
47. 40 = ______
48. 400 = ______
49. 63 = ______
50. 200 = ______

Decimal Division

Exercise No. 1

1.-20. State the decimal that is one-half of the stated value.

Example: .05 x 1/2 = .025 or .05 x .5 = .025

1. .10 ______	8. .010 ______	15. 2.0 ______
2. .46 ______	9. .20 ______	16. .066 ______
3. .72 ______	10. .07 ______	17. .036 ______
4. .004 ______	11. .38 ______	18. .002 ______
5. .026 ______	12. .042 ______	19. .016 ______
6. 1.5 ______	13. .18 ______	20. .84 ______
7. .006 ______	14. .012 ______	

21.-40. State the decimal that is one-quarter of the stated value.

Example: .05 x 1/4 = .0125 or .05 x .25 = .0125

21. .06 ______	28. .008 ______	35. .010 ______
22. .004 ______	29. .32 ______	36. .30 ______
23. .82 ______	30. 2.4 ______	37. .44 ______
24. 1.6 ______	31. .60 ______	38. 3.6 ______
25. .120 ______	32. .08 ______	39. .68 ______
26. .036 ______	33. .40 ______	40. .88 ______
27. .76 ______	34. .90 ______	

41.-50. State the square root for the stated values.

Example: $\sqrt{400} = 20$

41. 49 ______	45. 10,000 ______	49. 1296 ______
42. 1600 ______	46. 22,500 ______	50. 25 ______
43. 36 ______	47. 100 ______	
44. 5184 ______	48. 14,400 ______	

Exercise No. 2

1.-20. State the decimal that is one-half of the stated value.

Example: .06 x 1/2 = .030 or .06 x .5 = .030

1. .30 ______	8. .040 ______	15. 1.0 ______
2. .86 ______	9. .70 ______	16. .016 ______
3. .92 ______	10. .09 ______	17. .076 ______
4. .006 ______	11. .88 ______	18. .004 ______
5. .046 ______	12. .052 ______	19. .026 ______
6. 1.6 ______	13. .78 ______	20. .64 ______
7. .14 ______	14. .012 ______	

21.-40. State the decimal that is one-quarter of the stated value.

Example: .06 x 1/4 = .015 or .06 x .25 = .015

21. .09 ______	28. .004 ______	35. .030 ______
22. .008 ______	29. .12 ______	36. .28 ______
23. .84 ______	30. 4.8 ______	37. .52 ______
24. 2.4 ______	31. .30 ______	38. 1.2 ______
25. .220 ______	32. .02 ______	39. .64 ______
26. .018 ______	33. .60 ______	40. .16 ______
27. .94 ______	34. .76 ______	

41.-50. State the square root for the stated values.

Example: $\sqrt{196} = 14$

41. 64 ______	45. 3600 ______	49. 4225 ______
42. 1521 ______	46. 19,600 ______	50. 81 ______
43. 441 ______	47. 5184 ______	
44. 1936 ______	48. 14,400 ______	

Scientific Notation

Exercise No. 1

1.-20. Place the following ordinary numbers into the scientific notation system.

Example 1: 8600 = 8.6×10^{3} Example 2: .0065 = 6.5×10^{-3}

1. 26 ________
2. 340 ________
3. 6700 ________
4. 82,000 ________
5. 163,000 ________
6. 790 ________
7. 84 ________
8. 7100 ________
9. 98,000 ________
10. 1,200,000 ________
11. .003 ________
12. .00045 ________
13. .01 ________
14. .000007 ________
15. .000605 ________
16. .0000083 ________
17. .0000500 ________
18. .00307 ________
19. .49 ________
20. .0095 ________

21.-40. State the ordinary number equivalent for those in scientific notation.

Example 1: 3.6×10^{3} = 3600 Example 2: 7.2×10^{-3} = .0072

21. 9.3×10^{3} ________
22. 3.2×10^{6} ________
23. 8.1×10^{4} ________
24. 6.02×10^{4} ________
25. 3.1×10^{2} ________
26. 5.5×10^{1} ________
27. 6.8×10^{3} ________
28. 8.7×10^{5} ________
29. 4.66×10^{2} ________
30. 7.3×10^{7} ________
31. 1.2×10^{-3} ________
32. 9.5×10^{-6} ________
33. 5.3×10^{-4} ________
34. 2.6×10^{-2} ________
35. 3.3×10^{-1} ________
36. 5.9×10^{-5} ________
37. 1.7×10^{-7} ________
38. 6.9×10^{-3} ________
39. 7.4×10^{-4} ________
40. 8.2×10^{-2} ________

41.-50. Give the English unit equivalents for the following metric system values.

41. 200 cm = ________ in.
42. 10Kg = ________ lb.
43. 30 litres = ________ qt.
44. 50 Km = ________ miles
45. 180 cm = ________ in.
46. 4 litres = ________ gal.
47. 100 Kg = ________ lb.
48. 100 cm ________ in.
49. 90 cm = ________ in.
50. 65Kg = ________ lb.

The following conversions have been used to calculate these problems.

2.5 cm = 1 inch
1 Kg = 2.2 pounds
1 Km = .6 miles
1 litre = 1 quart

Scientific Notation

Exercise No. 2

1.-20. Place the following ordinary numbers into the scientific notation system.

Example 1: 32000 = 3.2×10^{4} Example 2: .00037 = 3.7×10^{-4}

1. 17 ________	8. 6200 ________	15. .00008 ________
2. 250 ________	9. 89,000 ________	16. .000009 ________
3. 5800 ________	10. 2,300,000 ________	17. .61 ________
4. 73,000 ________	11. .004 ________	18. .009 ________
5. 172,000 ________	12. .0021 ________	19. .00089 ________
6. 970 ________	13. .018 ________	20. .00025 ________
7. 48 ________	14. .00038 ________	

21.-40. State the ordinary number equivalent for those in scientific notation.

Example 1: 7.2×10^{4} = 72,000 Example 2: 3.3×10^{-4} = .00033

21. 1.3×10^{2} ________	28. 3.4×10^{2} ________	35. 1.6×10^{-1} ________
22. 2.5×10^{4} ________	29. 9.9×10^{4} ________	36. 3.3×10^{-4} ________
23. 2.8×10^{5} ________	30. 8.4×10^{5} ________	37. 8.9×10^{-7} ________
24. 1.28×10^{3} ________	31. 4.1×10^{-3} ________	38. 2.2×10^{-3} ________
25. 7.6×10^{7} ________	32. 5.6×10^{-4} ________	39. 5.1×10^{-2} ________
26. 6.5×10^{6} ________	33. 7.3×10^{-6} ________	40. 1.3×10^{-3} ________
27. 5.6×10^{1} ________	34. 4.7×10^{-2} ________	

41.-50. Give the metric unit equivalents for the following English system values.

41. 36 in. = ________ cm	45. 60 in. = ________ cm	49. 5 qts. = ________cc
42. 110 lbs. = ________ Kg	46. 121 lbs. = ________ Kg	50. 18 ins. = ________ cm
43. 60 qts. = ________ litres	47. 10 gals. = ________ litres	
44. 87 miles = ________ Km	48. 12 ins. ________ cm	

The following conversions have been used to calculate these problems.

2.5 cm = 1 inch

1 Kg = 2.2 pounds

1 Km = .6 miles

1 litre = 1 quart

mAs Conversion

Exercise No. 1

1. A radiograph is taken using 200 mA setting and .05 second exposure time. What mAs was used for this technique?

State the mAs that will result for the following mA and time value

	mA	Time	mAs		mA	Time	mAs
2.	100	.10	______	11.	300	.01	______
3.	100	.05	______	12.	300	.33	______
4.	100	.25	______	13.	300	.75	______
5.	100	.50	______	14.	400	.03	______
6.	200	.2	______	15.	400-	.40	______
7.	200	.03	______	16.	400	.05	______
8.	200	.10	______	17.	600	.50	______
9.	200	.30	______	18.	600	.15	______
10.	200	.25	______	19.	600	.30	______

20. A radiograph is taken using 40 mAs at 400 mA setting exposure. What time factor was used for this technique?

State the time factor that will result for the following mAs and mA values.

	mAs	mA	Time		mAs	mA	Time
21.	50	100	______	31.	15	150	______
22.	100	200	______	32.	25	400	______
23.	60	300	______	33.	50	1000	______
24.	20	600	______	34.	80	800	______
25.	20	800	______	35.	30	600	______
26.	5	400	______	36.	2	200	______
27.	50	500	______	37.	3	300	______
28.	30	300	______	38.	75	100	______
29.	25	100	______	39.	150	100	______
30.	40	200	______	40.	600	200	______

41. A radiograph is taken using 20 mAs at .05 second exposure time setting. What mA setting was used for this technique?

State the mA setting that will result for the following mAs and time values.

	mAs	Time	mA		mAs	Time	mA
42.	50	.10	______	47.	15	.01	______
43.	40	.05	______	48.	75	.75	______
44.	25	.20	______	49.	35	.35	______
45.	30	.05	______	50.	40	.25	______
46.	100	.08	______				

The following formulas have been used to solve these problems.

$\text{mAs} = \text{mA} \times \text{Time}$ $\text{mA} = \dfrac{\text{mAs}}{\text{Time}}$ $\text{Time} = \dfrac{\text{mAs}}{\text{mA}}$

mAs Conversion

Exercise No. 2

1. A radiograph is taken using 100 mA setting and .05 second exposure time. What mAs was used for this technique?

State the mAs that will result for the following mA and time values.

	mA	Time	mAs		mA	Time	mAs
2.	100	.15	______	11.	300	.04	______
3.	100	.35	______	12.	300	.012	______
4.	100	.005	______	13.	300	.008	______
5.	100	.75	______	14.	400	.024	______
6.	200	.30	______	15.	400	.46	______
7.	200	.02	______	16.	400	.018	______
8.	200	.20	______	17.	600	.60	______
9.	200	.60	______	18.	600	.44	______
10.	200	.15	______	19.	600	.10	______

20. A radiograph is taken using 40 mAs at 400 mA setting. What time exposure factor was used for this technique?

State the time factor that will result for the following mAs and mA values.

	mAs	mA	Time		mAs	mA	Time
21.	30	100	______	31.	100	100	______
22.	50	200	______	32.	25	200	______
23.	150	300	______	33.	250	1000	______
24.	30	600	______	34.	40	800	______
25.	20	600	______	35.	10	300	______
26.	15	300	______	36.	75	200	______
27.	60	600	______	37.	45	300	______
28.	30	900	______	38.	75	600	______
29.	75	300	______	39.	250	100	______
30.	80	400	______	40.	400	200	______

41. A radiograph is taken using 20 mAs at .05 second exposure setting. What mA setting was used for this technique?

State the mA setting that will result for the following mAs and time values.

	mAs	Time	mA		mAs	Time	mA
42.	100	.20	______	47.	225	.75	______
43.	20	.025	______	48.	70	.35	______
44.	60	.20	______	49.	48	.16	______
45.	96	.24	______	50.	20	.04	______
46.	200	.25	______				

The following formulas have been used to solve these problems.

$\text{mAs} = \text{mA} \times \text{Time}$ $\text{mA} = \dfrac{\text{mAs}}{\text{Time}}$ $\text{Time} = \dfrac{\text{mAs}}{\text{mA}}$

Exercise No. 1

1.-20. In the following equations, solve for the unknown quantity.

Example 1: $2a = 4 \quad a = \frac{4}{2} \quad a = 4 \div 2 \quad a = 2$ Example 2: $4a = 1 \quad a = \frac{1}{4} \quad a = 1 \div 4 \quad a = .25$

Example 3: $2a = 5x4 \quad a = \frac{5x4}{2} \quad a = \frac{20}{2} \quad a = 20 \quad a = 10$

1. $4a = 20$ a = ____
2. $6a = 24$ a = ____
3. $8y = 64$ y = ____
4. $3x = 36$ x = ____
5. $5x = 80$ x = ____
6. $2a = 12$ a = ____
7. $4a = 100$ a= ____
8. $8a = 2$ a = ____
9. $32a = 4$ a = ____
10. $16y = 4$ y = ____
11. $5x = 1$ x = ____
12. $25a = 10$ a = ____
13. $40a = 4$ a = ____
14. $50a = 25$ a = ____
15. $3a = 5x3$ a = ____
16. $4y = 16x4$ y = ____
17. $7y = 49x7$ y = ____
18. $ay = b$ y = ____
19. $xy = b$ x = ____
20. $ab = cd$ b = ____

21.-40. In the following equations, solve for the unknown quantity.

Example 1: $\frac{a}{4} = 8 \quad a = 8x4 \quad a = 32$ Example 2: $\frac{a}{3} = 12x2 \quad a = 12x2x3 \quad a = 72$

Example 3: $\frac{a}{6} = \frac{9}{3} \quad a = \frac{9x6}{3} \quad a = \frac{54}{3} \quad a = 54 \div 3 \quad a = 18$

Tip: The expression $\frac{a}{2}$ is mathematically the same as $\frac{1}{2}a$

21. $\frac{a}{6} = 3$ a = ____
22. $\frac{a}{4} = 2$ a = ____
23. $\frac{x}{9} = 4$ x = ____
24. $\frac{y}{3} = 7$ y = ____
25. $\frac{3}{4}a = 3$ a = ____
26. $\frac{2}{5}y = 40$ y = ____
27. $\frac{3}{7}x = 21$ x = ____
28. $\frac{a}{3} = 10x2$ a = ____
29. $\frac{a}{6} = 4x3$ a = ____
30. $\frac{x}{4} = 5x7$ x = ____
31. $\frac{y}{5} = 10x12$ y = ____
32. $\frac{2}{5}y = 6x3$ y = ____
33. $\frac{2}{3}b = 12x3$ b = ____
34. $\frac{3}{20}a = 7x6$ a = ____
35. $\frac{a}{2} = \frac{10}{4}$ a = ____
36. $\frac{a}{6} = \frac{24}{8}$ a = ____
37. $\frac{y}{5} = \frac{40}{8}$ a = ____
38. $\frac{x}{b} = c$ x = ____
39. $\frac{y}{x} = cb$ y = ____
40. $\frac{a}{y} = x$ a = ____

41.-50. In the following equations, solve for the unknown quantity.

41. $\frac{mAs}{Time} = mA$ mAs = ____
42. D x It = Log Io D = ____
43. mf x SOD = SID mf = ____
44. I x R = V I = ____
45. $\frac{E}{v} = h$ E = ____
46. D x F = W F = ____
47. $\frac{SOD}{FSS}$ x P = OID P = ____
48. kV x = 12.4 λ = ____
49. $\frac{Is\ Ns}{Ip} = Np$ Is = ____
50. Rad x QF = Rem Rad = ____

Exercise No. 2

1.-20. In the following equations, solve for the unknown quantity.

Example 1: $3a = 6 \quad a = \frac{6}{3} \quad a = 6 \div 3 \quad a = 2$ Example 2: $a = \frac{2}{8} \quad a = 2 \div 8 \quad a = .25$

Example 3: $3a = 6x5 \quad a = \frac{6x5}{3} \quad a = \frac{30}{3} \quad a = 30 \div 3 \quad a = 10$

1. $2a = 44$ a = ______
2. $4x = 28$ x = ______
3. $6y = 42$ y = ______
4. $3b = 81$ b = ______
5. $7y = 42$ y = ______
6. $8x = 72$ x = ______
7. $3a = 75$ a = ______
8. $7a = 28$ a = ______
9. $6b = 3$ b = ______
10. $3a = 1$ a = ______
11. $4x = 2$ x = ______
12. $30y = 5$ y = ______
13. $40x = 8$ x = ______
14. $80a = 10$ a = ______
15. $6a = 9x8$ a = ______
16. $5y = 25x8$ y = ______
17. $4b = 32x6$ b = ______
18. $3x = 27x9$ x = ______
19. $xyb = ac$ y = ______
20. $ax = cd$ x = ______

21.-40. In the following equations, solve for the unknown quantity.

Example 1: $\frac{a}{5} = 10 \quad a = 10 \times 5 \quad a = 50$ Example 2: $\frac{a}{6} = 3 \times 4 \quad a = 3 \times 4 \times 6 \quad a = 72$

Example 3: $\frac{a}{3} = \frac{18}{6} \quad a = \frac{18 \times 3}{6} \quad a = \frac{54}{6} \quad a = 9$

Tip: The expression $\frac{a}{2}$ is mathematically the same as $\frac{1}{2}a$

21. $\frac{a}{7} = 5$ a = ______
22. $\frac{b}{4} = 9$ b = ______
23. $\frac{x}{6} = 12$ x = ______
24. $\frac{b}{3} = 42$ b = ______
25. $\frac{2}{3}a = 30$ a = ______
26. $\frac{3}{4}y = 12$ y = ______
27. $\frac{4}{5}x = 20$ x = ______
28. $\frac{a}{3} = 6x4$ a = ______
29. $\frac{x}{7} = 3x7$ x = ______
30. $\frac{b}{6} = 4x8$ b = ______
31. $\frac{y}{6} = 9x11$ y = ______
32. $\frac{3}{4}a = 8x6$ a = ______
33. $\frac{4}{9}y = 6x2$ y = ______
34. $\frac{2}{3}b = 5x6$ b = ______
35. $\frac{a}{3} = \frac{10}{6}$ a = ______
36. $\frac{x}{4} = \frac{20}{2}$ x = ______
37. $\frac{y}{5} = \frac{50}{25}$ y = ______
38. $\frac{x}{a} = \frac{c}{b}$ x = ______
39. $\frac{xy}{a} = cb$ y = ______
40. $\frac{xa}{b} = \frac{c}{d}$ x = ______

41.-50. Solve the following equations for the stated quantity.

41. $\frac{PE}{D} = F$ PE = ______
42. $\frac{C}{2r} = \pi$ C = ______
43. $\frac{rem}{QF} = Rad$ rem = ______
44. $\frac{2A}{b} = h$ A = ______
45. $\frac{V}{R} = I$ V = ______
46. $\frac{Vs}{Vp} = \frac{Ns}{Np}$ Vs = ______
47. $\frac{mAs}{mA} = Time$ mAs = ______
48. $MF \times SOD = SID$ MF = ______
49. $Is \times Vs = IpVp$ Is = ______
50. $\frac{5}{9}C = (F° 32)$ C° = ______

Algebraic Principles (Squares)

Exercise No. 1

1.-20. In the following equations solve for the unknown quantity.

Example 1: $a^2 = 36 \quad a = \sqrt{36} \quad a = 6$

Example 2: $2a^2 = 8 \quad a^2 = \frac{8}{2} \quad a = \sqrt{4} \quad a = 2$

Example 3: $a^2 = 8 \quad a^2 = 8x2 \quad a = \sqrt{16} \quad a = 4$

1. $a^2 = 64$ a = ______	8. $2a^2 = 72$ a = ______	15. $\frac{a^2}{2} = 200$ a = ______
2. $x^2 = 81$ x = ______	9. $3a^2 = 243$ a = ______	16. $\frac{a^2}{3} = 27$ a = ______
3. $b^2 = 25$ b = ______	10. $2x^2 = 338$ x = ______	17. $\frac{y^2}{3} = 147$ y = ______
4. $y^2 = 25$ y = ______	11. $5y^2 = 245$ y = ______	18. $\frac{1}{4}x^2 = 81$ x = ______
5. $a^2 = 121$ a = ______	12. $6a^2 = 54$ a = ______	19. $\frac{1}{8}x^2 = 50$ x = ______
6. $x^2 = 169$ x = ______	13. $3b^2 = 48$ b = ______	20. $\frac{1}{4}y^2 = 64$ y = ______
7. $y^2 = 361$ y = ______	14. $7x^2 = 847$ x = ______	

21.-40. In the following equations solve for the unknown quantity.

Example 1: $a^2 = 4 \times 16 \quad a = \sqrt{4 \times 16} \quad a = \sqrt{64} \quad a = 8$

Example 2: $3a^2 = 4 \times 27 \quad a = \sqrt{4 \times 27 \div 3} \quad a = \sqrt{36} \quad a = 6$

Example 3: $\frac{a^2}{3} = 2 \times 6 \quad a = \sqrt{2 \times 6 \times 3} \quad a\sqrt{36} \quad a = 6$

21. $a^2 = 10x10$ a = ______	28. $2a^2 = 121x2$ a = ______	35. $a^2/2 = 648$ a = ______
22. $x^2 = 40x10$ x = ______	29. $3a^2 = 96x18$ a = ______	36. $a^2/3 = 60/5$ a = ______
23. $a^2 = 8x32$ a = ______	30. $2y^2 = 100x8$ y = ______	37. $a^2/8 = 224/7$ a = ______
24. $y^2 = 5x45$ y = ______	31. $5y^2 = 25x125$ y = ______	38. $1/2x^2 = 1014/3$ x = ______
25. $x^2 = 4x49$ x = ______	32. $4b^2 = 27x12$ b = ______	39. $1/4x^2 = 288/8$ x = ______
26. $b^2 = 2x72$ b = ______	33. $6x^2 = 225x24$ x = ______	40. $1/3x^2 = 735/5$ x = ______
27. $x^2 = 12x48$ x = ______	34. $4a^2 = 882x2$ a = ______	

Algebraic Principles (Squares)

Exercise No. 1

41.-50. Solve the following equations for the stated quanity.

41. $\frac{E}{C^2} = m$ — E = ______

42. $\frac{2E}{V^2} = m$ — E = ______

43. Is x Vs = Ip x Vp — Vs = ______

44. $\frac{\text{New SID}^2}{\text{Original SID}^2} = \frac{\text{Original Dose}}{\text{New Dose}}$ — New SID² = ______

45. $\frac{\text{New SID}^2}{\text{Original SID}^2} = \frac{\text{Original Dose}}{\text{New Dose}}$ — New SID = ______

46. $D^2 F = h\ Q1xQ^2$ — F = ______

47. $\frac{\text{New mAs}}{\text{New SID}^2} = \frac{\text{Original mAs}}{\text{Original SID}^2}$ — New mAs = ______

48. $\frac{\text{New SID}^2}{\text{Original SID}^2} = \frac{\text{New mA}}{\text{Original mA}}$ — New SID² = ______

49. $\frac{\text{New SID}^2}{\text{Original SID}^2} = \frac{\text{New Time}}{\text{Original Time}}$ — New SID² = ______

50. $\text{Original Time} \times \frac{\text{New SID}^2}{\text{Original SID}^2} = \text{New Time}$ — New SID = ______

Algebraic Principles (Squares)

Exercise No. 2

1.-20. In the following equations, solve for the unknown quantity.

Example 1: $a^2 = 16 \quad a = \sqrt{16} \quad a = 4$

Example 2: $2a^2 = 18 \quad a^2 = \frac{2}{18} \quad a^2 = 9 \quad a = \sqrt{9} \quad a = 3$

Example 3: $\frac{a^2}{2} = 8 \quad a^2 = 8 \times 2 \quad a^2 = 16 \quad a = \sqrt{16} \quad a = 4$

1. $a^2 = 49$ $\quad a =$ ______
2. $x^2 = 100$ $\quad x =$ ______
3. $b^2 = 144$ $\quad b =$ ______
4. $y^2 = 256$ $\quad y =$ ______
5. $a^2 = 324$ $\quad a =$ ______
6. $x^2 = 16$ $\quad x =$ ______
7. $y^2 = 441$ $\quad y =$ ______
8. $2y^2 = 162$ $\quad y =$ ______
9. $6a^2 = 150$ $\quad a =$ ______
10. $4z^2 = 64$ $\quad a =$ ______
11. $5x^2 = 45$ $\quad x =$ ______
12. $7x^2 = 700$ $\quad x =$ ______
13. $3a^2 = 363$ $\quad a =$ ______
14. $9y^2 = 1521$ $\quad y =$ ______
15. $\frac{x^2}{2} = 288$ $\quad x =$ ______
16. $\frac{a^2}{3} = 12$ $\quad a =$ ______
17. $\frac{y^2}{4} = 49$ $\quad y =$ ______
18. $\frac{1}{3} z^2 = 192$ $\quad a =$ ______
19. $\frac{1}{6} x^2 = 54$ $\quad x =$ ______
20. $\frac{1}{4} y^2 = 100$ $\quad y =$ ______

21.-40. In the following equations, solve for the unkonwn quanity.

Example 1: $a^2 = 3 \times 12 \quad a = \sqrt{3 \times 12} \quad a\sqrt{36} \quad a = 6$

Example 2: $2a^2 = \frac{2 \times 9}{2} \quad a = \frac{\sqrt{2 \times 9}}{2} \quad a = \sqrt{9} \quad a = 3$

Example 3: $\frac{a^2}{2} = 3 \times 6 \quad a = \sqrt{3 \times 6 \times 2} \quad a = \sqrt{36} \quad a = 6$

21. $a^2 = 4 \times 81$ $\quad a =$ ______
22. $a^2\ 8 \times 32$ $\quad a =$ ______
23. $y^2 = 4 \times 49$ $\quad y =$ ______
24. $x^2 = 12 \times 27$ $\quad x =$ ______
25. $b^2 = 2 \times 32$ $\quad b =$ ______
26. $a^2 = 3 \times 27$ $\quad a =$ ______
27. $z^2 = 12 \times 48$ $\quad z =$ ______
28. $2a^2 = 24 \times 3$ $\quad a =$ ______
29. $3x^2 = 48 \times 36$ $\quad x =$ ______
30. $2y^2 = 81 \times 8$ $\quad y =$ ______
31. $5b^2 = 20 \times 4$ $\quad b =$ ______
32. $2y^2 = 72 \times 4$ $\quad y =$ ______
33. $6b^2 = 450 \times 3$ $\quad b =$ ______
34. $4a^2 = 882 \times 2$ $\quad a =$ ______
35. $\frac{a^2}{2} = \frac{1568}{4}$ $\quad a =$ ______
36. $\frac{a^2}{3} = \frac{200}{6}$ $\quad a =$ ______
37. $\frac{b^2}{6} = \frac{162}{12}$ $\quad b =$ ______
38. $\frac{1}{4}x^2 = \frac{338}{8}$ $\quad x =$ ______
39. $\frac{1}{2}b^2 = \frac{968}{4}$ $\quad b =$ ______
40. $\frac{1}{3}y^2 = \frac{128}{6}$ $\quad y =$ ______

Algebraic Principles (Squares)

Exercise No. 2

41.-50. Solve the following equations for the state quantity.

41. $\frac{P}{I^2} = R$ P = ______

42. $\frac{A}{r^2} = \pi$ A = ______

43. $\frac{E}{M} = C^2$ E = ______

44. $\frac{A}{b} = h$ A = ______

45. $\frac{\text{New mA}}{\text{Original mA}} = \frac{\text{New SID}^2}{\text{Original SID}^2}$ New mA = ______

46. $\frac{\text{New SID}^2}{\text{Original SID}^2} = \frac{\text{New mA}}{\text{Original mA}}$ New SID^2 = ______

47. $\frac{\text{Original Dose}}{\text{New Dose}} = \frac{\text{New SSD}^2}{\text{Original SSD}^2}$ New Dose = ______

48. $\frac{\text{New SSD}^2}{\text{Original SSD}^2} = \frac{\text{Original Dose}}{\text{New Dose}}$ New SSD^2 = ______

49. $\frac{\text{New SSD}^2}{\text{Original SSD}^2} = \frac{\text{Original Dose}}{\text{New Dose}}$ New SSD = ______

50. $\frac{\text{Original mAs} \times \text{New SID}^2}{\text{Original SID}^2} = \text{New mAs}$ New SID = ______

Section I
Answer Key

Millisecond Conversion

Exercise No. 1

1. .8.	16. .350	31. 5	46. .008
2. .13.	17. .175	32. 90	47. .004
3. .08.	18. .035	33. 900	48. .002
4. .016.	19. .045	34. 170	49. .3
5. .005.	20. .001	35. 500	50. .75
6. .75.	21. 70.	36. 610	
7. .2.	22. 1.	37. 160	
8. .088	23. 800	38. 345	
9. .024	24. 30	39. 40	
10. .003	25. 4	40. 21	
11. .54	26. 650	41. .1	
12. .125	27. 310	42. .05	
13. .018	28. 95	43. .025	
14. .030	29. 20	44. .016	
15. .007	30. 60	45. .066	

Decimal Multiplication

Exercise No. 2

1. .18	16. 1.32	31. .96	46. 3600
2. .32	17. .98	32. .028	47. 1600
3. .024	18. .064	33. .38	48. 160,000
4. .014	19. .15	34. 3.28	49. 3969
5. .14	20. .004	35. 2.6	50. 40,000
6. .70	21. .16	36. .288	
7. 1.7	22. .40	37. .024	
8. .012	23. .092	38. .816	
9. 1.6	24. .016	39. .148	
10. .104	25. .1	40. .12	
11. .250	26. .48	41. 16	
12. .006	27. .5	42. 1024	
13. .12	28. .012	43. 19600	
14. .88	29. 3.2	44. 8464	
15. .072	30. .132	45. 32,400	

Millisecond Conversion

Exercise No. 2

1. .6	16. .13	31. 1	46. .06
2. .23	17. .275	32. 60	47. .004
3. .18	18. .075	33. 100	48. .25
4. .036	19. .035	34. 270	49. .375
5. .007	20. .005	35. 600	50. .666
6. .95	21. 90	36. 110	
7. .15	22. 3	37. 120	
8. .084	23. 500	38. 315	
9. .026	24. 80	39. 48	
10. .003	25. 7	40. 31	
11. .34	26. 250	41. .033	
12. .225	27. 530	42. .02	
13. .028	28. 75	43. .5	
14. .04	29. 40	44. .3	
15. .008	30. 20	45. .4	

Decimal Division

Exercise No. 1

1. .05	16. .033	31. .15	46. 150
2. .23	17. .018	32. .02	47. 10
3. .36	18. .001	33. .10	48. 120
4. .002	19. .008	34. .225	49. 36
5. .013	20. .42	35. .0025	50. 5
6. .75	21. .015	36. .075	
7. .003	22. .001	37. .110	
8. .005	23. .205	38. .9	
9. .10	24. .4	39. .17	
10. .035	25. .03	40. .22	
11. .19	26. .009	41. 7	
12. .021	27. .19	42. 40	
13. .09	28. .002	43. 6	
14. .006	29. .08	44. 72	
15. 1.0	30. .6	45. 100	

Decimal Multiplication

Exercise No. 1

1. .14	16. .66	31. .56	46. 1600
2. .30	17. .90	32. .004	47. 900
3. .024	18. .124	33. .30	48. 40,000
4. .008	19. .07	34. 2.48	49. 36
5. .10	20. .004	35. 3.40	50. 10,000
6. .50	21. .24	36. .368	
7. 1.5	22. .60	37. .032	
8. .018	23. .052	38. .416	
9. 1.2	24. .008	39. .068	
10. .064	25. .14	40. 1.84	
11. .250	26. 1.32	41. 4	
12. .010	27. .50	42. 1296	
13. .06	28. .02	43. 22,500	
14. .46	29. 1.6	44. 5184	
15. .16	30. .088	45. 32,400	

Decimal Division

Exercise No. 2

1. .15	16. .008	31. .075	46. 140
2. .43	17. .038	32. .005	47. 72
3. .46	18. .002	33. .15	48. 120
4. .003	19. .013	34. .19	49. 65
5. .023	20. .32	35. .0075	50. 9
6. .8	21. .0225	36. .007	
7. .07	22. .002	37. .13	
8. .020	23. .21	38. .3	
9. .35	24. .6	39. .16	
10. .045	25. .055	40. .04	
11. .44	26. .0045	41. 8	
12. .026	27. .235	42. 39	
13. .39	28. .001	43. 21	
14. .006	29. .03	44. 44	
15. .5	30. 1.2	45. 60	

Scientific Notation

Exercise No. 1

1. $2.6\ 10^{1}$
2. $3.4\ 10^{2}$
3. $6.7\ 10^{3}$
4. $8.2\ 10^{4}$
5. $1.63\ 10^{5}$
6. $7.9\ 10^{2}$
7. $8.4\ 10^{1}$
8. $7.1\ 10^{3}$
9. $9.8\ 10^{4}$
10. $1.2\ 10^{6}$
11. $3.0\ 10^{-3}$
12. $4.5\ 10^{-4}$
13. $1.0\ 10^{-2}$
14. $7.0\ 10^{-6}$
15. $6.05\ 10^{-4}$
16. $8.3\ 10^{-6}$
17. $5.0\ 10^{-5}$
18. $3.07\ 10^{-3}$
19. $4.9\ 10^{-1}$
20. $9.5\ 10^{-3}$
21. 9300
22. 3,200,000
23. 81,000
24. 60,200
25. 310
26. 55
27. 6800
28. 870,000
29. 466
30. 73,000,000
31. .0012
32. .0000095
33. .00053
34. .026
35. .33
36. .000059
37. .00000017
38. .0069
39. .00074
40. .082
41. 80 inches
42. 22 lbs.
43. 30 qts.
44. 30 miles
45. 72 inches
46. 1 gallon
47. 220 lbs.
48. 40 inches
49. 36 inches
50. 143 lbs.

Scientific Notation

Exercise No. 2

1. $1.7\ 10^{1}$
2. $2.5\ 10^{2}$
3. $5.8\ 10^{3}$
4. $7.3\ 10^{4}$
5. $1.72\ 10^{5}$
6. $9.7\ 10^{2}$
7. $48\ 10^{1}$
8. $6.2\ 10^{3}$
9. $8.9\ 10^{4}$
10. $2.3\ 10^{6}$
11. $4.0\ 10^{-3}$
12. $2.1\ 10^{-3}$
13. $1.8\ 10^{-2}$
14. $3.8\ 10^{-4}$
15. $8.0\ 10^{-5}$
16. $9.0\ 10^{-6}$
17. $6.1\ 10^{-1}$
18. $9.0\ 10^{-3}$
19. $8.9\ 10^{-4}$
20. $2.5\ 10^{-4}$
21. 130
22. 25,000
23. 280,000
24. 1280
25. 76,000,000
26. 6,500,000
27. 56
28. 340
29. 99000
30. 840,000
31. .0041
32. .00056
33. .0000073
34. .047
35. .16
36. .00033
37. .00000089
38. .0022
39. .0051
40. .0013
41. 90 cm
42. 50 kg
43. 60 liters
44. 145 km
45. 150 cm
46. 55 kg
47. 40 liters
48. 30 cm
49. 5000 cc
50. 45 cm

mAs Conversion

Exercise No. 1

1. 10 mAs
2. 10 mAs
3. 5 mAs
4. 25 mAs
5. 50 mAs
6. 40 mAs
7. 6 mAs
8. 20 mAs
9. 60 mAs
10. 50 mAs
11. 3 mAs
12. 99 mAs
13. 225 mAs
14. 12 mAs
15. 160 mAs
16. 20 mAs
17. 300 mAs
18. 90 mAs
19. 180 mAs
20. .1 sec.
21. .5 sec.
22. .5 sec.
23. .2 sec.
24. .033 sec.
25. .025 sec.
26. .125 sec.
27. .1 sec.
28. .1 sec.
29. .25 sec.
30. .2 sec.
31. .1 sec.
32. .0625 sec.
33. .05 sec.
34. .1 sec.
35. .05 sec.
36. .01 sec.
37. .01 sec.
38. .75 sec.
39. 1.5 sec.
40. 3 sec
41. 400 mA
42. 500 mA
43. 800 mA
44. 125 mA
45. 600 mAs
46. 1250 mA
47. 1500 mA
48. 100 mA
49. 100 mA
50. 160 mA

mAs Conversion

Exercise No. 2

1. 5 mAs
2. 15 mAs
3. 35 mAs
4. .5 mAs
5. 75 mAs
6. 60 mAs
7. 4 mAs
8. 40 mAs
9. 120 mAs
10. 30 mAs
11. 12 mAs
12. 3.6 mAs
13. 2.4 mAs
14. 9.6 mAs
15. 184 mAs
16. 7.2 mAs
17. 360 mAs
18. 264 mAs
19. 60 mAs
20. .1 sec.
21. .3 sec.
22. .25 sec.
23. .5 sec.
24. .05 sec.
25. .033 sec.
26. .05 sec.
27. .1 sec.
28. .033 sec.
29. .25 sec.
30. .2 sec.
31. 1 sec.
32. .125 sec.
33. .25 sec.
34. .05 sec.
35. .033 sec.
36. .375 sec.
37. .15 sec.
38. .125 sec.
39. 2.5 sec.
40. 2 sec.
41. 400 mA
42. 500 mA
43. 800 mA
44. 300 mA
45. 400 mA
46. 800 mA
47. 300 mA
48. 200 mA
49. 300 mA
50. 500 mA

Algebraic Principles

Exercise No. 1

1. 5
2. 4
3. 8
4. 12
5. 16
6. 6
7. 25
8. .25
9. .125
10. .25
11. .2
12. .4
13. .1
14. .5
15. 5
16. 16
17. 49
18. $\frac{b}{a}$
19. $\frac{b}{y}$
20. $\frac{cd}{a}$
21. 18
22. 8
23. 36
24. 21
25. 4
26. 100
27. 49
28. 60
29. 72
30. 140
31. 600
32. 45
33. 54
34. 280
35. 5
36. 18
37. 25
38. cb
39. cbx
40. xy
41. mA x Time
42. Log $\frac{IO}{IT}$
43. $\frac{SID}{SOD}$
44. $\frac{V}{R}$
45. hv
46. $\frac{W}{D}$
47. FSS x $\frac{OID}{SOD}$
48. $\frac{12.4}{Kv}$
49. $\frac{1pNp}{Ns}$
50. $\frac{REM}{QF}$

Algebraic Principles

Exercise No. 2

1. 22
2. 7
3. 7
4. 27
5. 6
6. 9
7. 25
8. 4
9. .5
10. .333
11. .5
12. .166
13. .2
14. .125
15. 12
16. 40
17. 48
18. 81
19. $\frac{ac}{xb}$
20. $\frac{cd}{a}$
21. 35
22. 36
23. 72
24. 126
25. 45
26. 16
27. 25
28. 72
29. 147
30. 192
31. 594
32. 64
33. 27
34. 45
35. 5
36. 40
37. 10
38. $\frac{ac}{b}$
39. $\frac{acb}{x}$
40. $\frac{cb}{da}$
41. /FD
42. π2r
43. Rad x QF
44. 1/2 bh
45. 1R
46. $\frac{Ns \times Vp}{Np}$
47. mA x Time
48. $\frac{SID}{SOD}$
49. $\frac{1pVp}{Vs}$
50. 9/5 (F^{o}-32)

Algebraic Principles (Squares)

Exercise No. 1

1. 8
2. 9
3. 5
4. 5
5. 11
6. 13
7. 19
8. 6
9. 9
10. 13
11. 7
12. 3
13. 4
14. 11
15. 20
16. 9
17. 21
18. 18
19. 20
20. 16
21. 10
22. 20
23. 16
24. 15
25. 14
26. 12
27. 24
28. 11
29. 24
30. 20
31. 25
32. 9
33. 30
34. 21
35. 36
36. 6
37. 16
38. 26
39. 12
40. 21
41. mc^2
42. $1/2mv^2$
43. $\frac{\text{Ip x Vp}}{\text{Is}}$
44. $\frac{\text{Original Dose x Original SID}^2}{\text{New Dose}}$
45. $\sqrt{\frac{\text{Original Dose x Original SID}^2}{\text{New Dose}}}$
46. $\text{h x}\frac{\text{Q1 x Q2}}{\text{D}^2}$
47. $\frac{\text{Original mAs x New SID}^2}{\text{Original SID}^2}$
48. $\frac{\text{New mA x Original SID}^2}{\text{Original MA}}$
49. $\frac{\text{New Time x Original SID}^2}{\text{Original Time}}$
50. $\sqrt{\frac{\text{New Time x Original SID}^2}{\text{Original Time}}}$

Algebraic Principles (Squares)

Exercise No. 2

1. 7
2. 10
3. 12
4. 16
5. 18
6. 4
7. 21
8. 9
9. 5
10. 4
11. 3
12. 10
13. 11
14. 13
15. 24
16. 6
17. 14
18. 24
19. 18
20. 20
21. 18
22. 16
23. 14
24. 18
25. 8
26. 9
27. 24
28. 6
29. 24
30. 18
31. 4
32. 12
33. 15
34. 21
35. 28
36. 10
37. 9
38. 13
39. 22
40. 8
41. I^2R
42. πr^2
43. mc^2
44. bh
45. $\frac{\text{New SID}^2\text{ x Original mA}}{\text{Original SID}^2}$
46. $\frac{\text{New mA x Original SID}^2}{\text{Original mA}}$
47. $\frac{\text{Original SSD}^2\text{ x Original Dose}}{\text{New SSD}^2}$
48. $\frac{\text{Original Dose x Original SSD}^2}{\text{New Dose}}$
49. $\sqrt{\frac{\text{Original Dose x Original SSD}^2}{\text{New Dose}}}$
50. $\sqrt{\frac{\text{New mAs x Original SID}^2}{\text{Original mAs}}}$

Section II
Radiation Protection and Biology

Eyes

Thyroid

Gonads

Radiation Biology

Chapter 1

1. The period of time after an acute exposure to radiation in which the effects of the exposure are seen is called the: 2:560

A. Latent period
B. Incident period
C. Manifest period
D. Exposure period

Pertaining to the survival curve answer questions 2 & 3

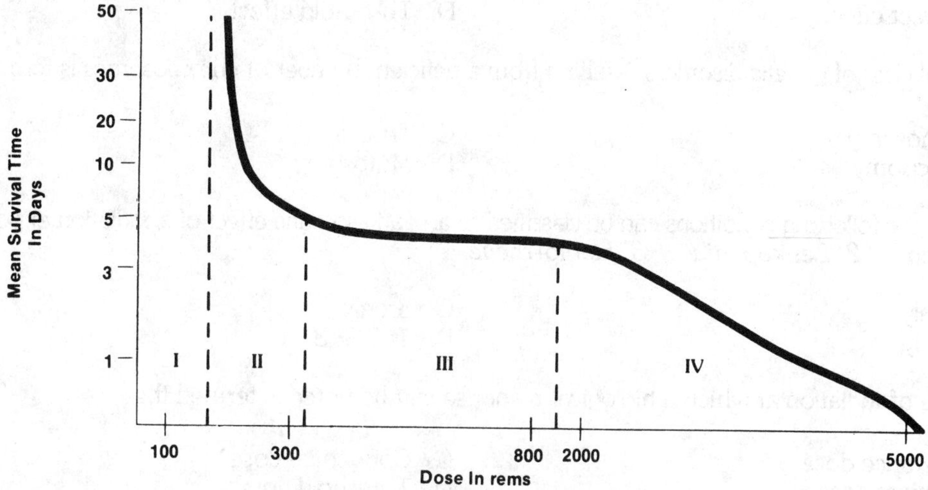

2. In the dose survival curve, the Roman numeral which corresponds to death from the gastrointestional tract is: 2:564

A. I
B. II
C. III
D. IV

3. According to the diagram at dose levels above 2000 Rems, death is most likely to occur in about: 2:564

A. 20-50 days
B. 10-20 days
C. 5-10 days
D. less than 2 days

4. Which of the following types of radiation-induced conditions is classified as a late somatic effect? 13:136
1. Cataract formation *2. Cancer* *3. Leukemia*

A. 1 only
B. 2 only
C. 3 only
D. 1, 2 & 3

5. A radiation-induced mutation that appears in the offspring of the exposed individual is termed a/an: 13:145

A. Acute mutation
B. Long term mutation
C. Genetic mutation
D. Somatic mutation

6. Approximately __________ of the genetic mutations that occur are considered harmful to the individual 7:45

A. 99%
B. 87%
C. 75%
D. 37%

7. The most radiosensitive part of a normal human cell is the: 2:572

A. Cell membrane
B. Centriole
C. Golgi Body
D. Chromosome

8. A large radiation exposure to the lens of the eye will most likely result in the formation of: 13:144

A. Glaucoma
B. Scleroma
C. Retinitis
D. Cataracts

9. Death that follows an acute exposure to radiation of about 500 rems will generally occur in about: 2:560

A. 3-5 hours
B. 3-5 days
C. 3-5 weeks
D. None of the above

10. The majority of damage to the body from radiation exposure results from the: 2:547

A. Direct effect
B. Indirect effect
C. Target effect
D. Threshold effect

11. A general class of genetic disorders resulting from a deficient number of chromosomes is termed: 14:905

A. Monosomy
B. Oconsomy
C. Trisomy
D. Mutosomy

12. Which of the following conditions can be classified as an early somatic effect of a radiation exposure? 13:108
1. Nausea *2. Leukopenia* *3. Hemorrhage*

A. 1 only
B. 2 only
C. 3 only
D. 1, 2 & 3

13. The dose of radiation at which a biological response can be noted is termed the: 13:136

A. Tolerance dose
B. Ambient dose
C. Congenital dose
D. Threshold dose

14. The loss of hair after exposure to large amounts of radiation is termed: 2:566

A. Epistaxis
B. Epilation
C. Bromidrosis
D. Erythema

15. The most radiosensitive tissue group listed below is the: 2:533

A. Central nervous tissue
B. Alimentary tract tissue
C. Muscular tissue
D. Cardiovascular tissue

16. The immediate symptoms that appear after an acute radiation exposure are termed the: 2:560

A. Latent syndrome
B. Proportional syndrome
C. Chronic syndrome
D. Prodromal syndrome

Pertaining to the diagram of the dose-response curves, answer questions 17-18.

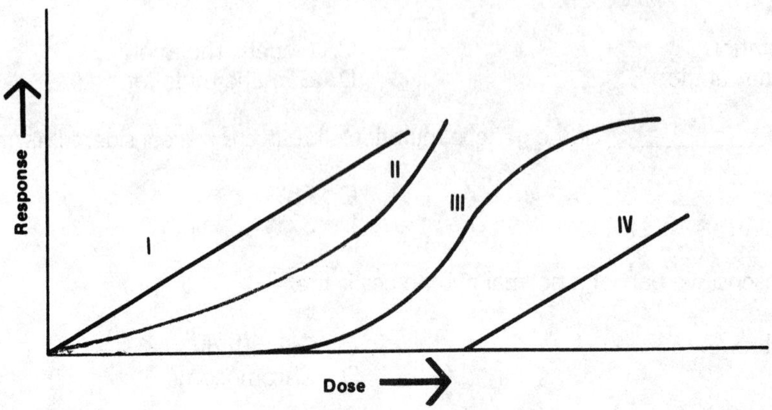

17. In the diagram, the linear non-threshold dose-response curve is represented by Roman numeral: 2:538

A. I
B. II
C. III
D. IV
E. None of the above

18. The curve represented by Roman numeral III is a: 2:538

A. Linear threshold curve
B. Non-linear threshold curve
C. Linear non-threshold curve
D. Non-linear non-threshold curve

19. The acute radiation syndrome associated with damage to the blood and blood-forming organs is called the: 2:561

A. Prodromal syndrome
B. CNS syndrome
C. G.I. syndrome
D. None of the above

20. The first biological effects of radiation in man to doses of approximately 25 rads are changes seen in the: 13:120

A. Blood count
B. Skin
C. Sperm count
D. Eyes

21. A noticeable skin reaction due to radiation exposures of over 150 R is called: 13:45

A. Hemodema
B. Leukogenesis
C. Erythema
D. Sponditis

22. In general, radiation damage increases with increases in the: 2:536
1. Oxygen enhancement ratio *2. Quality factor of the radiation* *3. Linear energy transfer*

A. 1 only
B. 2 only
C. 3 only
D. 1, 2 & 3 only

23. The laws of Bergonie and Tribondeau state that radiation damage to a cell is related to: 13:120
1. Size and weight *2. Differentiation* *3. Reproductive activity*

A. 1 & 2 only
B. 1 & 3 only
C. 2 & 3 only
D. 1, 2 & 3

24. The human fetus is most sensitive to radiation in the: 2:593

A. First trimester
B. Second trimester
C. Third trimester
D. Equally in all trimesters

25. Which of the following is a suspected type of heredity mutation? 13:147

A. Uremia
B. Mongolism
C. Ulcerative colitis
D. Epilation

26. The first sign of the acute radiation syndrome followed by a large exposure of radiation is usually: 13:130

A. Diarrhea
B. Incontinence
C. Nausea
D. Excessive bleeding

27. The most radiosensitive cell of the body listed below is the: 2:569

A. Neuron
B. Chondrocyte
C. Lymphocyte
D. Osteocyte

28. Which of the following effects may be observed if the fetus of a pregnant woman is exposed to radiation? 2:593
1. Spontaneous abortion 2. Mental retardation 3. Cancer

A. 1 & 2 only
B. 1 & 3 only
C. 2 & 3 only
D. 1, 2 & 3

29. The property of x-rays that accounts for their effect on biological systems is that they: 2:521

A. Have different energies
B. Have no electric charge
C. Cause ionization in matter
D. Travel at the speed of light

30. Trisomy is any class of genetic disorders resulting from a cell with: 14:1491

A. Extra chromosomes
B. Too few chromosomes
C. Twisted chromosomes
D. Mutated chromosomes

31. An x-ray photon passes into a cell causing a point mutation on a DNA molecule. This is classified as a/an: 13:108

A. Direct effect
B. Indirect effect
C. Pinhole effect
D. Threshold effect

32. Acute radiation exposure to the skin may result in erythema or an ulcerating condition called: 2:565

A. Desquamation
B. Vasculitis
C. Epilation
D. Isolysis

33. Which of the following tissue groups possess the lowest radiosensitivity? 13:123

A. Reproductive and hemopoietic
B. Muscular and nervous
C. Gastrointestinal
D. Endocrine and dermis

34. An adverse effect on the production of white blood cells is most likely after radiation exposure to the: 2:570

A. Central nervous system
B. Urinary system
C. Biliary system
D. Skeletal system

35. The linear (non-threshold) curve of radiation dose-response relationships states that radiation damage is directly proportional to radiation dose, and that: 13:128

A. Only high doses cause damage
B. Any dose may cause damage
C. Damage occurs only after minimum dose
D. None of the above

36. The dehydration that accompanies the gastrointestinal syndrome is the direct result of radiation damage to the: 2:570

A. Blood cells
B. Stem cells
C. Nerve cells
D. Muscle cells

37. Which type of human blood cell has the lowest radiosensitivity? 13:121

A. Mature erythrocyte
B. Mature leukocyte
C. Immature leukocyte
D. Immature thrombocyte

38. The time between the initial exposure to radiation and the appearance of visible effect is termed the: 2:570

A. Chronic period
B. Latent period
C. Manifest period
D. Pre-exposure period

39. Which of the following macro-molecules has the greatest sensitivity to radiation? 2:544

A. Proteins
B. Mitochondria
C. Dioxyribonucleic Acid
D. Adenosinetriphosphate

40. The degree to which recovery is possible after exposure to radiation is related to: 2:536
1. Total dose received *2. Degree of protraction* *3. LET of the radiation*

A. 1 only
B. 2 only
C. 3 only
D. 1, 2 & 3

41. A common by-product resulting from the irradiation of water in a living cell is a bleach called: 2:483

A. Hydrogen peroxide
B. Deoxyribose
C. Quanine
D. Ozone

42. The process by which energy from ionizing radiation is transferred from a directly ionized molecule to another molecule that is not directly ionized is called the: 2:547

A. Direct hit theory
B. Random interaction theory
C. Potential free radical theory
D. Indirect hit theory

43. The developing human is most sensitive to radiation during the ________ development stage. 2:593

A. Embryonic
B. Fetal
C. Neo-natal
D. Post-natal

44. The central nervous system syndrome that causes death in a matter of hours is only apparent at doses of radiation of approximately: 13:134

A. 300 rad (3 Gy)
B. 500 rad (5 Gy)
C. 1000 rad (10 Gy)
D. 5000 rad (50 Gy)

45. The study of a response to radiation or other stimulus occurring to tissues outside of the body (eg. test tube) is termed an ________ response. 2:542

A. In vino
B. In vento
C. In vitro
D. In vivo

46. A biologic tissue is least sensitive to radiation damage during: 2:566

A. Reproductive activity
B. An hypoxic state
C. Rapid cell growth (cell division)
D. None of the above

47. With large radiation exposures to the hemopoietic system, the following symptom is likely to occur: 13:132
1. Lowered blood count *2. Increased susceptibility to infection* *3. Longer coagulation time*

A. 1 only
B. 2 only
C. 3 only
D. 1, 2 & 3

48. As a general rule, somatic damage to radiation is approximately ________ reparable when small exposures are received in fractionated doses: 7:19

A. 90%
B. 60%
C. 30%
D. 10%

49. The time between a lethal radiation exposure and death is termed: 2:572

A. Mean survival time
B. Mean threshold time
C. Mean recovery ratio
D. Latent-manifest ratio

50. Which of the following is the most sensitive cell of the body to radiation exposure? 13:124

A. Chondrocytes
B. Mature red blood cells
C. Spermatogonia
D. Nerve cells

Pertaining to the diagram of the relative radiosensitivity graph by age, answer questions 51-52

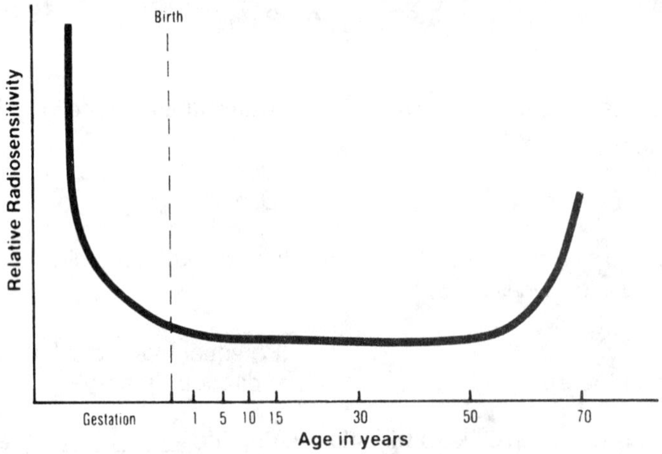

51. According to the diagram, the greatest sensitivity to radiation after a person is born is at the approximate age of: 2:537

A. 1 year
B. 10 years
C. 30 years
D. 70 years

52. In general, from ages 5-50 it can be said that radiosensitivity: 2:537

A. Increases with age
B. Decreases with age
C. Remains relatively constant with age
D. Varies greatly with age

53. Death resulting from radiation damage to the stem cells of the gastrointestinal tract is most likely to occur within: 13:133

A. 5 hours
B. 10 days
C. 4 weeks
D. 3 months

54. During the embryonic stage of development, much of the tissue differentiation takes place during the process called: 13:145

A. Organogenesis
B. Osmosis
C. Hemostasis
D. Oogenesis

55. A large (5 gray) exposure to the hemotologic system will result in the reduction of ________ in the circulating blood. 2:570
1. Leukocytes 2. Thrombocytes 3. Erythrocytes

A. 1 only
B. 2 only
C. 3 only
D. 1, 2 & 3

Interactions of Radiation With Matter

Chapter 2

1. The reduction in the number and energy of photons as radiation passes through matter is termed: 2:186

   A. Irradiation
   B. Deflection
   C. Photoelectrolysis
   D. Attenuation

2. A common nuclear emission, the beta particle, is identical in mass and charge with the: 2:49

   A. Positron
   B. Proton
   C. Electron
   D. Neutron

Pertaining to the diagram of the photoelectric probability curve, answer questions 3-4.

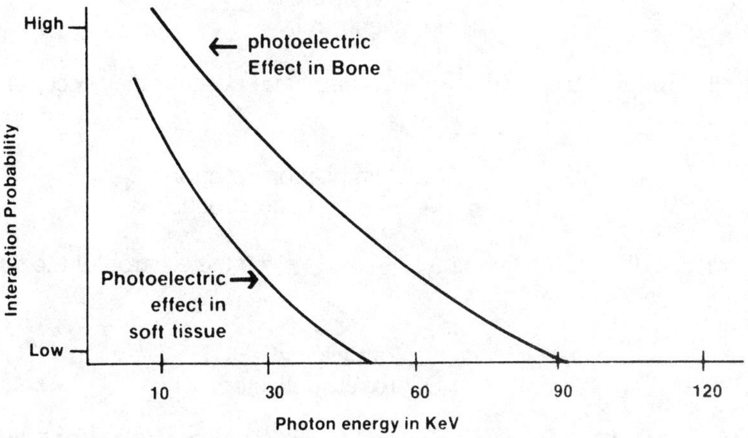

3. According to the diagram, the probability of a photoelectric interaction increases as: 2:148
   1. Atomic number increases *2. Photon energy increases* *3. Atomic mass decreases*

   A. 1 only
   B. 2 only
   C. 3 only
   D. 1, 2 & 3

4. According to the diagram the probability of a photoelectric interaction in soft tissue approaches zero with photon energies: 2:148

   A. Below 10 keV
   B. About 30 keV
   C. Above 60 keV
   D. Above 90 keV

5. The production of scattered radiation when x-radiation strikes an object principally results from the: 2:172

   A. Compton event
   B. Brem's event
   C. Characteristic event
   D. Photoelectric event

6. The type of interaction, in diagnostic kV range, which is responsible for the radiographic contrast between bone and soft tissue is termed: 2:182

   A. Compton effect
   B. Coherent scattering
   C. Pair production
   D. Photoelectric effect

7. At photon energies of 100 keV to 1 MeV, radiation is absorbed in soft tissue mainly by: 13:33

A. Pair production
B. Compton effect
C. Photoelectric effect
D. Coherent scattering

8. The most common attenuation reaction occurring with exposure to a 60 KeV diagnostic x-ray is: 2:178

A. Coherent scattering
B. Photoelectric effect
C. Compton scattering
D. Pair production

9. An unmodified (Thomson) scattering is most likely to occur in the energy range of: 13:28

A. 10-30 keV
B. 50-70 keV
C. 100-200 keV
D. 1-5 MeV

10. The alpha particle is most similar to a: 2:48

A. Helium nucleus
B. Positron
C. Negatron
D. Neutron

11. At photon energies above 1.022 MeV, the creation of a negatron and positron may occur in an event termed: 2:179

A. Photo disintegration
B. Pair production
C. Annihilation reaction
D. Van de Graff production

12. During a photoelectric interaction, the incident photon is _________ by the electron of the target atom. 2:176

A. Partially absorbed
B. Totally absorbed
C. Partially deflected
D. Totally deflected

13. The interaction of x-rays with matter that is most common in the energy range between 150 keV and 3 MeV is (the): 13:33

A. Photoelectric effect
B. Pair production
C. Compton scatter
D. Triplet production

14. The common emission from an unstable nucleus with the highest ionization potential is: 2:52

A. Alpha
B. Beta
C. Gamma
D. X-ray

15. The photoelectric interaction most commonly occurs when the incident photon strikes a/an: 2:176

A. Inner shell electron
B. Outer shell electron
C. Proton
D. Neutron

16. Which of the following interactions of x-ray in matter results in ionization of the target atom? 13:29
1. Pair production *2. Thomson scattering* *3. Compton scattering*

A. 1 only
B. 2 only
C. 3 only
D. 1, 2 & 3

17. Characteristic radiation emitted by the irradiated tissue following the photoelectric event: 13:31

A. Escapes totally from tissue
B. Is partially absorbed in tissue
C. Escapes partially from tissue
D. Is totally absorbed in tissue

18. The formation of positive and negative ions (ion pair) when radiation strikes matter requires 34 eV. What is the maximum number of ion pairs that could be formed by a 102 keV photon? 2:16

A. 2000
B. 3000
C. 3468
D. 4468

19. The Compton effect, when radiation strikes an atom, is known primarily as a/an _______ effect: 2:172

A. Scattering
B. Absorption
C. Nuclear
D. Filtration

20. A patient has received a radiographic examination of the chest. The technologist employed 5 mAs at 90 kV. The dose this patient receives is dependent upon: 13:29
1. Compton effect *2. Photodisintegration* *3. Photoelectric effect*

A. 1 & 2 only
B. 1 & 3 only
C. 2 & 3 only
D. 1, 2 & 3

21. The interaction of a x-ray photon and an orbital electron, which results in a partial transfer of energy, is termed: 2:172

A. Photoelectric effect
B. Pair production
C. Compton effect
D. Thomson effect

22. Pair production is a common reaction occurring in the photon energy range of: 2:179

A. 30-100 keV
B. 150-300 keV
C. 500-900 keV
D. 1-10 MeV

23. The common radioactive emissions, alpha, beta and gamma, all have the same origin in the atom being the: 2:52

A. Orbital shells
B. Compton area
C. Nucleus
D. Valence area

24. A 70 KeV photon strikes an inner shell electron, and all of the photon energy is transferred to the electron. This describes: 13:29

A. Coherent scattering
B. Compton scattering
C. Photoelectric effect
D. Photo-disintegration

25. An incident x-ray photon of 37 keV strikes an electron with a 40 keV binding energy. A scattered photon results with 37 keV of energy. This describes a: 13:28

A. Compton event
B. Brems event
C. Photoelectric event
D. None of the above

26. If a 60 keV electron strikes an inner shell electron with a binding energy of 26 keV, the resulting photoelectron will have a: 2:44

A. 60 keV
B. 34 keV
C. 32 keV
D. 26 keV

27. An incoming photon of 1 MeV interacts with an orbital electron causing ionization of the atom and a .9 MeV scattered photon. This describes an event called: 13:33

A. Pair production
B. Photoelectric interaction
C. Compton scattering
D. None of the above

28. An alpha particle consists: 2:48
1. Two neutrons 2. Two protons 3. Two electrons

A. 1 & 2 only
B. 1 & 3 only
C. 2 & 3 only
D. 1, 2 & 3

29. The number of Compton interactions is primarily dependent upon the ________ of the target atom: 5:66
1. Atomic number 2. Angle of interaction 3. Electron density

A. 1 only
B. 2 only
C. 3 only
D. 1, 2 & 3

30. The amount of energy absorbed by an irradiated object is primarily controlled by: 2:181
1. Speed of the incident photon 2. Atomic number of the tissue 3. Energy of incident photon

A. 1 & 2 only
B. 1 & 3 only
C. 2 & 3 only
D. 1, 2 & 3

31. At energies above 10 MeV, x-ray photons may interact directly with the atomic nucleus causing the emission of nuclear fragments in a process termed: 2:180

A. Photo-disintegration
B. Annihilation reaction
C. Triplet production
D. Magnetic resonance

32. The mass-energy equivalency of a positron or electron at rest velocity is: 2:180

A. .255 MeV
B. .511 MeV
C. 1.0 MeV
D. 1.022 MeV

33. When mass is converted into energy, which of the following equations is used to express their convertible natures? 2:71

A. $E = hv$
B. $E = mc^2$
C. $\lambda = \overline{kV}$
D. $C = v\,\lambda$

34. Which of the following members of the electromagnetic spectrum will possess the greatest mass-energy equivalency? 2:72

A. 20 MeV x-ray photon
B. 3 MeV gamma photon
C. 700 eV ultraviolet photon
D. 100 eV microwave

35. The incoming photon before any interaction with matter takes place is called the: 2:172

A. Incident photon
B. Scattered photon
C. Inherent photon
D. Recoil photon

Detection and Measurement of Radiation

Chapter 3

1. The linear energy transfer is a measure of the rate at which energy is transferred from ionizing radiation into: 2:469

   A. Air
   B. Soft tissue
   C. Bone
   D. Any matter

2. The principal active component of a thermoluminescent dosimeter is a special crystal called: 13:255

   A. Silver bromide
   B. Sodium iodide
   C. Lithium fluoride
   D. Calcium tungstate

3. The location of a lost radioactive source is most easily accomplished by using a: 13:257
   1. Survey meter *2. Pocket dosimeter* *3. Scintillation counter*

   A. 1 only
   B. 2 only
   C. 3 only
   D. 1, 2 & 3

4. When film badges are employed for radiation monitoring, which of the following factors may effect the reading of the badge: 13:257
   1. Exposure to excessive heat *2. Submersion of the badge in fluids* *3. Damage to the film wrapper*

   A. 1 only
   B. 2 only
   C. 3 only
   D. 1, 2 & 3

5. The roentgen is a unit of measurement that specifies the __________ of x-ray or gamma in air. 13:50

   A. Quality
   B. Quantity
   C. Absorption
   D. Characteristic nature

6. The rate of decay of a radionuclide is effected by: 2:49

   A. Chemical bonding
   B. Temperature changes
   C. Pressure changes
   D. None of the above

7. The most sensitive device for the detection of gamma radiation is a special type of radiation device called a/an: 2:640

   A. Scintillation counter
   B. Ionization chamber
   C. Film badge
   D. Pocket dosimeter

8. The amount of x-ray or gamma exposure resulting in the production of 2.58^4 x 10 coulombs in 1 Kg of air defines the: 2:15

   A. Rad
   B. Seivert
   C. LET
   D. Roentgen

9. A person is exposed to 100 rads of beta having a quality factor (QF) of 1. This person's dose equivalency is: 2:16

   A. 100 Curie
   B. 100 Becquerel
   C. 100 Rem
   D. 10 Roentgen

10. Which of the following is not a basic component of a film badge? 2:659

A. Plastic holder
B. Various metal filters
C. Light-tight pocket
D. Self-reading meter

11. A patient is exposed to 70 rads of 1 MeV neutrons having a quality factor (QF) of 5. This person's dose equivalency is: 13:52

A. 14 rem
B. 28 rad
C. 135 rad
D. 350 rem

12. The primary portion of most types of survey instruments is a gas filled chamber called a/an: 13:260

A. Cloud chamber
B. Cathode ray chamber
C. Ionization chamber
D. Valve chamber

13. The unit of absorbed dose, the rad (gray), is employed in the measurement of: 2:16
1. Alpha *2. Neutrons* *3. Gamma rays*

A. 1 only
B. 2 only
C. 3 only
D. 1, 2 & 3

14. The primary difference between a Geiger-Muller detector and cutie pie detector is: 2:637

A. Basic operating principle
B. Method of display
C. Method of detection
D. Operating potential

15. Because of environmental factors, the manufacturers recommend _________ as the maximum period of time that the film badge may be worn for personnel monitoring: 13:252

A. Two weeks
B. One month
C. 3 months
D. 6 months

16. Due to its high threshold to radiation, which type of radiation monitoring is considered unacceptable? 13:121

A. Photographic
B. Chemical
C. Biologic
D. Scintillation

17. The dose of ionizing radiation that, as far as we know, a person can absorb without sustaining appreciable bodily injury is termed: 13:252

A. ALARA
B. UNSCEAR
C. NCRP
D. MPD

18. When lead aprons are employed, a radiation monitoring device should be worn at the level of the: 13:244

A. Waist outside of the apron
B. Waist under the apron
C. Collar outside of the apron
D. Collar under the apron

19. A film badge can be used for the measurement of (a) ___________ source(s): 2:659
1. Alpha *2. Beta* *3. Gamma*

A. 1 & 2 only
B. 1 & 3 only
C. 2 & 3 only
D. 1, 2 & 3

20. The radiation unit best suited for measuring the output of a x-ray tube is the: 2:15

A. roentgen
B. rad
C. rem
D. gray

21. The instrument used for absolute exposure dose measurements is the: 7:63

A. Baldwin-Farmer ionization chamber
B. Victoreen condenser chamber
C. Geiger-Muller counter
D. Direct reading dosimeter

22. A person is exposed to 700 mR of x-rays is during a series of radiographic studies in a two week period of time. This individual's dose equivalent is approximately: 2:16

A. .7 rem (.07 Sv)
B. 14 becqueral (.14 mCi)
C. 35 rad (.35 Gy)
D. 1.4 rem (.14 Sv)

23. The smallest detectable dose of a standard film badge is approximately: 13:247

A. 1.0 mrem
B. 10 mrem
C. 50 mrem
D. 100 mrem

24. The only personal monitoring device which will enable a radiographer to determine there own dose immediately following a procedure is a: 13:25

A. TLD
B. G-M Counter
C. Film badge
D. Pocket dosimeter

25. Personnel monitoring is used to help insure that radiation: 13:245

A. Sources are properly shielded
B. Exposures are kept below effective dose limits
C. Exposures are maintained at the effective dose limit
D. All exposures are eliminated

26. The smallest radiation exposure that can be detected by a biological method is approximately: 2:559

A. 5 mrem
B. 100 mrem
C. 1,000 mrem
D. 25,000 mrem

27. The radiation unit derived when a rad is multiplied by the quality factor results in the unit of dose equiv9alency termed the: 2:16

A. roentgen
B. LET
C. rem
D. rep

28. Which type of radiation survey instrument is best suited for the detection of a high intensity radiation field: 13:260

1. G-M counter *2. Scintillation counter* *3. Cutie pie*

A. 1 only
B. 2 only
C. 3 only
D. 1, 2 & 3

29. Which of the following is classified as a rate meter? 13:260

A. Geiger-Muller counter
B. Thermoluminescent dosimeter
C. Pocket dosimeter
D. Film badge

30. The roentgen is most closely related to which other type of exposure? 2:15

A. Primary radiation
B. Secondary radiation
C. Remnant radiation
D. Secondary scatter radiation

31. A patient receives a dose of 2 gray of alpha having a quality factor (QF) of 20. This patient's dose equivalency is: 2:16

A. 10 rem
B. 20 seivert
C. 40 seivert
D. 40 rem

32. The amount of energy that is transferred to a material per unit length of travel is termed: 13:49

A. RBE
B. LET
C. rem
D. rad

33. The short interval of time during which a radiation counter is unable to respond to radiation is called: 2:536

A. Interval time
B. Spot time
C. Dead time
D. Recycle time

34. A patient receives a dose of 3 gray of fast moving neutrons with a quality factor (QF) of 10. The total dose equivalency of this patient is: 2:16

A. 30 rem
B. 30 seivert
C. 300 rem
D. 300 seivert

35. Pocket ionization chambers are not normally employed in a radiology personnel monitoring program because: 13:254
1. They require daily attention and recording *2. Provide only short term measurements*
3. They do not keep a permanent record

A. 1 only
B. 2 only
C. 3 only
D. 1, 2 & 3

36. Which is a commonly employed device used for the radiation monitoring of a radiographer? 13:255

A. Scintillation counter
B. Geiger-Muller counter
C. Thermoluminescent dosimeter
D. Victoreen R meter

37. The sensitivity of G-M counter is directly related to its: 2:637

A. Starting potential
B. Operating potential
C. Threshold potential
D. Discharge potential

38. A quenching agent is used in a G-M counter to improve the: 10:72

A. Resolving time
B. Hermatic seal integrity
C. Spectroscopic sensitivity
D. Pulse height discrimination

39. A 300 rad exposure is equivalent to a: 2:16

A. 0.3 gray
B. 3 gray
C. 30 gray
D. 300 gray

40. The linear energy transfer of a x-ray or gamma photon is low because of its: 2:534

A. Low energy
B. High penetration
C. Negative charge
D. Positive charge

41. A person receives an exposure to the following radiations. Which will result in the highest dose equivalency? 13:53

A. 4 gray of x-ray
B. 4 gray of beta
C. 4 gray of gamma
D. 4 gray of alpha

42. Which of the following nuclear emissions possesses the greatest threat to the internal organs from external exposures? 2:52
1. Alpha *2. Beta* *3. Gamma*

A. 1 only
B. 2 only
C. 3 only
D. 1, 2 & 3

43. The unit of absorbed dose is equivalent to an energy transfer of 100 ergs per gram of any matter and is termed the: 13:51

A. Rem
B. Rad
C. Roentgen
D. Seivert

44. The standard international unit for measurement of absorbed dose for all types of ionizing radiations is the: 13:51

A. Gray
B. Seivert
C. Becquerel
D. Curie

45. The curie is the unit used in the measurement of the quantity of: 2:16

A. Radioactivity
B. Ionization
C. Absorbed dose
D. Exposure dose

Pertaining to the ionization chamber voltage curve, answer questions 46-49.

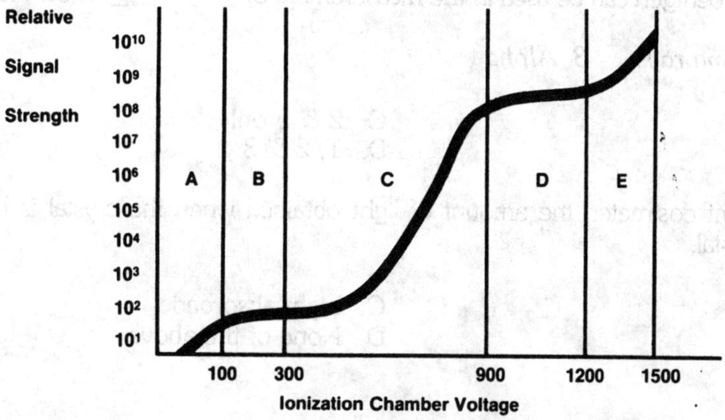

46. The area of the chamber voltage curve, letter D, represents the primary operating region for the: 10:26

A. Proportional counter
B. Geiger-Muller counter
C. Ionization chamber
D. None of the above

47. Ionization chambers require a voltage in which no recombination of ion pairs take place. This region is identified by: 10:26

A. Letter A & B
B. Letter C
C. Letter D
D. Letter E

48. In region C of the chamber voltage curve, the number of secondary ion pairs is proportional to the primary number of ion pairs produced and is termed the: 10:26

A. Threshold region
B. Recombination region
C. Proportional region
D. Glow region

49. In region E, the voltage is sufficient to cause spontaneous ionization of the chamber gases and is known as the region of: 10:26

A. Partial recombination
B. Maximum sensitivity
C. Minimum sensitivity
D. Continuous discharge

50. The calibration of an ionization type meter can be made by a comparison with a __________ radiation source. 10:83

A. High intensity
B. Short-lived
C. Long-lived
D. Alpha emitting

51. Which type of radiation detection devices will require the use of a photo multiplier tube to detect the light emitted from the crystal source: 2:642

1. Scintillation counter *2. Thermoluminescent dosimeter* *3. Pocket dosimeter*

A. 1 & 2 only
B. 1 & 3 only
C. 2 & 3 only
D. 1, 2 & 3

52. An ionization-type survey meter can be used in the detection of: 13:257

1. Primary radiation *2. Modulation transfer function* *3. Battery voltage*

A. 1 only
B. 2 only
C. 3 only
D. 1, 2 & 3

53. The most commonly employed inorganic crystal in a scintillation type counter is: 10:41

A. Calcium tungstate
B. Sodium iodide
C. Silver bromide
D. Calcium fluoride

54. The radiation unit the roentgen can be used in the measurement of ________ below the energy of 3 MeV. 13:49

1. X-rays *2. Gamma rays* *3. Alpha*

A. 1 & 2 only
B. 1 & 3 only
C. 2 & 3 only
D. 1, 2 & 3

55. In a thermoluminescent dosimeter, the amount of light obtained when the crystal is heated is related to the ________ by the crystal. 10:93

A. Energy absorbed
B. Mass absorbed
C. Light absorbed
D. None of the above

Protection of Personnel

Chapter 4

1. Primary barriers are those located perpendicular to the line of travel of the primary beam, and in the normal diagnostic range should have a thickness of: 13:217

   A. 0.5mm lead (1/40")
   B. 0.75mm lead (1/32")
   C. 1.5mm lead (1/16")
   D. 2.5mm lead (1/10")

2. Which is a type of gonadal shielding device currently available? 13:177
   1. Flat contact shield 2. Shadow shield 3. Shaped contact shield

   A. 1 only
   B. 2 only
   C. 3 only
   D. 1, 2 & 3

3. After a radioisotope is injected into the body, the rate at which is expelled by the body is termed the: 2:49

   A. Physiologic half-life
   B. Biologic half-life
   C. Effective half-life
   D. Affected half-life

Pertaining to the diagram of the effects of filtration on quantity of X-rays, answer questions 4-6.

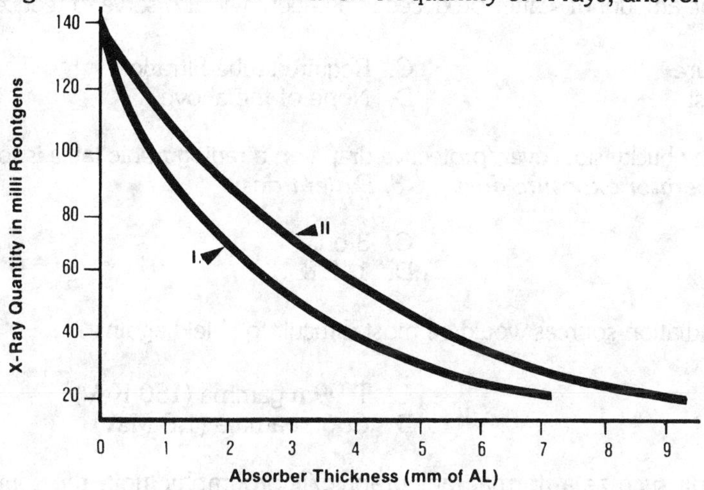

4. The half valve layer for curve I is approximately: 2:167

   A. 2 mm Al
   B. 3 mm Al
   C. 4 mm Al
   D. 6 mm Al

5. In order to reduce the intensity of curve II to 25% of its original value will require about: 2:167

   A. 3 mm Al
   B. 6 mm Al
   C. 8 mm Al
   D. 9 mm Al

6. Comparing the curves I and II, which would represent the beam with the highest intensity: 2:167

   A. Curve I
   B. Curve II
   C. Unable to determine

7. The principal source of scatter when performing a radiographic examination is the: 13:215

   A. Primary shutters of the collimator
   B. Patient being radiographed
   C. Floor of the radiographic room
   D. Radiographic table top

8. A primary protective barrier must extend upwards from the floor to a height of: 13:217

A. 3′ (1 meter)
B. 5′ (1.5 meters)
C. 7′ (2.1 meters)
D. 10′ (3 meters)

9. Which of the following technical factors or equipment will best reduce the gonadal dose? 13:184
1. Low mAs techniques *2. Radiographic grids* *3. Collimation*

A. 1 & 2 only
B. 1 & 3 only
C. 2 & 3 only
D. 1, 2 & 3

10. During a radiographic procedure if the patient dose is increased by a factor of two, the dose to the radiographer should:

A. Increase 2 times
B. Increase 3 times
C. Decrease by 2 times
D. Radiographer's dose is not effected

11. During fluoroscopy, a protective drape or sliding panel must be used to intercept scattered radiation from the patient. This should have a lead equivalency of at least: 13:225

A. .25mm
B. .50mm
C. .75mm
D. 1.5mm

12. Any device that reduces the amount of scatter produced in a patient will also serve to reduce: 13:213

A. Occupational exposure
B. Radiographic contrast
C. Required tube filtration
D. None of the above

13. The primary purpose of the bucky slot cover/protective drape on a radiographic table is to reduce: 13:225
1. Back scatter *2. Operator exposure dose* *3. Patient dose*

A. 1 only
B. 2 only
C. 3 only
D. 1, 2 & 3

14. Which of the following radiation sources would be most difficult to shield against?

A. 50 KeV x-ray
B. 250 KeV x-ray
C. Tc99m gamma (150 KeV)
D. Co60 gamma (1.5 MeV)

15. In a fixed fluoroscopic unit such as a normal radiographic/fluorographic unit, the source-to-skin distance shall not be less than: 20:10

A. 12″ (30 cm)
B. 15″ (38 cm)
C. 20″ (52 cm)
D. 25″ (64 cm)

16. The length of time it takes for a radionuclide to reduce its intensity to one-half of its original intensity by a normal decaying process is called the: 2:49

A. Half-value layer
B. Effective half-life
C. Biological half-life
D. Physical half-life

17. A secondary protective barrier must have a lead equivalency of at least: 2:633

A. .75mm Pb (1/32″)
B. 1.5mm Pb (1/16″)
C. 2.5mm Pb (1/8″)
D. 5mm Pb (1/4″)

18. When fluoroscopy is performed, aprons should be worn by the operator and must have an equivalency of at least: 20:18

A. .25mm lead
B. .5mm lead
C. 1.5mm aluminum
D. 2.5mm aluminum

19. The half-life of a radioactive substance: 2:49
1. Varies with time 2. Is characteristic of the isotope 3. Is constant

A. 1 & 2 only
B. 1 & 3 only
C. 2 & 3 only
D. 1, 2 & 3

20. Which of the following represents the amount of protective lead required around the average x-ray tube? 13:223

A. 1.5mm
B. 2.0mm
C. 2.5mm
D. 3.5mm

21. In order to minimize the radiographer's exposure while operating fixed radiographic equipment, at least how many times should scattered x-rays scatter before they can enter the shielded booth where the operator stands? 7:111

A. Two times
B. Three times
C. Five times
D. Ten times

22. Which of the following would lower a radiographer's dose more during a fluoroscopic study in which their presence in the room is required? 2:67

A. Having flouroscopic kilovoltage setting increased by 15%
B. Doubling the distance the radiographer stands from the table
C. Reducing the flouroscopic output by 30%
D. Reducing the flouroscopic time by 40%

23. In order to prevent an occupational worker from potentially exceeding the annual effective dose limit, the shielding for a controlled area must reduce the exposure to less than:

A. 5mR/week
B. 10mR/week
C. 50mR/week
D. 100mR/week

24. A fixed cone or spacer of not less than 30 cm (12") must be provided on ________ to insure this minimum source-to-skin distance: 20:11
1. Stationary radiographic units 2. Mobile radiographic units 3. Stationary flouroscopic units

A. 1 only
B. 2 only
C. 3 only
D. 1, 2 & 3

25. Which of the following would serve as the best method for protection from a high energy-gamma source? 2:602

A. Wearing a lead apron
B. Wearing a rubber glove
C. Increasing the distance from the source
D. Increasing the time you are exposed

26. The percentage of scattered radiation emanating from a patient during fluoroscopy is generally considered to be ________ of the primary beam when measured at one meter from the patient: 2:633

A. 0.1%
B. 0.5%
C. 1.0%
D. 10.0%

27. A diagnostic-type protection housing should reduce leakage radiation when a tube is operated at its leakage technique to an air kerma level less than ________ at a distance of one meter. 20:11

A. 5 rad (5cGy)/hr.
B. 1 rad (5cGy)/hr.
C. .5 rad (.5cGy)/hr.
D. .1 rad (.1cGy)/hr.

Pertaining to the diagram of the half-life of this radioactive isotope, answer question 28.

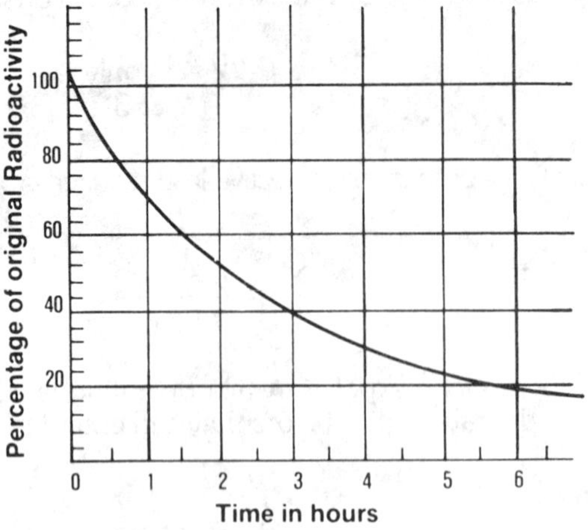

28. In the diagram, the amount of time it takes the radionuclide to decay to 50% of its original value is approximately:

A. 30 minutes
B. 1 hour
C. 2 hours
D. 4 hours

29. A scattered photon at 90 degrees and one meter from an object reduces intensity by about 1000 times. A beam that undergoes two scatterings will reduce the original intensity by about: 2:633

A. 10^{2} times
B. 10^{3} times
C. 10^{4} times
D. 10^{6} times

30. If the activity of a radioactive source is 400 mR/hr. and has a half-life of 6 hours, what will the activity of this source be in one day? 2:49

A. 133 mR/hr.
B. 25 mR/hr.
C. 12.5 mR/hr.
D. 6.25 mR/hr.

31. A radioactive source has a half-life of eight hours. How long would it take to reduce a 200 millicurie/hr. source to a level of 12.5 millicurie/hr.? 2:49

A. 8 hours
B. 16 hours
C. 24 hours
D. 32 hours

32. A radioactive source has a half-life of six hours. How long would it take to reduce a 200 millicurie/hr. source to a level of 12.5 millicurie/hr.? 2:49

A. 24 hours
B. 18 hours
C. 12 hours
D. 6 hours

33. If the exposure rate of a x-ray beam is initially 50 R/min., and it requires 2mm Al to decrease the exposure rate, to 25% of the original value, the half-value of this beam is: 2:175

A. 1mm Al
B. 2mm Al
C. 4mm Al
D. None of the above

34. In regard to the three important factors for radiation protection, the lowest exposure is possible if the: 2:602
1. Distance from a source is as short as possible *2. Time of exposure is as short as possible*
3. Shielding is as thin as possible

A. 1 only
B. 2 only
C. 3 only
D. 1, 2 & 3

35. For the purpose of radiation protection, a building can be divided into controlled and uncontrolled areas. Which area is generally considered a uncontrolled area? 13:221

1. Radiology waiting room *2. The darkroom* *3. File room*

A. 1 & 2 only
B. 1 & 3 only
C. 2 & 3 only
D. 1, 2 & 3

36. If it is necessary to restrain a patient during a radiographic examination, the least acceptable person to do this is a/an: 20:9

A. 50-year-old female radiology secretary
B. 25-year-old male technologist
C. 28-year-old pregnant friend of the patient
D. 18-year-old orderly

37. As the atomic number of a substance increases, the half-value layer thickness: 2:167

A. Increases
B. Decreases
C. Remains the same

38. The intensity of a beam is measured at 72 mR for a 200 mA, 150 ms, 77 kVp exposure. If all other factors remain unchanged, the approximate intensity at 300 mA, 200 ms, and 77 kVp will be: 2:653

A. 199 mR
B. 168 mR
C. 144 mR
D. 126 mR

39. Lead is usually used in the walls of diagnostic x-ray rooms instead of concrete because::

A. A smaller total mass of lead is required
B. Mass for mass, lead walls are cheaper than concrete
C. Lead is a good structural material
D. The protection calculations are easier

40. Most new fixed radiographic units have a short or fixed exposure switch. This is to help insure that the operator: 20:20

1. Will not use excessive prep times *2. Will use the proper technique*
3. Must remain behind the shielded booth

A. 1 only
B. 2 only
C. 3 only
D. 1, 2 & 3

41. The use of finger badge monitoring is of particular importance to those persons involved with the: 10:92

A. Holding of infants
B. Positioning of patients
C. Injections of contrast medias
D. Injections of radionuclides

42. In order to maintain the radiation exposure of a nonoccupational worker to within effective dose limit the walls, floors etc. for an uncontrolled area must reduce the air kerma to less than: 10:166

A. .01 cGy (.01 rad)/week
B. .1 cGy (.1 rad)/week
C. .5 cGy (.5 rad)/week
D. 1 cGy (1 rad)/week

43. The factors that must be considered in the determination of the thickness for a radiation barrer is/are: 10:169

1. Weekly workload *2. Use factors* *3. Occupancy factors*

A. 1 only
B. 2 only
C. 3 only
D. 1, 2 & 3

44. According to the ALARA concept of the N.C.R.P., the radiation exposure to an individual should be: 13:65

A. Kept at permitted MPD's
B. Carefully monitored at all times
C. Kept as low as possible
D. Kept on all children

45. The fraction of time that a radiation beam is directed at a specific barrier or area is termed the: 10:169

A. Use factor
B. Workload factor
C. Occupancy factor
D. Attenuation factor

In the following place an "A" next to a factor which is principally to protect a patient, a "B" next to a factor which is principally used to protect the radiographer or "C" next to a factor which is applicable to both persons.

46. ________ The use of diagnostic filtration

47. ________ The use of collimation during a fluoroscopic procedure

48. ________ The use of primary protection barriers

49. ________ The use of a lead apron by a patient

50. ________ The use of a protective curtain during fluoroscopy

Patient Protection

Chapter 5

1. Gonadal shielding should be employed if the: 13:173
   1. Person has a reasonable reproductive potential 2. Gonads lie within or close to the exposure field 3. Shielding does not interfere with the objectives of the study

   A. 1 only
   B. 2 only
   C. 3 only
   D. 1, 2 & 3

2. Patient dose in fluoroscopy may be minimized by: 13:90

   A. Leaving the shutters as wide open as possible
   B. Increasing fluoro mA
   C. Restricting field size
   D. Reducing SID

3. A x-ray machine operating at potentials above 70 kVp require at least ________ of tube filtration: 13:172

   A. 1.5mm aluminum
   B. 2.5mm aluminum
   C. 1.5mm lead
   D. 2.5mm lead

   Pertaining to the radiographic image, answer question 4.

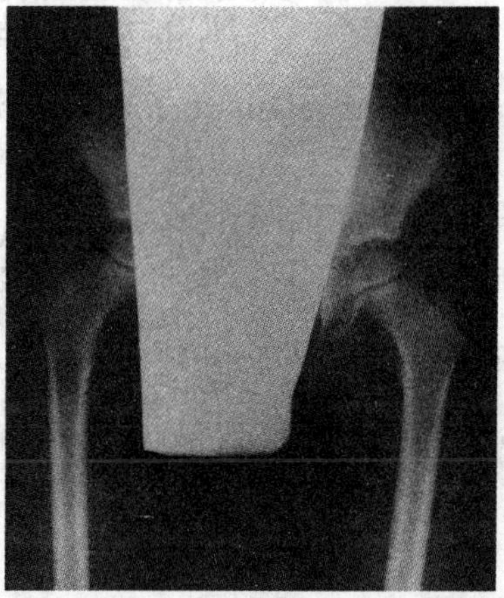

4. During the radiographic examination of the hips of this male patient, choose the statement that best describes the shielding employed. 2:655

   A. The gonads are properly shielded
   B. No shielding should be employed
   C. The shield covers important anatomy
   D. All required structures are seen

5. When possible, women of childbearing ages should be radiographed during the ________ of the menstrual cycle: 2:612

   A. First third
   B. Middle third
   C. Last third
   D. Any of the above

6. If the filtration for a radiographic unit were inadvertently omitted for a series of chest exposures the greatest increase in dose would occur in the: 10:564

   A. Diaphragm
   B. Skin of the chest
   C. Mediastinul structures
   D. Pulmonary vessels

7. The principal reason that diagnostic filtration must be employed for almost all radiographic studies is to: 2:167

A. Provide a higher contrast
B. Reduce skin exposures
C. Decrease beam quality
D. Reduce scatter production

8. Which of the following exposure and technical factors will result in a patient receiving the lowest dose? 2:157

A. 300 mA, 30 ms, 90 kVp 200 speed screen
B. 200 mA, 30 ms, 95 kVp 100 speed screen
C. 400 mA, 30 ms, 85 kVp 50 speed screen
D. 100 mA, 30 ms, 85 kVp 200 speed screen

9. The use of which device for the reduction of scattered radiation will result in an increased patient exposure? 13:185

A. Aperture diaphragm
B. Collimator
C. Grid
D. None of the above

10. Whenever possible a PA projection of the thorax instead of an AP projection is preferred to: 20:8

A. Reduce involuntary motion
B. Improve anatomic relationships
C. Decrease skin-to-image distance
D. Reduce exposures to the breast

11. The principal advantage of image intensified fluoroscopy as compared to conventional fluorography is a substantial reduction of: 13:190

A. Patient exposure
B. Operator exposure
C. Both of the above
D. Neither of the above

12. To protect the unborn fetus of a pregnant technologist, the following recommendation is generally made: 2:608

A. Quit work immediately
B. Do not work in the x-ray department
C. Work in the low radiation areas only
D. No special precautions are required

13. The number of repeated radiographs can be reduced by employing: 13:187
1. Effective communication skills *2. Careful selection of exposure factors*
3. Proper immobilization techniques

A. 1 only
B. 2 only
C. 3 only
D. 1, 2 & 3

14. Which of the following film-holder combinations will best reduce exposure during a radiographic examination? 2:221

A. 50 speed film-screen system
B. 200 speed film-screen system
C. 400 speed film-screen system
D. Non screen system

15. Which of the following would help in reducing the radiation expsoure to a child? 2:227
1. Use of gonadal shields *2. Use of high speed screen* *3. Use of a tightly collimated beam*

A. 1 only
B. 2 only
C. 3 only
D. 1, 2 & 3

16. The principal advantage of a rare earth imaging system as compared to a calcium tungstate imaging system is: 2:227

1. Less patient exposure *2. Less distortion* *3. Improved resolution*

A. 1 only
B. 2 only
C. 3 only
D. 1, 2 & 3

17. The use of image intensified fluoroscopy increases the brightness of an image with a substantial reduction in _________ as compared to conventional fluorography. 2:354
1. Resolution *2. Patient dose* *3. Visual acuity*

A. 1 only
B. 2 only
C. 3 only
D. 1, 2 & 3

18. The total amount of radiation received by a patient is effected by the: 13:191
1. Size of the field *2. Amount of beam filtration* *3. mAs*

A. 1 & 2 only
B. 1 & 3 only
C. 2 & 3 only
D. 1, 2 & 3

19. Which of the following techniques should be utilized if a reduction of patient dose is desirable? 2:157

A. 100 mA, 50ms, 80 kVp
B. 400 mA, 25ms, 80 kVp
C. 400 mA, 8ms, 90 kVp
D. 200 mA, 24ms, 90 kVp

Pertaining to the radiographic image, answer question 20.

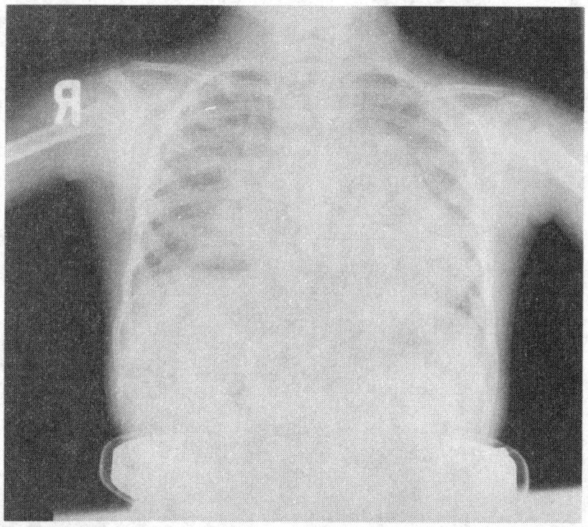

20. In the radiograph, the low density area in the bottom of the film indicates the use of:

A. Leaded waist apron
B. Shaped contact shield
C. Compensating bolus
D. Excessive collimation

21. Which of the following type of spot films will result in the lowest patient dose for a single exposure? 2:199

A. 70 mm photo (cine) spot film
B. 105 mm photo (cine) spot film
C. 8-10" cassette type spot film
D. 9x9" cassette type spot film

22. Which of the following exposure and technical factors will result in a patient receiving the lowest dose? 2:157

	mA	Time	kVp	Screen Speed	Grid
A.	100	100ms	87	250	12:1
B.	200	80ms	88	200	12:1
C.	100	150ms	84	50	6:1
D.	200	20ms	86	400	5:1

23. A diagnostic fluorographic unit operating at 2 mA, 80 kVp with 2.5mm Al filtration has an output of 150 roentgens per hour of operation at the table top. How long would it take this unit to deliver 375 roentgens? 12:189

A. 1.5 hours
B. 2.0 hours
C. 2.5 hours
D. 3.0 hours

24. In the following radiographic examinations list from greatest to least the expected fetal dose for a pregnant female. 2:614

1. AP hip image *2. Kidney, ureter, bladder (KUB)* *3. AP thoracic spine image*

A. 1, 2, 3
B. 2, 1, 3
C. 3, 2, 1
D. 1, 3, 2

25. A radiographic exposure of 200 mA at 100ms., 70 kVp is made registering 50 mR per exposure. Assuming no other changes are made, the exposure should produce about _______ at a 600 mA, 200 ms, 70 kVp technique. 2:650

A. 12.5 mR
B. 25 mR
C. 100 mR
D. 300 mR

26. When gonad shielding is employed during fluoroscopy, the radiographer should place the shield:

A. On the lateral surface of the table
B. Under the patient
C. Over the patient
D. No shielding can be used

Pertaining to the radiographic image, answer question 27.

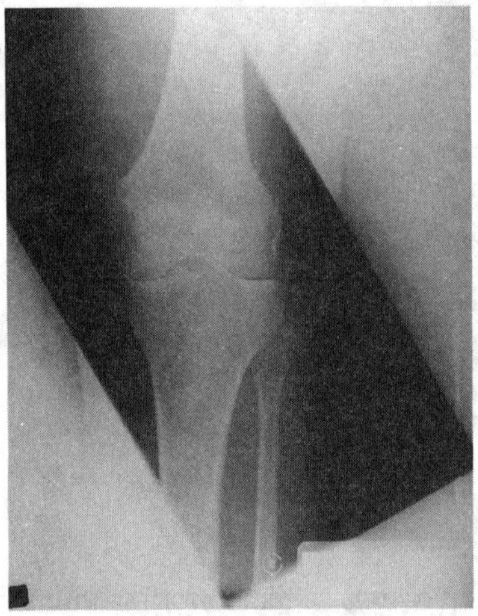

27. In the radiograph, what technical error caused the unnecessary radiation exposure to the patient: 2:243

A. Improper focal spot selection
B. Removal of the inherent filtration
C. Improper collimator orientation
D. Failure to collimate

28. During a right PA oblique (RAO) position of the stomach the area of the body receiving the greatest exposure is the: 2:633

A. Skin on the left dorsal surface
B. Internal organs of the right abdomen
C. Skin on the right ventral surface
D. Internal organs of the left abdomen

29. Whenever possible, women of childbearing ages should have radiographic examinations scheduled during the __________ of the menstrual cycle: 2:612

A. First 10 days
B. 10th - 20th day
C. Last 10 days
D. Any of the above

30. Which of the following factors will have the least effect on the radiation exposure to the patient if the mAs and kVp remain constant. 13:191

A. Amount of filtration
B. Source-to-image distance
C. The size of the exposure field
D. Focal spot size

Radiation Regulations

Chapter 6

1. According to the NCRP recommendations, the guidance level for cumulative exposures should be calculated using the formula: 19:4

   A. Age x 50 mSv
   B. 5 mSv (age - 18)
   C. 10 mSv x age
   D. $(\text{age x } 2\text{ mSv})^2$

2. The effective dose limits are based on this type of dose relationship. 2:603

   A. Linear, non-threshold
   B. Non-linear, threshold
   C. Linear, energy transfer
   D. Non-linear, energy transfer

3. When calculating the annual effective dose for a technologist, which of the following exposures should be considered? The exposure received: 13:75

   A. From your own thyroid uptake study
   B. From your own exposure while helping in a fluoroscopic room
   C. From nuclear fallout
   D. From your own dental radiographs

4. The effective dose limit are generally stated in terms of dose equivalents and are expressed in the units of: 2:16

   A. Roentgen or Seivert
   B. Rem or Becquerel
   C. Rem or Seivert
   D. Rad or Curie

5. Most of the currently employed radiation standards are acquired from the: 13:69

   A. Radiologic Committee for Protection
   B. National Radiation Protection Committee
   C. National Council on Radiation Protection and Measurement
   D. American Council for Protection of Radiation

6. The NCRP recommends that the effective dose limit to individuals in the general population be limited to _________ the occupational limit. 13:73

   A. 1/100 of
   B. 1/10 of
   C. 2 times
   D. 10 times

7. The smallest amount of artificial exposure to the general population is derived from: 2:6

   A. Nuclear fallout
   B. Medical x-rays
   C. Cosmic radiation
   D. Terrestrial gamma sources

8. It is recommended that the fetus should not exceed a dose equivalent limit of _________ for the entire gestation period. 19:5

   A. 1 mSv (.1 rem)
   B. 5 mSv (.5 rem)
   C. 15 mSv (1.5 rem)
   D. 30 mSv (3 rem)

9. The effective dose limit for the stochastic effects for an occupational radiation worker in a one year period is: 19:4

   A. .05 mSv (5m rem)
   B. 10 mSv (1 rem)
   C. 50 mSv (5 rem)
   D. 250 mSv (25 rem)

10. According to the NCRP, the air kerma at the level of the patient of a fluoroscopic unit shall not exceed: 20:16

A. 1 cGy (1 rad) per min
B. 3 cGy (3 rad) per min
C. 5 cGy (5 rad) per min
D. 10 cGy (10 rad) per min

11. On rare occasions when essential radiation related tasks must be performed by an occupational worker, a single emergency occupational exposure limit of _________ may be permitted. 19:44

A. .5 Sv (50 rem)
B. 1.5 Sv (150 rem)
C. 35 Sv (300 rem)
D. 55 Sv (500 rem)

12. An occupational worker receives 80 mSv/yr. while working in a controlled area, 30 mSv from yearly x-rays, and 110 mSv from all other environmental sources. This person's occupational dose is: 13:11

A. 30 mSv
B. 80 mSv
C. 140 mSv
D. 170 mSv

13. The largest estimated average annual whole body radiation exposure from natural sources excluding radon are from: 2:6

A. Cosmic rays from the sun
B. Internal terrestrial sources
C. External terrestrial sources
D. Atmospheric ultraviolet rays

14. In the calculation of an occupational workers effective dose limit, consideration shall be given to: 19:21

A. Internal exposures
B. External exposures
C. Both of the above
D. Neither of the above

15. According to the NCRP recommendations, the guidance level for cumulative effective dose for a 28 year-old occupational worker with three years of experience is: 19:4

A. 28 mSv
B. 76 mSv
C. 280 mSv
D. 350 mSv

16. According to the NCRP recommendations, the annual effective dose limit for deterministic (somatic) effects to all tissues other than the eye should not exceed: 19:4

A. 50 mSv (5 rem)
B. 150 mSv (15 rem)
C. 300 mSv (30 rem)
D. 500 mSv (50 rem)

17. A reasonable effective dose limit for a radiographer under ALARA might by: 13:196

A. 150 mSv (15 rem) per year
B. 80 mSv (8 rem) per year
C. 5 mSv (.5 rem) per year
D. Any of the above

18. A somatic effect which increases in severity with increasing radiation close above the threshold is called a: 19:8

A. Deterministic effect
B. Committed effect
C. Optimization effect
D. Reference effect

19. A student radiographer is admitted into training six months prior to his/her 18th birthday. This person's effective dose limit until the age of 18 is: 19:53

A. 1 mSv (.1 rem)
B. 3 mSv (.3 rem)
C. 5 mSv (.5 rem)
D. 10 mSv (1 rem)

Refer to the statement below in answering questions 20-21:
A radiation worker received an occupational dose of 3 mSv, a natural background dose of 2 mSv, and had 50 mSv of medical exposures during the year.

20. The total occupational dose of radiation this person received during the year is: 13:11

A. 2 mSv
B. 3 mSv
C. 52 mSv
D. 53 mSv

21. The total dose of radiation this person received during the year is: 13:11

A. 2 mSv
B. 3 mSv
C. 52 mSv
D. 55 mSv

22. An annual effective dose equivalent limit of 5 mSv (.5 rem) will apply to: 19:5
1. A pregnant radiographer *2. Embryo-fetus* *3. CT technologist*

A. 1 only
B. 2 only
C. 3 only
D. 1, 2 & 3

23. A thirty-year-old occupational worker has received a 6.5 mSv whole body exposure in the first month of employment. What is the maximum effective dose limit this person can receive in the next 11 months? 19:4

A. 3.5 mSv
B. 15.5 mSv
C. 23.5 mSv
D. 43.5 mSv

24. A pregnant radiographer receives a monthly badge report showing 70 mrem for the badge worn at the waist level. The approximate dose equivalent to the fetus is: 2:609

A. 170 mrem
B. 70 mrem
C. 62 mrem
D. 23 mrem

25. A high radiation area sign should be posted in an area where the exposure rate exceeds: 12:522

A. 10 mR/hr.
B. 100 mR/hr.
C. 100 mR/day
D. 100 mR/week

26. According to the NCRP, the annual effective dose limit for deterministic (somatic) effects for the lens of the eye is: 19:4

A. 10 mSv (1 rem)
B. 30 mSv (3 rem)
C. 50 mSv (5 rem)
D. 150 mSv (15 rem)

27. Once a pregnancy becomes known, the exposure to an embryo-fetus shall be no greater than ________ per month: 19:5

A. .5 mSv (.05 rem)
B. 5 mSv (.5 rem)
C. 18 mSv (1.8 rem)
D. 36 mSv (3.6 rem)

28. A member of the public will have a maximum annual effective dose limit of ________ is recommended for infrequent annual exposures. 19:46

A. 50 mSv (5 rem)
B. 30 mSv (3 rem)
C. 10 mSv (1 rem)
D. 5 mSv (.5 rem)

29. The guidance level for cumulative exposures of a 36 year old occupational worker with 10 years of experience is: 19:48

A. 130 mSv (13 rem)
B. 360 mSv (36 rem)
C. 460 mSv (46 rem)
D. 900 mSv (90 rem)

30. Which of the following is considered to be a type of natural source of radiation exposure? 2:6
1. Nuclear fallout 2. Internal gamma sources 3. Medical x-rays

A. 1 only
B. 2 only
C. 3 only
D. 1, 2 & 3

31. The average annual dose equivalent for all monitored radiation workers was found to be about: 19:33

A. 46 mSv (4.6 rem)
B. 32 mSv (3.2 rem)
C. 19 mSv (1.9 rem)
D. 2.1 mSv (.21 rem)

32. A member of the public is accidentally exposed to 4 mSv (.4 rem). This person has exceeded the annual effective dose limit. 19:48

A. True
B. False

33. An occupational radiation worker is exposed to and average 675mrem per month. How many months can this individual work without exceeding the annual effective dose equivalent limit. 19:48

A. 2 months
B. 4 months
C. 6 months
D. 7 months

34. In order to remain within the permitted effective dose equivalent limits for an occupational worker, an individual should not receive more than: 19:48

A. 1 mSv/week
B. 5 mSv/week
C. 10 mSv/week
D. 20 mSv/week

35. Which of the following is normally considered a type of stochastic effect:
1. Cataract formation 2. Cancer 3. Genetic conditions

A. 1 & 2 only
B. 1 & 3 only
C. 2 & 3 only
D. 1, 2 & 3

For each of the following workers place an A if they are classified as an occupational or a B if they are classified as a non-occupational.

36. ________ Ultra sonographer with no general radiographic responsibilities.

37. ________ Darkroom technologist in the radiology department.

38. ________ CT technologist with no general radiographic responsibilities.

39. ________ Staff radiographer with general radiographic responsibilities.

40. ________ Mammographer with no general radiographic responsibilities.

Exercise No. 1

1. A radiographer receives an exposure of 100 mR/week at a 200 cm SSD. What would this person's exposure have been at a 100 cm SSD?

2.-10. State the new exposure that would be received by a radiographer for the following source-skin distances and original exposure factors.

	Original SSD	New SSD	Original Exposure	New Exposure
2.	40 inches	80 inches	100 mR/week	______
3.	72 inches	36 inches	50 mR/week	______
4.	40 inches	60 inches	10 mR/week	______
5.	72 inches	60 inches	100 mR/week	______
6.	100 cm	200 cm	50 mR/week	______
7.	180 cm	90 cm	10 mR/week	______
8.	200 cm	50 cm	100 mR/week	______
9.	100 cm	140 cm	50 mR/week	______
10.	60 cm	180 cm	10 mR/week	______

11. A radiographer receives an exposure of 5 mR/hr. standing at a 200 cm distance. In order to reduce this exposure to 1.25 mR/hr., the radiographer should move to a distance of _________.

12.-16. State the new source-to-skin distance which will result from the use of the following source-to-skin distance and exposure.

	Original Exposure	New Exposure	Original SSD	New SSD
12.	40 mR/hr.	160 mR/hr.	72 inches	______
13.	20 mR/hr.	5 mR/hr.	60 inches	______
14.	5 mR/hr.	45 mR/hr.	100 cm	______
15.	2 mR/hr.	16 mR/hr.	100 cm	______
16.	100 mR/hr.	11 mR/hr.	180 cm	______

17. The time it takes a patient to receive a dose of 1 gray is 10 minutes at a 50 cm SSD. How long would it take to receive a dose of 2 gray at a 100 cm SSD?

18. If the time to deliver a dose of 100 mrad at a 100 cm SSD is 5 minutes, what is the dose rate per minute at this distance?

19. If the time it takes to deliver a dose of 200 mrad is 1 minute at 100 cm SSD, what dose would be received in 3 minutes at a 200 cm SSD?

20. A patient receives an exposure of 100 mR at a 180 cm SSD. If the SSD is doubled with no other changes made, the exposure will be reduced by:

   A. 25%
   B. 50%
   C. 75%
   D. 200%
   E. 400%

The following formulas have been used to solve these problems.

$$\text{New Exposure} = \text{Original Exposure} \times \frac{\text{Original Distance (SSD)}^2}{\text{New Distance (SSD)}^2}$$

$$\text{New Distance}^2 = \text{Original Distance (SSD)}^2 \times \frac{\text{Original Exposure}}{\text{New Exposure}}$$

Exercise No. 2

1. A radiographer receives an exposure of 200 mR/week at a 300 cm SSD. What would this person's exposure have been at a 100 cm SSD?

2.-10. State the new exposure that would be received by a radiographer for the following source-skin distances and original exposure factors.

	Original SSD	New SSD	Original Exposure	New Exposure
2.	30 inches	90 inches	100 mR/week	________
3.	72 inches	60 inches	50 mR/week	________
4.	40 inches	80 inches	20 mR/week	________
5.	36 inches	72 inches	40 mR/week	________
6.	200 cm	100 cm	50 mR/week	________
7.	130 cm	90 cm	30 mR/week	________
8.	200 cm	50 cm	70 mR/week	________
9.	100 cm	150 cm	50 mR/week	________
10.	80 cm	200 cm	20 mR/week	________

11. A radiographer receives an exposure of 10 mR/hr. standing at a 200 cm distance. In order to reduce this exposure to 2.5 mR/hr., the radiographer should move to a distance of ________.

12.-16. State the new source-to-skin distance which will result from the use of the following source-to-skin distance and exposure.

	Original Exposure	New Exposure	Original SSD	New SSD
12.	30 mR/hr.	120 mR/hr.	72 inches	________
13.	60 mR/hr.	15 mR/hr.	36 inches	________
14.	50 mR/hr.	450 mR/hr.	100 cm	________
15.	3 mR/hr.	36 mR/hr.	240 cm	________
16.	200 mR/hr.	22 mR/hr.	180 cm	________

17. The time it takes a patient to receive a dose of 2 gray is 1 minutes at a 60 cm SSD. How long would it take to receive a dose of 4 gray at a 120 cm SSD?

18. If the time to deliver a dose of 10 mrad at a 200 cm SSD is 10 minutes, what is the exposure rate per minute at this distance?

19. If the time it takes to deliver a dose of 300 mrad is 3 minutes at 200 cm SSD, what dose would be received in 6 minutes at a 100 cm SSD?

20. A patient receives an exposure of 40 mR at a 90 cm SSD. If the SSD is doubled with no other changes made, the exposure will be reduced by:

   A. 25%
   B. 50%
   C. 75%
   D. 200%
   E. 400%

The following formulas have been used to solve these problems.

$$\text{New Exposure} = \text{Original Exposure} \times \frac{\text{Original Distance (SSD)}^2}{\text{New Distance (SSD)}^2}$$

$$\text{New Distance}^2 = \text{Original Distance (SSD)}^2 \times \frac{\text{Original Exposure}}{\text{New Exposure}}$$

Section II

Explanation & Answer Keys

Radiation Protection and Biology

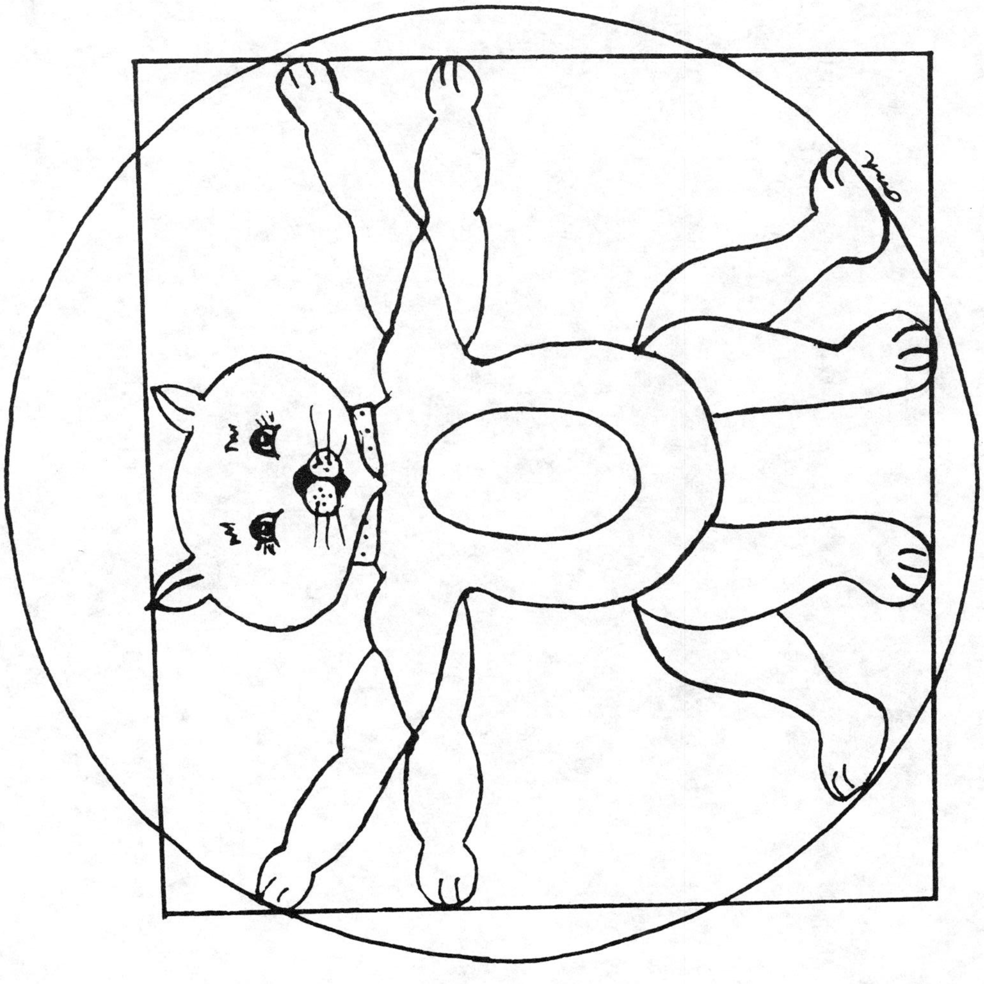

Radiation Biology

Chapter 1

1. C — Radiation exposures and their visible effects on an irradiated individual can be divided into two categories. Manifest effects, which are readily apparent in the person or latent (hidden) effects, which may remain dormant for many years.

2. C — Serious malabsorption and digestive problems result from dose equivalents in the 300-800 rem, (3-8 Sv) range. These gastrointestinal symptoms are often severe enough to cause death in a period of just 3-5 days.

3. D — A large dose equivalent in the range of 2-10,000 rem (2-10 Sv) will cause significant damage to the brain and spinal cord causing high intracranial pressure and death within 1-2 days.

4. D — Those effects which take many months or years to manifest themselves are called long term effects. Most forms of leukemia, cancer, or cataracts take between 4-20 years to develop.

5. C — Radiation induced mutations from damage to the DNA, genes, or chromosomes in the germ cells (sperm or egg), which appear in the offspring are called genetic effects. Radiation damage which effects only the exposed individual is classified as somatic.

6. A — Nearly all of the radiation or chemically induced genetic mutations are characterized by their negative impact on the viability of the exposed organism.

7. D — The genetic information of the cell is carried on a long chain of deoxyribonucleic acid (DNA) which is coiled into the subunits called chromosomes. DNA appears to be the most sensitive and important molecule to the action of chemical toxins or radiation.

8. D — Radiation exposures exceeding 150 R (150 cGy) to eye will substantially increase the rate of cataractogenisis in the exposed individual. Cataracts refer to the opacification of the lens of the eye or its capsule.

9. C — An acute dose equivalent of 500 rem (5 Sv) usually does not reach the gastrointestinal or central nervous system thresholds and so exposed individuals are most likely to die of damage to the hematologic organs which can take 20-60 days.

10. B — Because it comprises nearly 80% of the body's mass, water is the most common molecule effected by radiation. The harmful byproducts produced by its disassociation when struck by radiation such as hydrogen peroxide and hydroxyl ions constitute most of the harmful chemicals associated with the indirect effect.

11. A — Any condition resulting from a reduction in the normal number (46) of chromosomes is termed monosomy.

12. D — Among the first somatic effects which appear in an irradiated individual are nausea, vomiting, erythema, diarrhea, epilation and a reduction in all types of blood cells which can cause anemia, bleeding disorders and low immunocompetence.

13. D — The visible manifestation of a biologic response is called the threshold or minimum dose threshold.

14. B — At dose equivalents above 200 rem (2 Sv) to the skin, a common response is a temporary or permanent loss of hair called epilation. The condition becomes more severe with increasing dose, therefore is nonstocastic in nature.

15. B — The rapid proliferation of the cells in the gastrointestinal tract make them highly sensitive to the effects of irradiation.

16. D — The immediate symptoms of nausea, vomiting, diarrhea and leukopenia that occur at acute doses above 100 rem (1 Sv) are often called prodromal syndrome. These symptoms may last for a few hours at higher doses, to a couple of days at lower doses.

17. A — Linear curves distinguished by their straight line appearance are used when the biologic response is directly proportional to the dose. All nonthreshold curves originate at the junction of the dose response axii and indicate that any dose regardless of its magnitude may be harmful.

18. B — Nonlinear or sigmoid curves are characterized by a response that is not directly proportional to dose. All threshold curves require a minimal dose to cause a response and originate at some point on the dose axis away from the dose-response junction.

19. D — The blood and blood forming tissue are included as part of the hematologic or hematopoietic system which include spleen and bone marrow tissues.

20. A — The dose equivalent threshold for a reduction in the number of white blood cells is about 25 rem (.25 Sv) whole body. Erythema (reddening of the skin) requires a minimum dose equivalent of about 150 rem (1.5 Sv) with substantially higher values required to cause noticeable changes in the eyes or to lower the sperm count.

21. C — A sunburn-like condition noted by a reddening of the skin may occur following a relatively large radiation exposure to the skin is called erythema.

22. D — An increase in the amount of radiation damage is expected as the oxygen concentration, linear energy transfer and relative biologic effect numbers are increased. The age of the cell, sex of the exposed individual, and rate of radiation exposure also play an important role in the individual's response to radiation.

23. C — The classic laws of cellular sensitivity as proposed by Bergonie and Tribondeau relate cellular response to the cellular maturity (age), metabolic activity (O_2 consumption) and rate of cellular division (proliferation).

24. A — In utero irradiations have been found to be most harmful in the 2 week - 2 month development stage of the first trimester in which the organs develop (organogenesis). The relative risk in the period is about 5-6 times that of the last two trimesters.

25. B — In utero irradiation is thought to be associated with a higher incidence of mongolism, hydrocephalus, microcephaly, exencephalia, as well as a long list of other potential hereditary mutations. At the present time no human data is available which substantiates this prediction.

26. C — Nausea and vomiting are often the first sign of acute radiation injury and are caused by the high sensitivity of the linings of gastrointestinal tract organs.

27. C — The short life span of most lymphocytes makes them more sensitive to the harmful effects of irradiation than the longer lived thrombocytes or erythrocytes.

28. D The effects of a large irradiation to developing fetus are highly variable depending upon the total dose and the developmental stage. A radiation exposure above 50 rem(.5 Sv) to the embryo-fetus has been implicated in a number of bone, central nervous system disorders, as well as leukemia and many other types of cancers. Larger doses may lead to a spontaneous abortion or any number of serious growth defects to the fetus.

29. C The principal means by which radiation transfers energy to the biologic system is called ionization defined as the disassociation of compounds into their associated ions.

30. A Mongolism (Down's Syndrome) also called trisomy 21 is one of a number of hereditary disorders involving extra chromosomes which may be linked to in utero exposures.

31. A Because of its sensitive molecular structure, deoxyribonucleic acid (DNA) is prone to damage from a direct interaction with ionizing radiation.

32. A The denuding of the skin surface in response to large irradiation may result in the uncomfortable dry desquamation or more serious burn like moist desquamation of the skin.

33. B The high degree of specialization of nerve and muscle tissue tends to make them more resistant than most other cells of the body to the effects of radiation or chemical toxins.

34. D The red bone marrow which is found in the ends of all developing long bones and nearly all adult flat bones is responsible for the production of red blood cells, white blood cells, and platelets.

35. B A curve which originates at the dose-response junction indicates that any exposure no matter how small may cause a potential response in the irradiated individual.

36. B Since it is the stem cells of the gastrointestinal tract that are responsible for the uptake of fluids, their destruction will lead to dehydration and the loss of fluid volume within the cells of the body.

37. A From the most to least sensitive are the immature or precursor cells, the mature lymphocytes, mature platelets and the mature erythrocytes.

38. B Most individuals receiving large doses of radiation will have a period of time during which the effects are not apparent, called the latent period.

39. C The relative sensitivity of these macromolecules in decreasing order are DNA, RNA, proteins, and ATP.

40. D Recovery of an organism is related to a number of factors, including fractionation, protraction, total dose received, the LET of the radiation, as well as a large number of biologic factors.

41. B The irradiation of water will result in the removal of electrons and a chemical recombination of the hydrogen and oxygen ions that remain. Through a chemical recombination of these ions a number of new compounds and free radicals may be formed including hydrogen radicals, hydroxyl radicals, and hydrogen peroxide.

42. D The majority of the radiation damage that occurs in the body is a result of the radiolysis of water. Since chemical changes that occur in water are responsible for the majority of the tissue damage, it is classified as an indirect effect.

43. A As a fetus develops, it goes through a number of sensitive stages. By far the most sensitive period is the embryonic stage, followed by the ovum, fetal and neonatal periods.

44. D Because of the relatively low sensitivity of the central nervous system, a severe lethal effect such as excessive intracranial pressure will only appear at dose equivalents exceeding 2,000 rem (20 Sv).

45. C Two classic types of experimental testing methods are available for use with living organisms, in vivo and in vitro testing. In vivo testing accomplished within a living organism, though preferred, is often more difficult because a test subject may be subjected to potentially dangerous chemical or radiation stimuli. In vitro testing involves the removal of the tissue from the body and it's reaction to a stimulus.

46. B It is possible to reduce the radiosensitivity of a cell by the lowering of the cell's oxygen content through the use of antioxidants or radioprotectors such as some of the sulfhydryl compounds.

47. D The numbers of all blood cells are reduced following a large irradiation. The loss of the erythrocytes (RBC) will cause anemia and reduce the blood's ability to carry oxygen. A reduced leukocyte (WBC) count may negatively impact the immune function while a deficiency in the thrombocyte (platelet) count will lengthen the coagulation or clotting times.

48. A In an otherwise healthy individual, sublethal exposures are estimated to be about 90% repairable within just a few days following the exposure.

49. A The survival time following a supralethal exposure to radiation is primarily dependent upon the total dose, and time over which the exposure is received. The larger the dose, the shorter the survival time for the exposed individual.

50. C The stem cells of the male reproductive system, the spermatogonia, the female egg cells and the lymphocytes are considered the three most radiosensitive cells of the body.

51. D In a graph of the relative sensitivity of an individual over their life, two periods of increased sensitivity are noted. The first occurs before birth in the embryonic development stage, and the second in the more easily compromised elderly patient.

52. C The relative radiosensitivity for an individual remains nearly constant throughout childhood, adolescence and adulthood.

53. B Death caused by the excessive fluid loss, electrolytic imbalance and secondary infections seen with the gastrointestinal syndrome normally occurs within 5 - 10 days of the exposure.

54. A The development of the different organ systems from the undifferentiated cells occurs during the most sensitive portion of the development process called organogenesis. Organogenesis normally occurs in the embryonic stage between 2-8 weeks of intra-uterine development.

55. D All types of blood cells are dramatically reduced following a large irradiation. See explanation number 47 of this section.

Interactions of X Rays with Matter

Chapter 2

1. D — When photons interact with matter, 3 events may occur: penetration, scatter production and absorption (attenuation). In this last event, all of the energy of the photon is absorbed by the orbital elections of the target tissue.

2. C — Beta particles and electrons both have a mass of 1/2000 amu and negative charge making them physically identical with each other. It is the point of origin in the nucleus that distinguishes the beta particle from its orbital shell cousin.

3. A — The photoelectric event is energy dependent, occurring most frequently when lower energy photons interact with materials posing a high atomic number.

4. C — The probability of a photoelectric interaction drops to near zero in soft tissue at energies above 60 keV and in bone tissues at energies above 90 keV.

5. A — The vast majority of scattered radiation in the diagnostic or therapeutic regions result from a process called Compton's scattering. During this process, the incident photon collides with an outer shell electron, resulting in its ionization and remission of a scattered x-ray photons in all directions.

6. D — The different attenuation properties of tissue and their corresponding radiographic densities are primarily due to the photoelectric effect.

7. B — The energy dependent photoelectric effect drops to near zero in all substances at energies above 100 keV. As a result, the Compton's effect becomes the only interaction occurring in the energy range from 100 keV to 1 MeV.

8. B — At low photon energies, the majority of photons are attenuated by a process called the photoelectric effect, in which incident photons transfer all their energy to a tightly bound inner shell electron resulting in its ionization. This inner shell vacancy is quickly filled by the higher energy electrons occupying adjacent orbital shells called characteristic x-ray production.

9. A — Thompson's (coherent) scattering refers to the interaction of a low energy photon with a tightly bound electron of an atom. Since insufficient energy is available to cause the ionization of the affected electron, the net result is the re-emission of the incident photon energy in a new direction..

10. A — An alpha particle with a mass of 4 amu,and an atomic number of 2 is identical to a helium nucleus in all respects other than its absence of orbital electrons.

11. B — The creation of an electron-positron pair from a high energy photon which interacts with the columbic field of the nucleus is termed pair production. This interaction in which energy is converted into matter requires a photon to have a minimum of 1.022 MeV of energy.

12. B — The only by-product from a photoelectric event is the ionized orbital electron (photoelectron) that has absorbed the entire energy of the incident photon and the resulting low energy characteristic photons created during the refilling of the orbital vacancies.

13. C The Compton's interaction, which predominates diagnostic and therapeutic energy regions, is responsible for the majority of scattered photons produced in the tissue exposure to x ray or gamma radiation between 70 keV and 10 meV.

14. A Its relatively high mass and double positive charge makes alpha about 20 more times likely to cause ionization in a given tissue than an x-ray or gamma photon.

15. A The photoelectric event occurs when an incident photon strikes a tightly bound K shell electron and transfers all of its energy to the orbital electron.

16. C When an x-ray photon enters the tissue of the body, it may cause ionization of the outer shell electrons by the Compton's effect or ionization of the inner shell electron by the photoelectric effect.

17. D The low atomic number of most human tissues will produce only low energy characteristic x-ray photons that rarely have sufficient energy to escape from the tissues from which they were formed.

18. B If the average energy to form an ion pair is 34 eV, then a (102 KeV) photon could result in the formation of as many as 3,000 ion pairs.

19. A During the Compton event, the incident photon is absorbed, resulting in the ionization of the (Compton) electron and the formation of scattered photons with energies close to that of the primary beam.

20. B A polychromatic beam (x-ray beam), with a peak energy of 90 kVp will result in a beam with a quality (average energy) of about 75 keV. Photons produced in this energy region will interact with matter by the photoelectric and Compton events.

21. C The partial energy transfer that occurs during the Compton's effect is most likely when the incident energy level of the photon far exceeds the binding energy of the affected electrons. Most of this excess energy is re-emitted in the form of a scattered photon.

22. D Pair production is a high energy interaction in which a photon with at least 1.022 MeV of energy interacts with the columbic field of the nucleus and forms an electron-positron pair.

23. C Alpha, beta and gamma are the three common emissions originating from the nucleus of an unstable nuclide.

24. C During a photoelectric event, all of the incident photon energy is absorbed by an inner shell orbital electron resulting in its ionization. This ionized photoelectron may have sufficient energy to cause secondary ionizations in the tissues of the body by a process called bremsstrahlung photon production

25. D The only interaction of a photon with matter that results in the same incident and scattered photon energies is the low energy Thompson or coherent scattering event.

26. B The energy of a photoelectron is found to be the difference between the energy of the incident photon and the binding energy of the target electron.

27. C The Compton's event is characterized by a partial transfer of the incident photon energy and the resulting Compton electron and a single scattered photon.

28. A The Alpha particle which has a mass of 4 amu and an atomic number of 2 consists of 2 neutrons and 2 protons and no electrons.

29. C The number of Compton interactions is mainly dependent upon the electron density of target material.

30. C The amount of energy attenuated by a tissue is dependent upon the incident energy of the photons, the atomic number of the target material, and the angle of incidence between the photon and the electron.

31. A The release of nuclear fragments following its collision with a high energy photon is termed photo-disintegration.

32. B The mass energy equivalent is the numerical value of energy assigned to a particle based on its mass. An electron with a mass of 1/2000 amu has an energy equivalent of .511 MeV. The proton being 2000 times heavier would have an energy of about 1 GeV.

33. B Einstein's most famous equation $E=mc^2$ indicates the relationship between mass, energy and speed.

34. A Because there is a direct relationship between energy and mass, the photon with the highest energy will have the highest mass equivalence.

35. A The name given to a particle or photon before it collides with matter is termed incident radiation.

Detection of Radiation

Chapter 3

1. B The linear energy transfer (LET) is described as the amount of energy transferred per micrometer of length in soft tissue and has the units of keV per micrometer.

2. C The active component of a thermoluminescent dosimeter (TLD) is a small chip of the crystalline material, lithium fluoride. This material has the unusual property of emitting a different intensity of light when heated, depending on the amount of ionizing radiation to which it has been exposed.

3. A The constant readout and audible signal produced by a rate or survey meter is useful in the location of lost radioactive sources. The cutie pie and Geiger-Muller counters are the most well known of the rate meters.

4. D Improper film badge readings may occur from exposure of the badge to excessive heat, fluids or visible light from a damaged packet.

5. B The unit of radiation intensity or exposure, the roentgen (R), is limited to the measurement of x rays and gamma photons below 3 MeV in air. The output of an x ray unit is most often recorded in roentgens.

6. D The rate at which a radionuclide decays is always constant and is not effected by changes in temperature, pressure or the age of the radioactive material.

7. A When precise measurements of photon radiations are required a scintillation counter is often employed. This type of device which can actually measure single photon event is most often used in nuclear medical imaging devices and laboratory analyzers. Scintillation type devices can be designed to detect both the quantity and quality of the radiation.

8. D The unit of radiation intensity or exposure, the roentgen (R), is equal to the amount of radiation that will produce 2.58 x 10^{-4} coulombs of charge in 1 kilogram of air.

9. C The unit of dose equivalence, the rem, is derived from the formula: rem = absorbed dose (rad) x quality factor (QF).

10. D The plastic film badge holder is designed with a series of filters and openings to selectively enable the detection of high and low energy gamma rays, as well as beta particles.

11. D See explanation number 9 of this section.

12. C A gas filled chamber containing two electrodes, called an ionization chamber, works by recording the voltage change within a closed chamber when the gas is exposed to radiation. The majority of the survey or rate meters use some version of this chamber to detect radiation.

13. D The unit of radiation absorbed dose, the rad, is often preferred over the roentgen because it can be used in the measurement of all types of radiation, including alpha, beta, gamma, x rays and neutrons.

14. D The main difference in the types of ionization chamber type detectors is their sensitivity, which is directly related to the operating voltage range.

15. B If film badges are worn for more than a month, they may show abnormally high dose readings caused by the film's increased sensitivity following its initial exposure to radiation.

16. C Most radiation detection devices use photographic, chemical or scintillation effects to detect the radiation and are capable of measurements in 1-10 mrem range. The minimum dose equivalent threshold for even the most sensitive biologic effect is about 25,000 mrem (.25 Sv).

17. D Most radiation codes are based on the assumption that a certain minimum dose is required before a substantial amount of bodily injury occurs. This so-called threshold concept can be used to set acceptable limits for various types of radiation exposures received by an individual.

18. C Most experts in radiation safety favor the placement of a radiation monitor badge at the collar level outside the protective apron. This position enables the recording of the maximum estimated skin exposure to the unprotected surfaces of the body. The lower reading obtained by placing it behind the apron may give an underestimation of the actual dose to both shielded and unshielded areas. The most recent recommendation is for the use of two radiation monitors; one at each of these locations.

19. C See explanation number 10 of this section.

20. A See explanation number 5 of this section.

21. D The measurement of an individual's absolute dose is most commonly determined with a device called a dosimeter. Three common types of dosimeters that are currently employed are the pocket dosimeter, thermoluminesent dosimeter, and the film badge.

22. A Patient exposures are most often expressed in the unit of the roentgen (R) while the dose equivalent is what accounts for biological effectiveness of an exposure is expressed in rems or seiverts. Because 1 R is nearly equal to 1 rem and exposure of 700 mR would nearly equal to 700 mrem, .7rem, or .07 seivert.

23. B Photographic film can indicate exposures as low as about 5 - 10 mR or dose equivalent levels of 5 - 10 mrem.

24. D A pocket dosimeter is the only type of radiation detection device that permits the evaluation of an exposure after the completion of a study. This is made possible by the self-reading meter incorporated into the design of this device.

25. B Radiation monitoring devices are employed by all radiology departments to help insure proper radiation protection practices are followed and to insure all workers remain within established annual effective dose equivalent limits.

26. D Blood testing is no longer considered acceptable as a means of determining radiation exposure because of its high threshold of about 25,000 mrem (.25 Sv).

27. C The unit of dose equivalent, the rem, can be obtained from the formula rem = rad x QF.

28. C The somewhat lower sensitivity of the cutie pie detector makes it more suitable for the evaluation of higher intensity radiation areas than a GM counter.

29. A See explanation number 12 of this section.

30. A The output of an x-ray tube (primary radiation) as measured in air is normally stated in the unit of roentgens (R).

31. C In the international system of units (SI), the gray (Gy) is the unit of absorbed dose and the sievert (Sv) is the unit of dose equivalence. They are related by the formula Sv = Gy x QF.

32. B See explanation number 1 of this section.

33. C The time during which a detection device is unable to respond to an additional exposure of radiation is called its dead time. Detection devices with long dead times cannot be used in the measurement of short-lived radiation events such as those associated with pulsed x-ray sources.

34. B See question number 31 of this section.

35. D Pocket dosimeters, though sensitive, are relatively expensive, fragile, require daily monitoring, and a manual recording of all readings by a qualified radiation officer.

36. C Many departments now employ thermoluminescent dosimetric monitors instead of the more familiar film badge monitor because they are reusable and radiation measurements can be made in house.

37. B The sensitivity of any type of ionization type chamber detector is related to its operating potential. The higher operating voltage of a Geiger-Muller counter is the main reason for this device's higher sensitivity to radiation.

38. A The use of a quenching agent such argon can substantially improve the resolving time of an ionization detector.

39. B The conversion from rads to gray is based on relationship that 1 gray is equivalent to 100 rad.

40. B — X ray and gamma photons have a high degree of penetrability and therefore deposit only small amounts of energy per length of travel. These radiations can be said to have a low linear energy transfer (LET) and low ionization potential.

41. D — The high quality factor of alpha emissions will result in the highest dose equivalency of these particles.

42. C — Because gamma rays are highly penetrating, they are most likely to contribute to a high internal exposure but a relatively low skin exposure.

43. B — By definition, the energy transfer of 100 erg per gram of any matter by any type of radiation defines 1 rad.

44. A — In an attempt to standardize all units of radiation, the international system of units has selected the gray (Gy) as the unit of absorbed dose.

45. A — The curie is a unit of activity and is equal to 3.7×10^{10} disintegrations per second of a radionuclide.

46. B — In the ionization chamber voltage curve, the region between 900-1200 volts, is called the Geiger region. Any counter operating in this voltage range is termed a Geiger-Muller (GM) counter.

47. A — In the ionization chamber voltage curve the region under 300 volts is characterized by a lack of ion pair recombinations and the failure to produce a measurable signal making this region unsuitable for the operation of a detection device.

48. C — The slope of region C indicates that the number of primary and secondary ion pairs produced are proportional to each other, hence the name the proportional region.

49. D — At ionization chamber voltages above 1500 volts, a single electron creates a cascade or continuous discharge within the chamber and therefore is useless from a detection standpoint.

50. C — The accurate calibration of a survey or rate meter is normally accomplished through the use of a long half lived radioactive source. To insure consistency, the test source must be precisely prepared and should meet the National Bureau of Standards (NBS) standards.

51. A — The small amount of light emitted by a scintillation crystal and the thermally stimulated emission of a thermoluminescent dosimeter must first pass through a photomultiplier tube to convert the light into a detectable electrical signal.

52. A — Ionization type detectors can be used to measure most high energy photon radiation (primary, secondary, and exit) as well as most particulate radiations (alpha, beta, neutrons) through the attachment of specialized probes.

53. B — The principal scintillator used in most scintillation detectors is hermetically sealed thallium activated sodium or cesium iodide crystal.

54. A — See explanation number 20 of this section.

55. A — In a thermoluminescent dosimeter (TLD), lithium fluoride undergoes a physical change when exposed to radiation. When subsequently heated, the energy absorbed from its radiation exposure is detected as a change in the light intensity that it emits. A careful analysis of this light indicates the amount of radiation the crystal has absorbed.

Protection of Personnel

Chapter 4

1. C — In order to provide the required radiation protection to persons working in or around a radiation area, all diagnostic x-ray rooms must have a primary barrier with a lead equivalent thickness of about 1.5 mm or 1/16".

2. D — Gonadal shielding which can be made of lead or other high atomic material, comes in one of three forms called flat contact, shaped contact, or shadow shields.

3. B — The amount of time it takes for 1/2 of the injected amount of a substance to be removed from the body is termed the biologic half-life.

4. A — Use the corresponding chart to derive the appropriate values.

5. B — Use the corresponding chart to derive the appropriate values.

6. B — The higher half value layer thickness of curve II is a prime indicator for the higher energy of this beam.

7. B — The principal scattering object during the radiographic procedure is the patient undergoing the examination.

8. C — To insure that even the tallest of individuals are shielded during radiographic procedure, the shielding must extend up from the floor to a height of 7 feet or 2.1 meters.

9. B — The dose to the gonads can be reduced by a combination gonadal shielding, high speed imaging systems, tight collimation and low mAs techniques.

10. A — The operator and patient dose can be shown to be directly proportional. In theory, due to the design of the room and barriers, the operator's exposure will always be at least 1 million times less than that of the patient.

11. A — The majority of all temporary protective drapes and panels will require a minimum lead equivalency of not less than .25 mm.

12. A — All beam limiting devices such as diaphragms, cones, cylinders, and collimators are effective at reducing the scattered radiation produced in the patient, thus lowering the patient and operator doses, as well as improving the contrast of the radiographic image.

13. B — The bucky slot cover which automatically covers the opening between the table top and the table support and the protective drape on a fluoroscope both serve to reduce the dose to the operator of the equipment and any other personnel that must remain in the room during a procedure. These devices have little effect on the dose that is received by the patient.

14. D — The photon source with the highest energy, the 1.5 MeV gamma of Co60, also has the greatest penetration and is therefore most difficult to shield against.

15. A — The regulations concerning stationary radiographic or fluoroscopic units, requires that the minimum source-to-skin distance (SSD) shall not be less than 30 cm (12 in.) and should not be less than 38cm (15 in.).

16. D The half life of a radionuclide which involves the normal decay process is termed its physical half life.

17. A Secondary protective barriers or those which are only exposed to leakage or scattered radiation must have a lead equivalency of .75 mm (1/32").

18. B Protective aprons which are designed to have a minimum of a .5 mm lead equivalence will normally attenuate between 80-90% radiation hitting the aprons. According to current regulations, all personnel required to be present during fluoroscopy shall wear a protective apron of this thickness during a procedure.

19. C The physical half life of a radionuclide is unique for a specific isotope and is unaffected by temperature, pressure, time or chemical bonding.

20. A Since the tube housing is exposed directly to primary radiation, it is classified as a primary barrier and therefore will require at least 1.5 mm of lead equivalency.

21. A Each scattering event lowers the radiation intensity by about 1000 times as compared to that in the directly exposed tissue. By locating the exposure booths in a position that requires at least two scattering events can reduce the operator's dose by at least 1 million times compared to that of the patient.

22. B The greatest reduction in dose would result from a doubling of the distance which lowers the radiographers dose by about four times. A 15% in the kVp will bring down the mA and exposure by approximately 50%.

23. D The shielding for a controlled area must be thick enough to reduce the exposure to less than 100 mR/week or 5000 mR per year or, and air kerma of .1 rad (.1 cGy) per week.

24. D All mobile and fixed radiographic and fluorographic units must be provided with a fixed cone collimator or spaced that insure a minimum SSD of not less than 30 cm.

25. C See explanation number 22 of this section.

26. A See explanation number 21 of this section.

27. D The maximum allowable limit for leakage of radiation from the tube is less than 100 mR/hr. or an air kerma of .1 cGy in one hour when operated at it's leakage technique.

28. C The half-life is found by following the curve to the 50% line at about 2 hours.

29. D The intensity of scatter radiation 1 meter from a patient is approximately 1000 times (10^3) less than the intensity of the useful beam at the surface of the patient. If the beam is scattered twice the original intensity, the level of the beam is reduced by 1 million (10^6) times.

30. B Since the intensity of this source is reduced by one half every 6 hours, the following values can be obtained: 200 mR/hr after 6 hours; 100 mR/hr after 12 hours; 50 mR/hr after 18 hours; and 25 mR/hr after 24 hours.

31. D See explanation number 30 of this section.

32. B See explanation number 30 of this section.

33. A Since the intensity of the beam is reduced to 25% of its original value by 2mm of Al or the equivalent of 2 half value layers. Therefore if 2 mm of Al = 2 HVL than 1 HVL = 1 mmAl.

34. B In order to keep one's exposure to radiation as low as possible, it is important to keep as much distance between the individual and the source of radiation as possible, reduce the time the individual is exposed, and place as much shielding between the individual and source as possible.

35. D Controlled areas which are occupied by monitored occupational workers include the x-ray rooms, NM scanning areas, hot labs, and CT scanning area. Non-controlled areas are frequented by nonoccupational workers, and include such areas as offices, waiting rooms, darkrooms, file room, and corridors.

36. C Though the holding of patients should always be avoided when possible, when necessary it shall not be accomplished by pregnant women or persons under 18 years of age. All persons used for this purpose shall be shielded for primary radiation.

37. B The higher the atomic number of the material, the more radiation it attenuates, thereby reducing its half value layer thickness.

38. C The intensity of the beam is directly related to the mAs when the kVp remains constant. Therefore if a 200 mA, 150 mA provides 30 mAs and an exposure of 72 mR at 300 mA, 200 mS, 60 mAs exposure will result in a new intensity of about 144 mR.

39. A Lead, though more expensive, has many times the stopping power of concrete and therefore requires less total mass or thickness to obtain the same protective equivalency.

40. C In all newer stationary units, the exposure switch is attached to the control console to prevent its activation while the operator is standing inside the x-ray room.

41. D The nuclear medical technologist or physician responsible for radionuclide injections, or the person responsible for handling injections during special procedures, should be assigned ring badge monitors in addition to the collar and waist badges normally recommended.

42. A The shielding of a non-controlled area must reduce the exposure to an individual standing behind the barrier to an air kerma level of less than .01 rad (.01 cGy) per week.

43. D When determining thickness required for a specific source, one must consider the distance from the source, quality of the beam, workload factor (beam on time), use factor (fraction of time the beam is directed at a barrier), and the occupancy factor for adjacent rooms or areas.

44. C ALARA is defined as low as is reasonably achievable.

45. A See explanation number 43 of this section.

46. A Diagnostic filtration reduces the number of low energy photon reaching the patient.

47. C Collimation reduces the amount of exposure tissue in the patient and lowers the amount of scattered radiation. These factors combine to reduce the exposure to both the patient and the operator of the radiographic unit.

48. B Primary barriers are intended to protect the radiographer.

49. A Lead aprons placed on a patient are for their protection only.

50. B Protective drapes and bucky slot covers are primarily designed to protect the operator of the equipment.

Patient Protection

Chapter 5

1. D When patients have a substantial reproductive potential, gonadal shielding should be used whenever the gonads are within 5 centimeters (2") of the collimated beam unless the placement of the shield will compromise the objectives of the study.

2. C There are a number of things a radiographer or physician can do to lower the patient's exposure during fluoroscopy. These include reduction of the field size, the use of a lower mA value, the use of a higher kilovoltage setting, a reduction in the time that the fluoroscope is activated and the application of protective shielding when possible.

3. B Diagnostic filtration, which can substantially lower the skin exposure to a patient, is required for all diagnostic fluoroscopic and x-ray tubes. Those x-ray units which operate at levels above 70 kVp potential have a minimum requirement of 2.5 mm aluminum equivalent.

4. C This image indicates the usage of the most commonly employed type of gonadal shielding device, the flat contact shield. In this radiograph, the shield selected was too large, covering excessive anatomy of the femur and pelvic bones.

5. A Since pregnancy is highly unlikely during menstruation and the four or five days following menses, this is the preferred time for elective radiographic examinations.

6. B The principal function of diagnostic filtration is the reduction of radiation exposure to the skin, though two other positive effects will occur, namely an increase in beam quality and an increase in the homogeneity of the beam.

7. B See explanation number 6.

8. D Any factor which enables the use of lower mAs and kVp values will enable a reduction in the patient exposure. The use of screens with high speed numbers can substantially lower exposure to the patient.

9. C Any device that reduces scatter radiation before it reaches the patient such as beam limiting devices; i.e., cone, cylinder, aperture diagram, and collimator, will reduce patient exposures. Grids which eliminate scatter after it leaves the patient requires an increase in the incident beam intensity to maintain the desired radiographic density.

10. D During a radiographic procedure, the body itself will act like a filter significantly lowering the dose on the side of the body on which the beam exits. Because the breasts appear to be more sensitive than the skin of the back, a PA projection can be used to reduce the exposure to the breast.

11. C The electronic amplification which occurs within the image intensifier tube substantially increases the brightness level of the image, thus improving visual acuity of the eye while at the same time enabling about a 50% reduction in patient and operator exposure doses.

12. D The most recent regulations concerning pregnant radiation workers are designed to avoid possible discrimination of the woman. Because it is extremely unlikely that any trained worker is likely to exceed either their own or the fetal permissible dose equivalent limits, other than providing additional monitoring devices, no special precautions are advised or permitted.

13. D One of the largest contributing factors to the patient exposure dose is the number of repeat images that may be required. The repeat rate can be reduced by the proper maintenance of the equipment, use of immobilization techniques, careful positioning, careful selection of the technical factors and the use of appropriate communication skills.

14. C The higher (400) speed rare earth screens which are commonly employed in nearly all modern radiology departments must be used in conjunction with a specifically matched screen film to provide its optimum speed. In this question the 400 speed imaging system would be associated with the lowest patient exposure when a similar contrast and density are assumed.

15. D A radiographer can help reduce patient exposures in a number of ways, including the use of higher speed image receptors, proper shielding, tight collimation, and the selection of the appropriate technical factors.

16. A The main advantages of the rare earth imaging systems are an increased speed enabling the use of less radiation thus reducing the patient's exposure. The spatial resolution and radiographic contrast of 400 speed rare earth and 100 speed calcium tungstate screen systems are nearly identical.

17. B See explanation number 11 of this section.

18. D Among the numerous factors that effect the dose a patient receives are mAs, screen speed, grid ratio, SID, field size, filtration, and kVp.

19. C See explanation number 8 of this section.

20. A The use of protective (lead) aprons to limit gonadal exposures during chest radiography is common in modern radiography. The low density artifact seen in this image is the plastic waistband of one of these aprons.

21. A A photo (cine) spot film is associated with a significant reduction in the exposure to a patient. A further reduction will occur when a smaller format of 70 mm film system is employed.

22. D See explanation number 8 of this section.

23. C The time required to deliver a given dose is found by dividing the dose to be delivered by the dose rate.

24. B The estimated entrance exposure (mR) and fetal dose (mrad) for the three examinations in the question for a 400 speed imaging system are 110 mR and 35 mrad for a KUB, 110 mR and 25 mrads for a hip and 90 mR and 1 mrad for the thoracic spine respectively.

25. D When the kVp and other equipment factors remain unchanged, the intensity of the beam is directly proportional to the mAs value. Therefore, if 20 mAs is proportional to 50 mR; then 120 mAs is proportional to 300 mR.

26. B Since the fluoroscopic tube is normally located under the table top, in order to be effective any shielding must be placed on top of the table between the patient and the tube.

27. C The image was obtained with improper orientation of the collimator to the body part.

28. A In a right PA oblique (RAO) position the central ray will enter at the left dorsal surface of the body causing the greater entrance exposure on this surface.

29. A Because a pregnancy is highly unlikely during menstruation and the few days following, radiographic examinations, when possible, should be performed during this 10 day period of time.

30. D Though nearly all the technical and equipment factors will have an effect on the radiographic density and patient dose, the size of the focal spot is related to image resolution and under identical exposure conditions will not effect the radiographic density or dose to the patient.

Radiation Regulations

Chapter 6

1. C The NCRP handbook No. 116 has recommended the elimination of the old 5(n-18) formula in favor of the age x 10 mSv (age x 1 rem) formula for the cumulative effective dose limit.

2. A Annual effective dose limit, formerly the maximum permitted dose (MPD) are based on a linear nonthreshold relationship at low or acceptable risk levels.

3. B Occupational exposures are intended to reflect only those exposures received as part of one's employment or participation in work related activities in a controlled area. Exposures from your own thyroid study, nuclear fallout and your own dental x-rays are classified as nonoccupational.

4. C The units of dose equivalence are the rem and sievert. They are related by the relationship 1 Sv=100 rem.

5. C The vast majority of the recommendations concerning radiation and its control are provided by the National Council Radiation Protection and Measurement (NCRP).

6. B Untrained individuals working in or around a radiation area are not considered different than any other member of the general population and therefore can receive an annual effective dose equivalent limit of 5 mSv (500 mrem) or 1/10th of that of an occupational worker.

7. A The following are the estimated annual whole body dose equivalents received by an individual from artificial radiation sources. Diagnostic x-rays .4-.9 mSv, radiopharmaceuticals .1-.14 mSv, nuclear fallout .01-.03 mSv.

8. B The effective dose equivalent limit for the gestation period is 5 mSv (.5 rem) per year or .05 rem (.5 mSv) per month.

9. C An occupational worker has the following effective dose limit for the stochastic effects: 1 mSv (100 mrem) per week or 50 mSv (5 rem) per year.

10. D Under normal operating conditions, the kerma rate at the center of the point where the beam enters the patient for a fluoroscopic unit should not exceed 5 cGy (5 rad) per min. A rate of up to 10 cGy (10 rad) min is permitted when a high level control including a continuous audible signal is provided with the fluoroscopic unit.

11. A When life saving action justifies acute exposures they may be permitted to exceed annual effect dose limitations using the following guidelines. The use of volunteers is most desirable when an older worker with low lifetime accumulated effective doses are selected first. The maximum recommended effective dose be limited to .5 Sv (50 rem) or and equivalent dose of 5 Sv (500 rem) to the skin.

12. B Only the doses received in a controlled area should be considered in the calculation of the occupational dose. Diagnostic x-rays, therapeutic x-rays and all other environmental sources are nonoccupational in nature.

13. A Though a number of environmental factors may effect these values, the average whole body dose for natural sources are (.25-.40 mSv) for cosmic radiation, (.25-.35 mSv) for internal gamma and (.25.-35 mSv) for external gamma sources.

14. C All occupational exposures, whether internal or external, should be considered in the calculation of the effective dose equivalent limit.

15. C According to the current age formula in explanation number 1, the answer is 280 mSv (28 rem).

16. D The annual effective dose limit for the deterministic (somatic) effect to tissues, other than the eye, is 500 mSv (50 rem) per year.

17. C All radiation exposures should be kept to levels which are as low as reasonable achievable (AIARA). Unlike effective dose limits which are mandated absolute limits, AIARA includes economic and social factors to be considered. The AIARA limits enable a degree of flexibility which may differ significantly from the ideal. In general a reasonable recommendation is to maintain a radiation workers exposure to that of that of a non-occupational worker or 5 mSv (.5 rem) annually.

18. A A deterministic effect formally called a nonstochastic effect is defined as a somatic effect which increases in severity as a radiation dose increases above the dose threshold. Deterministic effects may be early, i.e., erythema, or late as with lens opacification, decreased fertility or organ atrophy.

19. A Any student under 18 years of age is allowed an annual effective dose equivalent limit of 1 mSv (.1 mrem) until his/her 18th birthday.

20. B See explanation number 12 of this section.

21. D See explanation number 12 of this section.

22. B A pregnant radiographer and a CT technologist are both classified as an occupational worker and would have an annual effective dose equivalent limit of 50 mSv (5 rem) only the embryo-fetus will have a dose equivalent limit of 5 mSv (.5 rem).

23. D Since an occupational worker has a yearly effective dose limit of 50 mSv, the 6.5 mSv dose is subtracted, leaving a 43.5 mSv balance for the next 11 months.

24. D According to the source listed it is possible to estimate fetal exposures based on monitor badge readings at the collar or waist. For a collar badge the fetal dose is estimated at about 3% of the collar reading. The fetal dose because of the attenuation of maternal tissues is about 30% of the waist badge reading.

25. B Two main radiation area signs are commonly employed in or around a controlled area. A radiation area sign is used when the rate is between 5-100 mR/hr and a high radiation area sign is used when the exposure rate exceeds 100 mR/hr.

26. D The specific annual effective dose limit for deterministic effects to the lens of the eye is 150 mSv (15 rem).

27. A See explanation number 8 of this section.

28. D All members of the public (non-occupational) workers have an effective dose limit of .1 mSv (10 mrem) per week, or 5 mSv (500 mrem) per year.

29. B According to the current age formula in explanation number 1, the answer is 36 x 10 mSv = 360 mSv (36 rem).

30. B Natural sources of radiation include cosmic radiation, internal and external gamma sources and radon gas.

31. D Though all radiation workers may receive up to 50 mSv (5000 mrem) per year, the average annual dose is normally many times lower at dose equivalent of about 2.1 mSv (210 mrem).

32. B Since a member of the public (nonoccupational) worker has a yearly dose equivalent of 5 mSv (500 mrem), this exposure has not exceeded this individual's yearly limit.

33. D This problem is solved by dividing annual limit by the monthly dose to give the number of months that this person can work without exceeding the permitted annual effective dose equivalent of 5000 mrem (50 mSv).

34. A See explanation number 9 of this section.

35. C A stochastic effect is a radiation induced injury which is assumed to have no dose threshold. Stochastic effects are normally limited to cancer and genetic effects that are considered by some to be all or nothing responses.

36. B Occupational workers are classified through their work with ionizing radiation. Radiographers, CT technologists, NM technologists, and mammographers, are all classified as occupational workers. Ultrasonographers, MRI technologists, darkroom technologists are classified as nonoccupational workers because they are not normally exposed to diagnostic x-rays.

37. B See explanation number 36 of this section.

38. A See explanation number 36 of this section.

39. A See explanation number 36 of this section.

40. A See explanation number 36 of this section.

Inverse Square Law

Exercise No. 1

1. 400 mR/week
2. 25 mR/week
3. 200 mR week
4. 4.4 mR/week
5. 144 mR/week
6. 12.5 mR/week
7. 40 mR/week
8. 1600 mR/week
9. 25.5 mR/week
10. 1.11 mR/week
11. 400 cm
12. 36 inches
13. 120 inches
14. 300 cm
15. 35.4 cm
16. 542.7 cm
17. 80 minutes
18. 20 mrad/min
19. 150 mrad
20. C

Inverse Square Law

Exercise No. 2

1. 1800 mR/week
2. 11.1 mR/week
3. 72 mR/week
4. 5 mR/week
5. 10 mR/week
6. 200 mR/week
7. 62.60 mR/week
8. 11.2 mR/week
9. 22.25 mR/week
10. 3.2 mR/week
11. 400 cm
12. 36 inches
13. 72 inches
14. 33.3 cm
15. 69.3 cm
16. 542.7 cm
17. 8 minutes
18. 1 mrad/min
19. 2400 mrad/min
20. C

Section III

Equipment Maintenance and Operation

Radiographic Graffiti II

Can you identify these common patient or technical related artifacts? (Answers appear at the bottom of the page.)

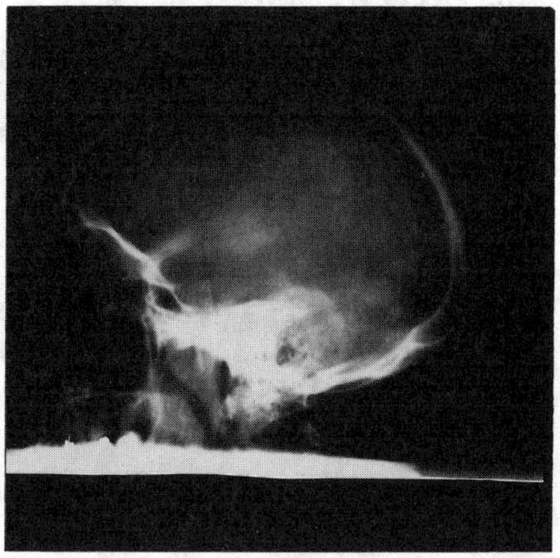

A

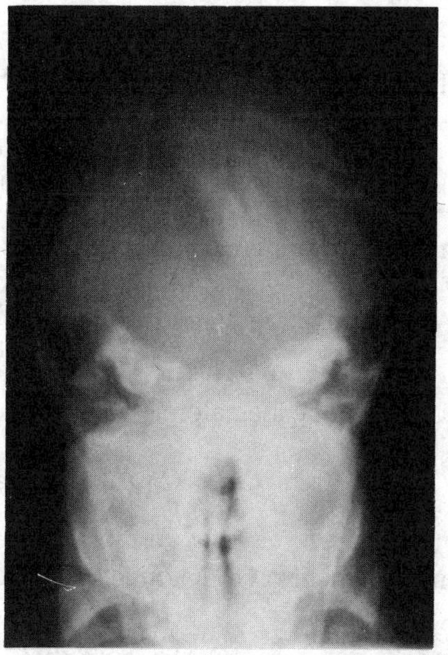

B

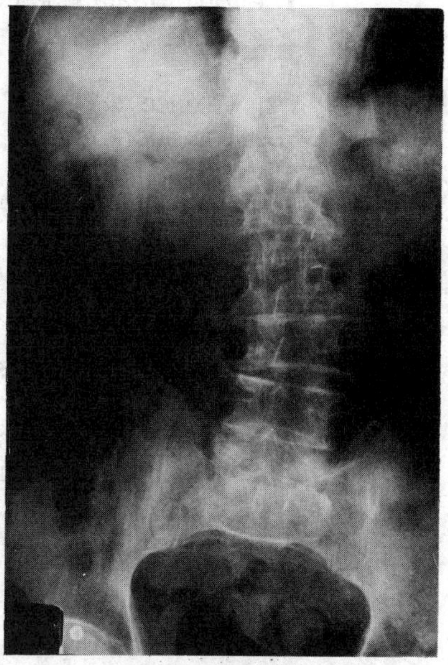

C

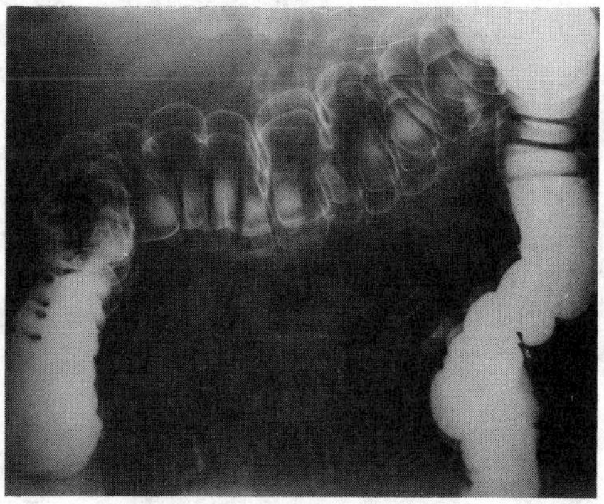

D

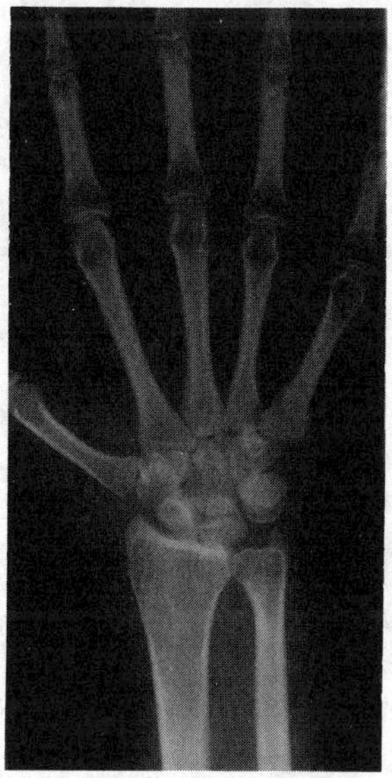

E

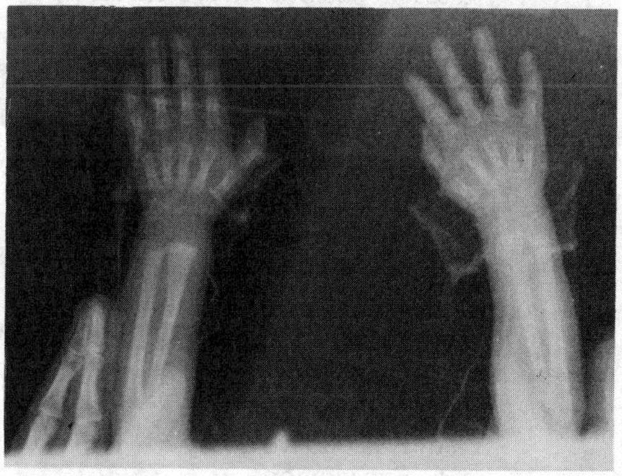

F

Answers

Film A. The faint low density areas are due to tightly wound hair braids (corn rows); **Film B.** The low density streaks are from a pillow used in restraining the head; **Film C.** The linear striation is caused by the transfer sheet under the patient; **Film D.** Superimposed images due to an inadvertent double exposure; **Film E.** Plastic patient identification bracelet around the wrist; **Film F.** Pajama sleeves and cloth tape restraints (missing: lead gloves on mothers hands).

Atomic Structure

Chapter 7

1. The force of attraction between the nucleus and an orbital electron is called the electron's ____________ ____________. 2:44

A. Valence
B. Transition number
C. Binding energy
D. Valence number

2. The schematic display most often employed to represent the structure of the atom was developed by: 3:13

A. Charles Dalton
B. Neils Bohr
C. Michael Crookes
D. Charles Darwin

3. The chemical properties of an atom are primarily controlled by its: 12:50

A. Number of protons
B. Number of neutrons
C. Atomic weight
D. Valence number

4. The orderly arrangement of the element by their atomic number and chemical properties is found in a: 12:45

A. Emission spectrum
B. Table of contents
C. Auger table
D. Periodic table

5. The principle unit of energy measurement on the atomic scale is the: 2:46

A. Electron volt
B. Neutron
C. Coulomb
D. Farad

6. The atomic number of the atom is determined by those particles which have an atomic mass unit of 1 and a single positive charge called: 12:43

A. Electrons
B. Positrons
C. Protons
D. Neutrons

7. The majority of the mass of the nucleus is derived from: 12:42

A. Protons and electrons only
B. Protons and neutrons only
C. Protons, neutrons and electrons only
D. Alpha, beta, gamma only

8. An atom that loses an orbital electron from the electrical influence of the nucleus has undergone: 3:14

A. Radioactive decay
B. Subluxation
C. Ionization
D. Covalent bonding

9. The number of outer shell electrons is known as the: 13:50

A. Electron quota
B. Orbital energy
C. Valence
D. Quantum mottle

10. In the symbol $^{60}_{27}C$, the number in the lower left-hand corner indicates the: 2:47

A. Mass number
B. Atomic number
C. Isotope number
D. Ionization level

11. In the previous question, the number in the upper left-hand corner indicates the: 2:47

A. Mass number
B. Atomic number
C. Isotope number
D. Ionization level

12. A neutral atom that loses an electron by ionization is termed a: 13:50

A. Negative atom
B. Positive atom
C. Negative ion
D. Positive ion

13. The smallest division of a substance possessing the same chemical and physical properties of the substance as a whole is called a/an: 3:13

A. Atom
B. Element
C. Molecule
D. Isotope

14. The tiny particles possessing a negative charge of 1.6 x 10^{-19} Coulombs that orbit the nucleus of an atom are termed: 3:14

A. Electrons
B. Protons
C. Neutrons
D. Photons

15. The centralized portion of an atom, the nucleus is primarily composed of particles called the: 12:42
1. Neutrons 2. Electrons 3. Protons

A. 1 & 2 only
B. 1 & 3 only
C. 2 & 3 only
D. 1, 2 & 3

16. The photons produced during the orbital transitions of electrons from a higher to lower energy are termed: 3:14

A. Characteristic radiations
B. Corpuscular radiations
C. Vacancy radiations
D. Field radiations

17. The formula $2n^2$ is used to calculate: 2:41

A. Minimum number of electrons in an orbital shell
B. Maximum number of electrons in an orbital shell
C. Total number of electrons in the atom
D. Number of outer shell electrons in an atom

18. The atomic mass of an atom can be estimated by adding the atomic masses of the: 2:40
1. Protons 2. Neutrons 3. Electrons

A. 1 & 2 only
B. 1 & 3 only
C. 2 & 3 only
D. 1, 2 & 3

19. The highest bonding energy of an atom is found in the electron's ____________ level. 2:44
1. First energy orbital 2. Second energy orbital 3. Third energy orbital

A. 1 only
B. 2 only
C. 3 only
D. 1, 2 & 3

20. A given atom has the following binding energies: K-shell - 1000 Ev, L-shell - 70 Ev, M- shell - 10 Ev. During a M-L transition, the photon energy emitted will be: 2:46

A. 930 Ev
B. 80 Ev
C. 60 Ev
D. Unable to determine

21. The chemical bond formed when two or more atoms share electrons is termed: 12:51

A. Ionic bonding
B. Polar bonding
C. Covalent bonding
D. Valence bonding

22. The interconvertability of mass and energy first proposed by Einstein is expressed by the equation: 3:16

A. $E= 1/2\ mv^2$
B. $E = hu$
C. $E = mc^2$
D. $E = 2v^2$

23. The octet rule states that the most chemically stable atoms have an outer shell configuration with ______ electrons. 12:41

A. Two
B. Four
C. Seven
D. Eight

24. The natural state of matter with the highest energy due to the movement of atoms is the: 12:38

A. Gaseous state
B. Liquid state
C. Plastic state
D. Solid state

25. About 99% of the atom's mass is located within the: 2:41

A. Orbital shell
B. Valence shell
C. Nucleus
D. Alpha particle

26. Ionization of an atom may occur from exposing the atom to: 12:53
1. Heat 2. Electric current 3. Radiation

A. 1 & 2 only
B. 1 & 3 only
C. 2 & 3 only
D. 1, 2 & 3

27. The electron binding energy is dependent upon the: 2:44
1. Valence number 2. Amount of protons 3. Electron to nucleus distance

A. 1 & 2 only
B. 1 & 3 only
C. 2 & 3 only
D. 1, 2 & 3

28. Which of the following atoms would possess the highest binding energy for its K shell electron? 2:44

A. $^{16}_{8}O$
B. $^{23}_{11}Na$
C. $^{63}_{29}Cu$
D. $^{126}_{53}I$

29. An atom that has seven outer shell orbital electrons belongs to a group of elements known as: 12:49

A. Inert gasses
B. Noble gasses
C. Halogens
D. Rare earth elements

30. Two or more chemicals with the same chemical formula, but having different chemical properties, are termed: 2:48

A. Isomeres
B. Isotopes
C. Isobars
D. Isotones

31. The formation of molecules can be accomplished through the interaction of valence electrons by a process termed: 12:48

A. Ionic bonding
B. Covalent bonding
C. Both of the above
D. Neither of the above

32. The mass of an orbital electron is about _________________ the mass of a proton. 2:40

A. 1/2000
B. 1/10
C. 20 times
D. 2000 times

33. Which of the following is true concerning the K-shell electrons of two different atoms? 2:44
   1. *The electron binding energy is the same for both atoms*
   2. *The electron binding energy is characteristic of each atom*
   3. *The heavier element will have more K-shell electrons*

A. 1 only
B. 2 only
C. 3 only
D. 1, 2 & 3

34. In a neutral atom the electrical neutrality is achieved by maintaining a balance between the: 12:42

A. Protons and electrons
B. Neutrons and positrons
C. Electrons and positrons
D. Protons and neutrons

35. The atomic particle which has a weight of 1 amu and carries no charge is called a/an: 12:42

A. Proton
B. Electron
C. Neutron
D. Negatron

36. The vertical columns of the periodic table contain groups of elements related by the same: 12:45

A. Atomic mass
B. Atomic number
C. Nuclear energy
D. Valence number

37. The physical state of matter is related to the degree of molecular attraction. Which state corresponds to the highest degree of molecular attraction? 12:38

A. Solid
B. Liquid
C. Plastic
D. Gas

38. A neutral atom that gains an electron will have a: 12:50

A. Neutral charge
B. Negative charge
B. Positive charge

39. The chemical bonding of two atoms by the transfer of an electron is a form of __________ bonding. 12:49

A. Ionic
B. Covalent
C. Mutual
D. Polar

40. Characteristic radiation emission occurs when: 3:14

A. Electrons move from lower to higher orbital shells
B. Electrons move from higher to lower orbital shells
C. The atom has an unstable nucleus
D. The atom undergoes nuclear fission

41. The mass energy equivalent value for an electron is: 3:16

A. .60 meV
B. .147 meV
C. .391 meV
D. .511 meV

42. An electron with a lowest binding energy is most likely located in the: 2:44

A. K shell
B. L shell
C. M shell
D. It is the same for all electrons

43. Which of the three fundamental particles has the lowest atomic mass? 2:40

A. Electron
B. Photon
C. Proton
D. Neutron

44. The most chemically stable elements having a filled outer shell configuration are termed: 12:49

A. Alkali metals
B. Alkaline metals
C. Nobel gases
D. Halogens

45. The physical characteristics of the atom are controlled by the atom's: 12:43

A. Atomic number
B. Atomic mass
C. Valence number
D. Electron binding energy

46. The stability of a nucleus is highly dependent upon the ________________ of the nucleus. 3:15

A. R/N ratio
B. $e^- - c^+$ ratio
C. N/Z ratio
D. All of the above

47. The nuclear particle possessing a single positive charge with a mass of 1.67×10^{-24} gm is termed the: 3:14

A. Neutron
B. Positron
C. Proton
D. Electron

48. The term applied to the raising of an electron to a higher energy orbit within a given atom is called: 3:14

A. Excitation
B. Ionization
C. Stabilization
D. Neutralization

49. The maximum number of inner or K-shell electrons in any atom is: 12:41

A. 1
B. 2
C. 4
D. 8

50. The smallest part of an element that retains all the characteristics properties of that element in bulk is called a/an: 12:38

A. Molecule
B. Atom
C. Coulomb
D. Nuclide

51. The following atoms are: $^{12}_{6}C$ $^{13}_{6}C$ $^{14}_{6}C$ 2:48

A. Isotopes
B. Isobars
C. Isotones
D. Isomers

Question 52 - 69. fill in the appropriate value for the following neutral atoms.

	$^{23}_{11}Na$	$^{7}_{3}Li$	$^{35}_{17}Cl$
State their atomic number	52 ____	58 ____	64 ____
State their atomic mass	53 ____	59 ____	65 ____
State their valence	54 ____	60 ____	66 ____
State their number of protons	55 ____	61 ____	67 ____
State their number of neutrons	56 ____	62 ____	68 ____
State their number of electrons	57 ____	63 ____	69 ____

70. In the three elements above, which will have the similar chemical properties?
1. Lithium 2. Sodium 3. Chlorine

A. 1 & 2 only
B. 1 & 3 only
C. 2 & 3 only
D. 1, 2 & 3

71. The principle force involved with nuclear binding is the: 3:15

A. Electromagnetic force
B. Weak force
C. Gravitational force
D. Strong force

72. Which of the following is a decay process by which an unstable nucleus tries to become stable? 3:20
1. Alpha decay 2. Beta decay 3. Positron decay

A. 1 only
B. 2 only
C. 3 only
D. 1, 2 & 3

73. Atoms with different types of nuclear configurations are termed: 3:13

A. Elements
B. Nuclides
C. Molecules
D. Mixtures

74. The term used to describe the process in which light nuclei combine to form a heavier nucleus is: 3:18

A. Fusion
B. Fission
C. Nuclear capture
D. De-excitation

75. Nuclides having the same atomic number but different atomic masses are termed: 2:48

A. Isotopes
B. Isotones
C. Isobars
D. Neutrinos

76. The rate of radioactive decay is greatly influenced by: 12:466

A. Chemical bonding
B. Pressure changes
C. Temperature changes
D. None of the above

77. After alpha decay, the atomic number (Z) of the parent nuclide will: 12:450

A. Increase by 2
B. Increase by 4
C. Decrease by 1
D. Decrease by 2

78. The atomic number of a daughter nuclide following a beta- decay will: 3:20

A. Increase by 1
B. Decrease by 1
C. Increase by 2
D. Decrease by 4

79. The fission process is best described as the: 3:17

A. Combining of two small nuclides
B. Combining of two large nuclides
C. Splitting of a large nucleus
D. None of the above

80. The most likely radionuclides undergoing spontaneous beta- decay will possess an unfavorably high N/Z ratio with an excess of: 3:20

A. Neutrons
B. Protons
C. Positrons
D. Electrons

Electricity and Magnetism

Chapter 8

1. The two types of electrical charges are derived from their corresponding charged atomic particles called the: 12:56

   A. Electrons and protons
   B. Photons and neutrons
   C. Quarks and Lepton
   D. Nucleons and quantum

2. The rubbing of fur and amber or silk with glass result in electrification by a process called: 12:57

   A. Hydrolysis
   B. Induction
   C. Friction
   D. Transference

3. Any materials which tend to oppose the flow of electrons are termed: 12:70
   1. Non-conductors 2. Conductors 3. Insulators

   A. 1 & 2 only
   B. 1 & 3 only
   C. 2 & 3 only
   D. 1, 2, & 3

4. The transfer of a like charge to an uncharged object by placing the two objects in contact with each other is termed: 12:57

   A. Friction
   B. Conduction
   C. Empedence
   D. Induction

5. According to the first law of electrostatics, two objects with opposite charges will tend to: 2:75

   A. Repel each other
   B. Attract each other
   C. Both of the above
   D. Neither of the above

6. An object with a net positive charge can be described as having a deficiency of: 12:56

   A. Protons
   B. Photons
   C. Neutrons
   D. Electrons

7. The electrical discharge that occurs between two oppositely charged objects when the insulating properties of the transfer medium is exceeded is called: 12:59

   A. Crosstalk
   B. Static
   C. Characteristic discharge
   D. Grounding

8. Electrification may occur from which of the following processes? 12:56
   1. Spontaneity 2. Conduction 3. Friction

   A. 1 & 2 only
   B. 1 & 3 only
   C. 2 & 3 only
   D. 1, 2 & 3

9. In a solid conductor, current consists of a flow of moving: 2:73

   A. Photons
   B. Neutrons
   C. Protons
   D. Electrons

10. The electrostatic force between two charged objects is _______________ proportional to the product of their magnitudes. 2:89

A. Directly
B. Inversely
C. Indirectly
D. None of the above

11. Which of the following are most often classified electrical conductors? 2:79
1. Copper   2. Rubber   3. Water

A. 1 & 2 only
B. 1 & 3 only
C. 2 & 3 only
D. 1, 2 & 3

12. According to the principle stated in Coulomb's law, the electrostatic force between two charged objects is___________ proportional to the _______________ of the distance between them. 2:77

A. Inversely, prime
B. Directly, prime
C. Inversely, square
D. Directly, square

13. The electrical resistance which occurs in a conductor is most often manifested as the amount of _____ created. 12:8

A. Heat
B. Electrical noise
C. X-rays
D. Magnetism

14. The force within an electrical circuit most responsible for the movement of current is termed the: 2:79

A. Resistance
B. Wattage
C. Amperage
D. Potential difference

15. A circuit which has 6.3×10^{18} electrons flowing per second is said to have a current of: 2:81

A. One ampere
B. One ohm
C. One volt
D. One watt

16. The type of electricity associated with the flow of current in a single direction is termed: 2:83

A. Alternating current
B. Direct current
C. Three-phase current
D. Repulsive current

17. A major source of electrical current which is produced by a chemical process is a: 2:79

A. Battery
B. Generator
C. Transformer
D. Capacitor

18. Which of the following class of materials have a high electrical resistance? 12:70
1. Metals   2. Rubber   3. Wood

A. 1 & 2 only
B. 1 & 3 only
C. 2 & 3 only
D. 1, 2 & 3

19. The laws which relate to the relative amperage, voltage and resistance in a direct current circuit is: 12:71

A. Coulomb's law
B. Joule's law
C. Boyle's law
D. Ohm's law

20. The resistance of a wire is related to its: 12:71
1. Length   2. Temperature   3. Cross-sectional area

A. 1 & 2 only
B. 1 & 3 only
C. 2 & 3 only
D. 1, 2 & 3

21. An electromagnetic device which converts mechanical into electrical energy is termed a: 12:68

A. Transformer
B. Generator
C. Electroscope
D. Autotransformer

22. Which of the following is a type of device used for storage of electricity? 12:83
1. Capacitor 2. Condenser 3. Battery

A. 1 & 2 only
B. 1 & 3 only
C. 2 & 3 only
D. 1, 2 & 3

23. Select the copper wire with the least electrical resistance. 12:71

A. 100 cm 20 gauge
B. 200 cm 20 gauge
C. 100 cm 10 gauge
D. 200 cm 10 gauge

24. The physical contact of two or more non-insulated wires carrying current will result in a/an: 12:82

A. Open circuit
B. Short circuit
C. Closed circuit
D. Relay circuit

25. The power loss as heat in a conductor is expressed by the formula P=: 12:84

A. IR
B. IV
C. VR^2
D. I^2R

26. The amount of electrical potential (potential difference in an electrical circuit) is measured by a unit called the: 2:79

A. Volt
B. Amp
C. Ohm
D. Forad

27. The power of a circuit is equal to the product of current and voltage and is measured by the unit: 2:86

A. Joule
B. Erg
C. Watt
D. Newton

28. The resistance of a wire will increase as its: 12:71
1. Length increases 2. Temperature increases 3. Diameter decreases

A. 1 & 2 only
B. 1 & 3 only
C. 2 & 3 only
D. 1, 2 & 3

29. An electromagnetic device consisting of a single iron core surrounded by a wire carrying current is termed a/an: 2:95

A. Voltic pile
B. Electroscope
C. Ammeter
D. Electromagnet

30. How many wet cell batteries must be connected in series to provide the 220 volt power supply of a battery powered mobile unit? 12:70

A. 55
B. 110
C. 220
D. 440

31. The production of electricity by the rotation of a coil of wire within a stationary magnetic field is termed: 2:96

A. Electromagnetic irradiation
B. Electric impedance
C. Magnetic fluxuation
D. Electromagnetic induction

32. If an x-ray unit has a maximum output of 120 kilovolts and 600 milliamperes, the generator must have a power rating of at least: 2:85

A. 63 kilowatts
B. 72 kilowatts
C. 120 kilowatts
D. 600 kilowatts

33. Which of the following is classified as a type of natural magnet: 2:87
1. Alnico 2. Lodestone 3. The earth

A. 1 & 2 only
B. 1 & 3 only
C. 2 & 3 only
D. 1, 2 & 3

34. The strong magnetic dipole of a hydrogen atom is due in large part to the: 2:86
1. Unpaired orbital electron 2. Imbalance of the protons and neutrons
3. Alternating nuclear spin

A. 1 only
B. 2 only
C. 3 only
D. 1, 2 & 3

35. The magnetic dipole resulting from the proton spin in a hydrogen nucleus is termed the; 2:86

A. Magnetic moment
B. Random domain
C. Unifield field
D. Attractive force

36. In a magnet the individual magnetic domains are aligned in: 2:86

A. Many different directions
B. Single direction

37. The magnetic force lines or lines of flux in a magnetic field flow: 12:90

A. Within the magnet itself
B. Outside the magnet
C. Both of the above
D. Neither of the above

38. Which of the following are units used in the measure of magnetic flux density? 2:87

A Watt or ohm
B. Erg or Joule
C. Newon or compass
D. Tesla or gauss

39. A commonly used device used in the detection of a magnetic field is a: 2:91

A. Monopole
B. Compass
C. Solenoid
D. Spectroscope

40. Magnetic fields are a phenomenon associated with: 2:86

A. Moving neutrons
B. Moving charges
C. Stationary monopoles
D. Stationary photons

41. The laws of magnetism are similar to the laws of electrostatic because they both: 2:89
1. Cause photons to deflect 2. Have the same atomic configuration
3. Are produced by the same physical process

A. 1 only
B. 2 only
C. 3 only
D. 1, 2 & 3

42. Which of the following will have an associated magnetic moment (field)? 2:86
1. A Hydrogen atom 2. An orbital electron 3. Iron atom

A. 1 & 2 only
B. 1 & 3 only
C. 2 & 3 only
D. 1, 2 & 3

43. The ability of a magnetic material to resist demagnetization is termed: 12:92

A. Retentivity
B. Grounding
C. Permeability
D. Excitation

44. Which of the following is an appropriate form of Ohm's law? 2:81
1. $V = IR$ 2. $R = V/I$ 3. $I = V/R$

A. 1 only
B. 3 only
C. 2 & 3 only
D. 1, 2, & 3

45. Pertaining to the following series of circuits answer questions 45 thru 49.

A series circuit with a current of 10 amperes contains three separate resistive elements with values of $R_1 = 5$ ohms, $R_2 = 10$ ohms = and $R_3 = 20$ ohms.

45. What is the total resistance for this circuit? 2:81

A. 3 ohms
B. 5 ohms
C. 15 ohms
D. 35 ohms

46. The total voltage for this circuit is: 2:81

A. 350 volts
B. 220 volts
C. 110 volts
D. 100 volts

47. What will the voltage measure across R_1? 2:81

A. 20 volts
B. 50 volts
C. 100 volts
D. 200 volts

48. The voltage across R_2 will measure: 2:81

A. 50 volts
B. 100 volts
C. 200 volts
D. 350 volts

49. The voltage across R_3 will measure: 2:81

A. 20 volts
B. 110 volts
C. 220 volts
D. None of the above

50. The total current through each resistor is the same for all parts of the series circuit. 2:81

A. True
B. False

Pertaining to the following parallel circuit, answer questions 51 thru 55.

A parallel circuit carrying a 120 voltage contains four separate resistive elements with values of $R_1 = 30$ ohms, $R_2 = 40$ ohms, $R_3 = 20$ ohms and $R_4 = 30$ ohms.

51. What is the total resistance for this circuit? 2:83

A. 111 ohms
B. 100 ohms
C. 17 ohms
D. 7 ohms

52. The total amperage found in this circuit is: 2:83

A. 17 amp
B. 21 amp
C. 43 amp
D. 57 amp

53. What will the current measure through R_1? 2:83

A. 1.5 amp
B. 2.5 amp
C. 3.6 amp
D. 4.0 amp

54. What will the current measure through R_2? 2:83

A. 1.2 amp
B. 1.6 amp
C. 3 amp
D. 4 amp

55. What will the current measure through R_3? 2:83

A. 1 amp
B. 3 amp
C. 4 amp
D. 6 amp

Basic X-Ray Circuitry

Chapter 9

1. The selection of voltage to the high tension transformer is most commonly made by adjusting the: 2:128

   A. Exposure timer
   B. Line voltage compensator
   C. Filament transformer
   D. Autotransformer

2. A device that can be used for the storage of electrical charge is called a: 12:83
   1. Transformer 2. Battery 3. Capasitor

   A. 1 & 2 only
   B. 1 & 3 only
   C. 2 & 3 only
   D. 1, 2 & 3

3. The high amperage necessary to cause thermionic emission in the filament circuit is developed from the use of a: 2:129

   A. Step-up transformer
   B. Step-down transformer
   C. Condenser
   D. Rectifier

4. The most common device used to create the high potential difference (kilovoltage) across the x-ray tube is the high tension: 2:133

   A. Autotransformer
   B. Variable resistor
   C. Saturable reactor
   D. Transformer

5. The prereading kV meter measures the potential difference on the: 2:128

   A. Primary side of the autotransformer
   B. Secondary side of the autotransformer
   C. Secondary side of the high tension transformer
   D. Across the x-ray tube electrodes

6. A step-down (Coolidge) transformer is most likely to be found in the: 12:245

   A. Primary circuit
   B. Secondary circuit
   C. Filament portion of the primary circuit
   D. None of the above

7. In a radiographic unit, the milliammeter will register during the _______ of the activation of the exposure switch. 12:247

   A. Boost (prep stage)
   B. Second (exposure stage)
   C. Both of the above
   D. Neither of the above

8. Fuses and circuit breakers are incorporated within electrical circuits to prevent: 10:244

   A. Short circuits
   B. Circuit overloads
   C. Voltage fluctuations
   D. Synchronism

9. To prevent possible shock hazards if a malfunction occurs, the high tension transformer is: 2:133
   1. Immersed in oil 2. Made with laminated steel plates 3. Used only in low voltage circuits

   A. 1 only
   B. 2 only
   C. 3 only
   D. 1, 2 & 3

10. The current necessary to create thermionic emission in the filament of a x-ray tube is about: 2:128

    A. 1-2 mA
    B. 3-6 mA
    C. 1-2 Amp
    D. 4-6 Amp

11. For safety reasons, the majority of components in a modern x-ray circuit are placed in the: 12:237

A. Primary circuit
B. Filament circuit
C. Accessory circuit
D. Secondary circuit

12. The type of current found between the secondary of the high tension transformer and the rectifiers is: 2:133

A. High voltage (kV) AC
B. Low voltage AC
C. High voltage (kV) DC
D. Low voltage DC

13. The type of meter used for measurement of the current (tube current) between the cathode and anode is a/an: 2:129

A. Kilovoltmeter
B. Milliammeter
C. Ammeter
D. Ohm meter

14. Most x-ray machines used in the United States are designed to operate on a 220 volt or 440 volt: 2:85

A. 60 hertz, alternating current
B. 60 hertz, direct current
C. 120 hertz, alternating current
D. 50 hertz, direct current

15. Which of the following current regulating devices is best described as a variable ratio transformer? 2:127

A. Rheostat
B. Autotransformer
C. High tension transformer
D. Filament transformer

16. The control for the amount of filament current in most modern radiographic units is directly related to the: 12:245

A. Timer setting selection
B. Kilovoltage selection
C. Milliampere selection
D. Focal spot size selection

17. In nearly all radiographic units, the high tension transformer is located in an oil-filled tank that also houses the: 2:133
1. Rectifiers *2. Filament transformer* *3. Autotransformer*

A. 1 & 2 only
B. 1 & 3 only
C. 2 & 3 only
D. 1, 2 & 3

18. In order to maintain a constant filament current as changes are made in the kilovoltage, a _______ is required: 12:246

A. Space charge compensator
B. Ballistics stabilizer
C. Filament boost circuit
D. Line voltage compensator

19. The milliamperage that is impressed across the x-ray tube will have a substantial effect on the: 2:129
1. Beam penetration *2. Filament heat* *3. Rate of photon production*

A. 1 & 2 only
B. 1 & 3 only
C. 2 & 3 only
D. 1, 2 & 3

20. In the filament branch of the primary circuit, the type of meter employed to measure this current is a/an: 12:245

A. Micro-ammeter
B. Milliammeter
C. Ammeter
D. Volt meter

21. The purpose of a ballistic-type milliampere-second meter is to measure current flow when: 12:247

A. High voltages are used
B. Low milliamperage settings are used
C. Rapid exposure time settings are used
D. Long exposure time settings are used

22. In most modern x-ray circuits, to overcome inconsistent voltage supplies which can vary by up to 5%, a ________ is incorporated to produce consistent voltage input to the machine. 2:127

A. Line compensator
B. Ballistic compensator
C. Wavetail compensator
D. Phase compensator

23. The selection of voltage to the primary of the high tension transformer in a three-phase unit is made by (a): 12:255

A. Single variac
B. Twin rectifier
C. Three autotransformers
D. Three circuit breakers

24. The x-ray tube has a direct electrical connections from the: 12:251
1. Filament circuit *2. Primary circuit* *3. Secondary circuit*

A. 1 & 2 only
B. 1 & 3 only
C. 2 & 3 only
D. 1, 2 & 3

25. The midpoint of the high tension transformer in a modern x-ray circuit is grounded. This serves to reduce the: 12:248

A. Number of rectifiers
B. Anode storage capacity
C. Reduce the insulation on the cables
D. Decrease tube vibrations

26. The autotransformer is most often located in the ________ portion of a modern x-ray circuit. 2:127

A. Primary
B. Filament
C. Accessory
D. Secondary

27. The pre-reading kV meter measures the ________ on the secondary side of the autotransformer: 2:129

A. Milliamperage
B. Kilovoltage
C. Wattage
D. Voltage

28. Which component is not generally located in the high voltage portion of a modern x-ray circuit? 2:130

A. Rectifiers
B. X-ray tube
C. Filament transformer
D. Milliampere meter

29. A resetable electromagnetic device, which is included in many electrical circuits to prevent damage from current surges, is termed a: 12:244

A. Circuit breaker
B. Line compensator
C. Saturable reactor
D. Safety relay switch

30. Which of the following is generally found on the operating console of a x-ray machine? 2:127
1. kV control switch *2. mA control switch* *3. Timer control switch*

A. 1 only
B. 2 only
C. 3 only
D. 1, 2 & 3

Transformers

Exercise No. 1

1. A transformer has 100 primary and 400 secondary turns of wire. What is the secondary voltage if 200 volts are placed on the primary coil (assuming no power losses occur)?

2.-10. Using a transformer, calculate the secondary voltage that will occur for the following number of primary and secondary turns and primary voltages, assuming no power losses occur.

	Primary Voltage	# Primary Turns	# Secondary Turns	Secondary Voltage
2.	220	100	1,000	
3.	110	200	20,000	
4.	220	1,000	10,000	
5.	110	10,000	100	
6.	440	5,000	500	
7.	200	2,000	20,000	
8.	90	300	30,000	
9.	380	30,000	300	
10.	80	880	8,800	

11. A transformer has 100 primary and 400 secondary turns of wire. What is the secondary current (amperage) if 100 amps are placed on the primary coil (assuming no power losses occur)?

12.-17. Using a transformer, calculate the secondary current that will occur for the following number of primary and secondary turns and primary currents.

	Primary Current	# Primary Turns	# Secondary Turns	Secondary Current
12.	100 amps	66	660	
13.	80 amps	120	12,000	
14.	50 amps	80	800	
15.	1 amp	10,000	100	
16.	10 amps	12,000	1,200	
17.	1 kiloamp	100	10,000	

18. If 220 volts are impressed on the primary side of an autotransformer that consists of 110 total turns, what is the secondary voltage if the 60th tapped turn is selected?

19.-25. Using an autotransformer, calculate the secondary voltage that will occur for the following primary voltages, total number and tapped turns for the autotransformer, assuming no power losses occur.

	Primary Voltage	Total # of Turns	# of Tapped Turns	Secondary Voltage
19.	100	100	50	
20.	200	400	100	
21.	110	220	110	
22.	220	110	55	
23.	440	880	110	
24.	80	550	250	
25.	140	500	300	

The following formula has been used in calculating these problems.

Transformer:

$$\text{Secondary Voltage} = \text{Primary Voltage} \times \frac{\text{\# of Secondary Turns}}{\text{\# of Primary Turns}}$$

$$\text{Secondary Amperage} = \text{Primary Amperage} \times \frac{\text{\# of Primary Turns}}{\text{\# of Secondary Turns}}$$

Autotransformers:

$$\text{Secondary Voltage} = \text{Primary Voltage} \times \frac{\text{\# of Tapped Turns}}{\text{Total Number of Turns}}$$

Exercise No. 2

1. A transformer has 600 primary and 1800 secondary turns of wire. What is the secondary voltage if 220 volts are placed on the primary coil (assuming no power losses occur)?

2.-10. Using a transformer, calculate the secondary voltage that will occur for the following number of primary and secondary turns and primary voltages, assuming no power losses occur.

	Primary Voltage	# Primary Turns	# Secondary Turns	Secondary Voltage
2.	220	200	20,000	______
3.	110	400	20,000	______
4.	220	2,000	10,000	______
5.	110	20,000	800	______
6.	440	6,000	600	______
7.	220	2,200	22,000	______
8.	90	330	330,000	______
9.	380	15,000	300	______
10.	80	20	8,000	______

11. A transformer has 100 primary and 1,000 secondary turns of wire. What is the secondary current (amperage) if 100 amps are placed on the primary coil (assuming no power losses occur)?

12.-17. Using a transformer, calculate the secondary current that will occur for the following number of primary and secondary turns and primary currents, assuming no power losses occur.

	Primary Current	# Primary Turns	# Secondary Turns	Secondary Current
12.	100 amps	160	3,200	______
13.	80 amps	100	24,000	______
14.	50 amps	80	1,800	______
15.	1 amp	14,000	200	______
16.	10 amps	12,800	3,200	______
17.	1 kiloamp	200	14,000	______

18. If 440 volts are impressed on the primary side of an autotransformer that consists of 110 total turns, what is the primary voltage if the 100th tapped turn is selected?

19.-25. Using an autotransformer, calculate the secondary voltage that will occur for the following primary voltages, total number and tapped turns for the autotransformer, assuming no power losses occur.

	Primary Voltage	Total # of Turns	# of Tapped Turns	Secondary Voltage
19.	220	400	50	______
20.	110	1,000	100	______
21.	800	880	110	______
22.	220	220	55	______
23.	440	880	220	______
24.	80	750	125	______
25.	200	450	150	______

The following formulas have been used in solving these problems:

Transformer:

$$\text{Secondary Voltage} = \text{Primary Voltage} \times \frac{\text{\# of Secondary Turns}}{\text{\# of Primary Turns}}$$

$$\text{Secondary Amperage} = \text{Primary Amperage} \times \frac{\text{\# of Primary Turns}}{\text{\# of Secondary Turns}}$$

Autotransformer:

$$\text{Secondary Voltage} = \text{Primary Voltage} \times \frac{\text{\# of Tapped Turns}}{\text{Total Number of Turns}}$$

Single and Three-Phase Rectification

Chapter 10

1. The process by which alternating current is converted into pulsating direct current is termed: 2:134

   A. Rectification
   B. Polarization
   C. Depolarization
   D. Electrification

2. In modern x-ray equipment, the device most often used for current rectification is a/an: 2:134

   A. Modified x-ray tube
   B. Thermionic valve tube
   C. Solid state diode
   D. Ionization chamber

Pertaining to the radiographic images, answer questions 3. & 4.

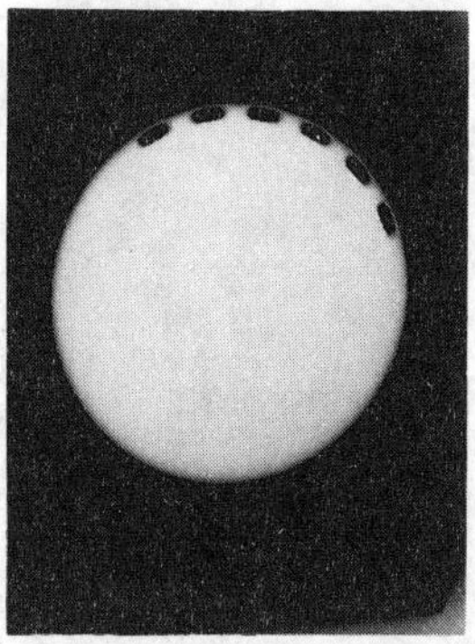

A

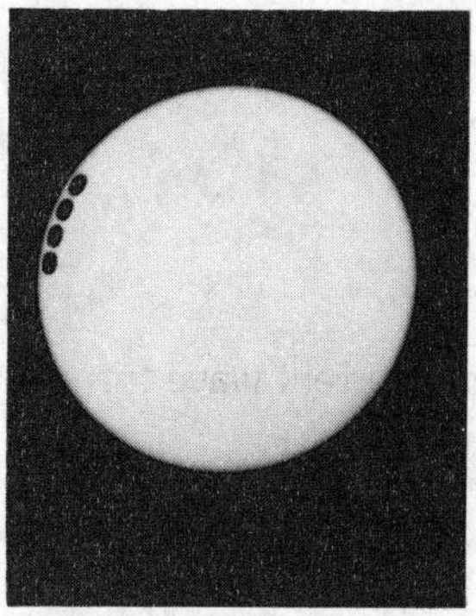

B

3. If the manual spin top test film represented by letter A was taken on a properly functioning single phase full wave rectified unit, what timer setting was employed? 2:132

   A. 200 milliseconds
   B. 100 milliseconds
   C. 50 milliseconds
   D. 10 milliseconds

4. If the manual spin top test represented by film B was taken at 33ms ($^1/_{30}$) second on a single phase full wave rectified unit, which statement best describes the unit? 2:134

   A. Timer is inaccurate
   B. Faulty valve tube
   C. Pitted anode
   D. Nothing is wrong

5. Which of the following types of single-phase x-ray equipment will permit the greatest instantaneous loading capacity? 2:139

   A. Self-rectified current
   B. Half-wave rectified current
   C. Full wave rectified current
   D. Rectification is not required

6. The small variation of the voltage waveform of three-phase rectified current is termed the: 5:50

   A. Constant potential
   B. Phase curve
   C. Helix effect
   D. Voltage ripple

7. The use of three-phase radiographic units has the advantage of ________ when compared to single-phase units. 2:139
   1. Lower exposure time settings *2. Higher milliamperage settings*
   3. More of the beam penetrability

   A. 1 only
   B. 2 only
   C. 3 only
   D. 1, 2 & 3

8. In a three-phase, twelve pulse unit, rectification is accomplished by employing: 5:50

A. 4 thermionic diodes
B. 6 solid state diodes
C. 12 solid state diodes
D. Delta and star windings

Pertaining to the voltage wave forms answer questions 9.-12.

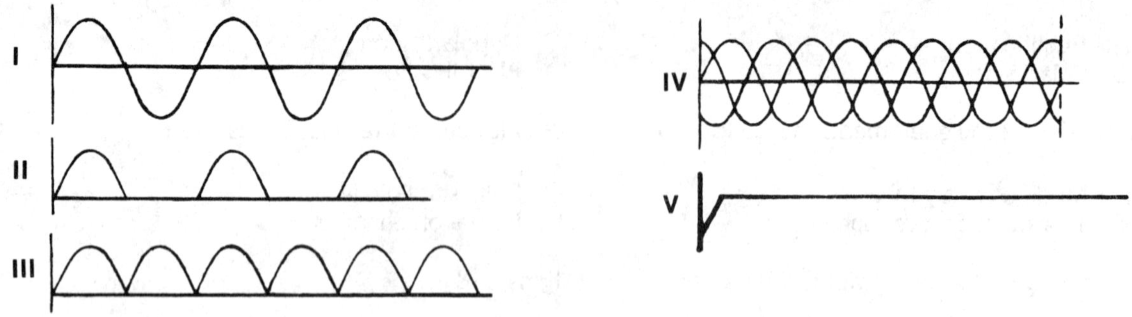

9. A full rectified wave form is represented by Roman numeral: 5:50

A. I
B. II
C. III
D. IV
E. V

10. A three phase alternating current wave form is represented by Roman numeral: 5:50

A. I
B. II
C. III
D. IV
E. V

11. Roman numeral V represents a: 5:50

A. 1/2 wave rectified curve
B. Alternating current curve
C. Three phase curve
D. Steady direct current curve

12. Which of the following represent direct current wave forms: 5:50

A. I, II & III
B. I & IV
C. II, III & IV
D. All of the above

13. The main advantage of the three-phase over single-phase (full wave) current is that in three-phase current, the voltage: 2:139

A. Never drops below the peak voltage
B. Never drops to a zero voltage
C. Is self-rectified
D. Is constant

14. The type of unit most likely to possess the smallest voltage ripple is a: 5:50

A. Full wave single-phase unit
B. Three phase unit
C. Self-rectified unit
D. Alternating current unit

15. The most common type of rectification for use in three-phase radiographic equipment is the: 5:46

A. Thermionic diode
B. Solid state diode
C. Rheostat
D. Valve tube

16. Rectification is used to suppress the inverse voltage associated with alternating current before it reaches the x-ray tube. This helps in the prevention of damage to the: 2:134

A. Filament
B. Anode
C. Bearing
D. Tube envelope

17. An exposure is made on a three-phase radiographic unit. In order to maintain the same radiographic density for a full-wave rectified unit, the mAs should be increased by about: 2:139

A. 5-10%
B. 15-25%
C. 30-40%
D. 50-55%

18. The rectification of high voltage alternating current into pulsating direct current occurs between the secondary of the high tension transformer and the: 2:134

A. Autotransformer
B. Solid state diodes
C. X-ray tube
D. Thermionic diodes

19. A three-phase, six-pulse unit has an effective kV of about ________ of the peak kVp. 5:50

A. 87%
B. 64%
C. 57%
D. 37%

20. The ripple value is the variation in voltage across the x-ray tub e as expressed as a percentage of the maximum value. With a single-phase system, a ripple value of ________ is normal. 5:50

A. 3%
B. 13%
C. 79%
D. 100%

21. A satisfactory radiograph is produced using a three-phase, 6-pulse unit. If the same technical factors are used on a single-phase unit, the radiograph would show: 2:139

A. Higher density
B. Lower contrast
C. Lower density
D. Lower detail

22. During the operation of an x-ray tube being supplied by full-wave rectified current in a 50 msecond exposure, the tube electrodes will change (reverse) polarity: 2:137

A. 24 times
B. 12 times
C. 6 times
D. None of the above

23. The most common method for obtaining a fully rectified tube current involves the use of: 2:136

A. Two diodes (valve tubes)
B. Four diodes (valve tubes)
C. Six diodes (valve tubes)
D. Twelve diodes (valve tubes)

24. The most common material used in the construction of solid state diodes is: 5:46

A. Arsenic
B. Gallium
C. Silicon
D. Tungsten

25. Compared to a single-phase x-ray unit, a three-phase unit using the same technical factors will create: 2:139

1. More photons *2. More heat units* *3. Higher energy photons*

A. 1 only
B. 2 only
C. 3 only
D. 1, 2 & 3

Pertaining to the syncronous spin top film, answer question 26.

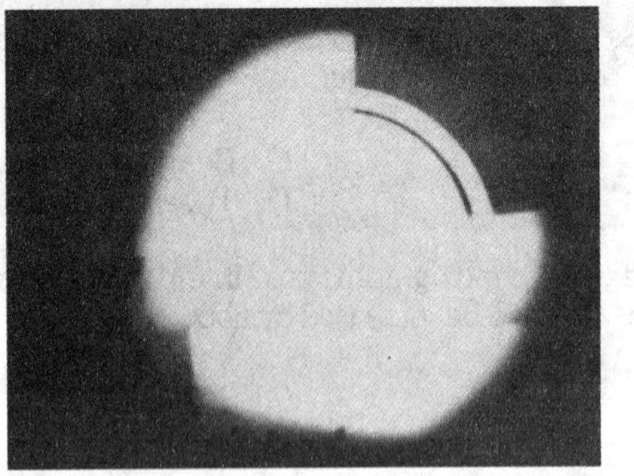

26. The syncronous spin top diagram represented by this film is an example of a 100 millisecond exposure taken on a: 2:133

A. Self rectified unit
B. 1/2 wave rectified unit
C. Full wave rectified unit
D. Three-phase rectified unit

27. A spin top test was performed on a full wave single phase unit and 24 dots appeared. Assuming nothing is wrong with the unit, what timer setting was employed? 2:133

A. 10 milliseconds
B. 100 milliseconds
C. 150 milliseconds
D. 200 milliseconds

28. A 50 millisecond exposure is taken on a single phase full wave rectified unit. The resulting film should show: 2:133

A. 3 dots
B. 6 dots
C. 9 dots
D. 12 dots

29. A 10 millisecond exposure is taken on a single phase full wave rectified unit. The resultant spin top test film shows 10 dots. This unit most likely has a/an: 2:133

A. Faulty valve tube
B. Inaccurate timer
C. Imbalanced anode
D. Nothing wrong with unit

30. A 50 millisecond exposure is taken on a single phase full wave rectified unit. The resulting spin top test shows 3 dots. This unit most likely has a/an: 5:48

A. Faulty valve tube
B. Inaccurate timer
C. Imbalanced anode
D. Nothing wrong with unit

31. Three phase current consists of three single-phase voltage wave forms separated by: 5:48

A. 30 degrees
B. 60 degrees
C. 120 degrees
D. 180 degrees

32. Solid state rectifiers as compared to thermionic diodes have/are: 2:135
1. More durable *2. More efficient* *3. A longer life*

A. 1 only
B. 2 only
C. 3 only
D. 1, 2 & 3

33.-35 In the following, state the number of dots that should appear on a spin top test film for a single phase full wave rectified unit. 2:132

	TIMER SETTING	EXPECTED NUMBER OF DOTS
33.	250 milliseconds	________________
34.	150 milliseconds	________________
35.	16 milliseconds	________________

Exposure Timers

Chapter 11

1. Which of the following types of radiographic units permits the lowest minimum exposure time settings? 15:208

   A. Half wave, single phase
   B. Full wave, single phase
   C. Three-phase (with AEC)
   D. Three-phase (without AEC)

2. Using an automatic exposure controlled (phototimed) unit, 50 speed screen cassette is mistakenly employed instead of a 100 speed screen cassette. Which of the following is likely to occur? 15:207

   A. The exposure time will be longer than expected.
   C. The exposure time will be shorter than expected.
   B. The radiograph will be too dark.
   D. The radiograph will be too light.

3. The timer accuracy is important for consistent radiographic density. Which of the following devices should be used in the determination of accuracy for a modern electronic timer of a three-phase unit? 2:133

   A. Oscilloscope
   B. Synchronous spintop
   C. Ballistics mAs meter
   D. Manual spintop

4. The majority of modern automatic exposure controlled units employ three or more thin ________ for the detection of remnant radiation. 5:58

   A. Photo-tubes
   B. Ionization chambers
   C. Fluorescent screens
   D. None of the above

5. Most modern three-phase radiographic units allow for exposure times as low as 1 ms of a second. The timer most commonly used for this purpose is a/an: 2:130

   A. Electronic timer
   B. Impulse timer
   C. Synchronous timer
   D. Thyratron

6. An automatic exposure controlled (phototimed) technique using 100 kVp is employed on a patient with intestinal gas from an obstruction. The most likely result is a/an: 2:130

   A. Overpenetrated radiograph
   B. Underpenetrated radio graph
   C. Radiograph with excessive density
   D. Radiograph with deficient density

7. With a grid controlled x-ray tube, the radiographic exposure is terminated when a preselected ________ is reached. 12:265

   A. Anode voltage
   B. Cathode voltage
   C. Grid voltage
   D. Filament amperage

8. Using an automatic exposure controlled exposure, which of the following factors will normally increase film density? 15:207

   A. Increased mA
   B. Increased kV
   C. Decreased distance
   D. None of the above

9. The use of an impulse or synchronous timer in a single-phase x-ray unit will permit accurate exposure times as short as: 2:130

   A. 1 ms (1/1000)
   B. 4 ms (1/240)
   C. 17 ms (1/60)
   D. 34 ms (1/30)

10. A radiographic exposure requires 24 mAs. What timer setting will result in the appropriate values using the 200 mA station? 2:132

A. 20 millisecond
B. 120 millisecond
C. 150 millisecond
D. 250 millisecond

11. Using an automatic exposure controlled unit for an abdomen exposure, in order to reduce motion, a/an _________ may be employed. 15:207
1. Shorter backup time *2. Increased SID* *3. Higher kVp*

A. 1 only
B. 2 only
C. 3 only
D. 1, 2 & 3

12. Using an automatic exposure controlled unit, a cassette with a 300 speed screen is mistakenly employed instead of a (100) speed screen cassette. Which of the following is likely to occur? 15:207

A. Overexposed radiograph
B. Underexposed radiograph
C. Longer than normal exposure time
D. Shorter than normal exposure time

13. For safety reasons, the timer circuits for most modern radiographic units are located in the:

A. Filament circuit
B. Primary circuit
C. Secondary circuit
D. High voltage circuit

14. The ionization detectors that are employed in most automatic exposure controlled units are located between the _________ and the _________ . 5:58

A. Tube/collimator
B. Collimator/table top
C. Table top/grid
D. Grid/Bucky tray

15. Using an automatic exposure controlled unit, an excessively long exposure time results. The possible cause is: 15:207
1. Improper centering *2. Excessive backup time* *3. Excessive kVp*

A. 1 only
B. 2 only
C. 3 only
D. 1, 2 & 3

16. A radiographic exposure utilizes a 300 millisecond timer setting. If you wish to reduce this by 40%, what new time should be employed? 2:132

A. 120 millisecond
B. 180 millisecond
C. 240 millisecond
D. 280 millisecond

17. In an automatic exposure controlled unit, the termination of the radiographic exposure is accomplished by a/an: 5:55

A. Ionization chamber
B. Photo multiplier
C. Thyratron tube
D. Fluorescent screen

18. A radiographic exposure requires 40 mAs. What timer setting will result in the appropriate values using the 800 mA station? 2:132

A. 50 ms
B. 100 ms
C. 500 ms
D. 750 ms

19. Using an automatic exposure controlled unit, a decrease in the exposure time will result from: 15:207
1. Decreasing backup time *2. Increasing kVp* *3. Decreasing SID*

A. 1 & 2 only
B. 1 & 3 only
C. 2 & 3 only
D. 1, 2 & 3

20. All exposure switches by law are required to terminate the exposure if not being depressed by the radiographer. The common name for this type of switch is: 20:20

A. Ballistics-type switch
B. Recycle-type switch
C. Deadman switch
D. Thyratron switch

21. An automatic exposure controlled exposure of an abdomen is taken. Which of the following is likely to occur after two cups of Barium are taken by this patient? 15:207

A. Film density will increase
B. Film density will decrease
C. Exposure time will increase
D. Exposure time will decrease

22. A radiographic exposure is taken using the 200 mA station and a 50 millisecond timer setting at 80 kV. This exposure will result in the production of: 2:132

A. 4 mAs
B. 10 mAs
C. 16 mAs
D. 40 mAs

23. An automatic exposure controlled (phototimed) unit is being used to make a radiograph of the abdomen. Which changes can be made to decrease the exposure time? 15:207

A. Use a higher speed screen
B. Increase grid ratio
C. Increase the kilovoltage setting
D. Use a shorter backup time

24. The modern electronic timer that is often used in newer radiographic equipment permits exposure times as short as 1/1000 sec. or:

A. 1 ms
B. 5 ms
C. 8 ms
D. 10 ms

25. The use of a backup timer is required in an automatic exposure controlled unit to: 3:506

A. Reduce patient exposure
B. Permit multiple exposures
C. Protect the unit
D. Check exposure times

26. What is the smallest mAs value obtainable for an AEC exposure using 300 mA 30 milliseconds if the unit has a minimum response time of 20 milliseconds? 15:207

A. 2 mAs
B. 3 mAs
C. 6 mAs
D. 9 mAs

27. A patient is to be radiographed using an AEC exposure at 600 mA and 80 kVp. If the minimum response time is 20 milliseconds, which of the following mAs values is possible? 15:207
1. 12 mAs *2. 24 mAs* *3. 36 mAs*

A. 1 only
B. 2 only
C. 3 only
D. 1, 2 & 3

28. An AEC unit is being utilized for a large patient. The backup timer is set at 400 mA 500 ms and 90 kVp. What is the largest permitted mAs value for this exposure? 15:207

A. 15 mAs
B. 75 mAs
C. 150 mAs
D. 200 mAs

29. When AEC exposure is made, what is the likely cause for backup time or overload light to appear? 15:207

A. Insufficient focal film distance
B. Insufficient collimation
C. Improper tube centering
D. Excessive patient movement

30. An AEC unit has a minimum response time of 33 ms (1/30) second and a backup time of 250 ms (1/4) second. During a radiographic exposure of 200 mA and 80 kVp, what is the smallest permitted mAs value for this unit? 15:207

A. 3 mAs
B. 7 mAs
C. 14 mAs
D. 50 mAs

31. A 6 mAs exposure is desired for a particular examination. An AEC exposure is used with 600 mA and 80 kVp. If the minimum response time of this unit is 20 ms (1/50) second, what timer setting will be indicated for the exposure? 15:208

A. 10 ms (1/100 sec.)
B. 20 ms (1/50 sec.)
C. 100 ms (1/10 sec.)
D. 200 ms (1/5 sec.)

32. A patient is to be radiographed using an AEC exposure at 400 mA and 88 kVp. If the minimum response time is 20 ms (1/50 sec.), which of the following mAs values can be obtained? 15:208
1. 4 mAs *2. 6 mAs* *3. 8 mAs*

A. 1 only
B. 2 only
C. 3 only
D. 1, 2 & 3

Pertaining to the following statement, answer questions 33.-35.

An AEC unit has a backup timer setting of 500 mA 500 ms (1/2 sec.). The unit is adjusted for a 12:1 bucky and a (400) speed cassette.

33. A patient is to be radiographed for a decubitus abdomen. Which of the following statements best indicates the required action in order to obtain an acceptable film? 15:209

A. Minimum response time must be shortened
B. Kilovoltage should be reduced
C. Backup time should be increased
D. The AEC should not be used

34. What is the largest permitted AEC mAs value that can be used for a supine abdomen on this patient? 15:209

A. 10 mAs
B. 60 mAs
C. 125 mAs
D. 250 mAs

35. The radiograph of the abdomen using an AEC exposure shows acceptable density and contrast. Which of the following actions would change this if a second exposure was attempted? 15:209
1. Decreasing the SID *2. Lowering the kVp to 75* *3. Using 100 speed cassette*

A. 1 only
B. 2 only
C. 3 only
D. 1, 2 & 3

X-Ray Tube

Chapter 12

1. The negative electrode of a x-ray tube is commonly called the: 2:117

A. Filament
B. Anode
C. Target
D. Diaphragm

2. The filaments in most modern x-ray tubes are surrounded by metal collars which serve to: 2:118

A. Increase filament heat
B. Compress the electron stream
C. Accelerate the electrons
D. Prevent shock hazards

3. The x-ray tube envelope is composed of a heat resistant material called: 2:117

A. Bakelite
B. Pyrex glass
C. Lead glass
D. Wratten 6B glass

4. The difference in charge between the cathode and anode during a x-ray exposure is termed the: 157

A. Power gradient
B. Space charge
C. Intensity factor
D. Potential difference

5. The temperature at which thermionic emission occurs is approximately: 5:11

A. 100 C°
B. 700 C°
C. 1270 C°
D. 2200 C°

6. The rotation speed of an anode in most single-phase x-ray units is about: 2:121

A. 200 rpm
B. 1000 rpm
C. 3000 rpm
D. 10,000 rpm

7. The temperature of the filament in a modern x-ray tube is directly related to the: 2:118

A. Exposure time
B. Quantity of x-rays
C. Speed of the electron stream
D. Speed of the anode rotation

8. Tungsten is the principal material used in the formation of the: 2:118
1. Target of a rotating anode *2. Target of a stationary anode* *3. X-ray tube filament*

A. 1 only
B. 2 only
C. 3 only
D. 1, 2 & 3

9. The reduction of focal spot size can be accomplished by: 2:124

A. Decreasing focus cup voltage
B. Using a smaller filament
C. Increasing mAs values
D. Increasing anode rotation speed

10. The thin, flattened surface of the x-ray tube envelope which allows for the minimum absorption of x-rays is termed the: 2:117

A. X-ray window
B. Primary diaphragm
C. Primary shutter
D. Focus cup

11. A modern dual focus x-ray tube contains: 2:119

1. One filament *2. Two filaments* *3. One anode*

A. 1 & 2 only
B. 1 & 3 only
C. 2 & 3 only
D. 1, 2 & 3

12. In a modern diagnostic x-ray tube, the majority of the energy of the electron stream is converted into _________ upon striking the anode. 2:119

A. Heat
B. Pressure
C. X-radiation
D. Gamma rays

13. The amount of space charge created at the cathode is primarily controlled by the: 5:11

A. Impressed amperage
B. Impressed kilovoltage
C. Tube current
D. Hysteresis

14. In a stationary anode tube, the target is imbedded in an anode made of: 5:14

A. Tungsten
B. Molybdenum
C. Copper
D. Rhenium

15. The speed of the electron stream across a x-ray tube is primarily controlled by the: 2:157

A. Applied kilovoltage
B. Vacuum material
C. Temperature of the filament
D. Focus cup voltage

16. The heating of the filament in a x-ray tube results in the liberation of electrons by a process termed: 5:11

A. Isotropic propagation
B. Thermionic emission
C. Thermoluminescence
D. Phosphorescence

17. In a modern x-ray tube, the positively-charged electrode that serves as the target for the electron stream is the: 5:14

A. Primary filament
B. Secondary filament
C. Anode
D. Cathode

18. The unit of power that the watt is derived from is the product of: 5:21

A. Amperage x time
B. Voltage x resistance
C. Resistance x amperage
D. Amperage x voltage

19. The measurement of the focal spot size in most diagnostic x-ray tube is based on the computation of the: 1:24

1. Actual focal spot *2. Effective focal spot* *3. Electronic focal spot*

A. 1 only
B. 2 only
C. 3 only
D. 1, 2 & 3

20. A modern radiographic tube with its hot filament was first developed in the early 1900's by: 2:11

A. Wilhelm Roentgen
B. William Coolidge
C. Neils Bohr
D. Madame Curie

21. The angle on the anode of a x-ray insures that all of the x-ray photons produced will be directed toward the patient. 2:124

A. True
B. False

22. If the angle on a x-ray tube anode is decreased from 12° down to 7°, it will result in a decrease in: 2:124
1. Focal spot size   2. Tube capacity   3. Film coverage

A. 1 & 2 only
B. 1 & 3 only
C. 2 & 3 only
D. 1, 2 & 3

23. In order to prevent the movement of electrons from the anode to the cathode during the exposure cycle, the type of current employed for x-ray tubes is: 2:134

A. Space charge limited current
B. Alternating current
C. Direct current
D. Three phase current

24. In a rotating anode, the molybdenum stem is a poor conductor of heat. This material helps in preventing: 5:11

A. Thermal expansion of the anode
B. Pitting of the anode
C. Bearing damage of the induction motor
D. Excessive prep time

25. Which is the proper sequence of materials travelled by the x-ray beam after leaving the anode of the x-ray tube? 2:115

A. Oil, aluminum, pyrex glass
B. Pyrex glass, oil, aluminum
C. Aluminum, pyrex glass, oil
D. Oil, pyrex glass, aluminum

26. The amount of heat storage capacity of a rotating anode is dependent upon the anode's: 2:121
1. Disk diameter   2. Rotational speed   3. Prep time

A. 1 & 2 only
B. 1 & 3 only
C. 2 & 3 only
D. 1, 2 & 3

27. Which of the following is an advantage that a rotating anode tube will have over a stationary anode tube? 5:16
1. Increased heat dissipation   2. Higher tube ratings   3. Reduced roughening of the target

A. 1 only
B. 2 only
C. 3 only
D. 1, 2 & 3

28. Which of the following factors will effect focal spot size? 5:13
1. Anode angle   2. Rotational speed   3. Cathode size

A. 1 & 2 only
B. 1 & 3 only
C. 2 & 3 only
D. 1, 2 & 3

29. As mA increases, blooming of the focal tract occurs and: 12:325

A. Effective focal spot size increases
B. Density decreases
C. Image blur decreases
D. Contrast increases

30. The material that is often added to a rotating anode disk to reduce surface defects and cracks is called: 5:17

A. Silver
B. Cesium
C. Lead
D. Rhenium

31. The main reason tungsten is employed as filament material in a x-ray tube is its: 2:117

A. Low phosphorescence
B. High degree of flexibility
C. High atomic number
D. High melting point

32. The discovery of x-rays was made by Wilhelm Roentgen in 1895 while experimenting with a partially evacuated glass tube called a/an: 2:8

A. Crooke's tube
B. Coolidge tube
C. Edison tube
D. Compton tube

33. The target of a modern rotating anode is bevelled to produce an incline to an angle of between: 2:118

A. 3°-7°
B. 7°-17°
C. 15°-25°
D. 25°-37°

34. The durability and efficiency of the tungsten filament is often improved by the addition of the metal: 2:111

A. Copper
B. Silver
C. Lead
D. Thorium

35. The anode of a modern rotating anode tube is generally composed of a combination of tungsten and: 5:17
1. Copper 2. Rhenium 3. Molybdenum

A. 1 & 2 only
B. 1 & 3 only
C. 2 & 3 only
D. 1, 2 & 3 only

36. The line-focus principle concerns the effect of target angulation and focal spot size. Which is a true statement of this principle? 2:118

A. The actual focal spot is larger than the effective focal spot.
C. As tube angulation increases, the effective focal spot decreases.
B. Radiation intensity is greatest on the cathode side of the tube.
D. More than one but not all the above.

37. The collective repulsion of the electron cloud around the cathode that creates an equilibrium with the rate of electron emission by the filament is termed the: 2:111

A. Space charge effect
B. Anode heel effect
C. Thermionic emission
D. Edison effect

38. The major reason that tungsten is the material chosen for the electrodes of a modern x-ray tube is its high melting point of: 5:14

A. 1073°C
B. 1283°C
C. 2330°C
D. 3370°C

39. The maximum anode heat storage capacity of a radiographic tube is related to the type and size of the: 5:15
1. Anode disk 2. Thermal cushion 3. Anode materials

A. 1 only
B. 2 only
C. 3 only
D. 1, 2 & 3

40. When a radiographer changes from a small to a large focal spot for an exposure, he/she is actually: 2:113

A. Using a larger filament
B. Increasing the filament-anode distance
C. Increasing the anode diameter
D. Increasing the anode angulation

41. Radiographic examinations requiring large tube currents and using serial exposures will often employ high speed rotating anodes with rotational speeds of about: 2:116

A. 400 rpm
B. 1,000 rpm
C. 3,000 rpm
D. 10,000 rpm

Pertaining to the diagram of the X-ray tube answer questions 42.-45.

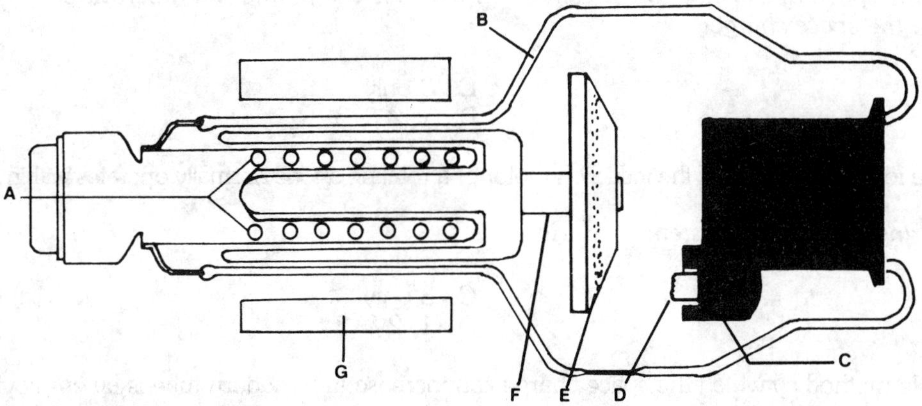

In the Diagram of the X-ray tube place the letter which corresponds to the following structures.

42. __________ Induction motor ball bearings

43. __________ Anode Stem

44. __________ Stator windings

45. __________ Focus cup

46. In a modern rotating anode, the rotator motion is accomplished by the use of a/an: 2:121

A. Mechanical motor
B. Induction motor
C. Synchronous motor
D. Solar motor

47. The steepness of the anode angle will effect the amount of the: 2:124
1. Anode heel effect *2. Field coverage* *3. Thermionic emission*

A. 1 & 2 only
B. 1 & 3 only
C. 2 & 3 only
D. 1, 2 & 3

48. Vibration of the x-ray tube during a radiographic exposure will most likely result in: 2:291
1. An increase in radiographic density *2. A decrease in radiographic contrast*
3. A decrease in radiographic definition

A. 1 & 2 only
B. 1 & 3 only
C. 2 & 3 only
D. 1, 2 & 3

49. The ball bearings which aid in the rotation of the induction motor require a special low vaporizing, lubricating material called: 12:211

A. Graphite
B. PCB's
C. Powdered silver
D. Grease

50. The general formula that is used to determine the anode heat units created with a fully rectified x-ray tube is: 5:20

A. HU = mA x kV x SID
B. HU = mA x Time x FSS
C. HU = mA x kV x Time
D. HU = mA x SID x FSS

51. A reduction in the kilovoltage setting during the operation of an x-ray tube will most likely: 5:198
1. Reduce the speed of the electron stream *2. Increase the filament temperature*
3. Decrease the space charge

A. 1 only
B. 2 only
C. 3 only
D. 1, 2 & 3

52. Which of the following portion of the induction motor of a rotating anode normally operates within the vacuum of the tube? 2:123
1. Ball bearings *2. Anode stem* *3. Rotor*

A. 1 only
B. 2 only
C. 3 only
D. 1, 2 & 3

53. The principle method by which the space charge can increase in a modern tube is by employing a higher: 5:11

A. mA station
B. kV setting
C. Timer setting
D. Screen speed

54. Which of the following events within the x-ray tube during the exposure (second) stage of the timer mechanism? 5:12
1. A high potential difference is developed *2. Filament heating begins*
3. Bucky movement begins

A. 1 only
B. 2 only
C. 3 only
D. 1, 2 & 3

55. The formation of the space charge in the filament of the x-ray tube is a accomplished during the ______ of the normal exposure cycle. 5:12

A. Prep or boost
B. Rectification stage
C. Exposure stage
D. Selection stage

Quality Assurance

Chapter 13

1. Nearly all causes of tube failure are related to ________ of the x-ray tube. 2:139

A. Mechanical characteristics
B. Electrical characteristics
C. Thermal characteristics
D. Manufacturing characteristics

2. A line pair or star phantom is often employed for determination of screen-film: 2:223

A. Resolution
B. Contrast
C. Sensitivity
D. Contact

3. In general, the proper method for warming up a x-ray anode may include exposures using: 12:223
1. A 30 second filament prep time *2. Short time, high kilovoltage techniques*
3. Long time, low kilovoltage techniques

A. 1 only
B. 2 only
C. 3 only
D. 1, 2 & 3

4. The determination of effective focal spot size can be accomplished by the use of a: 12:323

A. Spinning top tester
B. Wire mesh test
C. Pinhole camera
D. Penetrometer test

5. The degree to which mAs can be maintained by different combinations of mA and timer settings is termed: 2:630

A. Linearity
B. Reciprocity
C. Congruency
D. Reproducibility

6. The Wisconsin test tool and penetrometer are two commonly employed devices for the determination of:

A. Timer accuracy
B. Kilovoltage accuracy
C. Anode storage capacity
D. Circuit overload

7. The rotating anode of a modern x-ray unit requires about one second to accelerate to its designated speed. To avoid exposures during this time, rotation starts: 2:123

A. When the machine is turned on
B. When the technique is selected
C. During the prep stage of the exposure
D. During the exposure stage

8. Blurring of the radiographic image is often a result of: 2:232

A. Quantum mottle
B. Poor screen-film contact
C. Excessive drying temperature
D. Post-exposure fog

9. Even when proper warm-up procedures are followed, tube life is limited. A possible cause for this is: 2:140
1. Spintop leakage *2. Filament burnout* *3. Aperture diaphragm expansion*

A. 1 only
B. 2 only
C. 3 only
D. 1, 2 & 3

10. Good screen contact in cassette radiography is important because it reduces the amount of _____ from the screens: 2:232

A. Diffraction of light
B. Energy of light
C. Diffusion of light
D. Scattering of light

11. Most older radiographic tubes will appear to have a thin coating plated on the inside of the tube envelope. This is usually formed from vaporized: 2:140

A. Oil
B. Silver
C. Lead
D. Tungsten

12. A wire mesh test is the most common method for determination of: 2:232

A. Poor screen contact
B. Excessive phosphorescence
C. Opaque screen artifacts
D. Conversion efficiency

13. The induction motor of a rotating anode is damaged and stops before a series of multiple exposures is made. This will most likely result in: 12:221

A. Pitting of the anode
B. Cracking of the anode
C. Crazing of the anode
D. Cracking of the tube envelope

14. A stepwedge (penetrometer) is a device that can be used for: 2:269
1. Development of a characteristic curve *2. Calibrating x-ray equipment*
3. Determination of kVp

A. 1 only
B. 2 only
C. 3 only
D. 1, 2 & 3

15. Excessive heating of the tube may lead to vaporization of the filament or anode. Over a long period of time this may result in: 2:139
1. Arcing within the tube *2. Filtration of the beam* *3. Deterioration of the vacuum*

A. 1 only
B. 2 only
C. 3 only
D. 1, 2 & 3

16. A white or clear area appears on a series of films taken on the same cassette and in the same location on the films. The most likely cause is (a): 12:288

A. Opaque contrast residue on the table
B. Defect in the manufacturing of the film
C. Defect in the automatic processor
D. Stain on the intensifying screen

17. Which of the following events occurs in the first stage (prep) of a normal radiographic exposure? 12:216
1. Filament is heated *2. Production of x-rays* *3. Rotation of the anode*

A. 1 & 2 only
B. 1 & 3 only
C. 2 & 3 only
D. 1, 2 & 3

18. Which is not a common factor leading to the blurring of a radiographic image due to poor screen contact? 2:232

A. Improperly mounted screens
B. Broken cassette hinge
C. Warped cassette front
D. None of the above

19. If a radiographer does not employ the proper warm-up technique before making an exposure, which is most likely to occur? 12:219

A. Spark over
B. Cracking of the anode
C. Pitting of the anode
D. Melting of the anode

20. Surface irregularities, such as pitting of the anode, will often result in: 12:221

A. Excessive radiation output
B. Reduced radiation output
C. Blooming of the focal spot
D. Filament burnout

21. A radiograph comes out of the processor showing numerous areas of light density areas. The most likely cause is: 12:288

A. Improper film handling
B. Water on the feed tray
C. Excessive developer temperature
D. Dust on the screens

22. The timer of a three-phase radiographic unit is most often tested by using a/an: 20:12

A. Pinhole camera
B. Oscilloscope
C. Resolution phantom
D. Spintop tester

23. A loud, grinding noise is heard from a x-ray tube during the prep stage of the exposure. The most likely cause is: 2:139

A. A pitted anode
B. Filament burnout
C. Induction motor bearing damage
D. Cracked tube envelope

24. Which is a common method employed for extending the life of a x-ray tube? 2:139
1. Proper warm-up technique *2. Avoidance of prolonged prep time*
3. Observation of anode heat values

A. 1 only
B. 2 only
C. 3 only
D. 1, 2 & 3

25. A synchronous spintop tester is often employed for the determination of ________ on single-phase radiographic units.
1. Valve tube (diode) competency *2. Timer accuracy* *3. Beam quality*

A. 1 & 2 only
B. 1 & 3 only
C. 2 & 3 only
D. 1, 2 & 3

26. An exposure is attempted on a radiographic unit using the 100 mA station and an overload light results. If the same technique is permitted using the 300 mA station, the most likely cause is: 2:129

A. Small filament burnout
B. Anode pitting
C. Large filament burnout
D. Excessive heat storage capacity

27. The common indication of a gassy x-ray tube is ________ during the exposure.

A. Fluctuations in mA readings
B. Fluctuations in kVp readings
C. Vibration of the tube housing
D. No mA readings

28. The life of the filament can be dramatically reduced by employing greater ________ values than is generally required. 5:11

A. Potential difference (kVp)
B. Tube current (mA)
C. Focal spot
D. Focus cup

29. The principle piece of test equipment that is used to measure the radiation output of an x-ray unit is a: 2:630

A. Scintillation type meter
B. Pin hole camera
C. mAs/mR ionization type meter
D. Line pair phantom

30. What should be checked if no ready light appears when an exposure is attempted? 2:133
1. Bucky not in all the way *2. Door interlock systems* *3. Fluorocarriage properly parked*

A. 1 only
B. 2 only
C. 3 only
D. 1, 2 & 3

31. Smoke or fire is detected coming from the equipment cabinet. Your first action should be to: 8:57

A. Throw water on the cabinet
B. Turn off all power to the unit
C. Close the door to the room
D. Pull the fire alarm box

32. The principle agency involved with maintaining safety of radiation producing equipment is the: 13:64

A. BRH
B. AEC
C. CEC
D. EPA

33. Which of the following types of fire extinguishers are generally considered safe for use on electrical equipment? 8:58
1. Halon *2. Water* *3. CO^2*

A. 1 & 2 only
B. 1 & 3 only
C. 2 & 3 only
D. 1, 2 & 3

34. The output radiation intensity for a series of 10 radiographic exposures (reproducibility) must not exceed: 2:630

A. 40%
B. 20%
C. .5%
D. 2%

35. The most common cause for repeat radiographs is improper:

A. Positioning
B. Exposure Selection
C. Processing
D. Equipment Maintenance

36. The principal method of heat dissipation in a rotating anode x-ray tube is intended to be: 5:16

A. Conduction through the anode stem
B. Radiation to the tube envelope
C. Conduction through the anode bearings
D. Radiation to the focus cup

37. The careful observation of a tube rating chart may help to:
1. Prevent anode damage *2. Prevent filament damage* *3. Extend tube life*

A. 1 only
B. 2 only
C. 3 only
D. 1, 2 & 3

38. The accuracy of distance indicators and collimation devices is required to be within ________ of the ________. 2:630

A. 2%; SID
B. 2%; SOD
C. 10%; SID
D. 18%; SID

39. A radiographic (tube) rating chart is used in the determination of ________ for a tube. 2:140

A. Maximum exposure limits
B. Maximum heat capacity
C. Anode cooling characteristics
D. Housing cooling characteristics

40. Which is an important factor in the dissipation of heat away from the anode? 5:15
1. Amount of insulation oil *2. Temperature of the room* *3. Type of accessory cooling device*

A. 1 only
B. 2 only
C. 3 only
D. 1, 2 & 3

Anode Storage Capacity

Exercise No. 1

In the following exercise, all of the questions pertain to the anode cooling chart appearing below:

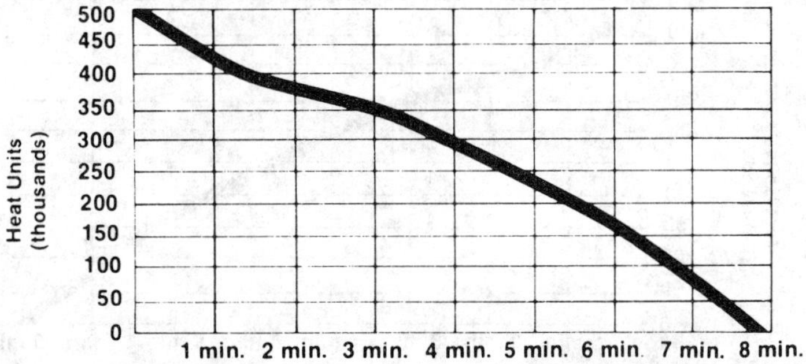

1. How many rapid full-wave rectified exposures at 100 mA, 250 ms 80 kVp can be made before exceeding the anode storage capacity of this tube?

2.-7. State the maximum number of full-wave rectified exposures at the following techniques which can be made without exceeding the anode storage capacity of this tube.

	mA	Time	kVp	Rectification	Maximum # of Exposures
2.	300	250 ms	90	Full-wave	________
3.	500	300 ms	80	Full-wave	________
4.	200	1.20 sec.	90	Full-wave	________
5.	400	600 ms	100	Full-wave	________
6.	100	1.5 sec.	95	Full-wave	________
7.	200	750 ms	105	Full-wave	________

8. How many rapid 3-phase, 6-pulse exposures at 100 mA, 250 ms, 80 kVp can be made before exceeding the anode storage capacity of this tube?

9.-14. State the maximum number of 3-phase, 6-pulse exposures at the following techniques which can be made without exceeding the anode storage capacity of this tube.

	mA	Time	kVp	Rectification	Maximum # of Exposures
9.	300	250 ms	100	3-phase, 6-pulse	________
10.	500	300 ms	90	3-phase, 6-pulse	________
11.	200	1.5 sec.	90	3-phase, 6-pulse	________
12.	400	750 ms	100	3-phase, 6-pulse	________
13.	800	50 ms	95	3-phase, 6-pulse	________
14.	600	150 ms	105	3-phase, 6-pulse	________

15. How many rapid 3-phase, 12-pulse exposures at 100 mA, 250 ms, 80 kVp can be made before exceeding the anode storage capacity of this tube?

16.-20. State the maximum number of 3-phase, 12-pulse exposures at the following techniques which can be made without exceeding the anode storage capacity of this tube.

	mA	Time	kVp	Rectification	Maximum # of Exposures
16.	400	70 ms	98	3-phase, 12-pulse	________
17.	300	250 ms	90	3-phase, 12-pulse	________
18.	600	250 ms	75	3-phase, 12-pulse	________
19.	800	270 ms	90	3-phase, 12-pulse	________
20.	300	330 ms	80	3-phase, 12-pulse	________

Correction factors used in this exercise for the 3-phase, 6-pulse unit is 1.35 and for the 3-phase, 12-pulse unit is 1.41.

Exercise No. 2

In the following exercise, all of the questions pertain to the anode cooling chart appearing below:

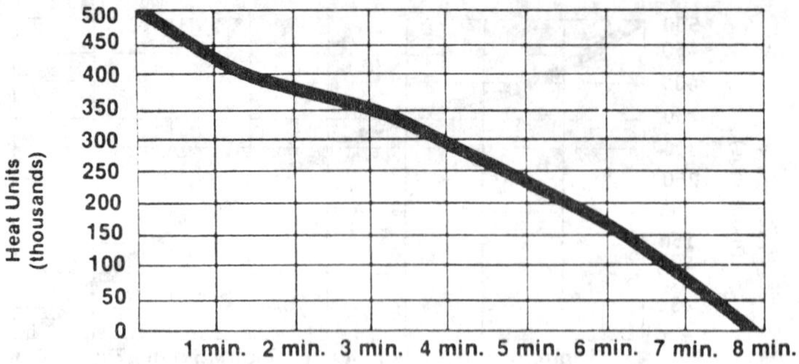

1. How many rapid full-wave rectified exposures at 200 mA, 250 ms 80 kVp can be made before exceeding the anode storage capacity of this tube?

2.-7. State the maximum number of full-wave rectified exposures at the following techniques which can be made without exceeding the anode storage capacity of this tube.

	mA	Time	kVp	Rectification	Maximum # of Exposures
2.	800	250 ms	80	Full-wave	
3.	300	300 ms	90	Full-wave	
4.	400	1.20 sec.	80	Full-wave	
5.	100	600 ms	110	Full-wave	
6.	200	1.5 sec.	95	Full-wave	
7.	400	750 ms	105	Full-wave	

8. How many rapid 3-phase, 6-pulse exposures at 200 mA, 250 ms, 80 kVp can be made before exceeding the anode storage capacity of this tube?

9.-14. State the maximum number of 3-phase, 6-pulse exposures at the following techniques which can be made without exceeding the anode storage capacity of this tube.

	mA	Time	kVp	Rectification	Maximum # of Exposures
9.	800	250 ms	100	3-phase, 6-pulse	
10.	600	300 ms	80	3-phase, 6-pulse	
11.	300	1.5 sec.	90	3-phase, 6-pulse	
12.	100	750 ms	110	3-phase, 6-pulse	
13.	300	50 ms	95	3-phase, 6-pulse	
14.	400	150 ms	105	3-phase, 6-pulse	

15. How many rapid 3-phase, 12-pulse exposures at 200 mA, 250 ms, 80 kVp can be made before exceeding the anode storage capacity of this tube?

16.-20. State the maximum number of 3-phase, 12-pulse exposures at the following techniques which can be made without exceeding the anode storage capacity of this tube.

	mA	Time	kVp	Rectification	Maximum # of Exposures
16.	200	70 ms	98	3-phase, 12-pulse	
17.	300	150 ms	60	3-phase, 12-pulse	
18.	400	250 ms	75	3-phase, 12-pulse	
19.	800	70 ms	90	3-phase, 12-pulse	
20.	600	330 ms	70	3-phase, 12-pulse	

Correction factors used in this exercise for the 3-phase, 6-pulse unit is 1.35 and for the 3-phase, 12-pulse unit is 1.41.

Tube Rating Chart

Exercise No. 1

In the following exercise, all of the questions pertain to the tube rating chart appearing below.

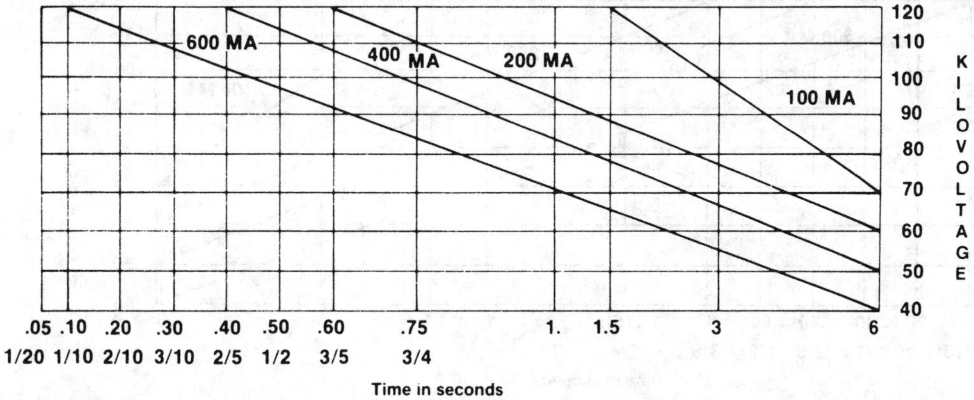

1. What is the maximum exposure time that can be used at 120 kVp for a 100 mA exposure?
2. What is the maximum exposure time that can be used at 120 kVp for a 400 mA exposure?
3. What is the maximum exposure time that can be used at 100 kVp for a 200 mA exposure?
4. What is the maximum exposure time that can be used at 100 kVp for a 600 mA exposure?
5. What is the maximum mA station that can be used for a .4 sec. exposure at 90 kVp?
6. What is the maximum mA station that can be used for a .75 sec. exposure at 120 kVp?
7. What is the maximum mA station that can be employed for a .2 sec. exposure at 120 kVp?
8. What is the maximum mA station that can be employed for a .05 sec. exposure at 120 kVp?
9. What is the maximum mA station that can be employed for a 1.5 sec. exposure at 90 kVp?
10. What is the maximum mA station that can be employed for a 3 sec. exposure at 60 kVp?
11. What is the maximum kilovoltage that can be used at 400 mA, .2 sec.?
12. What is the maximum kilovoltage that can be used at 600 mA, .5 sec.?
13. What is the maximum kilovoltage that can be used at 100 mA, .3 sec.?
14. What is the maximum kilovoltage that can be used at 200 mA, .75 sec.?
15. Will this unit permit a 400 mA, .5 sec., 80 kVp exposure?
16. Will this unit permit a 600 mA, .3 sec., 100 kVp exposure?
17. Will this unit permit a 100 mA, .5 sec., 120 kVp exposure?
18. Will this unit permit a 200 mA, .05 sec., 120 kVp exposure?
19. Will this unit permit a 400 mA, 1.0 sec., 100 kVp exposure?
20. Will this unit permit a 600 mA, 3.0 sec., 60 kVp exposure?

Tube Rating Chart

Exercise No. 2

In the following exercise, all of the questions pertain to the tube rating chart appearing below.

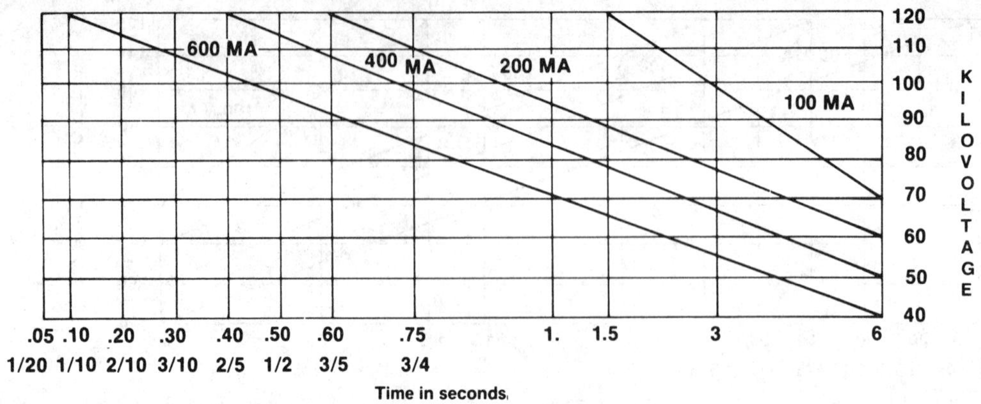

1. What is the maximum exposure time that can be used at 90 kVp for a 100 mA exposure?
2. What is the maximum exposure time that can be used at 100 kVp for a 400 mA exposure?
3. What is the maximum exposure time that can be used at 90 kVp for a 200 mA exposure?
4. What is the maximum exposure time that can be used at 120 kVp for a 600 mA exposure?
5. What is the maximum mA station that can be used for a .4 sec. exposure at 120 kVp?
6. What is the maximum mA station that can be used for a .75 sec. exposure at 110 kVp?
7. What is the maximum mA station that can be employed for a .3 sec. exposure at 120 kVp?
8. What is the maximum mA station that can be employed for a 1 sec. exposure at 100 kVp?
9. What is the maximum mA station that can be employed for a 3 sec. exposure at 70 kVp?
10. What is the maximum mA station that can be employed for a 1.5 sec. exposure at 70 kVp?
11. What is the maximum kilovoltage that can be used at 600 mA, .6 sec.?
12. What is the maximum kilovoltage that can be used at 400 mA, .75 sec.?
13. What is the maximum kilovoltage that can be used at 100 mA, 3 sec.?
14. What is the maximum kilovoltage that can be used at 200 mA, .1 sec.?
15. Will this unit permit a 600 mA, .5 sec., 80 kVp exposure?
16. Will this unit permit a 200 mA, .3 sec., 80 kVp exposure?
17. Will this unit permit a 100 mA, 1.5 sec., 120 kVp exposure?
18. Will this unit permit a 200 mA, .10 sec., 120 kVp exposure?
19. Will this unit permit a 600 mA, 1.0 sec., 100 kVp exposure?
20. Will this unit permit a 400 mA, 1.5 sec., 60 kVp exposure?

Filtration - HVL

Chapter 14

1. The use of diagnostic x-ray filtration has the following effect(s): 2:169
   1. Increased beam quality *2. Decreases skin dose*
   3. Decreases the number of long wavelength photons

   A. 1 & 2 only
   B. 1 & 3 only
   C. 2 & 3 only
   D. 1, 2 & 3

2. The thickness of absorbing material necessary to reduce the x-ray intensity to half of its original value is called: 2:165

   A. Half-life
   B. Half-wave reduction
   C. Half-penetration value
   D. Half-value layer

3. The inherent filtration of a radiographic machine is usually composed of the material: 2:168

   A. Pyrex glass
   B. Aluminum
   C. Bakelite
   D. Lead

4. The half-value layer of most diagnostic x-ray beams operating at a level of 95 kVp is: 2:165

   A. 3.5 mm Al (aluminum)
   B. 3.5 mm Cu (copper)
   C. 3.5 mm Sn (tin)
   D. 3.5 mm Pb (lead)

Pertaining to the diagram of the x-ray energy answer question 5.

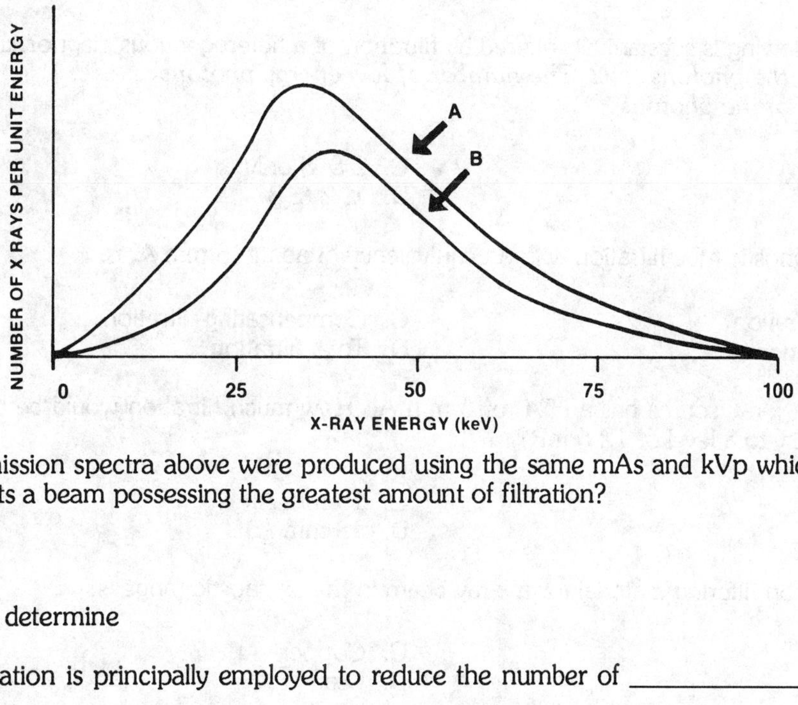

5. If the x-ray emission spectra above were produced using the same mAs and kVp which of the curve most likely represents a beam possessing the greatest amount of filtration? 2:158

   A. Curve A
   B. Curve B
   C. Unable to determine

6. Diagnostic filtration is principally employed to reduce the number of ______________ photons reaching the patient. 2:167
   1. Low energy *2. Long wavelength* *3. Scattered*

   A. 1 & 2 only
   B. 1 & 3 only
   C. 2 & 3 only
   D. 1, 2 & 3

7. A low energy gamma source is filtered by 4 mm Al which reduces it to 25% of its original intensity. What is the half-layer of this source? 2:167

A. 1 mm Al
B. 2 mm Al
C. 8 mm Al
D. 12 mm Al

8. Uniform film density during radiography of structures of unequal thickness can be accomplished with the use of a/an: 2:169

A. Thoraeus filter
B. Compensating filter
C. Compound filter
D. Inherent filter

Pertaining to diagram of thickness of aluminum absorber answer question 9.

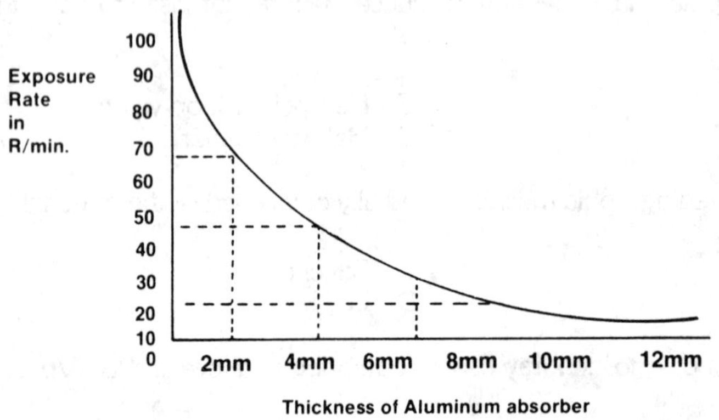

9. Pertaining to the diagram, approximately what is the half-value layer thickness for this beam? 2:167

A. 2 mm Al
B. 4 mm Al
C. 6mm Al
D. 8 mm Al

10. Which of the following is substantially altered by filtration of a heterogeneous diagnostic x-ray beam? 2:167
1. The speed of the photons *2. The number of low energy photons*
3. The quantity of the photons

A. 1 & 2 only
B. 1 & 3 only
C. 2 & 3 only
D. 1, 2 & 3

11. The normal diagnostic tube filtration with an equivalency of about .5 mm Al is: 4:87

A. Inherent filtration
B. Added filtration
C. Compensating filtration
D. Total filtration

12. A 100 mR/min. x-ray source has a HVL of 6 mm Al. How much filtration would be necessary to reduce the beam intensity to a level of 12.5 mR? 2:167

A. 9 mm Al
B. 12 mm Al
C. 15 mm Al
D. 18 mm A1

13. The most common filtering material for a x-ray beam in the diagnostic range is: 4:90

A. Thorium
B. Aluminum
C. Copper
D. Lead

14. The amount of diagnostic filtration can be determined by finding the sum of: 2:168

A. Inherent and compensating filtration
B. Compensating and added filtration
C. Added and compensating filtration
D. Inherent and added filtration

15. A x-ray beam has an intensity of 100 mR/min. unfiltered. If the addition of 10 mm Al reduces this level of 25 mR/min., the half-value layer thickness is about: 2:167

A. 2.5 mm Al
B. 5.0 mm Al
C. 7.5 mm Al
D. 15 mm Al

Pertaining to the diagram for beam filtration answer questions 16.-17.

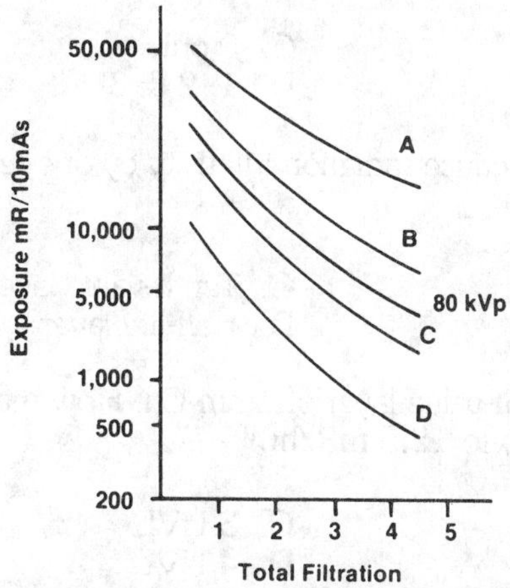

16. Pertaining to the diagram above, each of the lines A-D represents the intensity change due to filtration at different kVp levels. According to the graph, the highest kVp is represented by: 2:164

A. Curve A
B. Curve B
C. Curve C
D. Curve D

17. Pertaining to the diagram above, the greatest amount of attenuation is seen for values represented by: 2:164

A. Curve A
B. Curve B
C. Curve C
D. Curve D

18. A high energy gamma source requires 4 mm of copper to reduce the original intensity by 50%. This source has a half-value layer of: 2:167

A. 22 mm copper
B. 15 mm copper
C. 8 mm copper
D. 4 mm copper

19. The use of additional filtration in a diagnostic x-ray machine will: 2:167
1. Increase beam quality *2. Reduce anode heel-effect* *3. Increase film density*

A. 1 only
B. 2 only
C. 3 only
D. 1, 2 & 3

20. If it is desired to increase the amount of primary beam attenuation, the material used for the filter should have a __________ than the material it replaces. 2:167

A. Higher atomic number
B. Higher electron density
C. Lower atomic number
D. Lower electron density

21. Which of the following x-ray beams will possess the highest radiographic quality? 2:167

A. 80 kVp unfiltered
B. 90 kVp unfiltered
C. 80 kVp, 2.5 mm Al
D. 90 kVp, 2.5 mm Al

22. A single tenth-value layer will have about the same protective value as: 2:601

A. 2.4 half-value layers
B. 3.3 half-value layers
C. 5.2 half-value layers
D. 7.3 half-value layers

23. A beam that is filtered by two half-value layers of absorber is considered to be: 12:198
1. Polychromatic *2. Monochromatic* *3. Heterogeneous*

A. 1 only
B. 2 only
C. 3 only
D. 1, 2 & 3

24. The amount of material required to reduce radiation intensity by one-tenth of the original value is termed: 2:601

A. Filtration value layer
B. Half-value layer
C. Tenth-value layer
D. Half-life layer

25. A 100 mR/hr. x-ray source has a half-value layer of 1mm Cu. How many half-value layers are required to reduce the beam intensity to 12.5 mR/hr.? 2:167

A. 1 HVL
B. 2 HVL
C. 3 HVL
D. 4 HVL

Electromagnetic Radiation

Chapter 15

1. The wave-particle duality of radiation is described by (the): 2:69

   A. Coulomb's law
   B. Einstein's theory
   C. Ohm's law
   D. Quantum theory

2. The principal difference between a x-ray photon and a gamma ray is the: 2:64

   A. Wavelength of the photon
   B. Frequency of the photon
   C. Origin of the photon
   D. Speed of the photon

3. As the energy of electromagnetic radiation increases, the: 2:61
   1. Frequency decreases *2. Wavelength decreases* *3. Speed increases*

   A. 1 only
   B. 2 only
   C. 3 only
   D. 1, 2 & 3

4. The frequency (hertz) of a wave can be defined as the: 2:57

   A. Distance the wave travels
   B. Number of cycles per second
   C. Height of the wave
   D. Distance between the successive wave peaks

5. The shape of all electromagnetic radiations follow a form known as a/an: 2:59

   A. Shear curve
   B. Plank curve
   C. Attenuation curve
   D. Sine curve

6. Which of the following electromagnetic radiations is most likely to interact with matter as if it were a particle? 2:65

   A. X-ray
   B. Visible light
   C. Microwaves
   D. Infrared rays

7. The normal range of a diagnostic x-ray photon's wavelength is: 5:6

   A. 1-5 cm
   B. 100-150 cm
   C. 10-20 Å
   D. 0.1 - 0.5 Å

8. Which member of the electromagnetic spectrum will normally possess photon wavelengths longer than those of x-ray photons? 2:63
   1. Cosmic rays *2. Infrared rays* *3. Microwaves*

   A. 1 & 2 only
   B. 1 & 3 only
   C. 2 & 3 only
   D. 1, 2 & 3

9. Sunburn is a common reaction to exposure to a normal component of sunlight called: 2:63

   A. Ultraviolet radiation
   B. Cosmic radiation
   C. Ultrasonic radiation
   D. Monochromatic radiation

10. The relationship of energy keV and photon wavelength is expressed by the formula: 5:30

    A. $E = \lambda c$
    B. $E = \frac{12.4}{\lambda}$
    C. $C = \lambda v$
    D. $E = h c$

11. Which is not true of the members of the electromagnetic spectrum? They all have the same: 5:2

A. Velocity
B. Energy
C. Mass
D. Charge

12. Which of the following can be used to deflect the path of an electromagnetic photon beam? 5:2

A. Strong magnetic field
B. Strong electrostatic field
C. Weak gravitational field
D. None of the above

13. The member of the electromagnetic spectrum possessing a wavelength of 7000Å is: 2:63

A. Green light
B. Blue light
C. Yellow light
D. Red light

Pertaining to the following diagram answer questions 14. and 15.

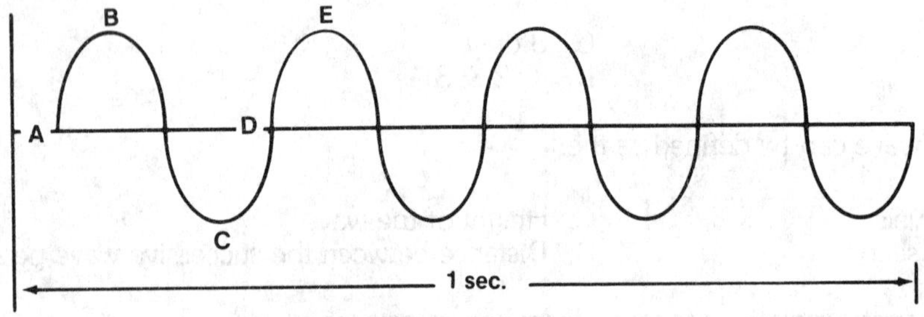

14. In the above diagram of an electromagnetic wave, which letters best define wavelength? 2:60

A. A-B
B. B-C
C. A-C
D. B-E
E. None of the above

15. In the diagram, this photon would have a frequency of: 2:60

A. Two cps
B. Three cps
C. Four cps
D. Eight cps

16. The energy of an electromagnetic radiation is directly proportional to the: 2:60
1. Wavelength of the radiation *2. Frequency of the radiation* *3. Speed of the radiation*

A. 1 only
B. 2 only
C. 3 only
D. 1, 2 & 3

17. Which of the following are included as members of the electromagnetic spectrum? 2:61
1. X-rays *2. Visible light* *3. Ultraviolet*

A. 1 & 2 only
B. 1 & 3 only
C. 2 & 3 only
D. 1, 2 & 3

18. The speed of all electromagnetic radiation in a vacuum is: 5:5

A. $3x10^{10}$ cm/sec.
B. $3x10^{10}$ m/sec.
C. $3x10^{10}$ km/sec.
D. Photons cannot travel in a vacuum

19. One of the principal dangers of exposure to radiation is the ability to cause ionization in the exposed tissue. Which type of radiation possesses sufficient energies to cause ionization? 2:63
1. Gamma rays 2. Infrared rays 3. X-rays

A. 1 & 2 only
B. 1 & 3 only
C. 2 & 3 only
D. 1, 2 & 3

20. Heat is a form of electromagnetic radiation produced at relatively low energies, and is also called: 2:59

A. Ultraviolet radiation
B. Gamma rays
C. Microwaves
D. Infrared radiation

21. What is the wavelength of a 248 KeV gamma photon 5:6

A. .05Å
B. .5Å
C. 1.5Å
D. 2.0Å

22. If the frequency of an electromagnetic photon is increased by a factor of 4, the wavelength will: 5:6

A. Increase 16 times
B. Increase 4 times
C. Decrease 16 times
D. Decrease 4 times

23. The unit used to measure the energy of a photon is the: 5:6

A. Angstrom unit
B. Quantum unit
C. Electron volt
D. Joule

24. The mass-energy equivalency of electromagnetic energy and matter was first predicted by: 12:36

A. Albert Einstein
B. Max Plank
C. Neils Bohr
D. Wilhelm Roentgen

25. Which formula is used in calculating the photon energy of electromagnetic radiations? 2:59

A. $E = mc^2$
B. $E = hv$
C. $E = 1/2\ mv^2$
D. $E = \frac{12.4}{keV}$

26. An individual packet of energy that interacts with matter in a specified manner according to the law of the electromagnetic forces is called a: 5:6

A. Vector
B. Photon
C. Spiral
D. Micron

27. Which of the following electromagnetic radiations are most likely to interact with matter as if it were a wave? 2:65
1. Visible light 2. X-ray rays 3. Gamma rays

A. 1 only
B. 2 only
C. 3 only
D. 1, 2 & 3

28. The wave length of photons are often measured by a unit of length corresponding to 10^{-8} cm called a/an: 5:5

A. Coulomb
B. Micron
C. Angstrom
D. Hertz

29. The energy of electromagnetic radiation is related to its wavelength by the relationshp E=: 5:6

A. hc

B. $\frac{hv}{\lambda}$

C. $\frac{hc}{\lambda}$

D. None of the above

30. The principle difference between electromagnetic radiation and sound waves is the: 5:6

A. Mass of the wave

B. Energy of the wave

C. Charge of the wave

D. Speed of the wave

Nature of X-Ray Photons

Chapter 16

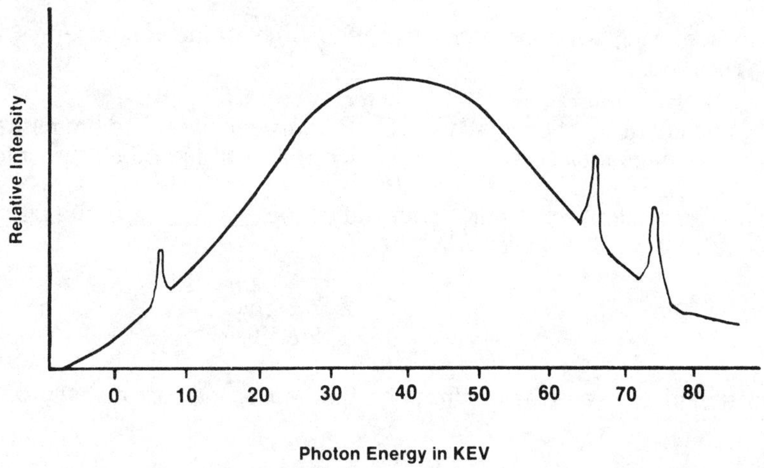

1. In the diagram of the x-ray spectral distribution curve, for a tungsten anode the three peaks seen represent: 2:149

A. K-edge absorption spectrum
B. Projectile electron energy
C. Brems x-ray production
D. Characteristic x-ray production

2. The collisions of x-ray tube electrons and outer shell orbit electrons of tungsten are principal causes of the __________ generated in the anode. 2:148

A. Microwaves
B. Ultraviolet photons
C. Infrared photons
D. X-ray photons

3. The x-radiation emission in a x-ray tube is considered to be __________ in nature. 12:160
1. Monochromatic *2. Heterogeneous* *3. Continuous*

A. 1 only
B. 2 only
C. 3 only
D. 1, 2 & 3

4. In the diagnostic range, the majority of the x-ray photons produced in the x-ray tube result from the: 2:151

A. Brems x-ray production
B. Compton x-ray production
C. Photoelectric x-ray production
D. Characteristic x-ray production

5. The penetrability or quality of a x-ray beam is effected by changes in: 2:165
1. Milliamperage *2. Filtration* *3. Kilovoltage*

A. 1 & 2 only
B. 1 & 3 only
C. 2 & 3 only
D. 1, 2 & 3

6. The energy of the photons produced in brems x-ray production is dependent upon the: 2:150
1. Projectile speed *2. Atomic number of the target* *3. Electron to nucleus distance*

A. 1 & 2 only
B. 1 & 3 only
C. 2 & 3 only
D. 1, 2 & 3

7. The loss of energy by a x-ray tube electron as it passes by a tungsten nucleus in the anode results in a photon being produced by an event termed: 5:30

A. Characteristic radiation
B. Compton interaction
C. Bremsstrahlung
D. Pair production

8. When the stream of fast-moving electrons interact with the target of the anode, x-rays are generated by two different processes. They are: 2:148

A. Bremsstrahlung and characteristic radiation
B. Characteristic and primary radiation
C. Primary radiation and bremsstrahlung
D. Compton and photoelectric interaction

9. The penetration of a x-ray photon is primarily controlled by the _________ of the photon: 5:6
1. Frequency *2. Weight* *3. Wavelength*

A. 1 & 2 only
B. 1 & 3 only
C. 2 & 3 only
D. 1, 2 & 3

10. When a high speed stream of x-ray tube electrons hit the anode, the vast majority of the electron's kinetic energy is converted into: 2:147

A. X-radiation
B. Infrared radiation
C. Ultraviolet radiation
D. Gamma rays

11. Increasing the speed of the electron stream across a x-ray tube will result in photons possessing a: 2:151
1. Shorter wavelength *2. Lower frequency* *3. Higher energy*

A. 1 & 2 only
B. 1 & 3 only
C. 2 & 3 only
D. 1, 2 & 3

12. The varying energies that occur during brems x-ray production are often referred to as: 12:160

A. Grenz radiation
B. Hertz radiation
C. White radiation
D. Discreet radiation

13. Which of the following factors will have a large effect on the photon energy of a characteristic x-ray photon? 1:49

1. Atomic number of the target *2. mA station employed* *3. Filament temperature*

A. 1 only
B. 2 only
C. 3 only
D. 1, 2 & 3

14. The frequency of a diagnostic x-ray photon directly proportional to the: 5:31
1. Quality of the beam *2. Beam penetration* *3. Energy of the beam*

A. 1 only
B. 2 only
C. 3 only
D. 1, 2 & 3

15. If one looks at a spectral distribution curve of an 80 kV x-ray exposure, the peaks seen at 59 and 69 KeV represent a: 5:33

A. Compton interaction
B. Bremsstrahlung photon production
C. Photoelectric interaction
D. Characteristic photon production

16. The efficiency of x-ray production can be calculated with the formula: 12:159

A. E = K x Z x kVp
B. $E = 1/2\ mv^2$
C. $E = mc^2$
D. E = kV x T x mA

17. A soft x-ray beam has a low penetrability. This beam is also said to possess a: 12:175

A. High quality
B. Low quality
C. High resolution
D. Low intensity

18. In a x-ray tube, the production of x-rays results from a rapid deceleration of fast moving ______ in the target material. 2:147

A. Electrons
B. Positrons
C. Protons
D. Neutrons

19. It would be possible to produce a homogeneous x-ray beam if the kilovoltage were held constant and all off-target radiations were prevented at the anode. 2:150

A. True
B. False

20. During x-ray production, the bremsstrahlung x-ray photons will comprise approximately _______ of the x-ray photons being created at 100 kVp. 2:151

A. 99%
B. 95%
C. 85%
D. 50%

Pertaining to the x-ray intensity distribution curve, answer questions 21. & 22.

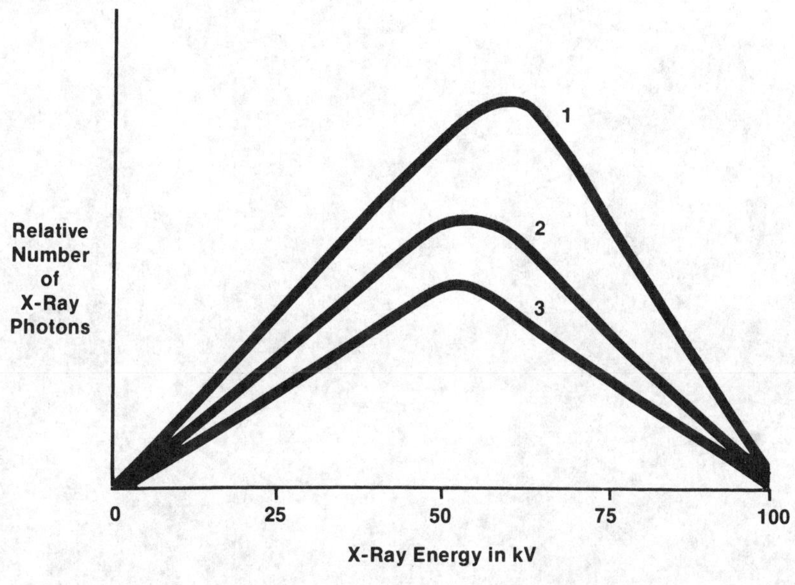

21. Which of the following curves resulted from the use of the lowest milliampere setting? 2:156

A. Curve 1
B. Curve 2
C. Curve 3

22. Curve 1 corresponds to the x-ray beam accomplished at a higher _________ than the other curves. 2:156
1. Distance 2. mA 3. kVp

A. 1 & 2 only
B. 1 & 3 only
C. 2 & 3 only
D. 1, 2 & 3

23. At a potential of 70 kVp the tube's electrons will reach a speed of approximately: 2:146

A. 9.1×10^6 m/s
B. 1.9×10^3 m/s
C. 3.0×10^8 m/s
D. 1.5×10^8 m/s

24. The quality or penetrability of an x-ray photon is affected by changes in the: 2:251
1. kVp 2. mA 3. SID

A. 1 only
B. 2 only
C. 3 only
D. 1, 2 & 3

25. In the diagnostic energy region, approximately what percentage of the electron stream energy is converted into x-rays? 5:33

A. 1%
B. 7%
C. 37%
D. 86%

Image Intensification

Chapter 17

1. The principal advantage of image intensified fluoroscopy compared to conventional radiography is that it allows for: 2:352
1. Higher spatial resolution *2. Lower patient dose* *3. Dynamic imaging*

A. 1 only
B. 2 only
C. 3 only
D. 1, 2 & 3

2. The time it takes for the human eye to adjust to the low light levels of conventional fluoroscopy is called: 2:355

A. Flicker time
B. Integration time
C. Adaptation time
D. Interlace time

3. A substantial increase in conventional fluoroscopic brightness levels is accomplished in most modern systems using a/an: 2:353

A. Television system
B. Image intensifier
C. Video tape unit
D. Magnification technique

4. When compared to conventional radiography, the amount of milliamperage employed for image intensified fluoroscopy is: 2:356

A. Significantly less
B. Significantly more
D. Approximately the same

5. The major factors effecting the amount of brightness gain in an image intensifier are: 2:357

A. Focal and scotopic vision gains
B. Adaptation and vignetting gains
C. Minification ratio and flux gains
D. Dispersion and retinal gains

6. The phosphor and photocathode layers, which convert the x-rays into electronic images in an image intensified tube, are located on the _________ of the tube. 2:356

A. Input side
B. Output side
C. Periphery
D. Focal tract

7. The output screen in an image intensifier serves to convert the electronic image into a: 2:357

A. Radiographic image
B. Visible light image
C. Video signal
D. Television frame

8. Most modern image intensifiers utilize an input fluorescent screen layer which is composed of: 2:356

A. Calcium tungstate
B. Cesium iodide
C. Barium lead sulfate
D. Zinc cadmium sulfide

9. The operating potential in most modern image intensifier tubes is approximately: 5:169

A. 2 kVp
B. 25 kVp
C. 140 kVp
D. 200 kVp

10. The squared ratio of the input to output screen diameter in an image intensifier tube is termed: 5:171

A. Magnification ratio
B. Conversion ratio
C. Modulation transfer ratio
D. Minification ratio

11. To aid the radiologist during fluoroscopy, most modern units will automatically change radiation levels for thicker or thinner body parts. This control is termed: 2:353

A. Automatic brightness gain
B. Automatic adaptation time
C. Automatic tomography
D. Automatic subtraction

12. Which of the following is a class of material found on the input of an image intensifier tube? 2:356
1. Incandescent screen *2. Fluorescent screen* *3. Photocathode*

A. 1 & 2 only
B. 1 & 3 only
C. 2 & 3 only
D. 1, 2 & 3

13. Which mode of a triple field image intensifier will have the highest gain factor (all other factors remaining the same)? 5:171

A. 4.5 inch
B. 6 inch
C. 9 inch
D. All possess the same gain factor

14. The normal milliamperage used for image intensified fluoroscopy is about: 2:353

A. 2-5 mA
B. 6-10 mA
C. 25-100 mA
D. 100-600 mA

15. The ICRU has recommended the use of a standardized method of evaluating the light emission of an image intensifier called: 5:170

A. Brightness gain
B. Relative aperture
C. Signal-to-noise ratio
D. Conversion factor

16. As an image intensifier ages the brightness gain of the device tends to: 5:171

A. Decrease
B. Increase
C. Remains the same

17. The brightness gain (image brightness) of an image intensifier can be increased by increasing the: 5:171
1. Exposure rate on the input screen *2. Minification ratio* *3. Acceleration potential of the tube*

A. 1 only
B. 2 only
C. 3 only
D. 1, 2 & 3

18. The principal advantage of the smaller field of a dual field image intensifier is: 2:358
1. Improved image quality *2. Less magnification* *3. Lower patient exposure*

A. 1 only
B. 2 only
C. 3 only
D. 1, 2 & 3

19. The improved ability to visualize small objects with image intensified fluoroscopy results from the human factor of vision associated with: 2:355
1. Peripheral vision *2. Photopic (cone) vision* *3. Scotopic (rod) vision*

A. 1 only
B. 2 only
C. 3 only
D. 1, 2 & 3

20. During an image intensified fluorographic study (manual mode), an increase in kilovoltage will: 5:171

A. Lower the penetration
B. Increase image brightness
C. Improve image contrast
D. Decrease image resolution

21. The automatic brightness (gain) control in most fluoroscopic systems serves to maintain image density as: 2:353
1. Part thickness varies *2. Kilovoltage varies* *3. Distance factors vary*

A. 1 & 2 only
B. 1 & 3 only
C. 2 & 3 only
D. 1, 2 & 3

22. Some of the increase in the light production by an image intensifier results from the acceleration of the electrons across the tube. This is referred to as: 5:171

A. Conversion gain
B. Minification gain
C. Flux gain
D. Lambert gain

23. The brightness of an image intensified fluoroscopic image is about ________ greater than a conventional fluorographic screen image. 2:358

A. 5-20 times
B. 20-40 times
C. 50-200 times
D. 5,000-20,000 times

24. The normal image intensifier tube is a glass envelope that has been:

A. Gas-filled
B. Air-filled
C. Gas-evacuated
D. Fluid-filled

25. The reduction of brightness at the periphery of an image intensifier is termed: 5:173

A. Vignetting
B. Diminution
C. Lateral dispersion
D. Modulation transfer

26. The reduction of field size in a dual field image intensifier is accomplished by: 5:173

A. Increasing the size of the input screen
B. Increasing the electrostatic lens voltage
C. Decreasing the size of the output screen
D. Decreasing the flux gain

27. The point in the image intensifier where the electrons crossover is termed the: 5:168

A. Optical focal spot
B. Input focal spot
C. Electronic focal spot
D. Flux focal spot

28. The greater brightness by an image intensifier tube is in large part due to the increase of electron energy, which results from the use of: 2:358
1. High potential difference across the tube *2. Photo-multiplier to increase signal strength*
3. Grid-controlled x-ray tube

A. 1 only
B. 2 only
C. 3 only
D. 1, 2 & 3

29. If the input screen of an image intensifier measures 20 cm and the output screen measures 2.0 cm, the minification ratio would be: 5:171

A. 2 times
B. 10 times
C. 50 times
D. 100 times

30. When viewing an image intensifier image on a large field of an image intensifier tube, the image brightness and resolution is greatest: 5:173
1. At the periphery of the image *2. Near the center of the image* *3. At higher dose rates*

A. 1 & 2 only
B. 1 & 3 only
C. 2 & 3 only
D. 1, 2 & 3

31. In a cesium iodide image intensifier the photocathode layer: 5:168

A. Is applied directly to the input phosphor
B. Is mixed with silver bromide
C. Serves to neutralize the input phosphor
D. Is applied directly to the output phosphor

32. A dual field Image Intensifier tube with automatic brightness (gain) control is being used in the smaller mode. This will result in: 2:356
1. More magnification *2. More contrast* *3. Less image noise*

A. 1 & 2 only
B. 1 & 3 only
C. 2 & 3 only
D. 1, 2 & 3

33. The output screen material of most modern image intensifiers is called zinc cadmium sulfide. This substance serves to convert: 3:357

A. X-rays into light
B. Electrons into x-rays
C. Electrons into light
D. Light into electrons

34. A thin layer of silver is coated on the inside of the image intensifier. When activated, this serves to focus the electron stream and is referred to as a/an: 2:355

A. Accelerating anode
B. Hooded anode
C. Photo-converter
D. Electrostatic lens

35. Using an image intensifier in the automatic brightness made, as the magnification of an image increases, the: 2:357
1. Image resolution increases *2. Contrast increases* *3. Minification ratio decreases*

A. 1 & 2 only
B. 1 & 3 only
C. 2 & 3 only
D. 1, 2 & 3

36. The ratio of the luminescence in candelas of the output phosphor to the input exposure rate in milliroentgens per second defines: 5:170

A. Brightness gain
B. Flux gain
C. Minification ratio
D. Conversion factor

Pertaining to the diagram of the Image Intensifier answer questions 37. - 40.

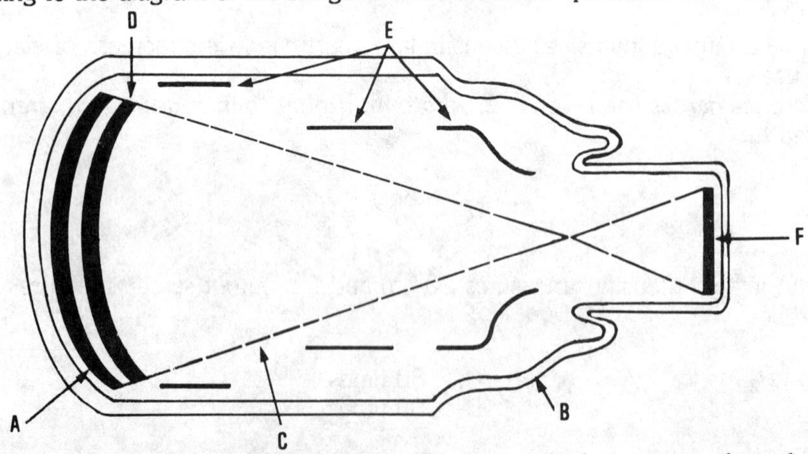

In the diagram of the Image Intensifier place the letter which corresponds to the following structures.

37. __________ Electrostatic lens 2:356
38. __________ Photocathode
39. __________ Output phosphor
40. __________ Input phosphor

Chapter 18

1. The vidicon and plumbicon are the two most commonly employed ________ for a modern radiography department. 2:359

   A. Image intensifier tubes
   B. Fluoroscopic x-ray tubes
   C. Cinefluorographic tubes
   D. Television camera tubes

2. The electronic signal that carries the information generated in the television camera to the TV monitor is termed the: 5:175

   A. Analog signal
   B. Synchronization pulse
   C. Video signal
   D. Precession signal

3. The scanning of the electronic image at the signal plate of a TV camera tube is accomplished with the grid controlled: 2:362

   A. Video signal
   B. Fiber optic bundle
   C. Electron gun
   D. Interlace framer

4. In order to accommodate human eye integration time, each frame of a television image must be presented in less than: 2:363

   A. 1/2 sec.
   B. 1/10 sec.
   C. 1/20 sec.
   D. 1/90 sec.

5. The conversion of the television output (video signal) into a digitalized form is the function of the: 2:392

   A. Analog-to-digital converter
   B. Signal-to-noise converter
   C. Digital-to-acquisition converter
   D. Matrix-to-digital converter

6. The weakest link in a televised image intensified chain is the resolution of the: 2:364

   A. Image intensifier
   B. Coupling devices
   C. Cinefluoroscopic system
   D. Television chain

7. All electronic devices possess a great deal of extraneous noise which serves to obscure the signal. Which of the following signal-to-noise ratios would be preferable for a high resolution system? 2:364

   A. 1:1
   B. 200:1
   C. 400:1
   D. 1000:1

8. The scanned electronic image on the signal plate of a vidicon television camera tube is converted into a/an: 2:360

   A. Visible image
   B. Video signal
   C. Gamma ray beam
   D. Audible signal

9. One frame of a standard closed circuit television system that is used in the United States will consist of: 5:182

   A. 262 scan lines
   B. 525 scan lines
   C. 625 scan lines
   D. 874 scan lines

10. The visible light image that impinges upon the target of a television camera tube is converted into a/an: 2:359

    A. Digital image
    B. X-ray image
    C. Audible signal
    D. Electronic image

11. The target or photo-conductive layer of a television camera is most often composed of a material called: 5:176

A. Cesium iodide
B. Lithium oxybromide
C. Antimony trisulfide
D. Calcium fluoride

12. In order to help eliminate image flicker during the viewing of a closed circuit television image, a technique called _________ is employed. 5:182

A. Depolarized light scanning
B. Image pre-amplification
C. Digital subtraction
D. Interlaced horizontal scanning

13. The common types of television tubes are classified as: 5:176

A. Airfilled tubes
B. Gasfilled tubes
C. Vacuum tubes
D. Solid state component tubes

14. The actual focusing of the electron beam in a television camera tube is accomplished by two sets of: 5:177

A. Focusing coils
B. Monitor chambers
C. Fluorescent screens
D. Ion traps

15. Which of the following components will result in the greatest acceleration of the electron beam? 5:180

A. Plumbicon camera tube
B. Vidicon camera tube
C. Orthicon TV camera tube
D. TV monitor tube

16. In a television picture tube, gas molecules that are ionized during image production are removed by a special component called a/an: 5:180

A. Electron gun
B. Ion trap
C. Control grid
D. Interlaced scanner

17. In order to coordinate the video signal between the camera and the television monitor, specialized electronic markers called _________ are added to the signal. 5:184

A. Synchronization pulses
B. Noise markers
C. Band width pulses
D. Magnetic markers

18. The resolution of a television system can be greatly improved by increasing the: 5:181
1. Bandwidth   2. Number of scan lines   3. Phospor crystal size

A. 1 & 2 only
B. 1 & 3 only
C. 2 & 3 only
D. 1, 2 & 3

19. When the televised image intensified image is viewed on a TV monitor, it will possess _______ than the image on a mirror optic viewing system. 2:359
1. Higher resolution   2. Higher contrast   3. Increased brightness

A. 1 & 2 only
B. 1 & 3 only
C. 2 & 3 only
D. 1, 2 & 3

20. When direct or TV viewing of the image intensified image and cine spot films are necessary, a/an _______ is used for image coupling. 2:361

A. Tandem mirror device
B. Video recorder
C. Interlace scanner
D. Fiber optic coupler

Pertaining to the diagram of the vidicon tube answer questions 12. - 25.

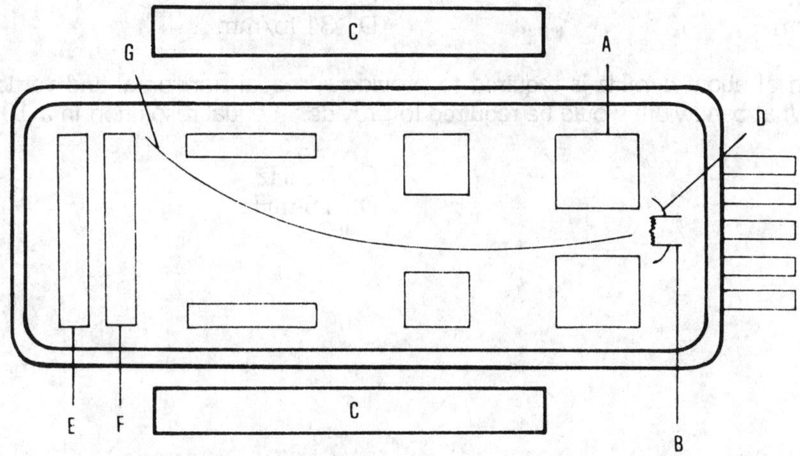

VIDICON

In the Diagram of the Vidicon television tube place the letter which corresponds to the following structures: 2:360

21. _________ Electron gun

22. _________ Signal plate

23. _________ Electron beam

24. _________ Accelerator grid

25. _________ Steering (deflection) coils

26. A small TV monitor is used to view an image, it will appear to have _________ than the same image viewed on a larger sized monitor. 5:182

    A. More vertical resolution
    B. More lag
    C. Less persistence
    D. Less horizontal resolution

27. The principle factor which effects the horizontal resolution is the: 5:183

    A. Number of vertical scan lines
    B. The voltage used by the TV monitor
    C. The bandwidth (band pass)
    D. The number of synchronization pulses used

28. The vertical resolution of a 525-line TV system will normally have a useable vertical resolution of about: 5:184

    A. 262 lines
    B. 370 lines
    C. 525 lines
    D. 1050 lines

29. Calculate the resolution in line pairs per millimeter for a 6" (152 mm) image intensifier if the TV monitor resolution is 185 line pairs/mm: 5:184

A. .8 lp/mm
B. 1.2 lp/mm
C. 28 lp
D. 31 lp/mm

30. If a bandwidth of about 4 mHz is required to provide an equal horizontal and vertical resolution in a 525 line system. What bandwidth would be required to provide an equal resolution in a 1024 line system. 5:183

A. 1 mHz
B. 4 mHz
C. 8 mHz
D. 16 mHz

Image Recording

Chapter 19

1. A beam-splitting mirror is required in the fluorography-television chain to allow for: 2:361

   A. Spot film cameras
   B. Cassette-loaded spot film
   C. Video recording
   D. Kinescopic recording

2. The greatest recorded detail is seen using: 2:364

   A. Cinefluorography spot film
   B. Video tape spot film
   C. Radiographic spot film
   D. Kinescopic spot film

3. The contrast enhancement of digital imaging techniques primarily results for the use of: 2:393

   A. The piezoelectric effect
   B. Tomographic sectioning
   C. Image subtraction
   D. Magnetic resonance

4. In recent years, new imaging techniques have been developed which do not require the use of ionizing radiations. Among these is: 2:449
   1. Ultrasonograpy *2. Nuclear magnetic resonance* *3. Thermography*

   A. 1 & 2 only
   B. 1 & 3 only
   C. 2 & 3 only
   D. 1, 2 & 3

5. A cinefluorographic unit will have the following advantage(s) as opposed to a video tape recording system. 5:187
   1. Improved spatial resolution *2. Lower patient dose* *3. Instant playback*

   A. 1 only
   B. 2 only
   C. 3 only
   D. 1, 2 & 3

6. During cinefluorography, a substantial reduction in patient exposure will result when a/an _______ is employed. 5:187

   A. Larger field size
   B. Pulsed x-ray beam
   C. Over-framing
   D. Faster frame rate

7. Because of the light emission from the output phosphor of most image intensifiers (Zn CdS:Ag), cinefluorographic films should be designed with a maximum sensitivity to the: 5:169

   A. Red spectrum
   B. Blue-violet spectrum
   C. Yellow-orange spectrum
   D. Yellow-green spectrum

8. In cinefluorography, the framing rate is directly related to the amount of patient dose received. 5:187

   A. True
   B. False

9. The pulsing of a x-ray beam during cinefluorography is often accomplished by the use of: 5:17

   A. Self-rectified diode
   B. Field emission x-ray tube
   C. Grid controlled x-ray tube
   D. Automatic exposure controlled x-ray tube

10. During televised image intensified fluoroscopy, permanent images can be produced and recorded using: 5:186
1. Cinefluorographic imaging 2. Conventional cassette spot film 3. Video tape recording

A. 1 only
B. 2 only
C. 3 only
D. 1, 2 & 3

11. During cinefluorography, in general, the greatest image quality would be found with the use of: 3:364

A. 16mm film
B. 35mm film
C. 70mm film
D. 105mm film

12. The magnetic recording of a televised image is often performed with the use of a: 5:191

A. Cine camera
B. Video tape recorder
C. Digital analog recorder
D. Kymographic recorder

13. Due to the ease of loading and unloading, the most popular of the magnetic recording systems is the: 5:191

A. Video tape recorder
B. Video cassette recorder
C. Video disc recorder
D. Cine tape recorder

14. The largest advantage that a video disc recorder has compared to other magnetic recording systems, is: 5:193

A. Random access
B. Instant processing
C. Longer recording times
D. Lower patient exposure

15. Which of the following imaging systems would possess the lowest spatial resolution? 2:364

A. Conventional fluorographic
B. Conventional radiographic
C. Intensified fluorographic
D. Televised fluorographic

16. The image quality in digital or computerized tomography is greatly improved by increasing: 2:399
1. The number of image detectors 2. Signal-to-noise ratio 3. Beam collimation

A. 1 & 2 only
B. 1 & 3 only
C. 2 & 3 only
D. 1, 2 & 3

17. The major components of a digital subtraction system include all of the following, except: 2:389

A. Image intensifier
B. Television camera
C. Computer
D. Circular detector array

18. Which system is capable of the highest resolution? 6:21

A. Human eye
B. Radiographic film
C. Intensifying screen
D. Fluoroscopic screen

19. The resolution of a computerized or digital image can be improved by increasing: 2:387
1. Pixel size 2. Matrix size 3. Area of the x-ray beam

A. 1 only
B. 2 only
C. 3 only
D. 1, 2 & 3

20. The principal advantage computed tomography has over conventional radiography is greatly improved: 2:425
1. Contrast resolution 2. Spatial resolution 3. Dynamic imaging

A. 1 only
B. 2 only
C. 3 only
D. 1, 2 & 3

Mobile Radiographic Units

Chapter 20

1. When taking a radiograph during an ongoing surgical procedure (such as a cholangiography), the machine should be located on the: 4:65

   A. Same side as the surgical field
   B. Opposite side of the surgical field
   C. Either of the above

2. The principle of terminating an exposure by the stoppage of the discharge of a capacitor at some pre-selected point is termed: 12:265

   A. Field emission switching
   B. Wavetail cutoff
   C. Three-phase switching
   D. Auto-rectification

3. In general, the type of unit with the lowest tube rating is a: 5:48
   1. Fixed installation (full wave) *2. Fixed installation (3-phase)* *3. Battery-operated mobile*

   A. 1 only
   B. 2 only
   C. 3 only
   D. 1, 2 & 3

4. After completion of a radiographic exposure using a capacitor discharge portable unit, the generation of x-rays is possible without exposure switch activation. This can be avoided by: 12:267

   A. Discharge the capacitor
   B. Unplugging the unit
   C. Moving the tube head
   D. Closing the collimator

5. A modern portable x-ray unit should have an exposure cord which permits the radiographer to be at least ________ from the tube assembly during the exposure. 21:50

   A. 2-4 feet
   B. 4-6 feet
   C. 6-12 feet
   D. None of the above

6. Because of the somewhat limited output of many types of mobile radiographic units, the use of ______ is/are generally recommended to overcome this difficulty. 4:59

   A. High ratio grids
   B. Longer exposure times
   C. Fast screen-film combinations
   D. No tube filtration

7. Which type of portable unit is best suited for use in a hospital with a limited number of standard voltage outlets? 4:60

   A. High output mobile unit
   B. Capacitor discharge unit
   C. Battery-powered mobile unit
   D. Field emission mobile unit

8. A mobile fluoroscope is being utilized for an examination of the lumbar spine. The minimum source-to-image distance that can be used is: 20:25

   A. 3 inches (7.5cm)
   B. 9 inches (23cm)
   C. 12 inches (30cm)
   D. None of the above

9. The principal advantage of capacitor discharge and battery-powered mobile units is that they permit the use of: 4:61

   A. Standard voltage sources
   B. Multiple rapid exposures
   C. Unlimited technical capacities
   D. Tomographic techniques

10. It is often recommended to use a moderately high kV when performing mobile or bedside exams. This tends to:
1. Reduce exposure times *2. Increase exposure latitude* *3. Insure adequate penetration*

A. 1 only
B. 2 only
C. 3 only
D. 1, 2 & 3

11. When radiographing patients with suspected spinal injuries, movement should be limited as much as possible but, when necessary, turning is accomplished using the ___________ method. 16:44

A. Log-rolling
B. Pole lift
C. Drawsheet pull
D. None of the above

12. Which type of mobile x-ray unit requires "charging" before each radiographic exposure is made? 4:61

A. Battery-powered mobile unit
B. Fluoroscopic C-arm unit
C. Field emission mobile unit
D. Capacitor-discharge unit

13. Because of problems in obtaining proper distances and centering, the grid that is most acceptable for mobile or bedside techniques is a: 4:63

A. Low ratio, parallel grid
B. High ratio, focus grid
C. Low ratio, crossed grid
D. High ratio, crossed grid

14. The termination of a x-ray exposure with a capacitor discharge mobile unit is often accomplished by a/an: 12:265

A. Grid-controlled x-ray tube
B. Synchronous timer
C. Impulse timer
D. Recycle timer

15. Manuel line voltage compensation adjustments are often necessary for proper exposures with many types of: 4:61

A. Fluorographic units
B. Cinefluorographic units
C. Mobile radiographic units
D. Tomographic units

16. Recently, a new class of portable fluoroscopic units have become widely available. Tube and image receptor are attached to a movable arm and are often referred to as: 4:66

A. Fixed arm units
B. C-arm units
C. B-scan units
D. T-arm units

17. A radiographer often encounters difficulties when performing portable radiographs. Therefore, it is acceptable to avoid using: 4:61

A. Gonadal shielding
B. Protective aprons
C. Proper collimation
D. None of the above

18. The current across the tube's electrons of a capacitor-discharge portable unit is most similar to that of a: 12:265

A. Half-wave rectified unit
B. Self-rectified unit
C. Full-wave rectified unit
D. Three-phase rectified unit

19. During a portable radiographic examination, a radiographer will receive the lowest amount of patient scatter by standing two meters: 7:113

A. Behind the patient
B. To the side of the patient
C. In front of the patient
D. None of the above

20. The most commonly employed power sources for a battery-powered mobile unit are: 4:60

A. Field emission batteries
B. Alkaline batteries
C. Parallel plate capacitors
D. Nickelcadmium cells

Computer Fundamentals

Chapter 21

1. The largest of the three main categories of computers is the: 2:369

   A. Microprocessor
   B. Microcomputer
   C. Minicomputer
   D. Mainframe computer

2. The physical components of a computer system are generally described as: 2:369

   A. Software
   B. Hardware
   C. Analog devices
   D. Digital devices

3. The control center of the computer, the central processing unit (CPU), consists of: 2:370
   1. Modem 2. Control unit 3. Memory unit

   A. 1 & 2 only
   B. 1 & 3 only
   C. 2 & 3 only
   D. 1, 2 & 3

4. Most computers operate on a simple number system consisting of only the digits 0 and 1. This is called the: 2:380

   A. Exponential system
   B. Duodecimal system
   C. Decimal system
   D. Binary system

5. The functional component of a computer involved with feeding information into the central processing unit is termed a/an: 2:373

   A. Input device
   B. Operating system
   C. Output device
   D. Disk drive

6. The mathematical computations or comparisons that can be made by a computer are performed by the: 2:372

   A. Cathode ray tube
   B. Read only memory
   C. Arithmetic/logic unit
   D. Random access memory

7. Thousands of individual circuit elements are necessary for the operation of a computer which are incorporated onto a small piece of silicon called a/an: 2:369

   A. Integrated circuit chip (IC)
   B. Floppy disk
   C. Byte chip
   D. Flip-flop

8. In order for a computer to understand a set of instructions, they must first be translated into a form that is understood by the computer. These are termed: 2:359

   A. Registers
   B. Languages
   C. Addresses
   D. Compilers

9. Information that is processed by a computer is most often displayed on a video terminal or: 2:373

   A. Integrated circuit chip (IC)
   B. Cathode ray tube (CRT)
   C. Analog to digital converter (ADC)
   D. Central processing unit (CPU)

10. The video display terminal that displays information as a series of letters, numbers or symbols is referred to as a/an: 2:373

    A. Graphic display
    B. Binary display
    C. Alphanumeric display
    D. Cryptic display

11. Permanent programs, dates and temporary information are normally stored in the ________ portion of the central proces sing unit. 2:374

A. Input
B. Output
C. Memory
D. Arithmetic/logic

12.-17. Match the following group of computer hardware to it appropriate classification. 2:373

A. Input device
B. Output device
C. Input/output devices

12. ________ Punch card

13. ________ Printer

14. ________ Storage disk

15. ________ Optical scanner

16. ________ Keyboard

17. ________ Video terminal with light pen

18. In computer terminology, a single binary digit is referred to as a: 2:380

A. Flip
B. Glitch
C. Bit
D. Chip

19. BASIC, FORTRAN, AND COBOL are three of the most commonly employed: 2:382

A. Input devices
B. Computer languages
C. Hard disk drives
D. Retrieval systems

20. Each character (letter or number) of the keyboard is normally encoded by 8 bits of binary information. This will then constitute the basic unit by which the capacity of a computer is measured called a/an: 2:380

A. Mega-bit
B. Octet
C. Byte
D. Word

21. Permanent instructions (programs) that cannot be modified are stored by the: 2:372

A. Read only memory (ROM)
B. Random access memory (RAM)
C. Both of the above
D. Neither of the above

22. The interpretation of the user's instructions and the transfer and retrieval to the appropriate location is the primary function of the: 2:371
1. Control unit 2. Memory unit 3. Arithmetic unit

A. 1 only
B. 2 only
C. 3 only
D. 1, 2 & 3

23. In order to identify and direct information to the proper memory location, a unique identifying label, called a/an ________, is employed. 2:371

A. Cursor
B. Terminal
C. Bug
D. Address

24. The invisible portions of a computer consists of languages and programs that are normally referred to as: 2:369

A. Software
B. Resistors
C. Graphics
D. Buffers

25. In computer terminology, a kilobyte is the equivalent of 1024 characters that consist of ________ bits of information. 2:380

A. 128
B. 1024
C. 8192
D. 102,400

26. When information is to be added, deleted or changed by the computer, the information is stored by the: 2:373

A. Read only memory
B. Random access memory
C. Arithmetic/logic memory
D. Multiformat memory

27. When large amounts of information or data are to be stored, a secondary memory device, such as a magnetic __________, may be employed. 2:374
1. Floppy disk *2. Hard disk* *3. Tape*

A. 1 & 2 only
B. 1 & 3 only
C. 2 & 3 only
D. 1, 2 & 3

28. Electrical data transmission over conventional phone lines from computer to computer may be accomplished with the use of a/an: 2:378

A. Assembler
B. Dot printer
C. Modem
D. Daisy wheel

29. The most frequently used device for the production of a hard copy image in radiology is a: 2:376

A. Line printer
B. Multiformat camera
C. Floppy disk
D. Display terminal

30. The use of computer has made the development of imaging modalities, such as ________ possible: 3:367
1. Ultrasonography *2. Digital imaging* *3. Computed tomography*

A. 1 only
B. 2 only
C. 3 only
D. 1, 2 & 3

Digital Imaging

Chapter 22

1. The organized arrangement of a digital image into columns and rows for its display is referred to as the: 2:387

A. Matrix
B. Signal plate
C. Noise pattern
D. Hybrid mode

2. The principle advantage that digital imaging systems have over conventional imaging system is in visualizing:

A. Electrical noise
B. High contrast resolution
C. Low contrast resolution
D. Compton's photons

3. The principle advantage of an area beam as compared to a fan beam is it: 2:387

A. Produces less scattered radiation
B. Enables shorter exposure setting
C. Provides a higher image contrast
D. It permits longer exposure times

4. The video system employed for a digital type imaging system requires a minimum signal to noise ratio of about: 2:391

A. 100:1
B. 200:1
C. 500:1
D. 1000:1

5. The specific device responsible for converting the video image into a digital form is called a/an: 5:397

A. CPU
B. Bit
C. ADC
D. H&D

6. Which of the following matrix size will be associated with the highest spatial resolution? 5:398

A. 128 x 128
B. 256 x 256
C. 512 x 512
D. 1024 x 1024

7. The brightness or gray scale of a digital image is most similar to the ________of an analog imaging system. 5:401

A. Image definition
B. Density scale
C. Noise factor
D. Spatial resolution

8. The windowing controls of a digital imaging system are provided to modify: 5:402

A. Image brightness
B. Image contrast
C. Both of the above
D. Neither of the above

9. The individual picture elements which contribute to the smallest segment of an image matrix are termed: 2:387

A. Pixels
B. Bits
C. Roms
D. Bytes

10. A digital technique which eliminates the majority of non vascular structures by enhancement of the vascular structures is best termed: 5:393

A. Photostimuable radiography
B. Scanned projection radiography
C. Digital subtraction angiography
D. Teleradiography

11. Compared to conventional fluoroscopy, the mA employed for digital fluoroscopy is higher to help: 5:411

A. Reduce patient exposures
B. Increase bit size
C. Reduce quantum mottle
D. Eliminate internal scattering

12. Which of the following is not a normal component of a digital radiographic system? 2:386

A. Image intensifier
B. Fan x-ray beam
C. Linear detector ray
D. High resolution monitor

13. The heart or brain of a digital imaging system is the: 5:396

A. A/D convertor
B. Multi format camera
C. Digital storage disk
D. Central control computer

14. The principle method which can be employed to overcome the motion which can occur between images in a digital subtraction technique is: 2:396

A. Hybrid subtraction
B. Noise smoothing
C. Reregistration
D. Edge enhancement

15. During an energy subtraction technique the increased absorption at different energy levels is principally due to the ________________ of the contrast media. 2:397

A. K absorption edge
B. Acquisition rate
C. Interlacing
D. Signal to noise ratio

16. The principle use of a dual collimated fan beam in a digital radiography system is a dramatic reduction in: 2:399

A. Exposure timer setting
B. Electronic noise
C. Interrogation time
D. Scattered radiation

17. Area beams are employed in all of the following techniques except: 2:386

A. Conventional radiography
B. Digital radiography
C. Digital fluoroscopy
D. Conventional fluoroscopy

18. The easiest way to reduce the noise of a digital imaging system is to: 5:409
1. Increase the exposure *2. Increase phosphor speed* *3. Decrease the time interval difference*

A. 1 only
B. 2 only
C. 3 only
D. 1, 2 & 3

19. A picture archiving and communication (PAC) system is best defined as a: 2:402

A. A computer assisted file and storage system for digital images
B. Digital method of recording radiographic images
C. Light stimulated method of image production
D. Satellite transmission network for analog images

20. During a digital subtraction angiogram the image preceding the appearance of the contrast media is called the: 2:394

A. Scroll image
B. Mask image
C. Interlace image
D. Progressive image

21. A picture archiving and communication system can be used for the storage of digital images such as those in: 2:402

1. CT *2. MRI* *3. DSA*

A. 1 & 2 only
B. 1 & 3 only
C. 2 & 3 only
D. 1, 2 & 3

22. High quality permanent copies of digitalized images can be provided through the use of a: 2:403

A. Automatic film processing
B. Photostimuable processor
C. Laser printer
D. Cathode ray terminal

23. The technique by which filed images may be transmitted over the phone lines for remote viewing is termed: 2:403

A. Computed radiology
B. Subtraction radiology
C. Teleradiology
D. Digital radiology

24. To employ the PACS system for the storage of convention radiographs, it will first require their: 2:402

A. Conversion into a analog format
B. Conversion into a dynamic format
C. Conversion into a digital image format
D. Conversion into a subtraction mask form

25. With a digital examination employing energy subtraction, the principle difference in appearance of the two images is the result of a dual ________________ setting. 2:393

A. mAs
B. kVp
C. Distance
D. Time

26. The adjustment of the image contrast with a digital unit is accomplished by using the: 3:611

A. Frame integration control
B. Window width control
C. Mask control
D. Time interval control

27. In order to retrofit a conventional fluoroscopic unit for digital fluoroscopy it will normally require a/an: 2:386

1. Digital image processing unit *2. Upgraded image intensifier* *3. Upgraded television system*

A. 1 only
B. 2 only
C. 3 only
D. 1, 2 & 3

28. The major source of noise in any electronic device is the: 2:391

A. Signal generator
B. Heated filament
C. Background radiation
D. Internal scattering

29. During a digital subtraction angiographic procedure the images for recording are composed of a preinjection mask and a ____________________. 2:393

A. Temporal mask
B. Digital mask
C. Analog image
D. Post injection image

30. The type of scan pattern required for the television monitor used in digital imaging systems is best described as a/an. 2:391

A. Interlaced scan
B. Skip scan
C. Progressive mode scan
D. Vertical mode scan

31. Which of the following is a type of digital image enhancement? 5:424

1. Filtering (smoothing)   2. Edge enhancement   3. Mask reregistration

A. 1 & 2 only
B. 1 & 3 only
C. 2, & 3 only
D. 1, 2, & 3

32. Most of the energy subtraction techniques are intended to demonstrate difference in the k absorption edge between: 2:397

A. Iodine and soft tissue
B. Water and spinal fluid
C. Oxygen and intestinal gas
D. Barium and air

33. A change in the window level of a digital display will greatly effect the __________ of the overall image: 5:401

A. Size distortion
B. Noise level
C. Shape distortion
D. Density or brightness

34. A failure to match up the images during the digital subtraction process will result in a: 2:396

A. Interrogation artifact
B. Misregistration artifact
C. Archiving artifact
D. Windowing artifact

35. Which of the following scintillation crystals is used in the detector or array of a digital radiographic system? 2:401

A. Calcium tungstate
B. Sodium bromide
C. Cadmium tungstate
D. Barium sulfate

Section III
Explanations & Answer Keys
Equipment Maintenance & Operation

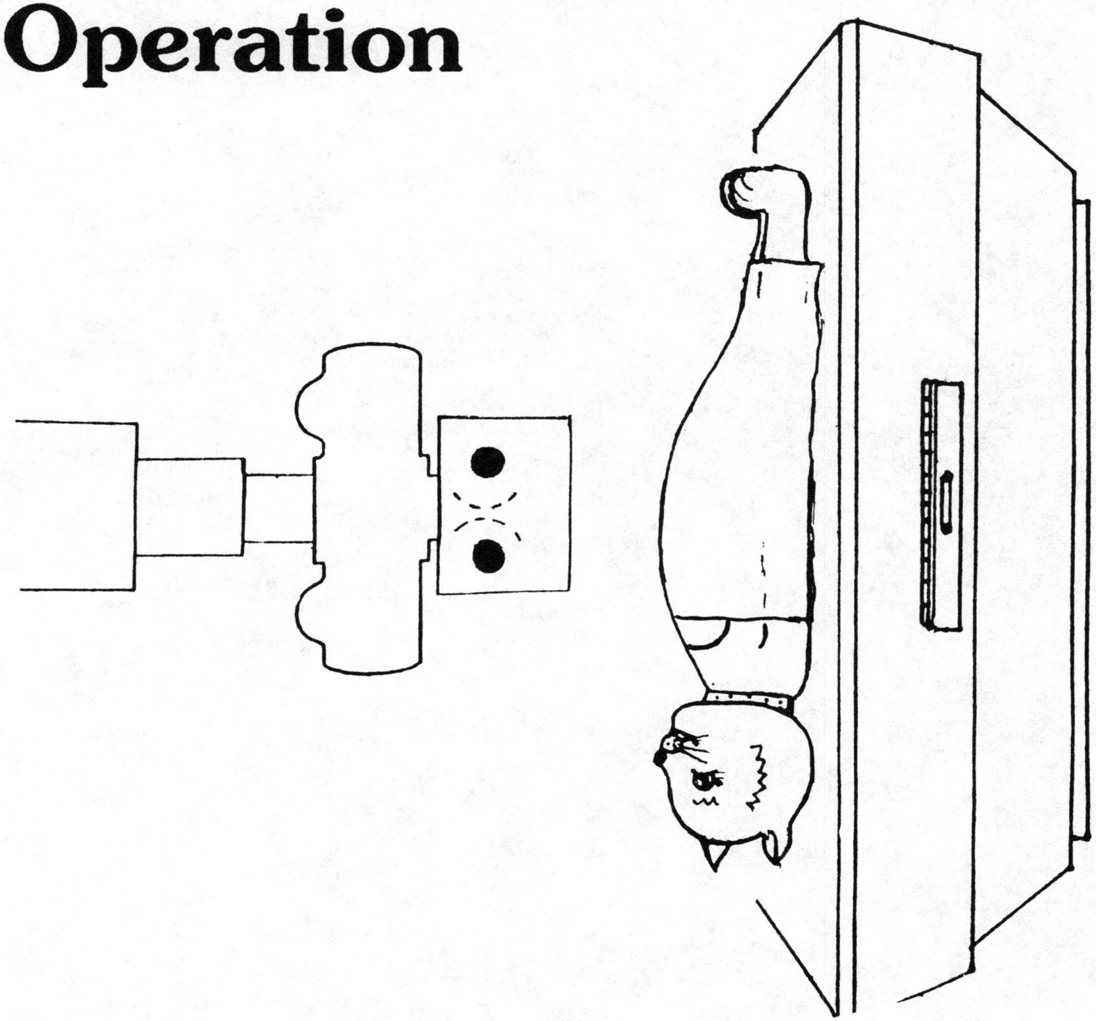

Atomic Structure

Chapter 7

1. C — The binding energy is the force by which the positively charged nucleus holds onto the negatively charged orbital electrons.

2. B — An atom is often said to resemble a miniature solar system with the nucleus of the atom representing the sun and the electrons appearing as the planets. This model which is still used today was first proposed by a Danish physicist Neils Bohr in 1913.

3. D — The chemical properties of an atom are a direct result of rearrangements in the number of the outer shell or valence electrons.

4. D — In the 1850's a Russian scholar Demitri Mendeleev described an arrangement of the known elements by their atomic mass and known chemical properties (valence) into what is now called the periodic table.

5. A — The basic unit of measurement of the atomic scale is the electron volt (eV).

6. C — The subatomic particle possessing a mass of one atomic mass unit (amu) and a single positive charge is the proton.

7. B — The largest portion of the atom's mass (atomic mass) is found in the nucleus of the atom. The nucleus is formed by two subatomic particles, the neutrally charged neutron and positively charged proton.

8. C — Following an atom's exposure to some forms of electromagnetic radiation, the orbital electrons can absorb enough energy to jump to higher orbital shells (excitation) or be completely removed from the atom's sphere of influence in a process called ionization.

9. C — The number of outer shell orbital electrons is called the valence and is directly related to the chemical properties of the atom.

10. B — The two numbers seen next to the atomic symbol or the (top) atomic mass (A) which indicates the total number of neutrons and protons and the (lower) atomic number (Z) equal to the number of protons in the nucleus.

11. A — See explanation number 10 of this section.

12. D — Since an electron carries a negative charge, its loss by ionization will leave a positively charged ion.

13. C — After a chemical bonding has occurred, a substance or compound is formed, the smallest subdivision of this compound is called a molecule.

14. A — The only particles which commonly exist in the orbital shells surrounding the nucleus are the negatively charged electrons.

15. B — See explanation number 7 of this section.

16. A — The binding energies of each atom are characteristic for a given element, when electrons move from higher to lower energy states or drop from higher to lower orbital shells. The photon energy created is identical to the difference of the shells binding energies.

17. B The maximum number of orbital electrons which can occupy a given orbital shell can be calculated by taking the squared value of the orbital shell number and multiplying that value by 2.

18. A The atomic mass (A) is found by adding the number of protons and neutrons.

19. A Because of their proximity to the nucleus, inner shell electrons possess the highest force of attraction or binding energy.

20. C See explanation number 6 of this section. 70eV - 10eV = 60eV.

21. C The two main types of chemical bonds are the ionic bond in which an outer shell electron is passed between the atoms and the covalent bond in which outer shell electrons are shared between two or more atoms.

22. C Perhaps the most well known and significant equation of Albert Einstein was $E = mc^2$ in which mass, energy and speed were found to be interrelated.

23. D The major driving force behind chemical interactions is the desire of an atom to have a filled valance (outer) shell consisting of 8 electrons often called the octet rule.

24. A Matter normally occupies one of the three normal states of matter as a solid, liquid or gas. The principal difference between these is in the energy of the atoms. In the lowest energy state or solid state, little atomic energy is found. By increasing the energy or heating the material it can pass into the liquid higher energy gaseous stages. Some material such as glass and most plastics reside is so-called quasi liquid or plastic state.

25. C Over 99% of the atom's mass resides in the nucleus with the protons and neutrons.

26. D The movement of electrons from lower to higher energy orbital shells (excitation) or the complete loss of the electrons (ionization) can be a result of an atom's exposure to the various forms of electromagnetic radiation including infrared (heat), visible light, ultraviolet, x-rays and gamma radiation.

27. C The force of attraction or binding energy between the nucleus and the orbital electrons is dependent upon the number of protons and the distance between the nucleus and the electrons.

28. D The highest binding energy for a given orbital shell is found in the atom with the highest atomic number.

29. C Among the most chemically reactive elements are those found in the seventh group of the periodic table called the halogens. These include the elements: fluorine, chlorine, bromine and iodine.

30. A The arrangements of atoms into compounds does not always result in identical compounds. Isomers are chemicals that have the same chemical formula but combine in different patterns to form compounds with different chemical natures. The sugars fructose and dextrose share the same C_6, H_{12}, O_6 formula but are chemically different.

31. C See explanation number 21 of this section.

32. A The mass of an electron is about 2000 times smaller than that of a proton or neutron.

33. B All atoms which share the same atomic number are called elements and will have binding energies which are characteristic of that element.

34. A — In order for an atom to remain neutral, it must have a balance between the number of negatively charged electrons and positively charged protons.

35. C — The principal uncharged nuclear particle is the neutron which has a mass of 1amu.

36. D — The groups of the periodic table share similar chemical properties because they posses the same number of outer shell or valence electrons.

37. A — See explanation number 24 of this section.

38. B — The addition of a negatively charged electron to a neutral atom will result in an excess charge and the renaming of this particle into a negative ion.

39. A — The specific type of chemical bond which involves transfer of an electron is called ionic bonding.

40. B — See explanation number 16 of this section.

41. D — All mass bearing particles according to Einstein's theory have an energy equivalence. The value for an electron is .511meV and about 1GeV for a proton.

42. C — The highest binding energy is found in an inner shell (K-shell) electron and the lowest in the loosely bound outer shell electrons.

43. A — Electrons are about 2000 times less massive than protons or neutrons.

44. C — Though no elements are entirely chemically inert, the noble gasses such as helium, neon, argon, krypton, and xenon are all extremely unreactive.

45. A — The physical properties such as appearance, color, texture, smell and taste are most closely related to the atomic number of the element.

46. C — The degree of nuclear stability is closely related to the neutron-proton (N/P) ratio. Elements with favorable ratios are found to be more stable than those with unfavorable N/P ratios.

47. C — The proton is the largest mass bearing particle carrying a positive charge.

48. A — See explanation number 8 of this section.

49. B — All atoms regardless of their atomic number can have a maximum of two inner shell electrons.

50. B — The fundamental form of matter that has all the characteristics of that matter is the atom.

51. A — By definition, any atoms which have the same atomic number and different atomic mass are termed isotopes.

52. 11

53. 23

54. 1

55. 11

56. 12

47. C — The proton is the largest mass bearing particle carrying a positive charge.

48. A See explanation number 8 of this section.

49. B All atoms regardless of their atomic number can have a maximum of two inner shell electrons.

50. B The fundamental form of matter that has all the characteristics of that matter is the atom.

51. A By definition, any atoms which have the same atomic number and different atomic mass are termed isotopes.

52. 11

53. 23

54. 1

55. 11

56. 12

57. 11

58. 3

59. 7

60. 1

61. 3

62. 4

63. 3

64. 17

65. 35

66. 7

67. 17

68. 18

69. 17

70. A Because sodium (Na) and lithium (Li) have the same number of valence electrons they will have similar chemical properties.

71. D Perhaps the strongest force in nature is that which binds the nucleus together called the strong force or nuclear glue.

72. D In an attempt to become stable or non-radioactive, a nucleus may emit a number of decay particles, among these are: alpha particles, beta particles, neutrons and positrons.

73. B The term nuclide is used to describe the nuclear configuration or the relative number of nucleons (protons and neutrons). The term nuclide is used to express any of the different nuclear configurations which may exist.

74. A The process by which our sun and most other stars emit energy is due to the fusion of hydrogen into the heavier element helium. This is the same process that enabled the development of the hydrogen bomb.

75. A See explanation number 51 of this section.

76. D The rate of radioactive decay occurs independently of all outside forces.

77. D Because the alpha particle consists of 2 protons and 2 neutrons its emission is associated with a decrease of 2 in the atomic number of the parent nuclide.

78. A During a beta decay a neutron is converted to a proton resulting in an increase of the atomic number of 1.

79. C Most nuclear reactors and the first atomic bombs utilized the fission process in which a nucleus of uranium or plutonium is split into two or more smaller atoms.

80. A Beta decay is most likely in nuclides possessing too few protons or an excess of neutrons.

Electricity and Magnetism

Chapter 8

1. A The two basic charges of all matter are the positive charge derived from the proton and negative charge obtained from electrons.

2. C The movement of certain materials across each other may cause the generation of charges during a process called friction.

3. B In order to protect an individual from electrical shocks most electrical components are surrounded by materials which oppose the flow of electricity called non-conductors or insulators.

4. B Conduction is defined as the transfer of a like charge by the physical contact of two bodies.

5. B The first law of electrostatics states that two charged objects with opposite charges will be attracted to each other and the objects with like charges will be repelled.

6. D In electrical terms a positive charge is normally caused by having a shortage of negatively charged electrons.

7. B A static discharge such as lightning occurs as electrons overcome the insulating properties of the surrounding air and travel from a negatively charged cloud to positively charged object.

8. C The three processes by which electrification can occur are conduction, induction, and friction.

9. D Though both positive ions and electrons can move in liquids only electrons can move in a solid conductor.

10. A — According to Coulomb's law, the magnitude of the electrostatic force is directly proportional to the product of the charges and inversely proportional to the square of the distance between the charged objects.

11. B — Electrical conductors which have a low resistance to the flow of electrons are usually made of copper or other metals. Tap water is also a relatively good conductor of electricity.

12. C — See explanation number 10 of this section.

13. A — The opposition to the flow of electrons (resistance) is most often manifested in the form of heat.

14. D — The driving force for the movement of electrical current is an electrical potential or potential difference that is caused by a difference in charge between the various components in a circuit.

15. A — By definition, 1 amp of current is equal to 1 coulomb of electrons flowing in 1 second.

16. B — The current from a battery, capacitor and DC generator all travel in a single direction and are referred to as types of direct current.

17. A — The major source of chemically produced electricity is a battery. At the present time a number of batteries are available which are classified as either wet or dry cell types.

18. C — The best insulations are usually made of rubber or plastic which have an extremely high electrical resistance. Certain types of oils are also very good insulators.

19. D — The behavior of electric currents in a circuit are best described by Ohm's law which is stated as V=RI where V=voltage, I=ampere, R=resistance.

20. D — In addition to the material from which the wire is made, the electrical resistance can be minimized by employing a larger cross sectional area (smaller gauge) lower conductor temperature, and reduced length of the wire.

21. B — One of the most important devices in the production of current electricity is an electromagnetic device called a generator which can convert mechanical into electrical energy.

22. D — Electricity can be stored for short periods of time in a capacitor or condenser and for longer periods of time in wet or dry cell batteries.

23. C — See explanation number 20 of this section.

24. B — Short circuits can be avoided by the proper insulation of the wires carrying electricity.

25. D — The heat loss in a conductor can be calculated using the power loss formula ($P=I^2R$) where P is the power loss as heat, I^2 is ampere squared, and R is the resistance of the wire.

26. A — There are 4 main units of electricity: the volt which is the unit of potential difference; the ampere, the unit of electrical quantity; the ohm, the unit of electrical resistance; and the watt or unit of power.

27. C — The unit of power, the watt, is obtained by multiplying the number of Amp x volts; i.e., P=IxV.

28. D — See explanation number 20 of this section.

29. D An electromagnet is a simple device consisting of an iron core surrounded by a coil of wire that is carrying current.

30. B Since voltage is an additive, it would take 110 separate 2 volt wet cell batteries to provide 220 volts of power supply.

31. D A generator which consists of a coil of wire which is rotated within a magnetic field, produces electricity by a process called electromagnetic induction.

32. B According to the power formula (P=IxV), the minimum rating for this device is 120 kVp x 600 mA = 72,000 watts.

33. C The earth and lodestone are the two common types of natural magnets. Alnico is a very strong type of artificial magnet.

34. A Because magnetism is always associated with movement of electrons, the unpaired orbital electron of hydrogen will make this matter magnetic.

35. A The induced magnetic field due to the nuclear spin of a hydrogen atom creates a magnetic field known as the magnetic movement.

36. B In a strong permanent magnet all the individual magnetic domains will be aligned in a single direction.

37. C The magnetic lines of force or magnetic flux move from the south magnetic pole to the north magnetic pole within the magnet and from north to south poles in the space surrounding the magnet.

38. D The tesla and gauss are the two main units for measuring magnetism. The earth's magnetic field measures about 1 gauss or .0001 tesla.

39. B The simplest device for the detection of a magnetic field is a compass which, due to its magnetic properties, aligns itself with any other external magnetic field.

40. B A magnetic field is associated with the movement of a charged particle; i.e., electron and proton or atoms with an imbalance in the number of orbital electrons, i.e,: hydrogen and iron.

41. C The laws of magnetism and the law of electrostatics are similar because they are created simultaneously with each other when charged particles are in motion.

42. D See explanation number 40 of this section.

43. A Permanent magnets are described as having two main properties; retentivity, or the ability of a material to resist demagnetization and permeability, the ability of a material to become magnetized.

44. D All are correct forms of Ohm's law.

45. D All are solved using the versions of Ohm's law in previous question and the formula $R_T=R_1+R_2+R_3$.

46. A See explanation number 45 of this section.

47. B See explanation number 45 of this section.

48. B See explanation number 45 of this section.

49. D See explanation number 45 of this section.

50. A See explanation number 45 of this section.

51. D All are solved using the versions of Ohm's law in question 44 and the formula $\frac{1}{R_T} = \frac{1}{R_1} + \frac{1}{R_2} + \frac{1}{R_3}$

52. A See explanation number 51 of this section.

53. D See explanation number 51 of this section.

54. C See explanation number 51 of this section.

55. D See explanation number 51 of this section.

Basic X-ray Circuitry

Chapter 9

1. D The incoming voltage to the x-ray unit can be modified into single volt increments by an electromagnetic device called an autotransformer. The kilovoltage selector switch on the control console has a direct connection to the autotransformer.

2. C Two main types of devices are available for the storage of electrical charges. The first is a capacitor consisting of a series of insulated plates on which charges can be stored temporarily and batteries which can both store and produce electrical charges.

3. B The release of electrons through the heating of a material is termed thermionic emission. This process is best accomplished by employing a high ampere, low voltage current. The use of a step-down transformer in the filament circuit is very effective at providing this type of current for the filament of the x-ray tube.

4. D The rapid acceleration of the electrons across the electrodes, filament, and anode of the x-ray tube is accomplished through the development of a high electrical potential. This potential difference is provided by the high tension (step-up) transformer located in the beginning of the secondary circuit.

5. B To ensure the appropriate voltage has been provided by the autotransformer, a voltmeter (pre-reading kVp meter) is connected to the secondary side of the autotransformer.

6 C The cathode or filament of an x-ray tube is placed in series with the mA selector switches, the rheostats and step down transformer of the filament portion of the circuit. The Cooledge (step-down) transformer provides the high amperage, low voltage current needed to heat the filament.

7. B The milliammeter which measures the amount of tube current or the electrical charge passing between the cathode and the anode of the x-ray tube only registers during the second phase of the radiographic exposure when the electron stream is accelerated across the tube.

8. B — All electronic equipment may experience unavoidable voltage drops or surges during its operation. Because some expensive components may be damaged by even small overloads, a fuse or circuit breaker can be installed to shut down the power to a component when an excessive voltage or amperage is detected in the circuit.

9. A — The extremely high and dangerous amounts of current utilized by the transformer could lead to a significant problem in the case of an electrical malfunction. The immersion of this transformer into an oil filled tank helps to both dissipate heat produced during its operation and provide an insulated barrier for prevention of electrical shocks.

10. D — The high heat necessary to boil off the electrons of the cathode or filament requires a temperature of about 2200°C. This is accomplished by the application of a current between 4-6 amps onto the filament.

11. A — The primary circuit of the radiographic unit carries current which is similar in form and voltage to normal household alternating current. Because this type of current is relatively safe to work with, the majority of accessory devices (motors, conveyers, lights) are placed into the primary circuit.

12. A — The high voltage (kilovoltage) alternating current which is developed on the secondary side of the high tension transformer needs to be converted into a direct current form before passing through the x-ray tube. The use of solid rectifiers accomplishes this conversion in a modern x-ray unit.

13. B — The tube current is measured across the cathode and anode of the x-ray tube by milliampere (mA) meter.

14. A — All of the power supplied to the electrical devices used in this country are in the form of a 60 cycle per second alternating current at voltages of 110, 220 or 440 volts depending on the power needs and nature of the equipment.

15. B — In all transformers it is the relative number in the turns of wire (ratio) that determines the effect on the voltage. Since an autotransformer enables a wide number of possible settings it is sometimes referred to as variable ratio transformer (variac).

16. C — In nearly all modern equipment, the selection of the filament temperature is tied into one of the preset mA selector buttons (stations). The higher the mA station selected, the higher the current and temperature achieved in the filament.

17. A — For safety reasons, any components which carry extremely large amounts of voltage or current when possible are placed in a highly insulated oil filled tank. The main components placed in this tank are the transformers (step-up and down) and the rectifiers.

18. A — Any change in the kilovoltage peak impressed on the electrodes of the x-ray tube will have an effect of the percentage of the electron cloud (space charge) moved across the tube. In order to provide a consistent exposure (mAs) at various kVp levels, the use of a space charge compensator is employed in a modern filament circuit.

19. C — The milliampere setting and the change in the filament temperature it causes (explanation 16) is directly related to the exposure rate of photon production and the resulting radiographic density.

20. C — In many older radiographic units, the selection of the filament temperature was a function of a filament control knob. An ammeter provided a measure of the actual current and means for estimating the filament heat employed. In most modern units, this meter has been replaced by preset mA stations.

21. C — In a radiographic unit which uses a meter to display the tube current or milliampere - seconds value, a rapid exposure time of less than 1/20th (.05) seconds is imperceptible to the human eye. A ballistic type of meter which lengthens or slows the swing of the indicator enables the operator to record the mAs values for the exposure.

22. A — A line (voltage) compensator is a type of electrical storage device called a capacitor which is able to store and release small amounts of current to compensate for small drops in the incoming voltage supply. Surge protectors often recommended for computers are similar in function to the line compensators employed in the x-ray circuits.

23. C — In a three phase circuit, each of the three voltage phases must be selected individually by its own autotransformer.

24. B — The cathode or filament of a modern radiographic tube has a dual electrical connection. The first filament connection is activated during the boost or prep stage which serves to pre-heat the filament. The second connection from the high tension circuit provides the potential difference across the cathode and anode in the tube.

25. C — By grounding the mid point of the secondary coil of the transformer, half of the selected voltage is below zero and one half above zero. In effect, each cable needs to be insulated for only this voltage, saving both cost and weight of the x-ray cables.

26. A — The autotransformer which selects the desired voltage to the primary side of the high tension transformer cannot be operated safely in the presence of high voltage and therefore is placed in the primary circuit.

27. D — See explanation number 5 of this section.

28. C — The inherent danger associated with the high voltage section can be minimized by placing only essential devices into this circuit; namely, the high tension transformer, rectifiers, insulated cables, milliameters and the x-ray tube.

29. A — Of the two types of overload protective devices, the fuse and circuit breaker, the most commonly employed device in a modern circuit is the electromagnetic circuit breaker which has the distinct advantage of being resetable.

30. D — The operating console of a modern radiographic unit will normally provide the monitoring and control of the mA, kVp, timer settings as well as any number of accessory systems such as AEC, tomographic, and fluoroscopic controls.

Transformers

Exercise No. 1

1. 880 V
2. 2200/2.2 kVp
3. 11,000 V/11 kVp
4. 2200 V, 2.2 kVp
5. 1.1 V
6. 44 volts
7. 2000 V, 2 kVp
8. 9000 V, 9 kVp
9. 3.8 V
10. 800 V
11. 25 amps
12. 10 amps
13. .8 amp, 800 mA
14. 5 amps
15. 100 amps
16. 100 amps
17. 10 amps
18. 120 V
19. 50 V
20. 50 V
21. 55 V
22. 110 V
23. 55 V
24. 36 V
25. 84 V

Transformers

Exercise No. 2

1. 660 V
2. 2200/22 kVp
3. 5500 V, 5.5 kVp
4. 1100 V, 1.1 kVp
5. 4.4 V
6. 44 V
7. 2200 V, 2.2 kVp
8. 90,000 V, 90 kVp
9. 7.6 V
10. 32,000 V, 32 kVp
11. 10 amps
12. 5 amps
13. .33 amps
14. 2.2 amps
15. 70 amps
16. 40 amps
17. 14.3 amps
18. 400 V
19. 27.5 V
20. 11 V
21. 100 V
22. 55 V
23. 110 V
24. 13.3 V
25. 66.6 V

Single & Three Phase Rectification

Chapter 10

1. A — The high voltage alternating current which emerges from the secondary side of the high tension transformer must be converted into direct current before reaching the x-ray tube. The use of solid state diode or the older value tubes provides the one way electrical conduction necessary to accomplish rectification.

2. C — Since the development of transistors in the 1960's, nearly all of the older type vacuum rectifiers have been replaced by the newer and more durable solid state diodes.

3. C — In fully rectified single phase radiograph units a 60 cps AC is converted by a 4 diodes arrangement to provide a 120 impulse per second direct current. When a high voltage direct current is passed into an x-ray tube, a short burst of x-rays is given off every 1/120 sec. (120 dots per second). When a metallic spin top with a single cutout is spun over radiographic film during an exposure, a series of dots representing the number of impulses is seen. Because U.S. power supplies are strictly maintained at 60 cps, the likely cause of a change in the number of expected dots is either a faulty x-ray timer or inoperative valve tube (diode). Image A shows 6 dots on a properly functioning unit. The timer setting is found by dividing: number of dots on the image by the number of dots per second (120) = timer setting. Since Image B was taken a 1/30 second, 4 dots should be visualized if the unit is operating within acceptable limits.

4. D — See explanation number 3 of this section.

5. C — The highest tube rating at the shortest exposure time is called the instantaneous loading capacity of a radiographic unit. This capacity is highest for three phase radiographic units. Single phase fully rectified, half wave and self rectified systems all fall well below three phase systems in their instantaneous loading capabilities.

6. D — The voltage wave form is found by plotting the relative voltage verse time (see diagrams in questions 9-12). In a fully rectified single phase unit (graph III) the voltage begins at zero volts and builds to a maximum voltage before returning to zero creating what is termed a 100% voltage ripple. In 3 phase 6 pulse units, the voltage never drops to zero and varies between 87 and 100% producing a 13% ripple and from 97 - 100% for a 3% ripple in a 3 phase 12 pulse type system.

7. D — The high instantaneous loading capacity of a three phase radiographic unit will enable shorter exposure settings and higher mA values. The ability of this system to provide a higher electrical potential to the electrons in the x-ray tube also improves the beam quality and penetrability of a three phase system.

8. C — Rectification is accomplished by the desired arrangement of a number of thermionic or solid state diodes. The type of rectification and the number of diodes required are as follows: self rectification - no diodes; half wave rectification - 2 diodes; full wave rectification - 4 diodes; 3 phase 6 pulse - 6 diodes; 3 phase 12 pulse - 12 diodes.

9. C — See corresponding curve.

10. D — See corresponding curve.

11. D — See corresponding curve.

12. C — Wave forms which alternate above and below the zero voltage lines are forms of alternating currents (graphs I & IV); direct current wave forms, which stay above line in a pulsed form (graphs II & III), are types of pulsating direct currents. The graph V represents a steady direct current wave form seen with batteries or capacitors.

13. B — The electrons in the charged cloud around the filament under the influence of a fully rectified electrical current are accelerated toward the anode as the voltage begins to build toward its maximum. When the voltage falls back to zero, the electron movement is slowed, causing x-ray photons that are produced at less than optimum energies.

14. B — See explanation number 6 of this section.

15. B — Until the 1970's, theriomic diodes or valve tubes were the principal type of rectifying devices. The advent of semiconductor technology allowed their replacement with less expensive, more durable, and longer lasting, solid state diodes.

16. A — The suppression of the inverse voltage with direct current enables x-ray tube electrons to pass in a single direction from cathode to the anode during the exposure. If this current were permitted to flow in both directions, the high energy electron stream would reverse its direction causing serious damage to the filament.

17. D — For a given set of technical factors, the output of a full wave rectified is lower in both quantity and quality than that of a three phase unit. In order to provide a similar density, a 50-55% increase in the mAs is required for the full wave technique.

18. C — The high voltage alternating current provided by the high tension transformer is directed into the rectifiers prior to its passage through the x-ray tube.

19. A — See explanation number 6 of this section.

20. D — See explanation number 6 of this section.

21. C — Because a three phase unit provides a higher quality and quantity of radiation, an identical set of exposure factors used on a single and 3 phase unit will result in a lower density and a slightly higher contrast for a single phase unit.

22. D — In direct current, no reversal or change in the electrical potentials of the cathode and anode occur.

23. B — See explanation number 8 of this section.

24. C — Most solid state diodes and semiconductors are made of crystalline silicon which has been modified (doped) with arsenic or gallium. These impurities are added to the silicon to adjust forward or reverse bias to the flow of electrons.

25. D — The higher efficiency of a three phase radiographic unit provides a tube current with more electrons per unit time and greater heat formation in the anode. In a three phase unit the mA x time x kVp heat unit formula requires a 1.35 (6 pulse) or a 1.41 (12 pulse) correction factor.

26. D — A synchronous spin top tester is motor driven to spin at a fixed rate of one revolution per second. Because of this fixed rotation, it can be used to test the timer settings of both single and three phase radiographic units.

27. D — See explanation number 3 of this section.

28. B — See explanation number 3 of this section.

29. B A failure of a spin top tester to show the correct number of dots is most likely due to an incorrectly adjusted timer switch which terminates the exposure after too long or short a period of time.

30. A A full wave rectified single phase unit requires the use of 4 valve tubes (two for each impulse of current). The loss of any valve tube will result in a 1/2 wave rectified wave form. A spin top test will record exactly 1/2 of the number of dots of a normally functioning unit.

31. C Three phase current consists of 3 single phase current wave forms which are shifted (phases) through an angle of 120 degrees with each other.

32. D The solid state silicon diodes used in most modern units have no filament to burn out, produce no low energy x rays, are more efficient and durable than the valve tubes they replace.

33. 30 See explanation number 3 of this section.

34. 18 See explanation number 3 of this section.

35. 2 See explanation number 3 of this section.

Timers

Chapter 11

1. D The shortest permissible timer setting allowed by a radiographic unit is related to its type of rectification and the timer design. The impulse timer used with nearly all fully rectified single phase units requires a minimum of one impulse or 1/120 sec. (8 ms) to terminate the exposure. Three phase units employ modern electronic (silicon rectifier) timers which enable exposures as short as 1 ms (1/1000 sec.). The automatic exposure controlled (AEC) circuits have timer systems that can permit exposures as low as 10 ms (1/100sec.).

2. D The operation of an automatic exposure control (AEC) system is based on the assumption that an appropriate screen, films, grids, degree of collimation, patient centering points and density control settings have been employed. Any failure to utilize the appropriate factors will result in a radiograph possessing an excessive or deficient density. The use of a slower speed screen which is not recognized by the AEC system results in an image possessing an insufficient radiographic density.

3. A The accuracy of any radiographic timing system can be tested by the use of the highly accurate oscilloscope or solid state radiation timing devices. In the past, determinations of the timer accuracy were made on single phase systems by the use of a spin top tester.

4. B — The sensors or radiation detectors of nearly all modern automatic exposure control (AEC) units consist of three 2"-4" ionization chambers which are incorporated into the bucky tray apparatus. The familiar  pattern enables the selection of any number of single or multiple sensors pattern based on the nature of the anatomical region. Most manufacturers recommend a single center sensor for non-contrast abdominal studies, skull positions, spine and skeletal structures and the lateral chest projections. All three sensors are selected for the majority of positive contrast studies of the abdomen and routine views of the pelvis. The two outer sensors are required to obtain the appropriate radiographic density of the lung during PA projections of the chest.

5. A — The most common type of device used in a modern three phase timer circuit is a silicon controlled rectifier (SCR) which is capable of exposures as short as 1 ms (.001 sec). Impulse, synchronous mechanical, and AEC timers are not normally capable of accurate exposures under 10 ms or .01 seconds.

6. D — Whenever one of the selected sensors is inadvertently placed under a gas pocket or gas filled loop in the intestine, radiation floods the sensor causing a premature termination of the exposure. This is one of the more common causes of AEC images providing an insufficient radiographic density. This is most frequently seen with obstructed patients and when double contrast techniques are employed.

7. C — In some mobile and cine radiographic systems, a special triode or grid controlled x-ray tube is used to terminate the exposure by the placement of a secondary charge on the grid. Compared to a more conventional system, these tubes can be rapidly turned on and off or pulsed to meet the requirements in cine radiography.

8. D — With a properly functioning automatic exposure control device, changes in the kVp, mA and SID will not normally effect the radiographic density because of the compensation in timer setting which will be made by the unit to adjust the exposure. Radiographic density modifications can best be accomplished through the adjustment of the density setting controls located on the control console of an AEC unit.

9. C — In older full wave rectified units timer selections were based on the number of impulses of current. In these units though maximum settings of 1/120 sec. or 8 ms was possible, in practice, 1/60 sec. or 17 ms setting was the shortest setting which provided consistent exposures.

10. B — By applying the time = mAs/mA formula the timer setting can be found.

11. C — During an automatic exposure controlled (AEC) technique, the main variable is the timer setting. In order to eliminate motion, high intensity radiation is required to reduce the time required for the exposure. This can be accomplished by the use of a higher mA setting, higher kVp setting, or shorter source-to-image receptor distance.

12. A — All automatic exposure control (AEC) units are adjusted for a specific image receptor combination for a given exposure. The use of an inappropriate image receptor which is not recognized by the AEC system results in a loss of density when slower speed combinations are employed or overexposed radiographic images if a higher speed system is used.

13. B — The actual physical devices which are incorporated into the radiographic unit to terminate the exposure are located in the low voltage or primary portion of the x-ray circuit.

14. C — The ionization type sensors or detectors used in nearly all modern automatic exposure control units are housed in a thin metal or plastic frame which sits directly above the radiographic grid, just below the top of the x-ray table.

15. A An excessive exposure time with an automatic exposure control (AEC) unit indicates that the intensity or exposure rate is low and that the sensors have not received adequate amounts of radiation to terminate the exposure. The main causes are insufficient kVp, low mA settings, excessive SID's and improper alignment of the tube to the sensor arrangement.

16. B In order to obtain the proper timer setting, the original setting is multiplied by 40% to obtain the required amount of correction. The answer is found by subtracting this value from the original timer setting.

17. C The physical device which triggers the termination of the x-ray exposure when the appropriate amount of radiation has been detected by the ionization chambers, in older AEC systems, was a triode tube called a thyratron. In more modern units, a solid state switch called a thyristor is employed.

18. A By applying the time = mAs/mA formula, the timer setting can be found.

19. C Any modification of the intensity (exposure rate) of the beam can be used to reduce the exposure time while maintaining the appropriate density. See explanation number 11 of this section.

20. C Any switch which requires a constant pressure to maintain its activation is classified as a dead man switch.

21. C During an automatic exposure controlled (AEC) technique an increase in the thickness or density of the patient will require a greater amount of incident radiation and a longer exposure time to obtain the appropriate radiographic density.

22. B The mAs can be determined by the application of the mAs = mA x time formula.

23. C See explanation number 11 of this section.

24. A See explanation number 5 of this section.

25. C In theory an automatic exposure control device has no upper limit for the duration of exposure. In practice, this limit is determined by the back-up timer. During an exposure the back-up timer is only activated when the selected maximum timer value is exceeded. The timer's principal function is to protect the unit from excessively long exposures.

26. C In an automatic exposure controlled (AEC) unit, the permitted range of exposures is found between the values obtained from multiplying the selected mA times, the minimum response time, the selected mA times, and the back-up timer setting. The shortest permissible exposure with an AEC unit is called the minimum response time. The longest permissible exposure value is related to the back-up timer setting.

27. D See explanation number 26 of this section.

28. D See explanation number 26 of this section.

29. C Though different manufacturers may employ different designations, all provide a light and/or audible sound signal which is activated when the maximum or back up time has been exceeded. Most units will display an overload, maximum output, or exposure reset light. These overloads are most frequently seen in radiography of obese patients and when the x-ray tube is improperly centered to the radiation sensors.

30. B See explanation number 27 of this section.

31. B Since this unit is incapable of exposures below the minimum response time, this will be the time indicated for the exposure.

32. C See explanation number 26 of this section.

33. D Unless an upright bucky device is available an automatic (AEC) exposure control cannot be used for decubitus projections. Decubitus exams are most commonly performed without the AEC device, by using manual exposure techniques.

34. D See explanation number 28 of this section.

35. C See explanation number 12 of this section.

X-Ray Tube

Chapter 12

1. A An x-ray tube contains two electrodes, the negatively charged tungsten filament (cathode) and the positively charged tungsten target called the anode.

2. B In order to compress and focus the electron cloud created around the filament of an x-ray tube, it is surrounded by a negatively charged collar called a focus cup.

3. B The transparent x-ray tube envelope is made of special heat resistant glass made from the oxides of boron called pyrex.

4. D The electrons which are freed by the heating of the filament of the x-ray tube are moved toward the anode by a difference in electrical potential measured by an electrical unit called the volt (kilovolt).

5. D The process by which metallic elements release free electrons when heated is called thermionic emission. In tungsten this emission occurs when the temperature reaches a level of about 2200° C or 4000° F.

6. C The speed at which the anode rotates is normally related to the type of rectification system employed. In most single phase units a rotation speed of 3000-3600 rpm is common. In three phase systems, the rotation speed is three times faster or about 10,000 rpm.

7. B The heat of the filament is directly related to the current (amperage) impression on it. The higher the current, the greater its temperature and the larger the number of electrons boiled off and x-ray photons produced.

8. D The high melting point of about 3400° C of tungsten makes it highly desirable for its use as filament material, as well as for the target (anode) of the electron stream. Tungsten is employed as the target material in both stationary and rotating anode type x-ray tubes.

9. B In most modern radiographic units, the selection of smaller filament is associated with the choice of mA stations with currents of less than 200 mA. During the manufacturing process, the use of a focus cup and steeper angled anodes will provide an x-ray tube with smaller effective focal spots.

10. A The pyrex glass envelope surrounding the x-ray tube causes some attenuation of the x-ray beam. A few tubes are designed with a thinner flattened area called the window to minimize the absorption of the x-rays by this portion of the tube envelope.

11. C The vast majority of modern diagnostic type x-ray tubes are designed with a small and large filament and a single rotating anode (target).

12. A The electron cloud surrounding the filament is drawn across the x-ray tube by the high potential difference created during the actual exposure. The majority of kinetic energy of this stream of electrons is converted into a low energy form of electromagnetic radiation called infrared rays or heat.

13. A The amount of free electrons produced at the cathode (space charge) is directly dependent upon the impressed amperage and the resulting filament heat.

14. C The target material in nearly all general purpose diagnostic type x-ray tubes is formed by a thin layer of the metal tungsten. In a stationary anode the majority of the anode is made of a solid block of copper.

15. A The speed developed by the electron stream as it passes from the cathode (filament) to the anode (target) side of the x-ray tube is directly proportional to the potential difference or the kilovoltage applied to the electrodes. The electrons in this 2-3 cm gap obtain speeds as high as 1.5×10^{10} cm/s or 1/2 the speed of light.

16. B See explanation number 50 of this section.

17. C The anode disk of a rotating anode x-ray tube is made of alloy of tungsten and rhenium with a stem made of the poor heat conducting molybdenum. This tungsten coated disk serves as the target for the high speed electron stream which originates at the heated filament.

18. D The watt is a unit used to express the amount of electrical power and is defined as the product of amperes and voltage; i.e., watt = amps x volts.

19. B The effective (projected) focal spot is directly related to the recorded detail of the radiographic image and is more important to image quality than the actual or electronic focal spot located on the surface of the anode.

20. B One of the first major advances which made modern radiography more practical was the heated filament made possible by William Coolidge through the development of the step down transformer that bears his name.

21. B Since x-rays are produced equally in all directions at the surface of the anode, many are directed into the anode disk and away from the patient. The angling of the anode helps direct a greater portion of unattenuated photons toward the image receptor by reducing the absorption of the photon in the heel of the anode.

22. B A reduction in the anode angle will provide for a smaller focal spot due to the line focus principle and a smaller field coverage because of the increase in the anode heel effect which occurs when the anode is more vertical.

23. C To ensure that the electron flow in the x-ray tube is always from the cathode to the anode, a type of direct current must be employed.

24. C The low heat conductivity of molybdenum makes this an ideal material for the stem of the anode disk. The heat which builds up on the anode during x-ray production is prevented from reaching the temperature sensitive ball bearings of the induction motor by this stem.

25. B After leaving the anode, the x-rays first encounter the pyrex glass of the tube envelope. Since most tubes are surrounded by a layer of insulating oil, the beam may pass through some oil before hitting the aluminum (filtration) prior to its encounter with the collimator mirror and plastic collimator cover.

26. A The heat storage capacity of a rotating anode tube can be increased by increasing the cross sectional area of diameter of the disk, using an anode with a high rotational speed, increasing the size of the thermal cushion (external housing) and by the application of an external cooling (fan) device.

27. D A rotating anode has a greatly expanded target area which serves to dissipate more of the heat produced during x-ray production. This reduces the damage to the target and enables higher instantaneous loading and tube rating capacities for the tube.

28. B See explanation number 9 of this section.

29. A As the mA station is increased, the heat of the filament and the amount of space charge expands. Since the electrons are striking a larger area on the anode, the effective focal spot is increased by a process called blooming. Image blur increases when a larger focal spot is employed.

30. D Pure tungsten is extremely brittle and is subject to surface cracks at high temperatures. The addition of rhenium increases both the thermal capacity and the resistance of the target to surface defects.

31. D The main characteristic required in a filament material is its need to resist melting at high temperatures. Pure tungsten or a mixture of thorium and tungsten provide wire filaments with melting points of about 3400° C or 6000° F.

32. A Sometimes called the first x-ray tube, the glass filled Crooke's-Hittorf tube provided Roentgen with the opportunity to make his important discovery of x-rays.

33. B The anode of most general purpose x-ray tubes is about 10-12°. The special fine focus tube needed for mammography employs angles as low as 7°.

34. D The addition of thorium to the brittle tungsten improves its ability to be coiled as well as improving its efficiency as an electron emitter; this will also increase its durability and longevity.

35. C See explanation number 17 of this section.

36. D Due to the line focus principle, the effective focal spot will appear smaller than the actual focal spot as long as the anode angulation is less than 45 degrees. The smaller the angle of the anode, the smaller the effective focal spot for a given filament size.

37. A When electrons are emitted by the filament in the absence of a large potential difference, they remain in the vicinity of the filament forming an electron cloud or space charge. This in turn limits the further emission of additional electrons by the so-called space charge effect.

38. D See explanation number 8 of this section.

39. D See explanation number 26 of this section.

40. A See explanation number 9 of this section.

41. D See explanation number 6 of this section.

42. A See accompanying diagram.

43. F — See accompanying diagram.

44. G — See accompanying diagram.

45. C — See accompanying diagram.

46. B — The induction motor of a rotating anode x-ray tube consists of a movable coil of wire (rotor) which is attached to the anode stem and disk. Two sets of ball bearings reduce the friction enabling the rotor to spin within the vacuum of the tube envelope. The stator (electromagnets) windings provide the driving force for the rotation of the anode disk surrounding the anode end of the tube outside the envelope.

47. A — The angulation of the anode is related to the size of the effective focal spot, the area that is covered by the beam and the amount of the anode heel effect.

48. C — Tube vibrations, though relatively rare, are almost always caused by an imbalanced induction motor assembly or a warped anode disk. Both will lead to inadequate rotation speed and a loss of recorded detail and contrast on the radiographic images.

49. C — Because the rotor of the induction motor is located inside the vacuum of the tube envelope, oil or graphite lubricants which vaporize at relatively low temperatures cannot be used. Powdered metals such as silver are successfully used as rotor lubricants of many modern x-ray tubes.

50. C — In a single phase fully rectified unit the number of anode heat units created for a given exposure is calculated using the formula: Anode heat units = MA x kVp x time. In a three phase 6 pulse unit a 1.35 correction factor is required. In a three phase 12 pulse system a 1.41 correction factor is required.

51. A — A lower tube potential (kVp) will decrease the speed and energy of the electron stream crossing the tube. This lower energy will decrease the quality of the beam by creating fewer photons with shorter wave lengths and higher frequencies.

52. D — See explanation number 46 of this section.

53. A — The filament temperature which has a direct relationship with the space charge can be modified by changes made in the mA setting.

54. A — In the first stage of an exposure, three events occur: the filament is heated, the bucky begins to move and the anode rotation commences. During the second or exposure stage, high potential difference moves the electron cloud toward the anode (target).

55. A — The heating of the filament which occurs in the first (prep or boost) stage of the exposure is directly related to the development of the space charge.

Quality Assurance

Chapter 13

1. C — The vast majority of the malfunctions associated with the operation of the x-ray tube are related to the high heat and thermal stresses associated with the x-ray production process.

2. A The resolution of an image receptor is most easily determined through the use of a resolution (phantom) consisting of precisely spaced lead strips and radiolucent spacers. The two types most commonly employed are called line pair and star phantoms. Examples of a star phantom and two line pair phantoms are found on page 43 at the beginning of section 2.

3. C Most modern radiographic tubes require a series of warm up techniques which serve to prevent uneven expansion of the anode and the development of heat stress cracks. All warm up techniques share 3 common properties: the use of a low kVp setting (70), large focal spot, and a long timer setting (500 ms).

4. C A piece of lead with a small pinhole is centered to a tube at an equal distance from the focal spot of the tube and the imaging receptor. A radiographic exposure will enable only a portion of the x rays to pass through the hole. This pinhole (camera) effect can be used to measure the size of the focal spot in the x-ray tube.

5. A One of the most important of the required equipment tests is in the determination of the consistency or linearity of a series of radiographic exposures. This test requires an exposure at each different mA station using the appropriate timer settings to provide the same mAs value. Using a mR/ms ionization type meter the intensity of each exposure is recorded and compared to other values. The linearity between exposures is considered acceptable if the variation is within 10% for all exposures at different mA stations.

6. B An estimation of the kilovoltage peak can be made radiographically by the use of a pentrometer (step wedge), Wisconsin kVp test cassette, Adren-Crooke's kVp test cassette. More recently, a number of manufacturers have developed kVp meters for the direct measurement of this important factor.

7. C To avoid accidental exposures to an anode which is stopped or rotating at less than its maximum speed, a relay or interlock switch is incorporated into the timer circuitry. This device requires a brief .5 sec. delay between the activation of the prep stage and the actual start of x-ray production process.

8. B In the recent past, the most common cause of image blurring was poor screen-film contact due to loose or low tension cassette straps. Though rarely seen with the new wafer cassettes, it can result from improperly mounted screens, faulty latches or hinges and warping of the cassette.

9. B All x-ray tubes are subject to a number of thermal and mechanical stresses. Though tube warm ups are critical for extending the life of the anode, little can be done to increase the longevity of the filament which is one of the more frequent causes of tube failure.

10. C Poor film-screen contact results in a rapid diffusion of the light from the screens before it reaches the film which noticeably blurs the radiographic image.

11. D The normal operation of the x-ray tube involves the production of extremely high temperatures of the filament and anode. Over time, this heat causes the vaporization of tungsten and the formation of a thin rust colored layer which coats the inside of the pyrex envelope called bronzing or tanning.

12. A The film-screen contact of an imaging system is best evaluated through the placement of a number 40 or 60 wire mesh over the cassette. Any irregularities shown will indicate a failure of the system to maintain the desired contact between the intensifying screen and the radiographic film.

13. A If a radiographic exposure is made on a stationary or slowly rotating anode, the heat is concentrated on a small area causing localized melting (pitting) of the anode.

14. D — The aluminum pentrometer (step wedge) is one of the oldest and widely used devices in the evaluation and testing of the imaging equipment. The step like design enables the production of a graduated scales of densities which can be employed to form a characteristic curve for the imaging components. A stepwedge can be used to compare and calibrate x-ray films, intensifying screens, exposure control settings, and automatic processing systems.

15. D — Any over heating of the x-ray tube can be associated with the vaporization of tungsten and the coating of the inside of the tube envelope. This layer can act as a third electrode, and additional filtration for the beam, and if it remains as a vapor, it can over time destroy the vacuum.

16. D — A low density artifact seen on a radiographic image indicates that either less radiation or light has reached the screen or film. Metallic objects in or on the patient, table, or positioning sponges may be visualized but are rarely found in the same location. Screen artifacts caused by dirty screens or screen damage will appear in the same region on all radiographs taken with the damaged cassette.

17. B — The first stage of the exposure, the prep or boost, is associated with three separate events: the preheating of the filament, the rotation of the induction motor and the activation of the bucky movement.

18. D — See explanation number 8 of this section.

19. B — A warm up technique is specifically designed to heat the anode of the x-ray tube in a manner which reduces the chance of a radical temperature change and uneven expansion which can crack or warp the anode disc.

20. B — Any irregularity on the surface of the anode disc such as pitting or crazing results in an increased attenuation of x-rays by the disc and a reduction of the radiation output for a given tube current.

21. D — See explanation number 16 of this section.

22. B — An evaluation of timer accuracy in a radiographic unit (single or three phase) can be made by a device called an oscilloscope.

23. C — The only moving component of a modern x-ray tube is the induction motor of the x-ray tube. The ball bearings incorporated into this motor help reduce the friction and noise associated with its operation. Bearing damage is commonly associated with noise during the prep or boost stage of the exposure.

24. D — Since the burn out of the filament is one of the most likely reasons for shortened tube life, the use of a lower mA station and shortened boost time will extend the life of the filament. The use of proper warm up times, observation of the heat units and tube rating limits will substantially reduce anode malfunctions.

25. A — A special motor driven spin top tester which spins at precisely one revolution per second can be used to test for valve tube competency in a single phase unit and timer accuracy on both single phase and three phase radiographic units.

26. A — A 25 mA, 50 mA and 100 mA selectors are all related to the choice of the small filament and focal spot. A faulty filament will negate all exposures attempted at any of these settings. Techniques selected on the large filament or high mA stations will be permitted it this filament may not be damaged.

27. A — As a tube ages, the tungsten from the filament and anode may begin to vaporize, destroying the vacuum within the x-ray tube. Since some of the tube's electrons will have difficulty reaching the anode and may cause fluctuations of the mA meter readings.

28. B Not unlike a light bulb, the life filament can be extended by limiting its use. The use of lower mA stations which require less space charge, lower filament temperatures, and the avoidance of the long boost and hold exposures, will maintain the longest possible life of the filament.

29. C The radiation output of an x-ray tube is measured in air by a milliroentgen/milliampere seconds meter. This modified ionization type meter is designed to accurately measure the short exposure bursts associated with diagnostic exposures.

30. D A ready light that is found on all newer radiographic units, indicate that the exposure parameters and the status of the accessory equipment are acceptable. A red indicator (fault) light will prevent exposures if any of the protective interlocks are open; i.e., bucky tray not pushed in, fluorocarriage not in park, bathroom or entrance doors opened or collimator not directed at the imaging receptor.

31. B The first and most important action of a radiographer faced with an electrical fire is the shut down of the main power to the unit. Secondly is the removal of the patient from the room. Third is to close all doors and lastly to pull the fire alarm box. Throwing water on an electrical fire can worsen the situation by increasing the potential for electrical shock.

32. A The main agency in charge of safety concerns of radiographic equipment is the Bureau of Radiologic Health (BRH).

33. B Electric fires require the use of a C type extinguisher. These include the dry chemical type and gaseous type devices like Halon or carbon dioxide.

34. C The reproducibility of radiographic units is critical in maintaining a consistent density for a given exposure. This can be tested by making a series of identical radiographic exposures and measuring the intensity of each using an mR/mAs meter. When properly adjusted, a unit shall provide exposures within a 5% reproducibility.

35. B A repeat analysis is likely to show the most common cause for a repeated image is improper exposure factor selection. This may be surprising in light of the wide spread use of automatic exposure control devices which are intended to eliminate this type of error.

36. B In a rotating anode tube excessive heat conduction through the stem would cause damage to the induction motor. The poorly conductive molybdenum stem of the anode encourages radiation of heat off the anode while resisting the conduction of the heat to the induction motor.

37. D A tube rating chart provides a means for determining if a given exposure can be safely permitted by the radiographic unit. In most modern units, these limits are programmed into the unit to protect the anode and cathode from excessive heat loads. A red light marked overload or maximum rating is normally located on the control console to indicate when unacceptable exposure settings have been selected.

38. A The recent recommendations of the NCRP requires that all new radiographic units posses distance indicators and collimation devices. Each must be accurate to within 2% of a given source-to-image distance; i.e., at a 100 cm SID the collimator must be accurate to within 2 cm.

39. A See explanation number 37 of this section.

40. D The maximum heat storage capacity or dissipation rate of the anode is effected by a number of intrinsic and extrinsic factors. In the anode itself its design, size and speed of rotation effect this rate. Externally, the size and design of the housing, amount of insulating oil and the presence or absence of cooling devices will also effect the spread of heat away from the anode.

Anode Storage Capacity

Exercise No. 1

1. 250	11. 13
2. 74	12. 12
3. 41	13. 97
4. 23	14. 39
5. 20	15. 177
6. 35	16. 129
7. 31	17. 52
8. 185	18. 31
9. 49	19. 18
10. 27	20. 44

Anode Storage Capacity

Exercise No. 2

1. 125	11. 9
2. 31	12. 44
3. 61	13. 259
4. 13	14. 58
5. 75	15. 88
6. 17	16. 258
7. 15	17. 131
8. 92	18. 47
9. 18	19. 70
10. 25	20. 25

Tube Rating Chart

Exercise No. 1

1. 1.5 sec.	11. 120 kVp
2. .40 sec.	12. 90 kVp
3. .75 sec.	13. 120 kVp
4. .40 sec.	14. 110 kVp
5. 600 mA	15. Yes
6. 100 mA	16. Yes
7. 400 mA	17. Yes
8. 600 mA	18. Yes
9. 100 mA	19. No
10. 400 mA	20. No

Tube Rating Chart

Exercise No. 2

1. 3 sec.	11. 90 kVp
2. .60 sec.	12. 90 kVp
3. 1 sec.	13. 90 kVp
4. .10 sec.	14. 90 kVp
5. 400 mA	15. Yes
6. 200 mA	16. No
7. 400 mA	17. Yes
8. 100 mA	18. Yes
9. 200 mA	19. No
10. 400 mA	20. Yes

Filtration - HVL

Chapter 14

1. D A minimum of 2.5 mm Al equivalent of diagnostic filtration is required for all radiographic units operating at above 70 kVp of energy. Though this amount of filtration will decrease the beam intensity by about 1/2, its primary purpose is the removal of the long wavelength (low energy) photons that increase the skin exposure of the patient without substantially effecting the radiographic image. Diagnostic filtration is estimated to reduce patient skin exposures by 30-50% compared to an unfiltered beam, while increasing the average energy (quality) of the beam by about 10%.

2. D The amount or thickness of a material required to reduce the intensity of the heterogeneous x-radiation beam to one half of its original value is called the half value layer.

3. A The total filtration required for a radiographic unit consists of an inherent layer formed primarily by the pyrex glass of the x-ray tube envelope and an added layer usually consisting of a thin sheet of aluminum. Some additional filtration is provided by the positioning mirror and plastic cover of the collimator.

4. A The half value layer of a diagnostic x-ray beam is highly variable depending on the kilovoltage peak selected. According to the NCRP 102 table 1.3 the HVL thickness at a 90 kVp is 2.6 mm Al for single phase units and 3.1 mm Al for three phase radiographic units. At 110 kVp, the HVL thickness for single phase and three phase units will be 3.1 mm Al and 3.6 mm Al respectively.

5. B The beam with the greatest filtration curve B has the lower intensity but higher beam quality shown by the curve being skewed to the higher energy portion of the graph.

6. A See explanation number 1 of this section.

7. B With a monochromatic (gamma) source, if 4 mm Al reduces the intensity by 25% of its original value than its half value layer would be 1/2 that amount or 2 mm Al.

8. B In some regions of the body; i.e., entire lower leg or entire spine, a single set of exposure factors will not provide the desired radiographic density in all areas. The use of specially designed (compensating) filters modifies the intensity of the beam thus compensating for these changes in tissue thickness or density, and therefore produces more uniform images of these structures.

9. B The correct value is obtained by following the curve from the 100 mR line to the 50 mR point and reading the thickness (4 mm) at that point.

10. C All x-ray beams produce a broad spectrum of photon energies and are referred to as being heterogeneous. The selective removal of the low energy photons during the filtering process will increase the average energy (quality) of the beam, reduce the intensity (quantity), and make the beam more homogeneous or monochromatic.

11. A The pyrex glass envelope which is designed to provide only a minimal resistance to the passage of the x-ray beam has an equivalent thickness of about only .5 mm Al.

12. D If a beam with a 100 mR/min. intensity has a half value layer (HVL) of 6 mm Al, the following reduction in intensity is seen: 50 mR/min. with 6 mm Al (1HVL), 25 mR/min. with 12 mm Al (2 HVL) and 12.5 mR/min. with 18 mm Al of filtration (3 HVL).

13. B Aluminum is employed as the principal filtering material in the diagnostic range because of its low cost. Special high density filters; i.e., gadolinium and holmium have been developed which may lower the doses and improve the contrast visualization during specific types of contrast studies.

14. D The total amount of diagnostic filtration is found by adding the amount of inherent filtration to the amount of added filtration.

15. B If 10 mm Al reduces the original beam intensity to one quarter of its original intensity, then half that amount or 5 mm Al would be the half value layer thickness for this beam.

16. A The curve with the highest kVp will undergo the least loss of intensity. According to the diagram the uppermost curve meets this criteria.

17. D Attenuation will increase as the beam energy decreases. The highest attenuation occurs in the curve of the lowest kVp value.

18. D By definition, the half value layer is the amount of material which will reduce the original intensity of the beam by 1/2 or 50%.

19. A The use of diagnostic filtration will increase the beam quality by the selective removal of the low energy photons. Since the beam intensity is reduced, the overall effect will be a reduction in the radiographic density.

20. A If higher degree of attenuation is desired, the atomic number of the filter should be increased.

21. D A beam with the highest quality is produced using a combination of the highest peak energy and the greatest amount of filtration.

22. B By definition, a tenth value layer reduces the beam intensity to 10% or 1/10th its original value. This is nearly identical to the effect of 3.3 half value layer thickness.

23. B After a beam has been filtered by two half value layer thickens of an absorber, the remaining wavelength of the photons are close enough in energy to be considered homogeneous in nature.

24. C A tenth value layer is defined as the amount of material which will reduce the original intensity of the beam by 1/10th of its original value.

25. C See explanation number 12 of this section.

Electromagnetic Radiation

Chapter 15

1. D One of the basic concepts of the quantum theory is the dual nature of the photons. This theory states that photons can either act as a wave or a particle. The penetration of light through glass or x rays through low density matter represents the weak interaction of the photon's wave nature. The particle's nature involves a substantial exchange in energy such as that between photons and electrons during excitation or ionization.

2. C If detected in free space, an x-ray photon and a gamma ray are identical in every measurable characteristic. Their classification is based entirely on their point of origin in the orbit shells for an x-ray photon and the nucleus of the atom for a gamma ray.

3. B All electromagnetic waves have a consistent velocity in a vacuum of $3x10^8$ meters per second (speed of light). Because their speed is constant, the relative energy is directly proportional to frequency and inversely proportional to the wavelength of the photon. In other words, an increased frequency and decreased wavelength are associated with a higher photon energy.

4. B — All electromagnetic waves if visible would appear as having a sine wave form, enabling it to be defined by the number of cycles passing by a given point each second (frequency). The frequency of a wave is directly related to the energy of the wave and is measured in the number of cycles per second called Hertz.

5. D — A sine wave is often conceptualized as an invisible piece of string of a fixed length that is being vibrated in an up and down fashion.

6. A — The particle nature of electromagnetic radiation is most frequently seen in waves possessing a significant amount of energy. The three principal waves which carry high amounts of energy are ultraviolet radiations, gamma rays and x rays.

7. D — The normal range of energy for diagnostic x rays is from about 25 keV to 125 keV according to the formula: $\min \lambda = 12.4/\text{keV}$. The wave length for this range of energies is found to be .1Å - .5Å where Å is the angstrom unit (10^{-8}cm).

8. C — X rays are a highly energetic form of electromagnetic radiation, possessing a short wavelengths and high frequencies. Therefore, all of the low energy members of the spectrum, radiowaves, microwaves, infrared radiation, visible light and ultraviolet radiation would all have wavelengths longer than x rays.

9. A — The ionization of the chemicals within the cells of the skin may cause a reddening to appear as the skin cells die. This damage when due to exposure to the sun (sunburn) is caused by one of the more energetic types of EMR called ultraviolet radiation (UV).

10. B — See explanation number 7 of this section.

11. B — Three common properties are shared by all electromagnetic radiations; namely, that they have no mass, no charge and travel at the same velocity in a vacuum (speed of light).

12. D — Because electromagnetic waves have no measurable charge or mass, they are unaffected by magnetic, electrostatic or weak gravitational fields. The extremely strong gravitational fields of black holes, neutron stars and our own sun are capable of bending photons in proportion to their masses.

13. D — The visible spectrum of light includes the various colors of light from the long wavelength red light (7000Å) to the shorter wavelength blue or violet (4000Å) light.

14. D — The wavelength of a photon is defined as the linear distance between the crests of two adjacent waves. As the wavelengths of a photon increases or becomes longer, its energy is decreased.

15. C — The frequency which is defined as the number of cycles or wave alterations passing through a given point in a second is determined by counting the number of peaks seen in the upper portion of the wave.

16. B — According to the formula $E=hv$ where h =planks constant and v = frequency of the wave, the energy of the wave is found to be directly proportional to its frequency.

17. D — The members of the electromagnetic spectrum include from lowest to highest energies: radiowaves, microwaves, infrared radiation, visible light, ultraviolet radiation, x-rays, gamma rays, and cosmic radiation.

18. A — The speed of all electromagnetic radiation in a vacuum is $3\text{x}10^{8}$cm/sec. or 186,000 miles/sec. This is not consistent and is dependent on the nature of the medium in which the waves are traveling.

19. B Many of the higher energy forms of electromagnetic radiation, including ultraviolet, x and gamma rays, are capable of removing (ionizing) electrons from the atoms exposed to this radiation.

20. D Though heat or infrared radiation is invisible to the human eye, its production is often associated with the inadvertent production of some form of visible light. The red color seen with a heating element of an electric stove is an example of this inadvertent light production.

21. A See explanation number 7 of this section.

22. D Since the frequency and wavelength of a photon are inversely proportional to each other, a four time increase in frequency results in a four time decrease in the photon wavelength.

23. C In our field it is most convenient to measure the energy of a photon in terms of its electron voltage (eV). The Joule is a unit by which photon energy is expressed in the SI system of measurement.

24. A One of the first concepts in Albert Einstein's theory of relativity was the equivalency he proposed between energy and matter. In recent years, the vast majority of this theory has been tested and found to be accurate.

25. B See explanation number 16 of this section.

26. B Not unlike the relationship of the atom as the smallest division of matter the photon is the smallest unit of electromagnetic radiation.

27. A Most low energy electromagnetic radiation such as radiowaves, microwaves, and infrared will interact with matter weakly as if they were waves. Higher energy waves which are responsible for ionizing effects of radiation act more like particles.

28. C The extremely small wavelength of diagnostic x-ray photons are measured by the units of angstroms (10^{-8} cm) or nanometers (10^{-9} cm).

29. C The energy of a photon can be found by the formula $E=hc/\lambda$ where h is planks constant, c is the speed of light, and λ is wave length of the photon.

30. D Sound differs from electromagnetic radiation in two major ways. The speed of sound is only 640 miles per hour in air and sound, unlike EMR, requires a propagation media and will not travel within a vacuum.

Nature of X-ray Photons

Chapter 16

1. D A graph of the x-ray distribution curve has a smooth bell curve appearance with three higher intensity spikes at the energies of 11, 59 and 69 keV. These represent the greater number of photons produced during characteristic x-ray production.

2. C The vast majority of high speed tube electrons which pass from the filament and interact with the atoms in the anode produce low energy infrared (heat) radiation.

3. B During the x-ray production process, the incident energy of the electron beam is converted into broad spectrums of low, medium and high energy photons. Because of the great variance in the nature of the beam, it is described as being heterogeneous or consisting of many different photon wavelengths.

4. A X rays are produced by two different processes following the collision of a high energy electron stream with the target material of the x-ray tube. Bremsstralung or brems radiation which results from the slowing down of the incident electron as it travels near the nucleus of the target atom accounts for about 70-85% of all the diagnostic quality photons in a beam. The remaining 15-30% of the photons are formed by a process called characteristic x-ray production which occurs following ejection of an inner shell electron by the incident electron x ray and other lower energy waves are produced during subsequent filling of the vacant orbital shells.

5. C The average energy of the x-ray photon in the beam existing from the tube is termed the beam quality. An increase in the potential difference (kVp) across the x-ray tube will produce a greater number of photons with a higher energy. The addition of 2-3 mm Al diagnostic filtration will selectively remove the majority of the low energy photons while increasing the average energy or quality of the beam.

6. D The brems radiation produced when an electron loses energy as it passes by the nucleus of a target atom is dependent upon three factors; namely, the original energy (speed) possessed by the incident (tube) electron, the atomic number of the target and the distance between projectile electrons and the nucleus of the target.

7. C Bremsstralung radiation is formed from the energy lost by a high speed electron as it passes near the positively charged tungsten nucleus. The loss of energy involved in this process appears in the form of an electromagnetic photon.

8. A See explanation number 4 of this section.

9. B The ability of an x-ray photon to penetrate through matter is dependent on both the nature of the matter and energy of the photon. The thicker or more dense the material the lower the degree of penetration by the photons. Higher energy (shorter wavelength) high frequency photons will have a higher degree of penetrability than their lower energy counterparts.

10. B The amount of energy converted into x rays during the x-ray production process is about 100 times less than the amount of energy formed as heat or infrared radiation.

11. B The use of a higher tube potential (kVp) will increase the speed (energy) of the electron stream across the x-ray tube. Since more energy is transferred to the atoms of the tungsten target, the photons produced will have higher frequency and a shorter wavelength than at lower energies.

12. C The Bremsstralung portion of the x-ray beam consists of photons consisting of virtually all photon energies, and is said to be polychromatic in nature. The term white radiation gives an accurate description for a heterogeneous x-ray beam which contains may different photon energies.

13. A Characteristic x rays are directly associated with the loss of energy resulting from the movement of electrons from a higher to lower orbital shell. In a given material, such as tungsten, these energies are highly consistent and depend primarily on the atomic number of the target.

14. D The energy of a photon is effected by its frequency and wavelength. Any change in this factor will also effect the quality, energy and the beam's ability to penetrate through matter. The frequency of the wave is directly proportional to these factors.

15. D See explanation number 1 of this section.

16. A The formula E=K x Z x kVp where K is a constant, Z = atomic number of the target material, kVp = potential difference applied across the tube, can be used to estimate the efficiency of the x-ray production process.

17. B A low quality beam has a low penetrability and is sometimes described as being soft.

18. A See explanation number 4 of this section.

19. B Though a reduction of off target radiations and a constant tube potential would help to improve the beam quality the basic heterogeneous nature of the bremsstralung process cannot be altered until the photons have exited from the tube.

20. C See explanation number 4 of this section.

21. C The lowest milliampere setting should correspond to the lowest intensity of curve 3.

22. C Curve 1 has more area under it than either of the other curves indicating a higher mA was employed. The curve is also shifted to the right (higher energy) indicating the use of a higher kVp setting.

23. D The electrons emitted from the filament of an x-ray tube are accelerated to speeds of about 1.5×10^8 meter per second or half the speed of light in the narrow gap separating the cathode and anode.

24. A The main factor that controls the quality and penetrability of the beam is the kVp setting.

25. A In the diagnostic energy range the percentage of the electron steam which is converted into x-rays is normally about .2 - .9%.

Image Intensification

Chapter 17

1. C Conventional radiography is limited to the production of still or static images. Fluoroscopic's main advantage is the ability to see the images in motion called dynamic or real time imaging.

2. C The human eye requires about 20 minutes to fully adjust or adapt to the low light levels of the old conventional fluoroscopic systems. Physicians preparing for these studies would sit in a darkroom or wear red goggles to prepare their eyes for the low light levels. One of the main advantages of image intensified (modern) fluoroscopy is that no light adaptation is required.

3. B The main advance which made modern fluoroscopy possible was the development of the image intensifier tube and its associated electronics. This device dramatically increases the brightness level of the fluoroscopic image.

4. A — During image intensified fluoroscopy the normal milliampere setting will vary between .5 - 5 mA, significantly less than the 25 - 1000 mA used in conventional radiography.

5. C — The increased brightness found with the image intensified fluoroscopy primarily occurs in the image intensifier (II) tube. The first of the two primary gain factors (flux gain) is related to the 25 kVp potential difference placed between the input and output sides of the tube which provides about a 100-140 time increase in the brightness of the image. The minification ratio occurs through the concentration of the electrons from the larger input screen being compressed onto a smaller (1") output screen. Quantitatively, this minification ratio is proportional to the squared ratio of the input screen to the squared ratio of the output screen diameter. The total gain is the product of the flux gain and minification ratio or about 8-10,000 in most modern systems.

6. A — After the imaging carrying (exit) radiation beam enters the input side of the image intensifier tube, a fluorescent screen (cesium iodide) is encountered which converts the x-rays into a visible light image. A photocathode layer immediately behind this screen transforms this visible image into a corresponding electronic image prior to its transit to the output side of the tube.

7. B — The output screen of an image intensifier consists of a 1" diameter fluorescent screen made of zinc cadmium sulfide which converts accelerated electronic images into a bright yellow-green colored visible image.

8. B — The concave input phosphor layer of the image intensifier is most often composed of a thin .1-.2 mm layer of sodium activated cesium iodide which absorbs a percentage of the exit beam and converts it into a visible light image.

9. B — In order to move the electrons comprising the electronic image on the input side of the image intensifier across to the output side of the tube requires a potential difference of about 25,000 volts or 25 kVp.

10. D — See explanation number 5 of this section.

11. A — In order to compensate for the various differences in patient thickness' and densities encountered during fluoroscopy, a system not unlike the automatic exposure control system in conventional radiography is employed by these units. These are most often called automatic brightness gain or control systems.

12. C — See explanation number 6 of this section.

13. C — See explanation number 5 of this section.

14. A — Most texts give the operating range of an image intensifier as between 2-5 mA. Recent improvements have further reduced this value to the point where .5-3 mA range may be more appropriate for the average sized patient.

15. D — In an attempt to standardize the unit of brightness gain, a second unit called a conversion factor has been developed. The conversion factor is defined as the ratio of luminescence in candela for the output phosphor to input intensity as measured in milliroentgen/seconds.

16. A — Over time the phosphor's and other components of the image intensified will degrade resulting in a loss of brightness gain. The deterioration for a modern II tube may be as high as 10% per year.

17. D — Brightness gain, though useful, can be modified by changes in the incident exposure rate, operating potential across the tube and ratio of input to output screen size (minification ratio).

18. A The use of smaller input field size with an image intensifier will provide images with a noticeable degree of magnification. Unlike radiographic magnification techniques which degrade image quality, a higher resolution and image quality are provided during fluoroscopic enlargement techniques. The main disadvantage of this technique is the substantial increase in the dose rate that is required to obtain the desired brightness level.

19. B The low light levels associated with conventional fluorography allows only the use of scotopic or rod vision. These cells which are active at low light levels lack the fine discrimination of the cone cells. The large increase in image brightness associated with the image intensifier enables the use of photopic (cone) vision and a substantial increase in visual activity.

20. B In the manual mode of a fluoroscopic unit (without automatic brightness control engaged), an increase in the kilovoltage setting will increase the intensity and brightness of the image. The increased penetration will decrease the amount of differential attenuation within the tissue resulting in a lower or longer scale contrast.

21. D The automatic control of a fluoroscopic system is designed to compensate for variations in part thickness, tissue density, and changes modification of the source-to-image receptor distance.

22. C The potential difference applied across the image intensifier serves to increase the energy of the electron stream by about 100-140 times.

23. D The total brightness gain of an image intensifier is dependent upon the product of minification ratio and the flux gain. In modern units this value normally falls in the range of 5000 for the small (6 inch) input magnification mode and as high as 20,000 for larger format (14 inch) tubes.

24. C In order to provide the electrons on the surface of the photocathode layer of the input screen an unimpeded flow to the output phosphor a vacuum is created in the image intensifier tube.

25. A Because of the curved input screen and the flat output screen in the image intensifier, a higher electron density occurs near the center of the image. This effect, called vignetting, provides more detail and brightness in the center of the image monitor than its periphery.

26. B The adjustment from a larger to smaller input size in an image intensifier tube is accomplished electronically by a modification of the voltages on the electrostatic lens.

27. C The electrons from the top of the input screen are focused to the bottom of the output screen while the electrons on the bottom of the input screen are focused to the top of the output screen. The point at which they cross is called the electronic focal spot.

28. A See explanation number 5 of this section.

29. D The solution is found by the application of minification factor formula where minification ratio factor = input screen diameter^2/output screen diameter^2.

30. C Due to vignetting (explanation 25), the image quality is best at the center of the image. A higher intensity or dose rate improves the image brightness, reduces image noise and improves quality of the image.

31. A The cesium iodide phosphor layer is directly applied to the glass of the input side of the tube.

32. D When a smaller input (magnification) mode is employed with an image intensifier, the reduction in the minification gain requires an increase in the intensity or dose rate to maintain the desired image brightness. This high dose rate decreases the noise of the image and provides a higher image contrast.

33. C See explanation number 7 of this section.

34. D A thin uniform layer of silver deposited on the inside of the image intensifier tube envelope and a series of internal metallic lenses within the tube are responsible for the proper focusing of the electron stream.

35. D See explanation number 32 of this section.

36. D See explanation number 15 of this section.

37. E See corresponding diagram.

38. D See corresponding diagram.

39. F See corresponding diagram.

40. A See corresponding diagram.

Television Imaging

Chapter 18

1. D A modern closed circuit TV chain, whether for fluorographic or digital applications, will most likely use a form of vidicon or plumbicon TV camera tube. Charge-coupled device (CCD) TV camera though more durable and sensitive than video tubes do not provide the resolution capabilities needed for most radiographic applications at this time.

2. C The electronic signal produced by a video or CCD TV camera is termed the video signal. When amplified, it is this signal that provides the instructions to the last portion of the TV chain, the monitor, for display of the image.

3. C In both vidicon and plumbicon TV camera tubes, the image on the signal plate is "read" by the use of a stream of electrons originating in the special designed cathode called the electron gun.

4. C When viewing a movie, each frame of the film must be presented in less than 1/20th of a second in order to see the images clearly without an annoying image flicker. In a similar fashion, each frame of a TV image must likewise be presented at speeds exceeding the eyes integration time of 1/20 second.

5. A The principal device used in the conversion of a non-digital (analog) signal to a digital form is an analog-to-digital (ADC) converter.

6. D Though the image intensifier and television systems have improved fluoroscopic imaging in a number of important ways, the absolute spatial resolution of the images is degraded with the addition of each new electronic component. The various components required in the TV chain give this system the poorest resolution of all of the conventional analog imaging systems.

7. D All electronic audio or imaging devices are associated with a large degree of background electrical noise which tends to degrade the image or sound quality of the original signal. This important imaging factor is normally expressed by the ratio of the maximum signal strength to the total electronic noise of the system, called the signal-to-noise ratio (SNR). In a conventional or home TV system, a SNR of 200:1 is considered acceptable. Most computer assisted or digital radiographic systems require a video system with a SNR of 500:1 or 1000:1 to maintain the image quality of these advanced techniques.

8. B With either a vacuum type video TV camera tube or charge-coupled device TV camera, the output is in the form of an electronic video signal which carries the information presented on the signal plate of the camera to the TV monitor.

9. B Most television monitors display an image consisting of 525 lines which are presented on the TV monitor. As with the signal-to-noise ratio, the TV systems required in most digital applications require an increase to 875 or 1024 lines to provide the increased resolution desired.

10. D The principal purpose of the target-signal plate assembly of the vidicon camera tube is the conversion of the visible image on the input screen into an electric image which can be transformed into the video signal. This photo-conductive target which is usually made of antimony trisulfide suspended in a mica matrix is the most important element of the TV camera.

11. C See explanation number 10 of this section.

12. D Image flicker, which may interfere with visualization of images from a television monitor can be eliminated by extremely rapid frame rates or by an electronic scanning pattern in which only every other line is scanned during each field called interlaced scanning.

13. C The use of a vacuum within both the TV camera and TV monitor eliminates most of the gases that would interfere with movement of electrons across these tubes.

14. A The electron beam (gun) in a video camera is steered to its desired location on the target assembly by two sets of electrostatic devices known as the focusing and deflection coils.

15. D The television camera tube will usually operate at a relatively low electrical potential difference of about 250 volts. In the larger television monitor the electrical potential is substantially greater at levels as high as 15,000 volts.

16. B Even with the best man-made vacuum some remaining gas and gasses formed by the imaging process are present. A charged ion trap located at the cathode end of the TV monitor helps to collect these gasses and maintain the quality of the image.

17. A To ensure the proper synchronization between the TV camera and monitor, a special electronic (synchronization) pulse is added to the video signal at the end of each scanning line.

18. A The two main factors that effect the image resolution of a TV system are the bandwidth and number lines per frame. The horizontal resolution is determined by a property called the bandwidth expressed in frequency (H_z) and describes the number of times the electron beam is changed (modulated). A value of 4.5 mHz is normal for most fluoroscopic systems while a high resolution system will have a value of 20mHz. The vertical resolution is determined by the number of scan lines in the system. See explanation number 9.

19. C Optical (mirror optics) monitoring directly off an image intensifier, though inexpensive, does not provide the increased contrast and brightness available with a TV chain system.

20. A Image coupling between an image intensifier and the television camera can be accomplished with a fiber optic coupler or tandem (beam splitting) mirror. The latter provides for remote recording by a cine spot film camera.

21. B See corresponding diagram.

22. E See corresponding diagram.

23. G See corresponding diagram.

24. A See corresponding diagram.

25. C See corresponding diagram.

26. A Since the distance between the scan lines is directly related to the vertical resolution, a larger TV monitor will display a greater scan line distance and a lower vertical resolution than a smaller TV monitor.

27. C The horizontal resolution of a video system is determined by the range of frequencies available in the video signal. The band pass (bandwidth) is the range of frequencies that the video system can transmit. A higher band pass will enable an improved horizontal resolution.

28. B The vertical resolution of a video system is dependent on the number of scan lines and the size of the monitor. A standard 10" monitor using a 525 line system will provide a vertical resolution of about 370 lines or 185 line pairs.

29. B In order to calculate the resolution (1p/mm) for an image intensifier will require the number of line pairs for a TV system to be divided by the size of the image display in millimeters In this example: $\frac{185 \text{ line pairs}}{152 \text{ mm}} = 1.21$ 1p/mm

30. D In order to provide the same vertical and horizontal resolution with a 1024 line system will require a bandwidth which is about 4 times larger than the 4 mHz required with a 525 line system.

Image Recording

Chapter 19

1. A When a cine spot film camera is to be used with an image intensified televised fluoroscopic chain, a tandem mirror is used as the coupling device between the image intensifier (II) and TV camera. The tandem mirror (beam splitter) consists of two optical lenses and special split mirror which enables the image from the image intensifier to go to both the TV and cine cameras simultaneously.

2. C Spot filming during fluoroscopy can be accomplished in a number of ways. The highest resolution can be obtained with the conventional or radiographic spot using the same film-screen combination employed in radiography. Cine (photo) spot films are photographic films taken off of the output phosphor of the image intensified and have a resolution limited by the relatively poor quality of the image intensifier tube. Video (Kine) spot images (video tape) recorded from the video signal have the poorest resolution due to the additional noise from the TV camera and magnetic tape recorder.

3. C In most of the first of the digital applications; i.e., digital subtraction angiography, the final image is first superimposed with its positive image counterpart taken before the introduction of a contrast agent. The net result is a final image in which most of the soft tissue and bony structures have been canceled out from the non-masked image. This and other digital techniques are discussed in chapter 28.

4. D Some of the new imaging techniques which do not employ ionizing radiation are listed with the source of energy used in the imaging process; ultrasonography (US); high frequency sound waves; magnetic resonance imaging (MRI); a strong magnetic field, and radio waves; thermography; infrared radiation (heat); diaphanography; high intensity visible light.

5. A Cine fluorography involves the photographic recording of images directly off of the image intensifier tube. The various forms of cine include the cine (photo) spot film or roll films (70-105 mm) and cine radiography (the recording of motion pictures during fluoroscopy). Compared to video (magnetic) tape images the advantage of cine imaging is their resolution. The disadvantage includes higher patient exposures and the need to process the photographic films prior to their display.

6. B The principal means for lowering the exposures in cine fluorography is by the use of a pulsed x-ray beam which is synchronized with the shutter opening to turn off the radiation while the film is being moved between frames.

7. D The roll type film used in cine fluorography is designed to possess a highest sensitivity to the yellow-green light emitted from the output screen of the image intensifier.

8. A Each frame in cine fluorography requires an exposure of about 100 microroentgens. As the framing rate increases there is the need to increase the number of x-ray pulses and patient dose rate.

9. C A special triode or grid controlled x-ray tube is used in cine fluorography to rapidly pulse the x-ray beam. In these tubes a separate power supply to a third electrode prevents the movement of the electrons toward the anode. By turning this voltage on and off, the electron stream and x-ray production can be controlled.

10. D In a televised fluoroscopic chain three types of spot image recording devices are available; the conventional radiographic spot image obtained from a cassette which is loaded into the fluorocarriage, the cine (photo) spot image obtained from the image intensifier and the video (magnetic) tape spots recorded from the video signal.

11. D During cine fluorography the greatest image quality is obtained for the film having the largest format. The disadvantage of the larger format film is the larger patient exposure required to obtain these images.

12. B See explanation number 10 of this section.

13. B At the present time video tape recorded on the VHS format are the most convenient and popular storage media for kinescopic images. A number of larger format tapes and hard disks have been used in radiology in the past.

14. A The video disc has been increasing in popularity since substantial increases in its storage capacity have been realized. The main advantage of the video disk is the rapid access it provides to all of the stored information.

15. D All electronic and photostimuable image systems introduce some distortion and noise onto the image. In terms of spatial resolution, systems with the most components; i.e., the televised fluoroscopic chain, will provide the poorest resolution.

16. D In computerized tomography, image quality is a function of size and number of the detectors, collimator design, number of scanning lines, signal-to-noise ratio of the TV chain and numerous aspects of the algorithms involved in image acquisition.

17. D A digital subtraction angiographic (DSA) system can be incorporated to many existing high resolution televised fluoroscopic chains through the addition of a digital computer and work station.

18. B The highest image resolution (100 Lp/mm) is found with a conventional radiographic film. Human vision is many times less acute than that of a radiographic film image.

19. B The resolution of all digital systems is dependent upon the size of the matrix and its individual picture elements (pixels). The larger the matrix and the smaller the pixels the higher the spatial resolution of the system.

20. A The main advantage of a computed (digital) system has over conventional film-screen imaging systems is a greatly improved contrast resolution.

Mobile Radiographic Units

Chapter 20

1. B During any surgical procedure the radiographic unit, whenever possible, is placed on the opposite side of the surgical field to avoid compromising the field's sterility.

2. B As a capacitor (condenser) discharges its voltage drops in a predictable manner. The termination of the exposure at some pre-selected spot on the curve is called wave tail cutting or cutoff.

3. C Because of their smaller size and smaller available power supplies all mobile radiographic units have power ratings well below those of a fixed installation.

4. A A residual charge may remain in a capacitor following a radiographic exposure. This must be eliminated prior to moving the unit to avoid current leakage and the inadvertent production of x rays without the activation of the exposure switch.

5. C In order to reduce the potential exposure to the operator during a mobile radiographic study the radiographer should wear a lead apron and stand as far from the patient and tube as possible. In order to accomplish this the exposure cord must be long enough to permit the operator to be at least 6" from the patient or the tube during the exposure.

6. C — The use of high speed (400-800) image system is advised for all mobile radiographic procedures to maintain low patient exposures and to provide acceptable images on large patients.

7. C — In many older hospitals the availability of standard 110 or 220 volt outlet receptacles is limited. Capacitor discharge and low output mobile units which will obtain their power supplies from these receptacles often encounter difficulties. The use of self contained battery powered mobile units developed in the 1980's have all but eliminated this problem.

8. C — The minimum source-to-image receptor distance allowable with any stationary or mobile radiographic or fluoroscopic unit shall not be less than 30cm (12") and should not be less than 38cm (15").

9. A — See explanation number 7 of this section.

10. D — The technical settings for most radiographic studies are similar to those used with fixed installations. With either unit the use of a high kVp value will increase the exposure latitude, ensure adequate penetration, and permit the use of shorter exposure times.

11. A — With suspected injuries to the spinal cord, movement should be limited as much as possible. The log rolling method is one of the few means for safely accomplishing this task.

12. D — A major drawback to the use of a capacitor discharge unit is the need to charge up the unit prior to making an exposure.

13. A — In order to reduce the amount of grid cutoff, a low ratio linear focus grid with a wide distance and centering latitude is preferred.

14. A — In addition to its use in cine fluorography a grid controlled (triode) x-ray tube may also be used to terminate the exposure in a capacitor discharge type mobile unit.

15. C — Some older mobile (capacitor discharge) units require a line voltage compensation control to maintain a consistent voltage to the radiographic unit.

16. B — Virtually all fluoroscopic units are designed with the tube and image receptor mounted on a C-shaped arm to ensure their alignment with each other.

17. D — Regardless of any difficulties a radiographer may encounter with a mobile examination of a patient, it is never acceptable to ignore the basic radiation safety practices which prevent unavoidable exposures.

18. D — Even though the voltage of a capacitor will drop off significantly after the first few milliseconds of the exposure, during the early part of the exposure the voltage wave form is most similar to that of three phase current.

19. B — The intensity of radiation produced in a scatting object is greatest behind the object, least to the sides of the object.

20. D — The rechargeable batteries most often used to power mobile radiographic units are made of nickel and cadmium compounds.

Chapter 21

1. D — Computers are made in three main sizes, the microcomputer (PC), the minicomputer used in many digital imaging applications and the main frame when extremely large data banks are required.

2. B — The physical components including housing, circuits and keyboard and image display devices are termed the hardware. Software refers to the programs, languages and data processing methods that make the computer useful.

3. C — The central processing unit (CPU) contains primary control center (microprocessor) which manipulates the data, performs mathematical operations, and stores the information for retrieval.

4. D — Most present day computers are only able to deal with a series of signals which translate information into on or off (flip-flop). A numerical system with only two digits (0 and 1) called the binary system is well suited for this type of operation.

5. A — Information can be introduced into the computer by way of an input device; i.e., keyboard, scanner. Floppy disks and hard drives can store or return information to the computer and are classified as input/output (I/O) devices. A printer is an example of a true output device.

6. C — The arithmetical unit of the central processing unit (CPU) contains the mathematical programs and logic circuits required for math operations.

7. A — Most modern computers have hundreds or thousands of tiny circuits which have been fused onto a small silicon block. These integrated circuits (IC) are normally called computer chips.

8. D — The use of high level computer languages will require their translation into a more usable form. This is accomplished by a compiler (interpreter).

9. B — The images related to the output are usually displayed on a cathode ray tube (CRT) or video monitor.

10. C — The alpha numeric display consists of the letters and numbers that appear on the television (video) monitor.

11. C — All of the new data which is input into the computer is either stored as temporary or main memory portions of the central processing unit (CPU).

12. A — Punch cards, scanners and keyboards are all input devices. Video monitors with touch screen or light pen capabilities, floppy and hard disks,and tape backup systems are both classified as input and output devices.

13. B — See explanation number 12 of this section.

14. C — See explanation number 12 of this section.

15. A — See explanation number 12 of this section.

16. A — See explanation number 12 of this section.

17. C See explanation number 12 of this section.

18. C The smallest quantity of computed information is carried by a single binary digit (bit). Since each alphanumeric character usually requires at least 8 bits or 1 byte of information, computer capabilities are usually based on the larger unit of a byte.

19. B A computer language provides the means for translating the more familiar alphanumeric system into the binary system which can be understood by the computer.

20. C See explanation number 18 of this section.

21. A The terms random access memory (RAM) and read only memory (ROM) are applied to the data which is temporarily stored in the computer or the more or less permanent memory.

22. A The principal control for the movement of data into other parts of a computer is the function of the control unit.

23. D The unique label which identifies data and enables its storage or movement without effecting all other data is called an address.

24. A See explanation number 2 of this section.

25. C If one Kilobyte is equivalent to 1024 x 8 bits and 1 character requires 1 byte then 1 kilobyte equals 8192 bits of information or 1024 characters.

26. B Random access memory is used to store data that must be periodically modified.

27. D In most computers a floppy disk, hard disk, hard drives or tape device can be used to store large amounts of data.

28. C The use of a modem enables the transmission of data from one computer to another over the phone lines. This device will soon enable transmission of virtually all types of diagnostic images to distinct viewing devices called teleradiology.

29. B Most of the digital images in the radiology department are transferred into a hard copy form by a multiformat camera or laser printer.

30. D Magnetic resonance imaging (MRI), computed tomography (CT), digital subtraction angiography (DSA), ultrasonography (US), computed radiography (CR), position emission tomography (PET), single photon emission computed tomography (SPECT), and diaphanography are among a rapidly growing list of imaging techniques that have emerged along with the modern computer.

Digital Imaging

Chapter 22

1. A A digital image is displayed as a series of tiny picture elements (pixels) which are arranged in vertical and horizontal columns (matrix) which are displayed on a cathode ray tube (video monitor).

2. C A digital imaging system compared to its analog counterpart can detect and expand the number of gray scales from about 30 shades to over 500 shades. This ability enables a significant improvement in the perception of contrast (contrast resolution) when digital image techniques are employed.

3. B The main advantage of an area beam; i.e., normal diagnostic imaging, is in its ability to provide a large area with an image forming dose of radiation in a short period of time. Fan beams like those used with most computed tomographic units reduce nearly all scattered radiation and therefore provide higher contrast images.

4. D The low signal strength associated with most types of digital systems requires a video chain with an exceptional signal to noise ratio of at least 1000:1.

5. C The analog signal from the television camera must first be converted into a digital signal by a device called an analog-to-digital converter (ADC). A digital-to-analog converter (DAC) is required to return the digital image back into an analog form the video monitor understands.

6. D The highest resolution on a television monitor is provided by a combination of the largest image matrix, the smallest TV monitor, and smallest pixel size.

7. B The term gray scale is used in a similar fashion as density and contrast in a film imaging system. Analog systems have only a narrow range of shades available. Digital image enables the use of greatly expanded numbers of gray scales as well as the ability to eliminate unwanted shades through post processing applications.

8. C The gray scale of a digital image can be modified by the adjustment of the window level and window width controls. A narrow window width tends to increase the relative difference in the gray scales and create images possessing higher contrast. The window level is used to bring different tissues into the visible range based on their attenuation properties.

9. A See explanation number 1 of this section.

10. C One of the first applications of digital imaging was in the enhancement of contrast filled vascular structures through the use of a digital subtraction angiographic (DSA) procedure.

11. C A higher mA value used in digital fluoroscopy or digital subtraction angiography provides less image noise and improves the visualization of small low contrast objects.

12. A The main components of a digital (computed) radiographic system are the x-ray tube, image detector or photostimulable phosphor and the associated TV image chain used in the display of the image.

13. D The most important component of a digital imaging system is the central control computer and its numerous functions related to the manipulation of the image quality.

14. C In digital subtraction angiography it is critical that the image (subtraction) mask and the post injection image be perfectly aligned. Failure to obtain this superimposition due to motion or movement will require the re-registration of images. In some advanced systems computer enhancement techniques may be available for improving the quality of images in which motion would normally be a problem.

15. A Temporal (energy) subtraction techniques involves the production and subtraction of two images taken at different energy (kVp) levels. The image results almost entirely from differences in the photoelectric effect of the K-shell electrons found in the bone tissue.

16. D The collimation of a fan beam at points near its origin and just prior to its interaction with the imaging detector enables the elimination of nearly all scattered radiation from the beam.

17. B The first application of digital radiography was termed scanned projection radiography. This required the use of a fan beam that moved with a single detector array from the head to toe direction during the exposure. This technique was slow and failed to live up to its early promise. New digital applications (computed radiography) using area beams and photostimulable phosphor plates have overcome many of the disadvantages of the early digital techniques.

18. A Noise in a digital imaging system is a result of two main factors: the electronic noise produced by the heated cathodes of the electrical TV cameras and monitors and the statistical fluctuations of the x-ray beam (quantum mottle) which becomes more noticeable at low exposure levels used in digital systems.

19. A The PACS, which is an acronym for picture archiving and communication system, is designed to file and store all types of digital images for later retrieval. PACS are sometimes referred to as electronic file rooms.

20. B The digital subtraction process is a three step process which starts with a mask image which is then digitally subtracted or turned into a positive image. When the positive mask is superimposed over the contrast containing image, The subtracted image will enhance the appearance of the contrast in the vessels through the suppression of overlying tissues. The subtracted image will enhance the appearance of the contrast in the vessels through the suppression of overlying tissues.

21. D PACS are ideally suited for the storage of images which are generated in digital form such as CT, CR, MRI, DSA, SPECT, PET and US.

22. C When hard copies of digital images are required, these can be obtained photographically by way of a multiformat camera or the use of a laser printer.

23. C One of the most practical applications of digital images is in their ability to be transmitted electronically in a process known as teleradiography.

24. C The standard hard copy analog radiographic image is not compatible with a digital storage system (PACS). In order to utilize the PAC system the radiograph must first be converted by a scanner or digitalizer into a digital form which can be understood by the computer.

25. B See explanation number 15 of this section.

26. B See explanation number 8 of this section.

27. D Though in principle a conventional televised fluoroscopic system could be converted over into a digital system, in practice the need for lower noise levels and improve display quality requires the replacement of both the image intensifier and the entire TV chain.

28. B Most electronic imaging devices; i.e., TV cameras and image intensifiers, are associated with high levels of electronic noise. The use of high potential differences and heated filaments in these devices will result in small but detectable secondary electrical interference which is seen as noise.

29. D See explanation number 20 of this section.

30. C The high resolution required in the television chain of a digital system is in part accomplished by the use of a progressive scanning mode rather than an interlaced mode.

31. D A large number of techniques are used in digital imaging to improve the image quality. They are sometimes referred to as post processing techniques because they involve manipulation of the image after it has been obtained. Among these are image smoothing which helps to eliminate image noise, edge enhancement for improving the appearance of small details and mask re-registration for reducing the motion or non-alignment between subtraction and post injection images during DSA.

32. A Temporal or energy subtraction is best suited for distinguishing between structures filled with iodinated contrast agents and soft tissue structures.

33. D See explanation number 26 of this section.

34. B See explanation number 14 of this section.

35. C A digital system may use a xenon gas-filled detection array or a solid state scintillation detector array consisting of cadmium tungstate, bismuth germanate, sodium iodide or cesium iodide.

Section IV
Image Production and Evaluation

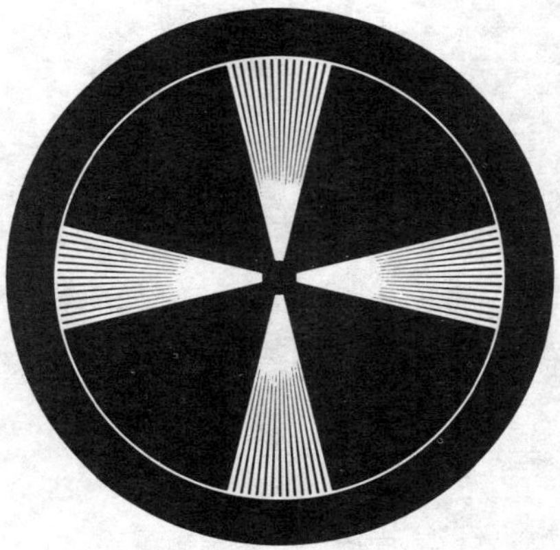

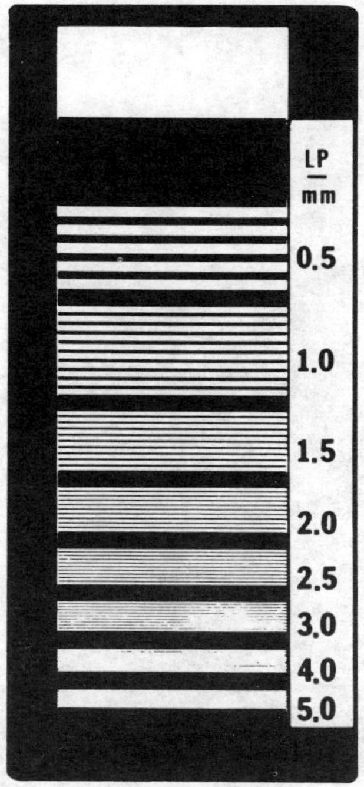

Radiographic Graffiti I

Can you identify these common patient related artifacts? (Answers appear at the bottom of the page.)

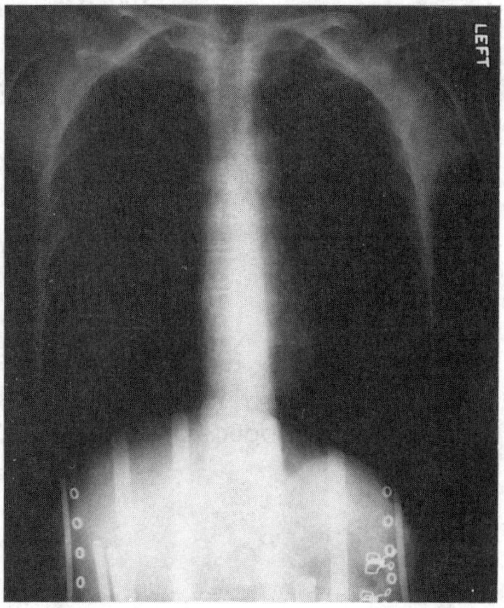

A

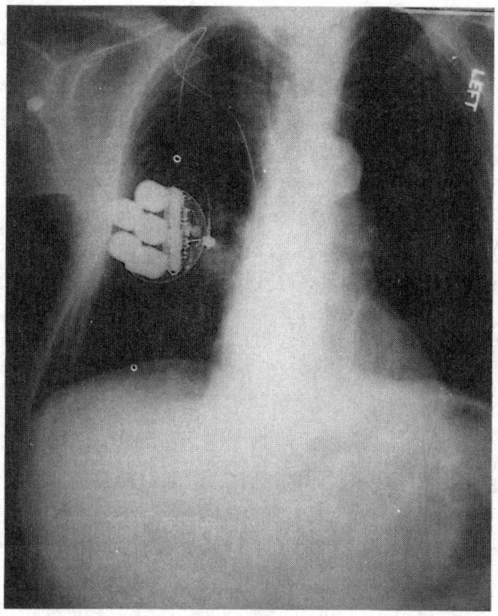

B

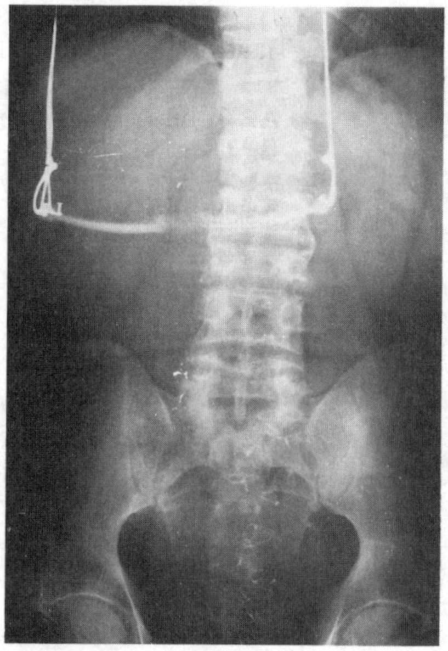

C

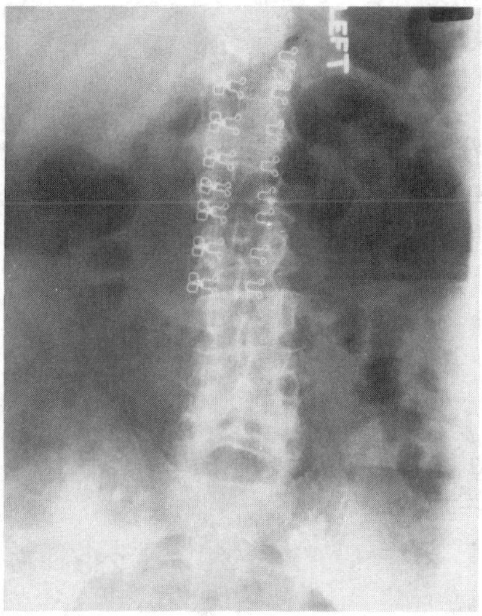

D

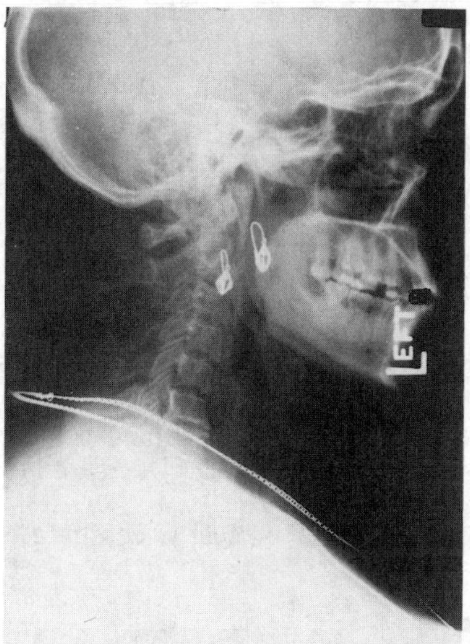

E

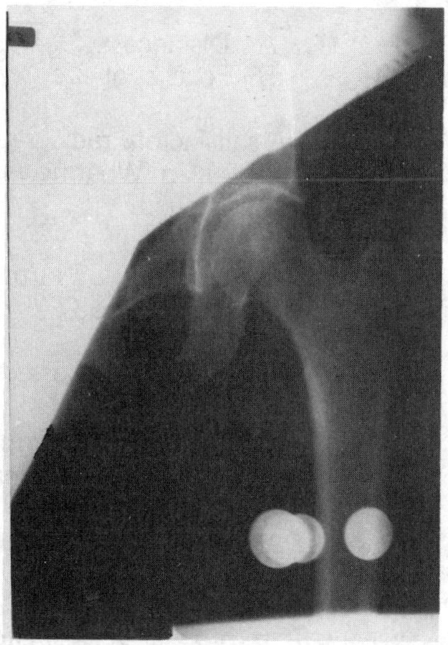

F

Answers

Film A. Girdle with plastic stayes and associated fasteners; **Film B.** Cardiac pacemaker and electrical leads; **Film C.** Patients glasses on abdomen - other metallic sutures; **Film D.** Bra fasteners; **Film E.** Patient earrings and necklace; **Film F.** Coins in the pocket of a patient shielded with a lead apron.

Radiographic Quality

Chapter 23

1. The amount of darkening from the black metallic silver deposited in the film is termed: 2:269

A. Contrast
B. Density
C. Attenuation
D. Base fog

2. Besides the variables of kV, time and mA, when preparing a technique chart, what other factors must be standardized? 15:216
1. Screen speed 2. Grid usage 3. Degree of collimation

A. 1 & 2 only
B. 1 & 3 only
C. 2 & 3 only
D. 1, 2 & 3

3. A technique of 400 mA 100 ms and 75 kVp is employed to produce an image with the desired radiographic density using the principles stated in the 15% kVp rule. What mA setting would produce a similar radiographic density at 50 ms sec. and 86 kVp? 15:153

A. 800 mA
B. 400 mA
C. 200 mA
D. 100 mA

4. The primary function of mAs during a radiographic exposure is to: 2:269

A. Improve resolution
B. Increase penetration
C. Provide a higher contrast
D. Regulate density

5. The primary exposure factor that is used for regulating radiographic contrast is: 2:290

A. Distance
B. Focal spot size
C. kVp
D. mAs

6. A satisfactory radiograph is made at a 160 cm (64") source-to-image receptor distance (SID) using 40 mAs at 87 kVp. What new factors will produce a similar density if a 80 cm (32") source-to-image distance (SID) is employed? 2:297

A. 5 mAs at 87 kVp
B. 10 mAs at 87 kVp
C. 20 mAs at 87 kVp
D. 40 mAs at 87 kVp

7. A film is said to possess excessive radiographic density. This would appear: 2:269

A. Lighter than normal
B. Darker than normal
C. To have excessive contrast
D. To have poor recorded detail

8. A reduction of scattered radiation on a radiographic film is likely to occur with an: 2:238

A. Increased kVp
B. Increased collimation
C. Increased part thickness
D. Decreased focal spot size

9. In general, an increase of 15% in kilovoltage will necessitate a ________ decrease in the mAs to maintain the same film density. 15:153

A. 15%
B. 30%
C. 50%
D. 90%

10. An example of a pathological change in a patient which requires an increase in penetration are: 15:241
1. Pagets disease 2. Acromegaly 3. Aseptic necrosis

A. 1 & 2 only
B. 1 & 3 only
C. 2 & 3 only
D. 1, 2 & 3

Pertaining to the diagram of the stepwedge film, answer questions 11 & 12.

11. In the diagram of the stepwedge film which Roman numeral corresponds to the film taken at the highest kVp 2:290

A. I
B. II
C. III
D. None of the above

12. Which of the stepwedge films best demonstrates a short scale contrast? 2:290

A. I
B. II
C. III
D. None of the above

13. An abdominal radiograph is taken of a patient measuring 21 cm. If all the same conditions are maintained and a second image is taken on a different patient measuring 29 cm, the image will most likely provide a lower: 15:111
1. Radiographic density 2. Radiographic contrast 3. Spatial resolution

A. 1 only
B. 1 & 2 only
C. 2 & 3 only
D. 1, 2 & 3

14. The dose a patient receives during a diagnostic exposure is directly proportional to the: 2:653
1. mAs employed 2. kVp employed 3. Focal spot size selected

A. 1 only
B. 2 only
C. 3 only
D. 1, 2 & 3

15. A decrease in scatter production will occur as: 2:234

A. OID decreased
B. SID decreases
C. kVp decreases
D. FSS decreases

16. Which of the primary exposure factors has the greatest effect on beam quality, attenuation, and exposure latitude? 2:165

A. mA
B. Time
C. mAs
D. kVp

17. A radiographic image is produced with insufficient contrast and excessive radiographic density. Which of the following changes in the technical factors will best correct for this error? 2:309

A. Increase the mAs, decrease the kVp
B. Increase the mAs, increase the kVp
C. Decrease the mAs and increase the kVp
D. Decrease the mAs and kVp

18. An increase in technique to maintain the radiographic density is often required for patients with: 15:241

A. Emphysema
B. Pulmonary edema
C. Osteoporosis
D. Leprosy

19. An increase in the number of x-ray photons reaching the radiographic film emulsion will result in a (an): 2:289

A. Higher image contrast
B. Increase in image density
C. Improved spectral match
D. Reduced exposure latitude

20. The normal human tissue with the lowest absorption coefficient to an x-ray photon in the 50-120 kVp range is found in: 5:75

A. Bone tissue
B. Connective tissue
C. Free air or gas
D. Muscle tissue

21. As the wavelengths of the photons in an x-ray beam are increased the: 2:65

A. Scatter produced by the beam increases
B. Beam penetration increases
C. The energy of the beam increases
D. None of the above

22. Radiographic contrast can be defined as the: 2:269

A. Difference in blackening between adjacent areas of the radiograph
B. Overall blackening of the radiograph
C. Base plus fog blackening of the radiograph
D. Ratio of film speed to screen speed

23. Which pair of tissue interfaces will possess the highest differential attenuation to diagnostic x-rays? 5:75

A. Muscle to bone
B. Water to muscle
C. Soft tissue to bone
D. Soft tissue to muscle

24. Many skeletal disorders will require an increase in technique due to an overgrowth of bone tissue. These are referred to as: 15:241

A. Osteoblastic conditions
B. Osteolytic conditions
C. Osteoporotic conditions
D. Osteonecrotic conditions

25. An example of a pathologic change in a patient which requires an increase in penetration is: 15:241
1. Pleural effusion *2. Empyema* *3. Ascites*

A. 1 & 2 only
B. 1 & 3 only
C. 2 & 3 only
D. 1, 2 & 3

26. The percentage of scatter reaching the radiographic film during a non-grid exposure of an abdomen comprises about ______________ of the image. 12:364

A. 10%
B. 20%
C. 50%
D. 90%

27. A technique of 10 mAs at 78 kVp is used at a 140 cm (56") source-to-image receptor distance (SID). What new source-to-image receptor distance (SID) should be used to decrease this exposure to 5 mAs at 78 kVp? 2:300

A. 70 cm (28")
B. 99 cm (40")
C. 198 cm (80")
D. 280 cm (112")

28. Most gasses such as nitrogen, oxygen and hydrogen which have a low attenuation for x-rays and are classified as: 2:181

A. Radiogeneous
B. Radiolucent
C. Radiopaque
D. Radioactive

29. The principal method of controlling voluntary motion is by the use of: 1:5

A. Sandbags
B. Proper instructions
C. Restraining devices
D. Long exposure times

30. Which of the following conditions is the most difficult to penetrate radiographically? 15:241

A. Empyema
B. Gout
C. Active osteomyelitis
D. Pneumothorax

31. A radiograph is produced of a high tissue density structure. The area of the film corresponding to this structure will show a: 2:289

A. High radiographic density
B. Low radiographic density
C. Tissue density in not related to radiographic density

32. A radiograph that shows a short scale of contract often results from the use of: 12:340

A. Breathing techniques
B. Tilted anode techniques
C. Low kilovoltage techniques
D. Falling load techniques

33. Which of the following sets of technical factors will result in a film with the least amount of involuntary motion? 1:5

	mA	Time	kVp
A.	200	100 ms	80
B.	400	.05 sec.	80
C.	800	.03 sec.	80
D.	800	75 ms	70

34. The utilization of high kVp (above 90 kVp) for radiographs of the abdomen will result in: 2:290

A. High exposure latitude and short scale contrast
B. High exposure latitude and long scale contrast
C. Low exposure latitude and short scale contrast
D. Low exposure latitude and long scale contrast

35. According to the law of reciprocity, which set of technical factors should give the same density as one using 200 mA, 40 ms at 80 kVp? 2:272

	mA	Time	kVp
A.	200	40 ms	80
B.	400	20 ms	80
C.	800	10 ms	80
D.	All of the above		

36. The appropriate degree of penetration for a small body part (finger-hand) is normally accomplished by using a kilovoltage setting in the range of:

A. 55-65 kVp
B. 85-95 kVp
C. 95-105 kVp
D. 115-125 kVp

37. Which of the following is likely to occur if kilovoltage is increased for a radiographic exposure? 12:340
1. Increased density *2. Lower penetration* *3. Longer scale contrast*

A. 1 & 2 only
B. 1 & 3 only
C. 2 & 3 only
D. 1, 2 & 3

38. A beam with a low penetrability is often referred to as a soft beam or: 2:165

A. High intensity beam
B. High quality beam
C. Low quality beam
D. High resolution beam

39. Which of the following substances possesses the highest degree of radiopacity? 12:340

A. Calcium
B. Aluminum
C. Lead
D. Oxygen

40. A radiograph that shows a small differential attenuation is best described as having a: 12:340

A. Low contrast
B. High density
C. High resolution
D. Poor definition

41. In order to see a noticeable change in radiographic density, a change of __________ in the mAs is required. 2:68

A. 5%
B. 10%
C. 15%
D. 30%

42. A child has been radiographed using 100 mA 20 ms at 74 kVp, but due to movement, the radiographic image is unacceptable. In order to reduce motion, a 60% reduction in time is desired. What new timer setting should be employed?

A. 8 ms
B. 12 ms
C. 60 ms
D. 120 ms

43. Which of the following sets of technical factors will result in a film with the greatest amount of involuntary motion? 1:5

	mA	**Time**	**kVp**
A.	200	.02 sec.	86
B.	300	10 ms	84
C.	400	.13 sec.	79
D.	800	50 ms	76

44. An increase in technique is usually required to maintain normal radiographic density for patients with: 15:241

A. Giant cell sarcoma
B. Leprosy
C. Osteoporosis
D. Acute kyphosis

45. When a substance is exposed to diagnostic x-rays, the amount of attenuation is primarily dependent upon the __________ of the attenuator. 5:75

A. Atomic number
B. Electron density
C. Viscosity
D. Proton-neutron ratio

46. Radiographic density corresponds to the amount of light from the view box that is __________ the radiographic film. 2:272

A. Transmitted through
B. Scattered by
C. Reflected by
D. Absorbed by

47. A satisfactory radiograph is made using 15 mAs at 75 kVp at a 140 cm source-to-image distance (SID). Assuming no other technical changes are made, what new mAs value will produce the same radiographic density at a 180 cm SID? 2:300

A. 5 mAs
B. 9 mAs
C. 19 mAs
D. 25 mAs

48. The use of 15% kilovoltage rule in the maintenance of radiographic density as mAs is halved offers the advantage of: 12:352
1. Smaller patient dose *2. Greater exposure latitude* *3. Shorter exposure times*

A. 1 only
B. 2 only
C. 3 only
D. 1, 2 & 3

49. A radiograph is produced of a homogeneous object: the resultant image will show a: 2:273

A. Long scale contrast
B. Short scale contrast
C. It will show no contrast

50. An acceptable radiograph is made at 10 mAs and 75 kVp on full wave radiographic unit. In order to maintain the same radiographic density on a 3-phase unit would require a technique of about: 2:137

A. 5 mAs, 68 kVp
B. 5 mAs, 75 kVp
C. 20 mAs, 75 kVp
D. 20 mAs, 85 kVp

51. Which set of technical factors would best be used on a small child's abdomen in order to limit motion? 1:5

	mA	Time	kVp
A.	100	.2 sec.	70
B.	200	100 ms	70
C.	400	.05 sec.	70
D.	800	20 ms	70

52. When developing a variable-kilovoltage technique chart, an increase of _______________ should be used for every centimeter of thickness difference. 15:223

A. 2 kVp
B. 15% in kVp
C. 39% of mAs
D. 50% of mAs

53. Any material that greatly reduces the number of incident x-ray photons passing through the material is classified as: 2:66

A. Isotropic
B. Radiopaque
C. Radiotransparent
D. Radiolucent

54. Which of the following pathologic conditions will normally require a decrease in technique? 15:241

A. Empyema
B. Emphysema
C. Pulmonary edema
D. Silicosis

55. An acceptable anterioposterior abdominal radiograph is taken using 50 mAs at 80 kVp. If the patient is placed in the oblique position, with no change in technique, the radiograph will possess:

A. Excessive radiographic density
B. Insufficient radiographic density
C. Excessive radiographic contrast
D. Insufficient radiographic contrast

56. A satisfactory radiograph is made using 400 mA, 100 ms, 83 kVp at an 100 cm SID. Assuming no other factors change, which of the following techniques will produce a radiograph showing a similar density? 12:352

A. 200 mA, 20 ms, 96 kVp
B. 400 mA, 20 ms, 83 kVp
C. 800 mA, 50 ms, 83 kVp
D. 800 mA, 50 ms, 96 kVp

57. A radiograph of a patient with ascites (fluid in the abdominal cavity) will often appear to have an insufficient radiographic density. This is most likely caused by the ______________ of the abdominal fluid 14:352
1. Excessive penetration *2. Excessive scatter* *3. Excessive absorption*

A. 1 only
B. 2 only
C. 3 only
D. 1, 2 & 3

58. The principal controlling factor for the number of photons produced during an exposure is: 2:290
1. Timer setting *2. Milliampere setting* *3. Kilovoltage*

A. 1 & 2 only
B. 1 & 3 only
C. 2 & 3 only
D. 1, 2 & 3

59. Which of the body tissues has the highest absorption coefficient when exposed to diagnostic x-rays? 2:183

A. Bone
B. Soft tissue
C. Muscle
D. Gas

60. A step wedge (penetrometer) film showing a small number of widely varying shades of density is desired. Choose the set of factors that will best demonstrate this appearance. 5:164

	mA	Time	kVp	FSS
A.	100	500 ms	80	1 mm
B.	100	250 ms	90	1 mm
C.	300	500 ms	60	2 mm
D.	300	250 ms	70	2 mm

61. An increase in the radiographic density can result from an increase in: 2:269
1. Source-to-image receptor distance *2. Kilovoltage* *3. Milliampere setting*

A. 1 & 2 only
B. 1 & 3 only
C. 2 & 3 only
D. 1, 2 & 3

62. A radiograph of an obstructed patient is performed showing an excessive radiographic density. The probable cause is due to ______________ of the intestinal gas. 12:340

A. Excessive scatter
B. Excessive absorption
C. Excessive density
D. Excessive penetration

63. An exposure of 40 mAs can be accomplished by using which of the following set of factors? 2:300

A. 200 mA, 100 ms
B. 10 mA, 250 ms
C. 400 mA, 100 ms
D. 400 mA, 50 ms

64. Most of the scattered radiation reaching the film during an x-ray exposure is produced at an energy: 12:373

A. Above that of the primary beam
B. Nearly the same as the primary beam
C. Substantially lower than the primary beam

65. If you halved the time of exposure, how much would you change the milliampere to obtain the same density on the radiograph? 2:300

A. Double the milliampere
B. Add 10 milliamperes
C. Use 4 times the milliampere
D. Use half the milliampere

66. When an air gap technique is employed, the amount of scatter reaching the film is decreased because: 2:263

A. Of the filtering effects of the air
B. More scatter misses the film
C. Less scatter is produced at a longer SID
D. Scatter consists only of primary radiation

67. The x-ray absorption characteristics of tissues are often variable. Which of the following may effect the degree of radiation attenuation? 15:241
1. Patient's age *2. Muscle development* *3. Pathological conditions*

A. 1 & 2 only
B. 1 & 3 only
C. 2 & 3 only
D. 1, 2 & 3

68. A larger percentage of scattered radiation is likely to occur if: 2:234

A. Thickness of a tissue decreases
B. Density of a tissue decreases
C. Kilovoltage employed decreases
D. None of the above

69. A technique of 160 mAs and 60 kVp has been employed in producing a satisfactory radiograph. In order to reduce this to 80 mAs, a new kVp of ______________ should be employed 15:153

A. 51 kVp
B. 69 kVp
C. 74 kVp
D. 86 kVp

70. An exposure of 500 mA, 50 ms at 84 kVp is made at an 100 cm SID. What new time setting can be used to maintain the film density at a 200 cm SID? 2:300

A. 25 ms
B. 100 ms
C. 200 ms
D. 400 ms

71. Which of the following pathologic conditions will generally require an increase in penetration? 15:241

A. Osteoporosis
B. Emphysema
C. Paget's disease
D. Pneumothorax

72. The principal controlling factor for scale of contrast is: 2:290

A. Kilovoltage
B. Milliampere
C. Focal spot size
D. Distance

73. A satisfactory radiograph is made using 15 mAs at 78 kVp at an 100 cm SID. Assuming no other technical changes are made, what new mAs value will produce the same radiographic density at an 180 cm SID? 2:300

A. 5 mAs
B. 8 mAs
C. 27 mAs
D. 49 mAs

74. A radiograph possessing a small number of widely varying density values is said to possess: 2:290

A. Low subject contrast
B. Low inherent contrast
C. Short scale contrast
D. Long scale contrast

75. The percentage of scattered radiation produced during a radiograph is dependent upon: 5:66
1. Kilovoltage employed *2. Part thickness* *3. Tissue density*

A. 1 & 2 only
B. 1 & 3 only
C. 2 & 3 only
D. 1, 2 & 3

76. A radiograph taken at a 200 cm source-to-image distance (SID) requires 32 mAs and 92 kVp. What new source-to-image distance (SID) can be employed to reduce the exposure to 8 mAs at 92 kVp? 2:300

A. 175 cm
B. 150 cm
C. 100 cm
D. 80 cm

77. A film-screen system is used to make the following exposures: 2:264

	mA	Time	kVp	Distance
1.	600	100 ms	70	100 cm (40')
2.	300	200 ms	70	100 cm (40")
3.	100	600 ms	70	100 cm (40")

Though the same density should be expected, film #3, shows a higher density. This is an example of the failure of the:

A. Inverse square law
B. Reciprocity law
C. Conversion efficiency law
D. K-shell law

78. To insure proper quality of a radiograph after the introduction of a positive contrast media (iodine), is necessary. 1:670

A. A decrease in kilovoltage
B. A decreased exposure time
C. An increase in collimation
D. An increase in milliamperage

79. A radiograph of the chest is normally performed without using a grid. The reason that this is possible is due to the _______________ of the structures within the chest. 4:162

A. High tissue density
B. Natural air gap
C. High subject contrast
D. Low subject contrast

80. The following condition(s) must be satisfied in order to use a technique chart for radiographic exposure: 14:237
1. X-ray machines must be calibrated *2. All patients are measured*
3. Processing must be standardized

A. 1 only
B. 2 only
C. 3 only
D. 1, 2 & 3

81. An exposure is taken at 30 mAs and 78 kVp using a small focal spot. The radiograph has high contrast and recorded detail but lacks sufficient density. Which of the exposure factors below should be employed to improve the radiographic density? 2:296

A. 15 mAs, 90 kVp
B. 60 mAs, 66 kVp
C. 42 mAs, 78 kVp
D. Any of the above

82. During a manual radiographic exposure, if the kilovoltage is increased from 80 kVp to 90 kVp, with no other changes made, an increase in the _______________ will occur. 2:290
1. Radiographic contrast *2. Exposure latitude* *3. Scatter produced*

A. 1 & 2 only
B. 1 & 3 only
C. 2 & 3 only
D. 1, 2 & 3

83. A film is taken that shows a long scale of contrast due to excessive penetration. Which technique change would best correct the error? 2:290

A. Raise the kVp
B. Lower the kVp
C. Raise the mAs
D. Lower the mAs

84. An x-ray generator has a maximum capacity of 800 mA. What is the shortest possible exposure time for a technique requiring 40 mAs at 78 kVp? 2:130

A. 10 milliseconds
B. 20 milliseconds
C. 50 milliseconds
D. 100 milliseconds

85. A chest radiograph is taken at 180 cm (72") source-to-image distance (SID) using a technique of 200 mA 20 ms 78 kVp. What new distance factor could be utilized to reduce the exposure to 2 mAs at 78 kVp? 2:300

A. 90 cm (36")
B. 127 cm (51")
C. 267 cm (107")
D. 360 cm (144")

86. Calibers are used to obtain measurements of: 15:237

A. Patient thickness
B. Film density
C. Screen thickness
D. Beam quality

87. A satisfactory lumbar spine radiograph is performed using 60 mAs at 85 kV. If a cone-down view of L3 is done using the same technique, the film density will ________ and contrast will be ________. 2:237

A. Increase/lower
B. Decrease/higher
C. Increase/higher
D. Decrease/lower

88. In order to maintain the same radiographic density, if the milliampere setting employed for a given exposure is halved, would require a doubling of the: 10:353
1. Exposure time *2. Kilovoltage* *3. Added filtration*

A. 1 only
B. 2 only
C. 3 only
D. 1, 2 & 3

Pertaining to the diagram, answer question 89.

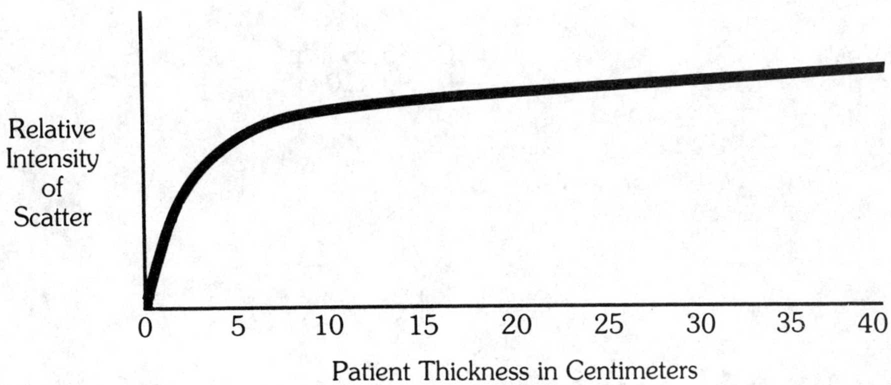

89. According to the graph, the intensity of the scattered radiation produced is ________ proportional to the ________. 2:237

A. Directly, radiographic density
B. Inversely, patient thickness
C. Inversely, radiation quality
D. Direct, patient thickness

90. Which of the following devices could be used to reduce the intensity of scattered radiation produced in a patient measuring 38 centimeters? 2:287
1. Slower speed screen *2. Compression band* *3. Smaller focal spot*

A. 1 only
B. 2 only
C. 3 only
D. 1, 2 & 3

Pertaining to the following statement answer questions 91 thru 96.

A technique of 300 mA, 100 ms and 75 kVp is employed to produce an image with the desired radiographic density. Using the principle stated in the 15% kVp rule, fill in the values below that will maintain the same radiographic density. 15:153

	mA	Time	kVp
91.	300	____	86
92.	600	____	64
93.	300	200 ms	___
94.	100	150 ms	___
95.	____	25 ms	86
96.	____	600 ms	64

According to the principles of the 15% kVp rule, all three exposures will produce a similar radiographic density.

	mA	Time	kVp
A.	300	50 ms	92
B.	100	300 ms	80
C.	600	100 ms	68

97. Which technique is associated with the lowest radiographic contrast?

98. Which technique is associated with the highest patient exposure?

99. Which technique should be employed to reduce the effects of involuntary motion?

100. When a barium type contrast agent is employed the desired degree of penetration can be accomplished by using a kilovoltage setting in the range of:

A. 50-60 kVp
B. 60-80 kVp
C. 75-85 kVp
D. 90-110 kVp

Time - Distance

Exercise No. 1

1. A radiograph is taken using 300 mA, 20 ms at 90 kVp at a 200 cm SID. What new time setting can be used to maintain the same radiographic density at a 100 cm SID, assuming all other factors remain unchanged?

2.-10. State the new timer setting that can be used to maintain the proper radiographic density given the original timer setting original SID and new SID, assuming all other factors remain unchanged.

	Original Timer Setting	Original SID	New SID	New Timer Setting
2.	40 ms	100 cm	50 cm	________
3.	160 ms	200 cm	100 cm	________
4.	150 ms	180 cm	100 cm	________
5.	250 ms	150 cm	100 cm	________
6.	300 ms	100 cm	180 cm	________
7.	50 ms	40 inches	30 inches	________
8.	1 sec.	200 cm	150 cm	________
9.	1.0 sec.	300 cm	50 cm	________

10. A radiograph is taken using 400 mA, 50 ms at 90 kVp at a 200 cm SID. What new mA setting can be used to maintain the same radiographic density at a 100 cm SID, assuming all other factors remain unchanged?

11.-16. State the new mA setting that can be used to maintain the proper radiographic density given the original mA, original SID and new SID, assuming all other factors remain unchanged.

	Original mA Setting	Original SID	New SID	New mA Setting
11.	400 mA	40 inches	20 inches	________
12.	100 mA	100 cm	300 cm	________
13.	800 mA	200 cm	100 cm	________
14.	200 mA	36 inches	72 inches	________
15.	600 mA	180 cm	90 cm	________
16.	900 mA	180 cm	60 cm	________

17. A radiograph is taken using 600 mA, 150 ms at 60 kVp at a 180 cm SID. What new mAs can be used to maintain the same radiographic density at a 90 cm SID, assuming all other factors remain the same?

18.-25. State the new mAs that can be used to maintain the proper radiographic density given the original mA, time and SID and the new SID, assuming all other factors remain unchanged.

	Original mAs	Original SID	New SID	New mAs
18.	5 mAs	100 cm	20 cm	________
19.	8 mAs	200 cm	50 cm	________
20.	12 mAs	180 cm	60 cm	________
21.	20 mAs	30 inches	40 inches	________
22.	25 mAs	72 inches	36 inches	________
23.	16 mAs	40 cm	60 cm	________
24.	30 mAs	100 cm	300 cm	________
25.	100 mAs	140 cm	180 cm	________

The following formulas have been used in solving these problems.

$$\text{New Timer Setting} = \text{Original Timer Setting} \times \frac{\text{New SID}^2}{\text{Original SID}^2}$$

$$\text{New mAs} = \text{Original mAs} \times \frac{\text{New SID}^2}{\text{Original SID}^2}$$

$$\text{New mA Setting} = \text{Original mA Setting} \times \frac{\text{New SID}^2}{\text{Original SID}^2}$$

Exercise No. 2

1. A radiograph is taken using 300 mA, 240 ms at 90 kVp at a 200 cm SID. What new time setting can be used to maintain the same radiographic density at a 100 cm SID, assuming all other factors remain unchanged?

2.-9. State the new timer setting that can be used to maintain the proper radiographic density given the original timer setting, original SID and new SID, assuming all other factors remain unchanged.

	Original Timer Setting	Original SID	New SID	New Timer Setting
2.	120 ms	120 cm	60 cm	________
3.	320 ms	400 cm	200 cm	________
4.	250 ms	90 cm	50 cm	________
5.	200 ms	150 cm	300 cm	________
6.	100 ms	100 cm	200 cm	________
7.	600 ms	40 inches	60 inches	________
8.	1.5 sec.	300 cm	150 cm	________
9.	1.0 sec.	300 cm	60 cm	________

10. A radiograph is taken using 400 mA, 50 ms at 90 kVp at a 200 cm SID. What new mA setting can be used to maintain the same radiographic density at a 100 cm SID, assuming all other factors remain unchanged?

11.-16. State the new mA setting that can be used to maintain the proper radiographic density given the original mA, original SID and new SID, assuming all other factors remain unchanged.

	Original mA Setting	Original SID	New SID	New mA Setting
11.	200 mA	20 inches	40 inches	________
12.	900 mA	300 cm	100 cm	________
13.	800 mA	400 cm	200 cm	________
14.	100 mA	36 inches	72 inches	________
15.	600 mA	60 cm	120 cm	________
16.	900 mA	150 cm	50 cm	________

17. A radiograph is taken using 800 mA, 250 ms at 60 kVp at a 180 cm SID. What new mAs can be used to maintain the same radiographic density at a 90 cm SID, assuming all other factors remain unchanged?

18.-25. State the new mAs that can be used to maintain the proper radiographic density given the original mA, time and SID and the new SID, assuming all other factors remain unchanged.

	Original mAs	Original SID	New SID	New mAs
18.	15 mAs	100 cm	200 cm	________
19.	80 mAs	400 cm	100 cm	________
20.	36 mAs	180 cm	60 cm	________
21.	120 mAs	40 inches	60 inches	________
22.	24 mAs	72 inches	36 inches	________
23.	8 mAs	30 cm	40 cm	________
24.	80 mAs	150 cm	250 cm	________
25.	10 mAs	140 cm	180 cm	________

The following formulas have been used in solving these problems.

$$\text{New Timer Setting} = \text{Original Timer Setting} \times \frac{\text{New SID}^2}{\text{Original SID}^2}$$

$$\text{New mAs} = \text{Original mAs} \times \frac{\text{New SID}^2}{\text{Original SID}^2}$$

$$\text{New mA Setting} = \text{Original mA Setting} \times \frac{\text{New SID}^2}{\text{Original SID}^2}$$

Advanced Time - Distance Exercise

Exercise No. 1

1. A radiograph is taken using 400 mA, 10 ms at 90 kVp at a 200 cm SID. What new timer setting of .04 sec, is desired. What new SID can be used to maintain the same radiographic density, assuming all other factors remain unchanged?

2.-9. State the new SID that can be used to maintain the proper radiographic density given the original timer setting and SID and new timer setting, assuming all other factors remain unchanged.

	Original Timer Setting	New Timer Setting	Original SID	New SID
2.	20 ms	80 ms	100 cm	____________
3.	50 ms	200 ms	200 cm	____________
4.	160 ms	40 ms	180 cm	____________
5.	300 ms	600 ms	40 inches	____________
6.	900 ms	100 ms	300 cm	____________
7.	1.5 sec	500 ms	30 inches	____________
8.	330 ms	990 ms	100 cm	____________
9.	400 ms	600 ms	200 cm	____________

10. A radiograph is taken using 100 mA, 350 ms at 80 kVp at a 60cm SID. A new mA setting of 400 mA is desired. What new SID can be used to maintain the same radiographic density, assuming all other factors remain unchanged?

11.-16. State the new SID that can be used to maintain the proper radiographic density given the original mA setting and SID and the new mA setting, assuming all other factors remain unchanged.

	Original mA Setting	New mA Setting	Original SID	New SID
11.	100 mA	25 mA	100 cm	____________
12.	400 mA	100 mA	200 cm	____________
13.	600 mA	1200 mA	80 cm	____________
14.	900 mA	100 mA	150 inches	____________
15.	300 mA	600 mA	40 inches	____________
16.	800 mA	200 mA	100 cm	____________

17. A radiograph is taken using 200 mA, 400 ms, 80 kVp at 100 cm SID. What new mAs of 20 mAs is desired. What new SID can be used to maintain the same radiographic density, assuming all other factors remain unchanged?

18.-25. State the new SID that can be used to maintain the proper radiographic density given the original mAs and SID and the new mAs, assuming all other factors remain unchanged.

	Original mAs	New mAs	Original SID	New SID
18.	160 mAs	40 mAs	100 cm	____________
19.	10 mAs	40 mAs	200 cm	____________
20.	30 mAs	270 mAs	100 cm	____________
21.	50 mAs	100 mAs	40 inches	____________
22.	80 mAs	50 mAs	180 cm	____________
23.	15 mAs	200 mAs	100 cm	____________
24.	20 mAs	30 mAs	150 cm	____________
25.	15 mAs	45 mAs	200 cm	____________

Advanced Time - Distance Exercise

Exercise No. 2

1. A radiograph is taken using 800 mA, 20 ms at 80 kVp at a 200 cm SID. A new timer setting of .08 sec, is desired. What new SID can be used to maintain the same radiographic density, assuming all other factors remain unchanged?

2.-9. State the new SID that can be used to maintain the proper radiographic density given the original timer setting and SID and new timer setting, assuming all other factors remain unchanged.

	Original Timer Setting	New Timer Setting	Original SID	New SID
2.	10 ms	40 ms	100 cm	
3.	5 ms	20 ms	300 cm	
4.	240 ms	60 ms	150 cm	
5.	800 ms	600 ms	60 inches	
6.	900 ms	100 ms	300 cm	
7.	1.0 sec	600 ms	40 inches	
8.	330 ms	660 ms	100 cm	
9.	500 ms	800 ms	200 cm	

10. A radiograph is taken using 200 mA, 150 ms at 80 kVp at a 60 cm SID. A new mA setting of 800 mA is desired. What new SID can be used to maintain the same radiographic density, assuming all other factors remain unchanged?

11.-16. State the new SID that can be used to maintain the proper radiographic density given the original mA setting and SID and the new mA setting, assuming all other factors remain unchanged.

	Original mA Setting	New mA Setting	Original SID	New SID
11.	25 mA	100 mA	100 cm	
12.	100 mA	400 mA	180 cm	
13.	300 mA	1200 mA	100 cm	
14.	900 mA	100 mA	40 inches	
15.	150 mA	300 mA	36 inches	
16.	300 mA	100 mA	100 cm	

17. A radiograph is taken using 400 mA, 200 ms, 90 kVp at 200 cm SID. A new mAs of 20 mAs is desired. What new SID can be used to maintain the same radiographic density, assuming all other factors remain unchanged?

18.-25. State the new SID that can be used to maintain the proper radiographic density given the original mAs and SID and the new mAs, assuming all other factors remain unchanged.

	Original mAs	New mAs	Original SID	New SID
18.	180 mAs	45 mAs	200 cm	
19.	25 mAs	100 mAs	200 cm	
20.	15 mAs	135 mAs	36 inches	
21.	60 mAs	100 mAs	40 inches	
22.	90 mAs	50 mAs	100 cm	
23.	15 mAs	105 mAs	50 cm	
24.	120 mAs	30 mAs	150 cm	
25.	25 mAs	65 mAs	200 cm	

Recorded Detail

Chapter 24

1. The true image shadow of a radiographic image is termed the: 2:287

   A. Blur edge
   B. Edge gradient
   C. Umbra
   D. Point source

2. Which of the following factors will effect the recorded detail of the radiographic image? 2:287
   1. OID   2. Focal spot size   3. Anode rotation speed

   A. 1 & 2 only
   B. 1 & 3 only
   C. 2 & 3 only
   D. 1, 2 & 3

3. The recorded detail (definition) possessed by a radiograph can be improved by increasing the: 2:223

   A. Film base thickness
   B. Crystal layer thickness
   C. Amount of collimation
   D. Phosphor crystal size

4. Which of the following sets of factors will produce a radiograph with the greatest unsharpness? 2:288

	mAs	kVp	SID	OID	FSS
A.	20	83	100 cm	20 cm	.3 mm
B.	40	78	180 cm	10 cm	1 mm
C.	15	87	100 cm	20 cm	2 mm
D.	35	74	180 cm	20 cm	.3 mm

5. The amount of magnification of a radiographic image can be determined by the formula:

   A. $\text{Mag.} = \frac{\text{SID}}{\text{SOD}}$
   B. $\text{Mag.} = \frac{\text{OID}}{\text{SOD}}$
   C. $\text{Mag.} = \frac{\text{SOD}}{\text{OID}}$
   D. None of the above

6. Elongation and/or foreshortening of a radiographic image will most likely occur from a/an: 2:285

   A. Increased source-to-image receptor distance (SID)
   B. Decreased object-to-image receptor distance (OID)
   C. Angulation of the central ray
   D. Increase in the anode-heel effect

7. Which of the following changes will increase the size distortion (magnification) that will occur in a radiographic image? 2:280
   1. A decreased SID   2. An increased FSS   3. An increased OID

   A. 1 & 2 only
   B. 1 & 3 only
   C. 2 & 3 only
   D. 1, 2, & 3

8. A radiograph is performed using a 150 mAs, 80 kVp exposure with a 100 cm (40") source-to-image receptor distance (SID), a 40 cm (16") object-to-image receptor distance (OID), and a .3 mm focal spot. The magnification factor for this image is: 2:280

   A. .09
   B. 1.66
   C. 2.5
   D. 3.3

9. Improper tube, object and image receptor alignment with no change in source-to-image receptor distance will result in: 2:283

A. Size distortion
B. Shape distortion
C. Motion distortion
D. Screen distortion

10. Which of the following set of factors will produce the greatest amount of geometric unsharpness? 2:288

	mA	Time	kVp	SID	OID	FSS
A.	300	.001	82	300 cm	20 cm	.3mm
B.	200	.003	84	100 cm	20 cm	2 mm
C.	600	.001	86	300 cm	20 cm	1 mm
D.	400	.002	85	150 cm	20 cm	1 mm

11. The image of a spherical shaped object which is exposed by the peripheral portion of an x-ray beam will appear as a/an _______________ shaped image. 2:283

A. Cylindrical
B. Circular
C. Elliptical
D. Irregular

12. During a macroradiographic (magnification) technique, grids are not generally employed because of the resulting scatter reduction due to the: 12:355

A. Increased SID
B. Increased OID
C. Smaller focal spot
D. Reduced exposure factors

13. The use of tube angulations in many radiographic examinations are useful for the visualization of: 2:283

A. Low density tissues
B. High density tissues
C. Superimposed tissues
D. None of the above

14. Which of the following sets of technical factors will produce a radiograph showing the greatest magnification? 2:280

	mAs	kVp	SID	OID	FSS
A.	25	86	140 cm	10 cm	2 mm
B.	35	86	180 cm	20 cm	1 mm
C.	15	93	180 cm	30 cm	.3 mm
D.	25	93	100 cm	30 cm	.3 mm

15. As the distance from a radiation source increases, the object receives less radiation. This occurs principally because x-ray photons: 12:355
1. Are absorbed by the air 2. Lose energy at long distances 3. Are distributed over a larger area

A. 1 only
B. 2 only
C. 3 only
D. 1, 2 & 3

16. An increase in source-image distance (SID) with no other changes made, will result in a radiograph showing: 2:295

A. Shorter scale of contrast
B. Lower radiographic density
C. Lower recorded detail
D. More magnification distortion

17. Magnification radiography is being performed using an 140 cm source-to-image receptor distance (SID), 40 cm object-to-image receptor distance (OID), and .3 mm focal spot. The magnification factor for this set of factors is: 2:280

A. 4.2
B. 3.5
C. 1.8
D. 1.4

18. The formation of the image unsharpness (blur) around a radiographic image is effected by changes in: 2:287

1. OID   2. SID   3. Degree of collimation

A. 1 & 2 only
B. 1 & 3 only
C. 2 & 3 only
D. 1, 2 & 3

19. If a heart measures 12.5 cm from side to side at its widest point, and its image on a chest radiograph measures 15.0 cm, the magnification factor is: 2:280

A. 1.2
B. 1.5
C. 1.8
D. 2.4

20. In order to maintain the same relative size of a radiographic image, if the object-to-image receptor distance is increased, a radiographer could: 2:280

A. Increase focal spot size
B. Decrease focal spot size
C. Increase source-to-image receptor distance
D. Decrease source-to-image receptor distance

21. Which of the following set of factors will result in the radiographic image possessing the greatest recorded detail? 2:288

	mAs	kVp	Grid Radio	Screen Speed	Focal Spot Size
A.	20	87	8:1	(100)	2 mm
B	50	83	6:1	(50)	1 mm
C.	40	74	6:1	(100)	1 mm
D.	70	73	12:1	(50)	2 mm

22. The image of a disk shaped object which is exposed on the central axis of the beam will appear as a/an shaped image: 2:283

A. Cylindrical
B. Circular
C. Elliptical
D. Irregular

23. Which of the following changes will increase the amount of distortion appearing on a radiographic image? 2:283

1. Decreasing the SID   2. Employing tube angulations   3. Increasing OID

A. 1 only
B. 2 only
C. 3 only
D. 1, 2 & 3

24. An object measuring 18 cm in length is radiographed at a 100 cm source-to-image receptor distance (SID) and a 10 cm object-to-image receptor distance (OID). How long will the resulting radiographic image be? 2:280

A. 16 cm
B. 20 cm
C. 24 cm
D. 34 cm

25. To insure an acceptable amount of image unsharpness (blur) during macroradiographic or mammographic studies, a/an ________________ should be employed. 12:352

A. Fractional focal spot
B. Large focal spot
C. Air gap technique
D. Subtraction technique

26. Which of the following set of factors will produce a radiographic image with the least amount of geometric unsharpness? 288

	mA	Time	kVp	SID	OID	FSS
A	100	50 ms	83	100 cm	12 cm	1 mm
B.	300	20 ms	76	300 cm	40 cm	.3 mm
C.	100	100 ms	64	150 cm	20 cm	1 mm
D	500	50 ms	79	100 cm	6 cm	.3 mm

27. Which of the following factors will effect the recorded detail of the radiographic image? 2:288
1. Source-to-image receptor distance 2. Object-to-image receptor distance 3. Focal spot size

A. 1 only
B. 2 only
C. 3 only
D. 1, 2 & 3

28. The amount of size distortion in an object is effected by changes in the: 2:287
1. SID 2. OID 3. Focal spot size

A. 1 & 2 only
B. 1 & 3 only
C. 2 & 3 only
D. 1, 2 & 3

29. Which of the following set of factors will produce a radiograph with the greatest recorded detail (definition)? 2:288

	mA	Time	kVp	SID	OID	FSS
A.	200	100 ms	85	300 cm	20 cm	1 mm
B.	400	50 ms	85	200 cm	10 cm	2 mm
C.	200	100 ms	85	200 cm	10 cm	1 mm
D.	400	50 ms	85	100 cm	20 cm	.5 mm

30. If an x-ray beam was produced from a single point source, the resultant image would consist of: 12:339

A. Umbra only
B. Penumbra only
C. Both of the above

31. A 10 cm object is radiographed using a 90 cm SID and 60 cm OID. The image of the object at the film plane should measure: 2:280

A. 7.5 cm
B. 12.5 cm
C. 23 cm
D. 30 cm

32. The recorded detail of a radiographic image can be improved by employing a: 2:280
1. Longer SID 2. Smaller FSS 3. Shorter OID

A. 1 & 2 only
B. 1 & 3 only
C. 2 & 3 only
D. 1, 2 & 3

33. The degree of target (anode) angulation will principally effect the amount of: 2:125

A. Film coverage
B. Thermionic emission
C. Secondary scatter production
D. Image magnification

34. Three identical thin objects are located on the same plane and at the same distance from the imaging system. The object showing the greatest magnification is located: 2:281

A. Toward the cathode side of the tube
B. Toward the anode side of the tube
C. On the central axis
D. None of the above

35. A change in which geometric factor will not effect the magnification of a radiographic image? 2:280

A. Source-to-image receptor distance
B. Source-to-object distance
C. Focal spot size
D. Object-to-image receptor distance

Magnification

Exercise No. 1

1. An object measuring 20 cm in width is radiographed and appears as a 50 cm in width radiographic image. What is the magnification factor for this image?

2.-8. State the magnification factors for the following objects when they are radiographed and result in the given image widths.

	Object Width	Image Width	Magnification Factor
2.	10 cm	40 cm	________
3.	15 inches	30 inches	________
4.	20 cm	30 cm	________
5.	30 inches	40 inches	________
6.	10 cm	15 cm	________
7.	25 cm	50 cm	________
8.	35 cm	105 cm	________

9. An object is radiographed at a 100 cm SID and 20 cm SOD. What is the magnification factor for this object?

10.-17. State the magnification factors which will result from the use of the following source-to-image receptor and source-to-object distances.

	SID	SOD	Magnification Factor
10.	40 cm	10 cm	________
11.	40 inches	20 inches	________
12.	40 inches	30 inches	________
13.	100 cm	40 cm	________
14.	100 cm	50 cm	________
15.	120 cm	60 cm	________
16.	180 cm	60 cm	________
17.	72 inches	36 inches	________

18. An object is radiographed at a 100 cm SID and 40 cm OID. What is the magnification factor for this object?

19.-25. State the magnification factors which will result from the use of the following source-to-image and object-to-image receptor distances.

	SID	SOD	Magnification Factor
19.	40 cm	10 cm	________
20.	40 inches	20 inches	________
21.	40 inches	30 inches	________
22.	100 cm	60 cm	________
23.	100 cm	40 cm	________
24.	180 cm	18 cm	________
25.	300 cm	60 cm	________

The following formulas have been used in solving these problems.

$$\text{Magnification Factor} = \frac{\text{Image width}}{\text{Object width}} \qquad \text{Magnification Factor} = \frac{\text{SID}}{\text{SOD}}$$

$$\text{SOD} = \text{SID} - \text{OID}$$

Magnification

Exercise No. 2

1. An object measuring 30 cm in width is radiographed and appears as a 60 cm in width radiographic image. What is the magnification factor for this image?

2.-8. State the magnification factors for the following objects when they are radiographed and result in the given image widths.

	Object Width	Image Width	Magnification Factor
2.	20 cm	40 cm	______
3.	10 inches	30 inches	______
4.	30 cm	80 cm	______
5.	30 inches	50 inches	______
6.	10 cm	25 cm	______
7.	25 cm	75 cm	______
8.	30 cm	90 cm	______

9. An object is radiographed at a 200 cm SID and 20 cm SOD. What is the magnification factor for this object?

10.-17. State the magnification factors which will result from the use of the following source-to-image receptor and source-to-object distances.

	SID	SOD	Magnification Factor
10.	30 cm	10 cm	______
11.	80 inches	20 inches	______
12.	60 inches	30 inches	______
13.	100 cm	60 cm	______
14.	150 cm	50 cm	______
15.	160 cm	80 cm	______
16.	180 cm	90 cm	______
17.	72 inches	18 inches	______

18. An object is radiographed at a 200 cm SID and 40 cm OID. What is the magnification factor for this object?

19.-25. State the magnification factors which will result for the use of the following source-to-image and object-to-image receptor distances.

	SID	SOD	Magnification Factor
19.	40 cm	30 cm	______
20.	60 inches	20 inches	______
21.	90 inches	30 inches	______
22.	180 cm	60 cm	______
23.	100 cm	60 cm	______
24.	100 cm	20 cm	______
25.	300 cm	50 cm	______

The following formulas have been used in solving these problems.

$$\text{Magnification Factor} = \frac{\text{Image width}}{\text{Object width}} \qquad \text{Magnification Factor} = \frac{\text{SID}}{\text{SOD}}$$

SOD = SID - OID

Advanced Magnification

Exercise No. 1

1. An object 35 cm in width is radiographed at 100 cm SID and at a 50 cm SOD. What is the width of the radiographic image of the object?

2.-6. State the length of the radiographic image which will result from using the following source-to-image receptor and source-to-object distances and object lengths. 2:280

	Object Width	SID	SOD	Image Size
2.	15 cm	100 cm	50 cm	____________
3.	20 inches	150 cm	50 cm	____________
4.	30 inches	200 cm	50 cm	____________
5.	30 cm	100 cm	40 cm	____________
6.	40 cm	100 cm	60 cm	____________

7. An object 10 cm in length is radiographed at 200 cm SID and at 100 cm OID. What is the length of the radiographic image of the object?

8.-12. State the length of the radiographic image which will result form the use of the following source-to-image receptor and object-to-image receptor distances and object lengths.

	Object Width	SID	OID	Image Size
8.	25 cm	100 cm	40 cm	____________
9.	30 cm	150 cm	50 cm	____________
10.	15 inches	200 cm	150 cm	____________
11.	20 inches	100 cm	60 cm	____________
12.	30 cm	180 cm	18 cm	____________

13. A radiograph is taken of an object 10 cm in width and the resulting radiographic image measures 20 cm in width. If a 100 cm SID was used, what was the source-to-object distance employed?

14.-18. State the source-to-object distance which will result for the following source-to-image receptor distance and object and images sizes. 2:280

	Object Width	Image Width	SID	SOD
14.	25 cm	50 cm	100 cm	____________
15.	15 cm	20 cm	180 cm	____________
16.	50 cm	75 cm	100 cm	____________
17.	5 cm	15 cm	300 cm	____________
18.	12 cm	20 cm	100 cm	____________

19. An object is radiographed at a 200 cm SID and 100 cm OID. What is the percentage of magnification for the resulting radiographic image?

20. An object is radiographed at a 200 cm SID and 150 cm SOD. What is the percentage of magnification?

21. An object undergoes a two time linear magnification. The area of the radiographic image covered is ____________ greater than the original object.

22. A 4 cm x 8 cm object undergoes a 1.5 magnification factor. What are the dimensions of the radiographic image?

23. A 6 cm diameter circular object undergoes a 2.5 magnification factor. What is the diameter of the radiographic image?

24. An 8 cm x 10 cm object undergoes a 2.0 magnification factor. What is the total area covered by the radiographic image?

25. A circular object covering 30 cm^2 undergoes a 1.5 magnification factor. What is the total area covered by the radiographic image?

Advanced Magnification

Exercise No. 2

1. An object 25 cm in width is radiographed at 100 cm SID and at a 50 cm SOD. What is the width of the radiographic image of the object?

2.-6. State the length of the radiographic image which will result from using the following source-to-image receptor and source-to-object distances and object lengths. 2:280

	Object Width	SID	SOD	Image Size
2.	25 cm	100 cm	60 cm	________
3.	30 inches	150 cm	30 cm	________
4.	10 inches	100 cm	50 cm	________
5.	80 cm	200 cm	40 cm	________
6.	10 cm	180 cm	60 cm	________

7. An object 20 cm in length is radiographed at 200 cm SID and at 150 cm OID. What is the length of the radiographic image of the object?

8.-12. State the length of the radiographic image which will result from the use of the following source-to-image receptor and object-to-image receptor distances and object lengths.

	Object Width	SID	OID	Image Size
8.	5 cm	100 cm	60 cm	________
9.	20 cm	250 cm	50 cm	________
10.	5 inches	300 cm	150 cm	________
11.	10 inches	200 cm	40 cm	________
12.	40 cm	180 cm	60 cm	________

13. A radiograph is taken of an object 10 cm in width and the resulting radiographic image measures 20 cm in width. If a 200 cm SID was used, what was the source-to-object distance employed?

14.-18. State the source-to-object distance which will result for the following source-to-image receptor distance and object and images sizes. 2:280

	Object Width	Image Width	SID	SOD
14.	15 cm	45 cm	100 cm	________
15.	25 cm	30 cm	180 cm	________
16.	60 cm	120 cm	40 cm	________
17.	5 cm	25 cm	300 cm	________
18.	12 cm	36 cm	100 cm	________

19. An object is radiographed at a 200 cm SID and 150 cm OID. What is the percentage of magnification for the resulting radiographic image?

20. An object is radiographed at a 300 cm SID and 150 cm SOD. What is the percentage of magnification?

21. An object undergoes a four time linear magnification. The area of the radiographic image covered is ____________ greater than the original object.

22. A 6 cm x 9 cm object undergoes a 2.5 magnification factor. What are the dimensions of the radiographic image?

23. A 12 cm diameter circular object undergoes a 1.5 magnification factor. What is the diameter of the radiographic image?

24. An 7 cm x 15 cm object undergoes a 3.0 magnification factor. What is the total area covered by the radiographic image?

25. A circular object covering 24 cm^2 undergoes a 1.5 magnification factor. What is the total area covered by the radiographic image?

Image Unsharpness

Exercise No. 1

1.-7. Calculate the amount of image unsharpness (focal-spot blur) given the focal spot size, object-to-image receptor distance and source-to-object distance.

	Focal Spot Size	OID	SOD	Image Unsharpness
1.	1 mm	50 cm	50 cm	________
2.	2 mm	30 cm	100 cm	________
3.	1 mm	30 cm	60 cm	________
4.	2 mm	10 cm	90 cm	________
5.	.3 mm	30 cm	70 cm	________
6.	2 mm	30 cm	70 cm	________
7.	.3 mm	30 cm	90 cm	________

8.-14. Calculate the amount of image unsharpness (focal-spot blur) given the focal spot size, object-to-image receptor distance and source-to-image receptor distance.

	Focal Spot Size	OID	SID	Image Unsharpness
8.	1 mm	50 cm	100 cm	________
9.	2 mm	15 cm	60 cm	________
10.	.3 mm	25 cm	100 cm	________
11.	2 mm	5 cm	60 cm	________
12.	1 mm	60 cm	100 cm	________
13.	.3 mm	10 cm	180 cm	________
14.	.5 mm	15 cm	150 cm	________

15.-20. Given the following pairs of factors, select the one which will show the least amount of image unsharpness (greatest recorded detail).

	SID	OID	FSS
15.	A. 100 cm	20 cm	1 mm
	B. 100 cm	20 cm	2 mm
	C. 100 cm	20 cm	.3 mm
16.	A. 200 cm	10 cm	1 mm
	B. 100 cm	10 cm	1 mm
	C. 180 cm	10 cm	2 mm
17.	A. 180 cm	5 cm	1 mm
	B. 180 cm	10 cm	1 mm
	C. 200 cm	5 cm	2 mm
18.	A. 180 cm	20 cm	.3 mm
	B. 100 cm	10 cm	1 mm
	C. 200 cm	10 cm	.3 mm
19.	A. 180 cm	40 cm	1 mm
	B. 100 cm	20 cm	2 mm
	C. 200 cm	80 cm	1 mm
20.	A. 200 cm	30 cm	2 mm
	B. 180 cm	60 cm	.3 mm
	C. 100 cm	30 cm	2 mm

The following formula has been used in calculating these problems.

Image Unsharpness = $FSS \times \frac{OID}{SOD}$ SOD = SOD - OID

Image Unsharpness

Exercise No. 2

1.-7. Calculate the amount of image unsharpness (focal-spot blur) given the focal spot size, object-to-image receptor distance and source-to-object distance.

	Focal Spot Size	OID	SOD	Image Unsharpness
1.	2 mm	100 cm	100 cm	______
2.	1 mm	60 cm	100 cm	______
3.	1 mm	20 cm	60 cm	______
4.	2 mm	20 cm	180 cm	______
5.	.3 mm	30 cm	70 cm	______
6.	2 mm	40 cm	70 cm	______
7.	.3 mm	35 cm	70 cm	______

8.-14. Calculate the amount of image unsharpness (focal-spot blur) given the focal spot size, object-to-image receptor distance and source-to-image receptor distance.

	Focal Spot Size	OID	SID	Image Unsharpness
8.	2 mm	50 cm	100 cm	______
9.	1 mm	20 cm	80 cm	______
10.	.3 mm	35 cm	135 cm	______
11.	1 mm	15 cm	60 cm	______
12.	2 mm	40 cm	100 cm	______
13.	.3 mm	100 cm	180 cm	______
14.	.5 mm	30 cm	300 cm	______

15.-20. Given the following sets of factors, select the one which will show the least amount of image unsharpness (greatest recorded detail).

	SID	OID	FSS
15.	A. 200 cm	20 cm	1 mm
	B. 200 cm	20 cm	2 mm
	C. 200 cm	20 cm	.3 mm
16.	A. 100 cm	20 cm	1 mm
	B. 200 cm	20 cm	1 mm
	C. 100 cm	10 cm	2 mm
17.	A. 180 cm	5 cm	1 mm
	B. 180 cm	10 cm	1 mm
	C. 90 cm	10 cm	1 mm
18.	A. 180 cm	20 cm	.3 mm
	B. 100 cm	10 cm	2 mm
	C. 200 cm	10 cm	.3 mm
19.	A. 180 cm	40 cm	1 mm
	B. 100 cm	20 cm	2 mm
	C. 100 cm	40 cm	1 mm
20.	A. 200 cm	30 cm	2 mm
	B. 180 cm	60 cm	.3 mm
	C. 100 cm	40 cm	2 mm

The following formula has been used in calculating these problems.

$$\text{Image Unsharpness} = \text{FSS} \times \frac{\text{OID}}{\text{SOD}} \qquad \text{SOD} = \text{SOD} - \text{OID}$$

Beam Lighting Devices

Chapter 25

1. The use of a beam limiting device (collimator) reduces patient exposure but to maintain proper radiographic density: 2:238

   A. mAs should be increased
   B. kV should be decreased
   C. FFD should be increased
   D. FSS should be increased

2. The variation in the exposure rate along the longitudinal axis of the x-ray anode is termed the: 2:124

   A. Ripple effect
   B. Heel effect
   C. Blooming effect
   D. Canting effect

3. Collimation is a term used to denote: 12:388

   A. Limiting the beam of the radiation to the required dimension
   B. Monitoring of personnel for exposure to radiation
   C. Pre-examination preparation of patient
   D. Excessive exposure

4. The anode heel effect, which results from absorption of x-rays by the anode itself, results in a reduction of beam intensity on the: 2:124

   A. Cathode side of the tube
   B. Anode side of the tube
   C. Stem of the tube
   D. Stator heel of the tube

5. A modern automatic collimator consists of: 12:389
   1. Parallel sheets of lead *2. Electronic sensors* *3. Motor driven shutters*

   A. 1 & 2 only
   B. 1 & 3 only
   C. 2 & 3 only
   D. 1, 2 & 3

6. As a general rule, in the utilization of the anode heel effect the thickest portion of the body is placed under the: 2:124

   A. Anode side of the tube
   B. Cathode side of the tube
   C. Bucky side of the table
   D. Back side of the table

7. The upper or entrance shutters in a collimator serve principally to reduce: 2:243
   1. Primary radiation *2. Off-focus radiation* *3. Scatter radiation*

   A. 1 only
   B. 2 only
   C. 3 only
   D. 1, 2 & 3

8. If the focal spot to mirror distance in a collimator is 5 cm, the light to mirror distance should be: 5:94

   A. 5 cm
   B. 7.5 cm
   C. 10 cm
   D. 15 cm

9. The use of beam limiting devices will result in a smaller field coverage that, in turn, will tend to produce a higher: 2:234
   1. Radiographic contrast *2. Radiographic density* *3. Beam intensity*

   A. 1 only
   B. 2 only
   C. 3 only
   D. 1, 2 & 3

10. If a reduction in patient dose is desired, which of the following fields will expose the least amount of tissue? 12:17

A. 12 cm diameter circular field
B. 8 cm x 8 cm square field
C. 10 cm x 7 cm rectangular field
D. 9 cm x 8 cm rectangular field

11. To best utilize the anode heel effect during upright chest radiography, the anode should be placed ______ to the cathode. 2:125

A. Laterally
B. Medially
C. Inferiorly
D. Superiorly

12. All types of currently employed beam limiting devices are located between the __________ and the ____________ . 2:238

A. Patient/film
B. Tube/patient
C. Patient/bucky tray
D. None of the above

13. The Bureau of Radiologic Health has mandated that all new radiographic units must have a type of collimation known as: 2:244

A. Fixed collimation
B. Incident collimation
C. Automatic collimation
D. Negative collimation

14. An acceptable full abdomen radiograph of the gallbladder is obtained on a 14" x 17" film. If a cone-down gallbladder view is taken at the same technique, the film will show a/an: 2:235

A. Higher density and contrast
B. Increased density and lower contrast
C. Decreased density and higher contrast
D. Higher density and contrast

15. Automatic collimation is required on all new radiographic units. These serve to decrease: 2:238
1. Patient exposures 2. Beam quality 3. Scatter production

A. 1 & 2 only
B. 1 & 3 only
C. 2 & 3 only
D. 1, 2 & 3

16. The anode heel effect is most pronounced: 5:19

A. At long source-to-image receptor distances
B. At short source-to-image receptor distances
C. When a small sized film is used
D. With a low grid ratio

17. A 20 cm x 8 cm aperture diaphragm is located 10 cm below the x-ray source at a 40 cm SID. What is the size of the exposed field? 5:97

A. 5 cm x 2 cm
B. 10 cm x 4 cm
C. 40 cm x 16 cm
D. 80 cm x 32 cm

18. If the diameter of the x-ray beam is 24 cm at 100 cm source-to-image receptor distance, what diameter would the beam cover at 200 cm? 5:97

A. 12 cm
B. 36 cm
C. 48 cm
D. 96 cm

Pertaining to the radiation Intensity answer question 19 & 20.

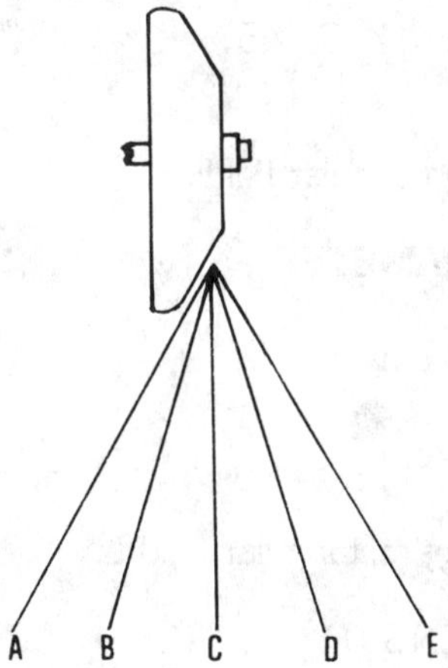

19. Pertaining to the diagram, which of the following letters corresponds to the region of the least radiation intensity? 2:124

A. Letter A
B. Letter C
C. Letter E
D. All possess the same intensity.

20. Pertaining to the diagram, the letter C corresponds to a relative exposure intensity rate of approximately: 2:124

A. 70%
B. 90%
C. 100%
D. 120%

21. The width of the image unsharpness (geometric blur) can be reduced by placing a beam limiting device as _________ the x-ray source as possible. 5:94

A. Close to
C. Far away from

22. When a light illuminated collimator is employed, which of the following components remains within the path of the x-ray beam. 5:95

A. Collimator light bulb
B. Positioning mirror
C. Motion trimmers
D. None of the above

23. The thickness of the collimator shutter leaves must have the same lead equivalency as that of a/the: 2:632

A. Lead apron
B. Secondary barrier
C. Protective housing
D. Diagnostic filtration

24. Which of the following devices is not classified as a type of beam limiting device: 5:94

A. Collimator
B. Apeture diaphragm
C. Hooded anode
D. Extension cylinder

25. In order to provide the appropriate alignment with the x-ray beam, the positioning mirror should be mounted at an angle of _________ to the central ray. 5:83

A. 20%
B. 30%
C. 45%
D. 90%

28-27. Calculate the area covered for the following circular fields.

	FIELD SHAPE	RADIUS	DIAMETER	AREA	
26.	Circular	10 cm	20 cm	_________	12:17
27.	Circular	23 cm	46 cm	_________	12:17

28-30. Calculate the area covered for the following rectangular fields.

	FIELD SHAPE	WIDTH	LENGTH	AREA	
28.	Rectangular	18 cm	25 cm	_________	12:16
29.	Rectangular	12 cm	30 cm	_________	12:16
30.	Rectangular	40 cm	15 cm	_________	12:16

Radiographic Grids

Chapter 26

1. The main advantage of a moving grid compared to a stationary grid of the same ratio and frequency is the: 5:110
1. The grid lines are blurred 2. Grid cutoff is reduced 3. Greater tube angulations are permitted

A. 1 only
B. 2 only
C. 3 only
D. 1, 2 & 3

2. The reduction of secondary radiation reaching the image receptor can be accomplished by using: 5:101
1. An increased OID 2. Radiographic grids 3. Beam Limiting devices

A. 1 & 2 only
B. 1 & 3 only
C. 2 & 3 only
D. 1, 2 & 3

3. Grid ratio is represented by the formula: 2:250

A. r. = h x D
B. $r. = \frac{25.4}{D + d}$
C. $r. = \frac{h}{D}$
D. $r. = \frac{D}{h}$

4. The distance at which a focus grid can be used is directly related to the: 5:101

A. Grid ratio
B. Grid frequency
C. Convergence line
D. Interspacing material

5. When the greatest reduction of scattered radiation is required, a/an __________ ratio grid should be employed. 2:253

A. 6:1
B. 8:1
C. 12:1
D. 16:1

6. Which type of grid pattern will result in a radiograph with the least amount of primary beam cutoff? 2:257

A. Focus grid
B. Parallel grid
C. Crossed grid
D. Rhombic grid

7. A Potter-Bucky diaphragm serves to: 5:110

A. Limit the field of radiation
B. Lower patient exposures
C. Move the grid during an exposure
D. Limit backscattered radiation

8. Which of the following set of technical factors will produce a radiograph with the least contrast? 2:252

	mA	Time	kVp	Grid Ratio
A.	200	25 ms	86	16:1
B.	200	20 ms	94	12:1
C.	400	10 ms	102	8:1
D.	300	10 ms	94	16:1

9. Which of the following focus grids would possess the greatest amount of centering and distance latitude? 5:106

A. 6:1 ratio
B. 8:1 ratio
C. 12:1 ratio
D. 16:1 ratio

10. In the selection of a grid with the proper ratio, the factor that must be considered is: 2:261
1. Positioning latitude 2. Degree of cleanup 3. Kilovoltage range

A. 1 only
B. 2 only
C. 3 only
D. 1, 2 & 3

Pertaining to the Radiographic image answer questions 11 & 12.

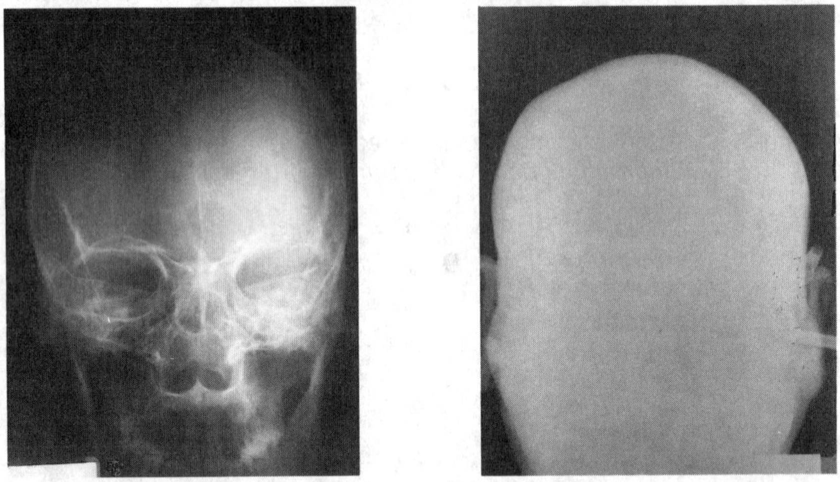

A B

11. In the radiographic image of film A, the decreased density seen in the lateral portion of the film mostly resulted from a/an: 2:261

A. Excessive collimation
B. Improperly aligned grid
C. Patient motion
D. Excessive focal film distance

12. In the radiographic image of film B, the decreased density seen throughout the film was caused by a/an: 2:253

A. Improper centering
B. Excessive bucky movement
C. Upside down gird
D. Improper technical compensation

13. In order to avoid the appearance of grid lines using a moving grid, the Bucky tray motion must commence: 12:384

A. Before the exposure begins
B. After the exposure begins
C. Grid movement is not necessary

14. A grid that is constructed with all the grid strips running parallel to each other is an example of a: 2:254

A. Linear focus grid
B. Linear parallel grid
C. Rhombic grid
D. Cross-hatched grid

15. When grids are utilized, the technical factors must be adjusted because: 2:253
1. Scatter production increases 2. Scatter absorption increases 3. SID must be increased

A. 1 only
B. 2 only
C. 3 only
D. 1, 2 & 3

Using the following diagram, answer questions 16. and 17.

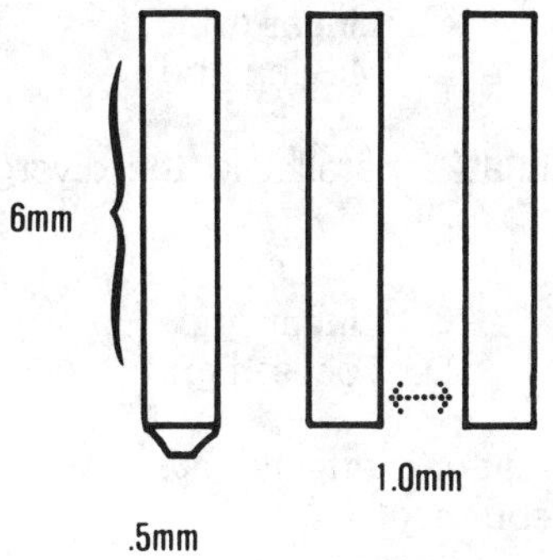

16. This grid will have a ratio of: 2:250

A. 16:1
B. 12:1
C. 8:1
D. 6:1

17. The frequency of the grid strips is: 5:101

A. 100 lines/in.
B. 66 lines/in.
C. 33 line/in.
D. None of the above

18. Which of the following set of factors will produce a radiograph with the greatest contrast? 5:104

	mA	Time	kVp	Grid Ratio
A.	300	30 ms	68	12:1
B.	400	25 ms	87	8:1
C.	200	50 ms	84	10:1
D.	600	15 ms	98	12:1

19. Which of the following factors will not result in the production of grid cut-off using a focus grid? 5:106

A. Exceeding distance latitude
B. Exceeding centering latitude
C. Angulating to tube along the grid strips
D. Angulating to tube across the grid strips

20. If grid ratio was increased, and not technical changes made, the radiograph would show an increase in: 2:248

A. Density
B. Magnification
C. Contrast
D. Distortion

21. High ratio grids are recommended for use when: 2:261

A. High kilovoltage is used
B. Distances are variable
C. Off-centering is necessary
D. Dose reduction is necessary

22. In a modern 8:1 focus grid, the majority of scatter reduction results from the: 2:251

A. Lead grid strips
B. Aluminum grid strips
C. Lead interspacing material
D. Aluminum interspacing material

23. A given grid has a grid strip thickness of 4 mm, a grid interspace thickness of 20 mm and height of 160 mm. What is the grid ratio? 2:250

A. 5:1
B. 6:1
C. 8:1
D. 16:1

24. The amount of primary beam cut off by grids can be greatly reduced by employing a: 2:257

A. Crossed grid
B. Focused grid
C. Linear grid
D. Moving grid

25. A type of grid in which the grid lines are situated parallel to the divergent rays of the x-ray beam is termed a: 2:257

A. Stationary grid
B. Parallel grid
C. Linear grid
D. Focused grid

26. The efficiency of a grid or its ability to reduce scatter is effected by: 2:253
1. Grid ratio 2. Grid radius 3. Grid frequency

A. 1 & 2 only
B. 1 & 3 only
C. 2 & 3 only
D. 1, 2 & 3

27. If a 12:1 ratio grid is substituted for an 8:1 ratio grid, and the same technical factors are employed, the image will possess a higher: 4:90
1. Radiographic density 2. Radiographic contrast 3. Percentage of scatter

A. 1 only
B. 2 only
C. 3 only
D. 1, 2 & 3

28. Tube angulation techniques are often required during many normal radiographic procedures. For this reason, the least commonly employed grid for a general purpose room is a: 2:254

A. Parallel grid
B. Crossed grid
C. Focused grid
D. Linear grid

29. The amount of scatter pickup (grid efficiency) will increase as: 2:252
1. Grid ratio increases 2. Frequency increases 3. Selectrivty decreases

A. 1 only
B. 2 only
C. 3 only
D. 1, 2 & 3

30. The appearance of grid cutoff when focus grids are employed is likely when the tube is: 2:260
1. Angled along the grid strips 2. Off-centered to the grid 3. Used at the wrong distance

A. 1 & 2 only
B. 1 & 3 only
C. 2 & 3 only
D. 1, 2 & 3

31. The advantage(s) involved with the use of a radiographic grid is/are: 2:258

A. Improved contrast
B. Reduced scatter production
C. Lower patient exposures
D. More than one but not all above

32. A grid is designed with .125 mm thick lead strips, 8 mm in height and separated by .5 mm aluminum spacers. The grid has a ratio of: 2:250

A. 4:1
B. 8:1
C. 12:1
D. 16:1

33. A radiograph is produced in a table bucky, and the tube was properly centered to the image receptor in the bucky tray yet grid lines were seen. A probable cause is: 2:260
1. Excessive grid speed 2. Insufficient grid speed 3. Insufficient collimation

A. 1 only
B. 2 only
C. 3 only
D. 1, 2 & 3

34. The man most often credited with the construction of the first practical radiographic grid is: 2:249

A. W.K. Roentgen
B. H. Potter
C. G. Bucky
D. J.P. Focus

35. The principal reason for using a radiographic grid in a diagnostic examination is: 2:247

A. Backscatter reduction
B. Reduction of scatter reaching the film
C. Reduction of patient exposure
D. Increase in radiation quality

36. A cross-hatched grid is formed by 8:1 and 6:1 linear grids. This combined grid will have distance and centering latitudes proportional to a/an: 2:256

A. 6:1 linear grid
B. 8:1 linear grid
C. 13:1 linear grid
D. 16:1 linear grid

37. The amount of scatter reduction when a 6:1 crossed grid is used is nearly equivalent to that of a/an: 2:256

A. 6:1 Fine Line grid
B. 12:1 linear grid
C. 8:1 linear grid
D. 5:1 focus grid

Pertaining to the radiographic images, answer questions 38 & 39.

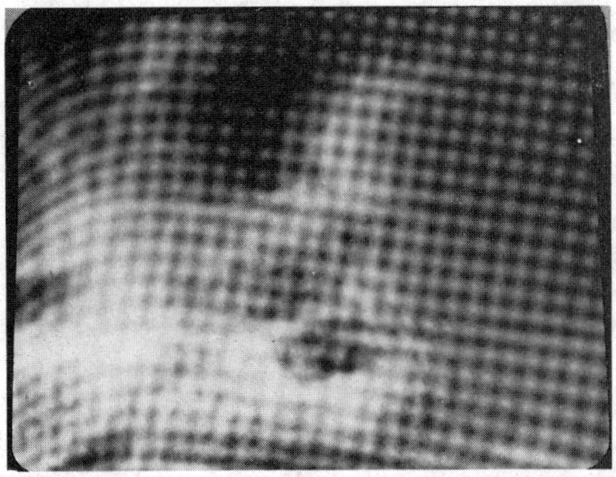

A

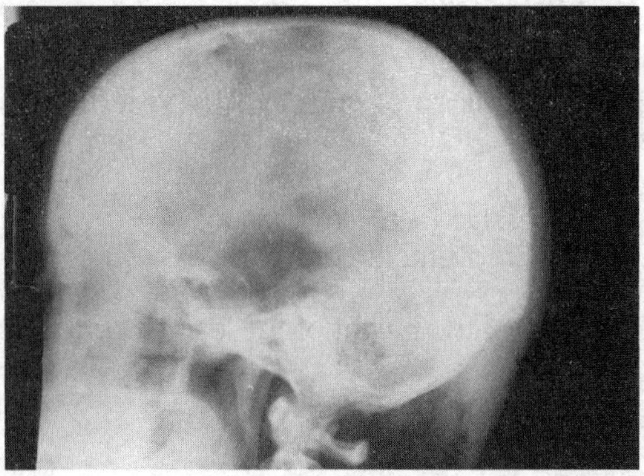

B

38. The radiographic image of film A demonstrates a grid artifact called the:

A. Moire effect
B. Quantum mottle effect
C. Rhomboid effect
D. Cross hatch effect

39. In the radiograph image of film B, the decrease of density in the lateral areas of the image were caused by a/an:

A. Off focus decentering
B. Upside down focus grid
C. Improperly aligned grid
D. Tube angulation across the grid

40. When an 8:1 grid is used in addition to a film-screen holder (cassette), in order to maintain film density, the technique is: 2:253

A. Increased by 1 1/2 times the mAs
B. Increased by 3 times the mAs
C. Decreased by 1 1/2 times the mAs
D. Decreased by 3 times the mAs

41. A radiograph is taken that shows peripheral grid cutoff on both sides of the film. The possible cause includes a/an: 2:61
1. Upside down focus grid *2. Excessive source-to-image distance*
3. Deficient source-to-image distance

A. 1 only
B. 2 only
C. 3 only
D. 1, 2 & 3

42. In general, when tube potential exceeds 90 kVp the minimum grid ratio that should be employed is: 2:262

A. 5:1
B. 6:1
C. 12:1
D. 16:1

43. Which of the following set of factors will produce a radiograph with the greatest contrast? 2:252

	mA	Time	kVp	Grid Ratio
A.	300	15 ms	110	8:1
B.	200	35 ms	90	5:1
C.	200	25 ms	100	6:1
D.	400	5 ms	84	8:1

44. The first moving grid was designed by: 2:258

A. H. Potter
B. T. Edison
C. G. Bucky
D. M. Pupin

45. The inclination of grid strips toward a central point is termed: 2:257

A. Interspacing
B. Crossing
C. Canting
D. Scattering

46. The appearance of grid lines using a focus grid in a bucky tray is unlikely under which of the following situations? 2:259

A. Excessive focal film distance
B. Grid movement before the exposure
C. Grid stoppage during the exposure
D. Improper centering

47. The ratio of the height of the lead strips to the distance between the lead strips defines: 2:250

A. Grid ratio
B. Grid radius
C. Grid latitude
D. Lines per inch

48. Which of the grid factors listed below will result in the greatest amount of scattered radiation attenuation: 2:252

A. 6:1 ratio/60 1p/in
B. 8:1 ratio/60 1p/in
C. 6:1 ratio 100 1p/in
D. 8:1 ratio 100 1p/in

49. The placement of two identical parallel grids on top of each other with the grid strips running in the same direction may cause: 4:40

A. Parallel effect
B. Strobe effect
C. Moire effect
D. Heel effect

50. The contrast improvement that can be obtained with a particular grid is dependent upon the: 5:93
1. Kilovoltage employed *2. Size of the field* *3. Patient thickness*

A. 1 only
B. 2 only
C. 3 only
D. 1, 2 & 3

Grid Conversion

Exercise No. 1

1. A radiograph is taken with medium (100) screens, non-bucky, using 10 mAs at 80 kVp. If a 6:1 grid is employed, what new mAs should be used to maintain the radiographic density?

2.-10. State the mAs value that should be employed to maintain the same radiographic density when the following grid ratio changes are made.

	Object mAs	Original Grid Ratio	New Grid Ratio	New mAs
2.	21 mAs	No grid	8:1	
3.	20 mAs	8:1	12:1	
4.	15 mAs	6:1	12:1	
5.	60 mAs	No grid	12:1	
6.	100 mAs	12:1	16:1	
7.	50 mAs	6:1	8:1	
8.	70 mAs	No grid	6:1	
9.	5 mAs	No grid	16:1	
10.	12 mAs	5:1	8:1	

11. A radiograph is taken with a rare earth (320) screen with 12:1 Bucky using 40 mAs at 80 kVp. If a 6:1 grid is substituted, what new mAs should be used to maintain the radiographic density?

12.-20. State the mAs value that should be employed to maintain the same radiographic density when the following grid ratio changes are made.

	Object mAs	Original Grid Ratio	New Grid Ratio	New mAs
12.	100 mAs	12:1	8:1	
13.	20 mAs	6:1	No grid	
14.	50 mAs	12:1	6:1	
15.	60 mAs	16:1	No grid	
16.	42 mAs	8:1	No grid	
17.	12 mAs	16:1	12:1	
18.	24 mAs	8:1	5:1	
19.	16 mAs	12:1	No grid	
20.	60 mAs	6:1	5:1	

All answers on this form were calculated using the following grid conversion factors:

5:1 = 1.5 x mAs
6:1 = 2 x mAs
8:1 = 3 x mAs
10:1-12:1 = 4 x mAs
16:1 = 5 x mAs

The following formula has been used in calculating these problems:

$$\text{New mAs} = \text{Original mAs} \times \frac{\text{New Grid Conversion Factor}}{\text{Original Grid Conversion Factor}}$$

Grid Conversion

Exercise No. 2

1. A radiograph is taken with medium (100) screens, non-Bucky, using 5 mAs at 80 kVp. If an 8:1 grid is employed, what new mAs should be used to maintain the radiographic density?

2.-10. State the mAs value that should be employed to maintain the same radiographic density when the following grid ratio changes are made.

	Object mAs	Original Grid Ratio	New Grid Ratio	New mAs
2.	30 mAs	No grid	12:1	______
3.	10 mAs	No grid	16:1	______
4.	25 mAs	6:1	16:1	______
5.	30 mAs	12:1	16:1	______
6.	7 mAs	No grid	8:1	______
7.	15 mAs	8:1	12:1	______
8.	12 mAs	6:1	16:1	______
9.	5 mAs	5:1	8:1	______
10.	100 mAs	No grid	12:1	______

11. A radiograph is taken with a rare earth (320) screen with 12:1 Bucky using 60 mAs at 85 kVp. If a 6:1 grid is substituted, what new mAs should be used to maintain the radiographic density?

12.-20. State the mAs value that should be employed to maintain the same radiographic density when the following grid ratio changes are made.

	Object mAs	Original Grid Ratio	New Grid Ratio	New mAs
12.	10 mAs	6:1	No grid	______
13.	5 mAs	16:1	No grid	______
14.	7 mAs	12:1	8:1	______
15.	16 mAs	8:1	No grid	______
16.	100 mAs	6:1	5:1	______
17.	24 mAs	12:1	6:1	______
18.	36 mAs	8:1	5:1	______
19.	80 mAs	12:1	No grid	______
20.	30 mAs	16:1	12:1	______

All answers on this form were calculated using the following grid conversion factors:

5:1 = 1.5 x mAs
6:1 = 2 x mAs
8:1 = 3 x mAs
10:1 or 12:1 = 4 x mAs
16:1 = 5 x mAs

The following formula has been used in calculating these problems:

$$\text{New mAs} = \text{Original mAs} \times \frac{\text{New Grid Conversion Factor}}{\text{Original Grid Conversion Factor}}$$

Density Analysis (Grid)

Exercise No. 1

In each of the following set of technical factors, select the technique that will result in the greatest radiographic density.

	mA	Time	kVp	Grid Ratio
1.	A. 300	30 ms	94	5:1
	B. 300	10 ms	89	12:1
	C. 300	20 ms	84	8:1
	D. 300	1 ms	86	16:1
2.	A. 200	15 ms	76	8:1
	B. 100	25 ms	74	12:1
	C. 300	20 ms	78	8:1
	D. 200	6 ms	78	12:1
3.	A. 600	50 ms	90	12:1
	B. 400	50 ms	92	8:1
	C. 800	50 ms	88	6:1
	D. 600	50 ms	90	No grid
4.	A. 600	35 ms	105	12:1
	B. 300	12 ms	103	6:1
	C. 300	33 ms	98	12:1
	D. 600	15 ms	95	8:1
5.	A. 200	250 ms	88	6:1
	B. 300	330 ms	98	12:1
	C. 100	500 ms	88	8:1
	D. 200	15 ms	86	6:1
6.	A. 300	500 ms	96	12:1
	B. 200	750 ms	92	16:1
	C. 100	1.5 sec	90	5:1
	D. 400	600 ms	88	8:1
7.	A. 400	1 ms	86	12:1
	B. 800	50 ms	89	No grid
	C. 400	70 ms	94	6:1
	D. 800	100 ms	94	No grid
8.	A. 300	50 ms	72	No grid
	B. 600	.15 s	80	16:1
	C. 100	.15 s	82	12:1
	D. 300	1 ms	75	8:1
9.	A. 200	3/10	92	6:1
	B. 100	2/5	92	12:1
	C. 200	3/4	88	8:1
	D. 100	3/8	82	6:1
10.	A. 800	1/40	86	5:1
	B. 600	1/30	84	8:1
	C. 400	1/20	92	12:1
	D. 600	1/120	86	8:1

Density Analysis (Grid)

Exercise No. 2

In each of the following set of technical factors, select the technique that will result in the greatest radiographic density.

	mA	Time	kVp	Grid Ratio
1.	A. 100	30 ms	84	6:1
	B. 100	10 ms	76	16:1
	C. 100	20 ms	82	8:1
	D. 100	10 ms	76	No grid
2.	A. 200	150 ms	86	12:1
	B. 400	330 ms	89	12:1
	C. 300	250 ms	82	8:1
	D. 100	150 ms	84	12:1
3.	A. 600	150 ms	92	6:1
	B. 300	70 ms	82	No grid
	C. 800	50 ms	84	12:1
	D. 1000	10 ms	74	6:1
4.	A. 300	15 ms	96	16:1
	B. 100	25 ms	92	8:1
	C. 100	33 ms	87	16:1
	D. 400	15 ms	76	12:1
5.	A. 200	10 ms	86	5:1
	B. 300	10 ms	89	8:1
	C. 300	10 ms	92	16:1
	D. 200	10 ms	84	12:1
6.	A. 800	50 ms	86	8:1
	B. 800	70 ms	86	12:1
	C. 600	150 sec	92	16:1
	D. 600	100 ms	88	16:1
7.	A. 300	25 ms	87	8:1
	B. 100	33 ms	76	No grid
	C. 600	33 ms	89	6:1
	D. 30	15 ms	86	8:1
8.	A. 400	7 ms	83	8:1
	B. 600	1 ms	92	No grid
	C. 400	100 ms	91	12:1
	D. 600	70 ms	70	16:1
9.	A. 800	3/20	76	12:1
	B. 400	3/4	82	16:1
	C. 300	2/3	94	16:1
	D. 800	3/8	94	8:1
10.	A. 100	1/60	82	6:1
	B. 300	1/60	84	8:1
	C. 200	1/60	82	8:1
	D. 100	1/120	82	6:1

Intensifying Screens

Chapter 27

1. In general, about __________ of a x-ray film's exposure is due to the light emitted by the screen's phosphors. 5:121

A. 2-5%
B. 10-13%
C. 47-50%
D. 95-98%

2. The principal advantage of direct exposure (cardboard) radiography compared to screen radiography is: 2:223

A. Decreased patient dose
B. Shorter scale of contrast
C. Increased resolution
D. Decreased technical factors

3. The emission of light from a phosphor after the incident radiation exposure is terminated is called: 2:220

A. Fluorescence
B. Solarization
C. Phosphorescence
D. Incandescence

4. Radiographic parallax can be reduced by using: 4:128

A. Smaller crystals
B. A thicker crystal layer
C. A thinner film base
D. A blue-tinted base

5. The amount of x-ray absorption of an intensifying screen phosphor is primarily dependent upon the: 2:228

A. Pair production effect
B. Photoelectric effect
C. Compton effect
D. Thomson effect

6. The effect of quantum mottle is most commonly seen when a __________ is employed. 2:266

A. Cardboard holder
B. Slow film-screen combination
C. Fast film-screen combination
D. Double emulsion film

7. Rare earth phosphors are a significant improvement over calcium tungstate screens. This is due in large part to their high: 2:229

A. Conversion efficiency
B. Spectral matching
C. Phosphorescence
D. Quantum mottle

8. As crystal size in an intensifying screen increases, recorded detail will tend to _________: 2:222

A. Increase
B. Decrease
C. Remain the same

9. The speed of a film-screen combination can be increased by making the following screen modification. 2:218
1. Larger crystals *2. Thinner crystal layer* *3. Reflective backing*

A. 1 & 2 only
B. 1 & 3 only
C. 2 & 3 only
D. 1, 2 & 3

10. The primary advantage of the use of rare earth phosphors in modern intensifying screens is: 2:227

A. Improved resolution
B. Reduced patient dose
C. Lower cost
D. Lower quantum noise

11. Some of the phosphors currently employed in intensifying and fluoroscopic screens were first discovered by: 15:66

A. Michel Pupin
B. Thomas Edison
C. Wilhelm Roentgen
D. George Eastman

12. As the photon energy hitting an intensifying screen increases: 2:222
1. Screen speed increases *2. Resolution increases* *3. Detail increases*

A. 1 only
B. 2 only
C. 3 only
D. 1, 2 & 3

13. Which film-holder combination will result in the highest intensification factor? 2:221

A. Non-screen film in a cardboard holder
B. Screen film in a cardboard holder
C. Screen film in a (50) speed cassette
D. Screen film in a (100) speed cassette

14. Which of the following photoemissive chemicals (phosphors) possess the highest conversion efficiency? 229

A. Gadolinium oxysulfide
B. Calcium tungstate
C. Silver bromide
D. Copper oxybromide

15. The resolution of an intensifying screen can be improved by employing: 2:217

A. Thicker crystal layer
B. Light-absorbing dyes
C. Larger crystal size
D. Reflective screen backing

16. In order to provide the maximum efficiency of the screen-film combination, the light emitted by the screens should correspond to the film's light sensitivity. This is termed: 2:226

A. Spectral matching
B. Homogeneity
C. Conversion efficiency
D. Synergism

17. Gadolinium and lanthanum are two common materials used in a new class of high speed screens called: 2:226

A. Rare earth phosphors
B. Parallax phosphors
C. Gamma phosphors
D. Quantum phosphors

18. The speed of an intensifying screen can be increased by employing: 2:217

A. Light absorbing dyes
B. A reflective backing
C. Smaller crystals
D. Thinner crystal layer

19. The emission of light equally in all directions when phosphors are exposed to a stimulus is termed: 5:122

A. Quantum mottle
B. Selectivity
C. Spectral scattering
D. Isotropic propagation

20. As the speed of an imaging system increases: 2:267
1. Image blur decreases *2. Quantum mottle increases* *3. Noise increases*

A. 1 & 2 only
B. 1 & 3 only
C. 2 & 3 only
D. 1, 2 & 3

21. The principal action of the crystals found in intensifying screens is the conversion of: 5:118

A. Visible light into electrons
B. X-rays into visible light
C. Visible light into x-rays
D. Electrons into visible light

22. Lead foil is incorporated into the back lid of most cassettes. This serves to: 2:225

A. Reduce backscatter
B. Focus the x-ray beam
C. Create an air gap
D. Improve resolution

23. The manufacturers of many slow speed intensifying screens add a pink or yellow dye to the crystal layer. This will: 2:222

A. Increase screen speed
B. Increase screen resolution
C. Increase contrast
D. Reduce scatter production

24. The major advantage(s) of using intensifying screens as compared to a direct exposure technique is/are: 2:225

1. Lower patient dose *2. More latitude* *3. Improved resolution*

A. 1 only
B. 2 only
C. 3 only
D. 1, 2 & 3

25. The resolution (resolving power of an intensifying screen) is generally measured in units of: 2:223

A. Lines pairs inch
B. Lines pairs millimeter
C. Candela per inch
D. Candela per millimeter

26. A series of uneven densities may be apparent on high sensitivity imaging systems due to the random distribution of x-rays, its effect is termed: 2:267

A. Quantum mottle
B. Isotropic propagation
C. Vignetting
D. Convergence ratio

27. Reflection from successive underlying layers of luminescent crystals in slow intensifying screens can be lessened by: 2:222

A. Using intensifying screens that are pure white
B. Adding a light-absorbing dye
C. Using a slower film
D. Placing dark paper between the intensifying screen and the film

28. The speed of an intensifying screen is effected by changes in: 2:222
1. Size of the crystals *2. Thickness of the crystals* *3. Temperature of the screen*

A. 1 & 2 only
B. 1 & 3 only
C. 2 & 3 only
D. 1, 2 & 3

29. The ratio of exposures required to produce the same film density with and without screens is termed the: 2:221

A. Film-screen ratio
B. Resolution
C. Minification ratio
D. Intensification factor

30. A high speed imaging system is used in the production of a radiograph with a low resolution. The principal cause for this low resolution is a high degree of radiographic: 2:67

A. Lucency
B. Noise
C. Opacity
D. Contrast

31. The process by which a crystalline material gives off visible light in response to x-radiation is termed: 2:219

A. Luminescence
B. Phosphorescence
C. Incandescence
D. Photo-electrolysis

32. The ratio of the x-ray energy absorbed by a phosphor to the visible light energy emitted is termed: 2:229

A. K-shell absorption edge
B. Spectral matching
C. Conversion efficiency
D. Intensification factor

33. The most common material used in a (100) speed screen cassette is a blue light-emitting phosphor called: 2:226

A. Calcium tungstate
B. Lanthanum oxybromide
C. Silver bromide
D. Sodium thiosulfate

34. The principal advantage that rare earth imaging systems have over calcium tungstate screens is a greater speed without an appreciable loss of: 2:229
1. Radiographic density *2. Radiographic contrast* *3. Resolution*

A. 1 only
B. 2 only
C. 3 only
D. 1, 2 & 3

35. Poor screen contact is determined by using a: 2:227

A. Pinhole camera
B. Spin top test
C. Wire mesh tester
D. Test for lag

36. The front of any type of film holder must be covered by a light-tight substance that is both _________ and to _________ x-radiation. 2:225

A. Homogeneous/radio-opaque
B. Heterogeneous/radio-opaque
C. Homogeneous/radiolucent
D. Heterogeneous/radiolucent

37. In a cassette, the thin layer of foam rubber (felt) behind each intensifying screen serves to: 2:225
1. Increase screen speed *2. Improve film-screen contact* *3. Decrease patient scattering*

A. 1 only
B. 2 only
C. 3 only
D. 1, 2 & 3

38. A calcium tungstate screen is to be manufactured with a thicker crystal layer. This will tend to increase the: 2:222
1. Screen speed *2. Resolving power of the screen* *3. Conversion efficiency*

A. 1 only
B. 2 only
C. 3 only
D. 1, 2 & 3

39. The light emitted by a phosphor during its exposure to x-radiation is termed: 2:219

A. Photoelectrolysis
B. Thermionic emission
C. Incandescence
D. Fluorescence

40. A non-screen (cardboard) requires a technique of 200 mA, 500 ms, 75 kVp. If a 100 speed screen is used at 100 mA, 50 ms, 75 kVp which gives the same radiographic density, the intensification factor is: 2:221

A. 5 times
B. 10 times
C. 20 times
D. 40 times

Screen Conversion Exercise

Exercise No. 1

1. A radiograph is taken with a 50 speed screen, non-bucky, using 40 mAs at 80 kV. If a rare earth (400) screen was substituted, what new mAs could be used to maintain a similar radiographic density?

2.-10. State the mAs value that should be employed to maintain the same radiographic density when the following screen speed changes are made.

	Object mAs	Original Screen Speed	New Screen Speed	New mAs
2.	10 mAs	50	300	
3.	20 mAs	100	200	
4.	15 mAs	100	400	
5.	60 mAs	50	200	
6.	100 mAs	200	400	
7.	50 mAs	160	480	
8.	70 mAs	50	100	
9.	5 cm	50	200	
10.	600 mAs	No screen	100	

11. A radiograph is taken with a 400 speed screen, non-bucky, using 2.5 mAs at 70 kV. If a slow (50) screen was substituted, what new mAs could be used to maintain a similar radiographic density?

12.-20. State the mAs values that can be employed to maintain the same radiographic density when the following screen speed changes are made.

	Object mAs	Original Screen Speed	New Screen Speed	New mAs
12.	24 mAs	200	100	
13.	12 mAs	100	50	
14.	30 mAs	320	160	
15.	20 mAs	200	100	
16.	70 mAs	200	50	
17.	50 mAs	320	80	
18.	90 mAs	480	160	
19.	15 mAs	180	60	
20.	10 mAs	400	100	

The following formulas have been used in solving these problems.

$$\text{New mAs} = \text{Original mAs} \times \frac{\text{Original Screen Speed Number}}{\text{New Screen Speed Number}}$$

Exercise No. 2

1. A radiograph is taken with a medium 100 speed screen, 12:1 Bucky, using 40 mAs at 90 kVp. If a (400) speed screen is substituted, what new mAs should be used to maintain a similar radiographic density?

2.-10. State the mAs value that should be employed to maintain the same radiographic density when the following screen speed changes are made.

	Object mAs	Original Screen Speed	New Screen Speed	New mAs
2.	40 mAs	300	100	______
3.	30 mAs	100	50	______
4.	25 mAs	50	200	______
5.	45 mAs	200	400	______
6.	100 mAs	50	100	______
7.	80 mAs	160	320	______
8.	5 mAs	100	400	______
9.	10 mAs	200	50	______
10.	15 mAs	50	300	______

11. A radiograph is taken with a 320 speed screen, non-Bucky, using 5 mAs at 60 kVp. If a (160) speed screen is substituted, what new mAs should be used to maintain a similar radiographic density?

12.-20. State the mAs values that can be employed to maintain the same radiographic density when the following screen speed changes are made.

	Object mAs	Original Screen Speed	New Screen Speed	New mAs
12.	30 mAs	200	100	______
13.	24 mAs	100	50	______
14.	100 mAs	300	400	______
15.	50 mAs	320	160	______
16.	35 mAs	200	400	______
17.	20 mAs	100	200	______
18.	270 mAs	50	300	______
19.	100 mAs	320	480	______
20.	40 mAs	50	100	______

The following formula has been used to calculate these problems.

$$\text{New mAs} = \text{Original mAs} \times \frac{\text{Original Screen Speed Number}}{\text{New Screen Speed Number}}$$

Density Analysis (Screens)

Exercise No. 1

In each set of technical factors listed find the technique which will result in the greatest radiographic density.

	mA	Time	kVp	Screen Speed
1.	A. 200	10 ms	92	100
	B. 200	5 ms	74	300
	C. 200	25 ms	84	50
	D. 200	25 ms	74	100
2.	A. 100	30 ms	84	100
	B. 100	300 ms	87	50
	C. 200	10 ms	74	200
	D. 300	1 ms	74	50
3.	A. 300	100 ms	96	320
	B. 200	150 ms	88	160
	C. 100	200 ms	90	200
	D. 300	200 ms	96	320
4.	A. 400	40 ms	90	100
	B. 800	40 ms	96	50
	C. 300	1.0 sec.	105	No screen
	D. 300	40 ms	94	No screen
5.	A. 100	150 ms	83	400
	B. 400	250 ms	85	200
	C. 600	50 ms	86	100
	D. 300	50 ms	88	100
6.	A. 300	1.5 sec.	72	No screen
	B. 200	1.0 sec.	64	50
	C. 400	750 ms	66	50
	D. 300	600 ms	72	No screen
7.	A. 200	50 ms	62	100
	B. 100	35 ms	74	50
	C. 200	40 ms	86	320
	D. 200	35 ms	74	320
8.	A. 200	1/2 sec.	76	50
	B. 300	500 ms	88	50
	C. 100	1/2 sec.	88	100
	D. 600	250 ms	86	100
9.	A. 400	1/2 sec.	88	200
	B. 800	125 ms	90	160
	C. 200	50 ms	89	320
	D. 400	1/8 sec.	88	100
10.	A. 300	1/100 sec.	82	320
	B. 200	1/50 sec.	91	100
	C. 300	1/60 sec.	86	50
	D. 200	1/20 sec.	74	50

Density Analysis (Screens)

Exercise No. 2

In each set of technical factors listed find the technique which will result in the greatest radiographic density.

		mA	Time	kVp	Screen Speed
1.	A.	150	100 ms	64	200
	B.	300	100 ms	68	100
	C.	600	100 ms	66	100
	D.	400	100 ms	66	50
2.	A.	600	150 ms	82	160
	B.	800	50 ms	87	100
	C.	1000	40 ms	93	320
	D.	800	15 ms	86	160
3.	A.	300	60 ms	91	50
	B.	200	150 ms	84	100
	C.	200	250 ms	88	100
	D.	300	10 ms	86	50
4.	A.	500	35 ms	87	320
	B.	300	60 ms	83	50
	C.	100	150 ms	83	100
	D.	300	5 ms	89	100
5.	A.	400	250 ms	92	50
	B.	800	150 ms	88	No screen
	C.	200	250 ms	82	100
	D.	600	150 ms	84	No screen
6.	A.	50	3.0 sec.	74	160
	B.	100	1.0 sec.	68	200
	C.	300	500 ms	70	100
	D.	500	10 ms	65	200
7.	A.	600	50 ms	94	100
	B.	600	10 ms	80	320
	C.	600	70 ms	92	100
	D.	600	20 ms	84	50
8.	A.	300	1.5 sec.	80	100
	B.	150	1.5 sec.	66	200
	C.	50	5.0 sec.	82	100
	D.	50	3.0 sec.	74	200
9.	A.	800	80 ms	79	100
	B.	400	160 ms	76	50
	C.	300	1/10 sec.	72	160
	D.	400	1/4 sec.	70	320
10.	A.	300	150 ms	77	320
	B.	200	1/16 sec.	87	200
	C.	200	660 ms	90	100
	D.	300	3/20 sec.	84	50

Grid and Screen Conversion

Exercise No. 1

Use the following information to answer questions 1.-7.

1.-7. A radiograph is taken at 30 mAs, 12:1 Bucky, using a 100 speed screen. What new mAs should be employed to maintain the radiographic density for the following screen and grid ratio changes?

	New Screen Speed	New Grid Ratio	New mAs
2.	50	6:1	______
3.	200	8:1	______
4.	300	5:1	______
5.	50	16:1	______
6.	200	No grid	______
7.	300	6:1	______
8.	400	12:1	______

Use the following information to answer questions 8.-14.

8.-14. A radiograph is taken at 40 mAs using 8:1 grid and 50 speed screen. What new mAs should be employed to maintain the radiographic density for the following screen and grid ratio changes?

	New Screen Speed	New Grid Ratio	New mAs
8.	50	No grid	______
9.	400	5:1	______
10.	100	8:1	______
11.	30	No grid	______
12.	100	12:1	______
13.	300	8:1	______
14.	200	No grid	______

Use the following information to answer questions 15.-20.

15-20. A radiograph is taken at 60 mAs with a 6:1 grid using 200 speed screen. What new mAs should be employed to maintain the radiographic density for the following screen and grid ratio changes?

	New Screen Speed	New Grid Ratio	New mAs
15.	400	No grid	______
16.	100	12:1	______
17.	50	6:1	______
18.	400	12:1	______
19.	100	16:1	______
20.	50	8:1	______

The following formula has been used in calculating these problems.

$$\text{New mAs} = \frac{\text{Original mAs} \times \text{Original Screen Speed No.}}{\text{New Screen Speed No.}} \times \frac{\text{New Grid Conversion Factor}}{\text{Original Grid Conversion Factor}}$$

Exercise No. 2

1.-7. Use the following information to answer questions 1.-7.

A radiograph is taken at 10 mAs, non-Bucky, using a 50 speed screen. What new mAs should be employed to maintain the radiographic density for the following screen and grid ratio changes?

	New Screen Speed	New Grid Ratio	New mAs
1.	100	6:1	______
2.	200	8:1	______
3.	100	12:1	______
4.	400	12:1	______
5.	200	8:1	______
6.	100	5:1	______
7.	No screen	No grid	______

Use the following information to answer questions 8.-14.

8.-14. A radiograph is taken at 30 mAs using 12:1 Bucky using a 400 speed screen. What new mAs should be employed to maintain the radiographic density for the following screen and grid ratio changes?

	New Screen Speed	New Grid Ratio	New mAs
8.	100	16:1	______
9.	50	6:1	______
10.	200	No grid	______
11.	200	6:1	______
12.	50	No grid	______
13.	200	8:1	______
14.	100	15:1	______

15-20. A radiograph is taken at 50 mAs using 6:1 grid using a 100 speed screen. What new mAs should be employed to maintain the radiographic density for the following screen and grid ratio changes?

	New Screen Speed	New Grid Ratio	New mAs
15.	50	5:1	______
16.	200	12:1	______
17.	300	12:1	______
18.	50	8:1	______
19.	400	16:1	______
20.	200	No grid	______

The following formula has been used in calculating these problems.

$$\text{New mAs} = \text{Original mAs} \times \frac{\text{Original Screen Speed No.}}{\text{New Screen Speed No.}} \times \frac{\text{New Grid Conversion Factor}}{\text{Original Grid Conversion Factor}}$$

Radiographic Films

Chapter 28

1. The majority of screen-type radiographic films consist of a film base and a: 2:190

A. Single film emulsion
B. Dual film emulsion
C. Multiple film emulsion
D. None of the above

Pertaining to the radiographic images, answer questions 2 & 3.

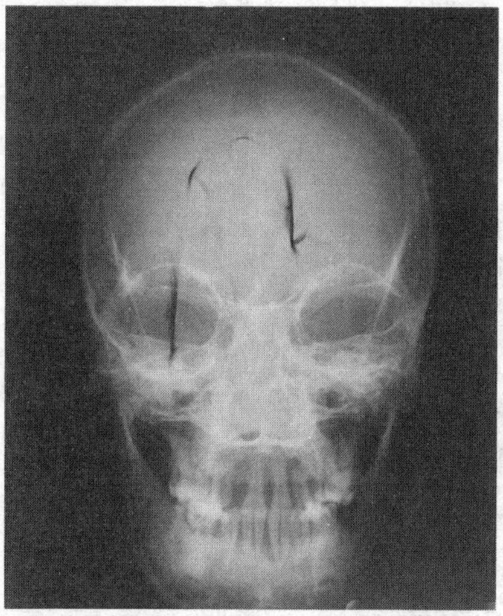

A

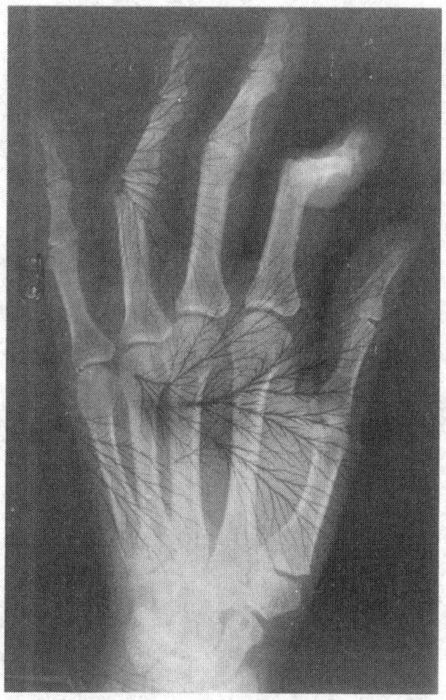

B

2. In the radiographic image in Film A, the increased density marks seen on the skull most likely resulted from: 2:445

A. Improper film handling
B. Dirty intensifying screen
C. A warped cassette
D. Improper safelight filtration

3. In the radiographic image of Film B, the increased density artifact was caused by: 2:445

A. Bending the film
B. Emulsion scratches
C. Static electricity
D. Water on the feed tray

4. The useful range of densities on a characteristic (H&D) curve is approximately: 2:272

A. 0.1-.25
B. 0.1-1.5
C. 0.2-2.5
D. 0.5-5.0

5. Many of the characteristic properties of a film were first described by: 2:269

A. Hurter and Driffield
B. Hinze and Downs
C. Hittorf and Day
D. None of the above

6. The major factor(s) that will cause a variation in film speed is/are: 2:195
1. Grain size *2. Emulsion thickness* *3. Base material*

A. 1 & 2 only
B. 1 & 3 only
C. 2 & 3 only
D. 1, 2 & 3

Pertaining to the diagram of the characteristic curve, answer questions 7 & 8.

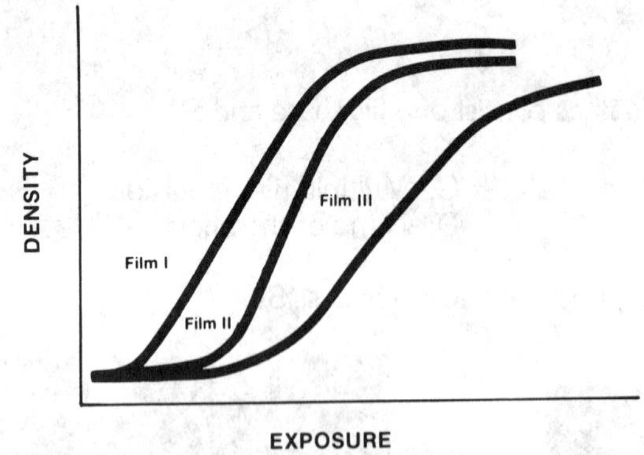

7. Which of the three films shows are greatest radiographic contrast: 2:275

A. Film 1
B. Film 2
C. Film 3
D. All are the same

8. Which of the three films has the greatest speed? 2:275

A. Film 1
B. Film 2
C. Film 3
D. All are the same

9. Any unwanted marks on a film are classified as: 2:275

A. Glitches
B. Pi lines
C. Artifacts
D. Scatter

10. The graphic relationship between the intensity of exposure to a film and the resulting density is demonstrated by a/an: 2:277

A. Characteristic curve
B. Threshold curve
C. Spectral curve
D. A to D curve

11. The latent image is the invisible image formed by the action of x-rays and the: 2:192

A. Film base
B. Gelatin of the film emulsion
C. Crystals in the film emulsion
D. Developer solution

12. Exceeding the film manufacturer's temperature recommendation levels during storage will cause the film to have a/an: 2:200
1. Reduced contrast *2. Increased speed* *3. Increased fog*

A. 1 & 2 only
B. 1 & 3 only
C. 2 & 3 only
D. 1, 2 & 3

13. A film is found to transmit 1/100 of the light from a viewbox. The density of this film is: 2:271

A. .01
B. .25
C. 1.0
D. 2.0

14. A normal screen-type radiographic film should have a base plus fog density of about: 2:271

A. 0.1
B. 0.5
C. 0.8
D. 1.6

15. The optical properties of a film base can be improved by the incorporation of blue dye into the: 2:190

A. Film emulsion
B. Film base
C. Gelatin layer
D. Protective coating

16. In general, radiographic film possesses the greatest sensitivity to light in the _________ spectrum. 2:192
1. Red-orange 2. Yellow-orange 3. Blue-green

A. 1 only
B. 2 only
C. 3 only
D. 1, 2 & 3

17. A characteristic (H&D) curve expresses the relationship between density and exposure and is used to compare the film factors of: 2:277
1. Opacity 2. Sensitivity 3. Contrast

A. 1 & 2 only
B. 1 & 3 only
C. 2 & 3 only
D. 1, 2 & 3

18. Black metallic silver is formed in a chemical reaction in which a/an ____ is added to the silver ions: 2:206

A. Photon
B. Proton
C. Neutron
D. Electron

19. A characteristic (H&D) curve is made from a film by plotting _________ and the resulting _________ 2:275

A. Log relative density/contrast
B. Log relative exposure/density
C. Log relative latitude/density
D. Log relative density/latitude

20. The inherent ability of film emulsion to react to radiation and record a range of densities is termed: 2:273

A. Radiographic contrast
B. Subject contrast
C. Exposure latitude
D. Film latitude

21. The clear, porous substance of the film that permits the uniform dispersion of the silver halogen crystals is called: 2:191

A. Polyester
B. Collogen
C. Gelatin
D. Cycon

Refer to the diagram of the characteristic curve answer questions 22.-24.

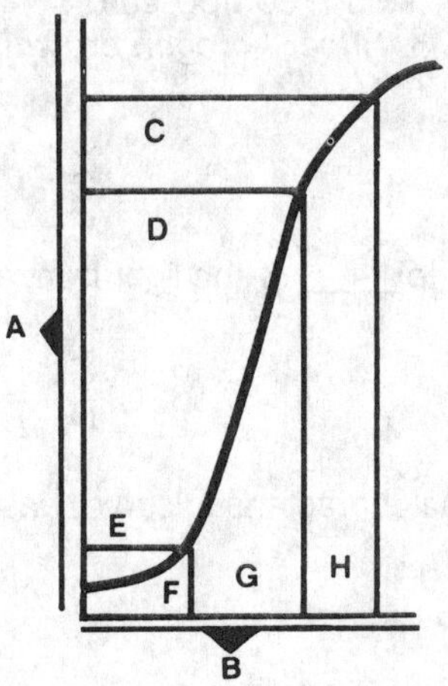

22. Which letter corresponds to the log relative exposure axis? 2:275

A. Letter A
B. Letter B
C. Letter D
D. Letter G

23. Which letter corresponds to the latitude of diagnostically useful density range? 2:275

A. Letter A
B. Letter B
C. Letter D
D. Letter G

24. Letter E corresponds to the area of the film possessing the least amount of: 2:275

A. Image blur
B. Radiographic contrast
C. Radiographic density
D. Sensitivity

25. A radiograph comes out of a processor with numerous dark, treelike markings. These most likely are caused by: 2:445

A. Bending the film
B. Static electricity
C. Light fog during processing
D. Water on the feed tray

26. Which method can be employed to reduce post-exposure fog in the darkroom? 12:294
1. Increase the safelight distance to the film *2. Increase the wattage of the bulbs*
3. Increase the number of safelights

A. 1 only
B. 2 only
C. 3 only
D. 1, 2 & 3

27. Small amounts of impurities (sulfur) are added to the silver halogen crystals in the film to aid in development. These are called: 12:299

A. Artifacts
B. Sensitivity specks
C. Reticulation specks
D. Surficants

28. The active light-sensitive components of a radiographic film are located in the ________ of the film. 2:191

A. Adhesive layer
B. Emulsion
C. Super-coating
D. Base

29. The slope or steepness of the characteristic curve of a radiographic film is used in the determination of its: 2:277

A. Radiographic contrast
B. Subject contrast
C. Film resolution
D. Recorded detail

30. A typical radiographic film emulsion is composed of: 2:191
1. Gelatin *2. Polyester* *3. Silver halogen crystals*

A. 1 & 2 only
B. 1 & 3 only
C. 2 & 3 only
D. 1, 2 & 3

31. Film density of 3.0 corresponds to _____ of the light from a viewbox being transmitted through the film. 2:271

A. 1/10
B. 1/30
C. 1/300
D. 1/1000

Pertaining to the diagram of the characteristic curve, answer questions 32.-34.

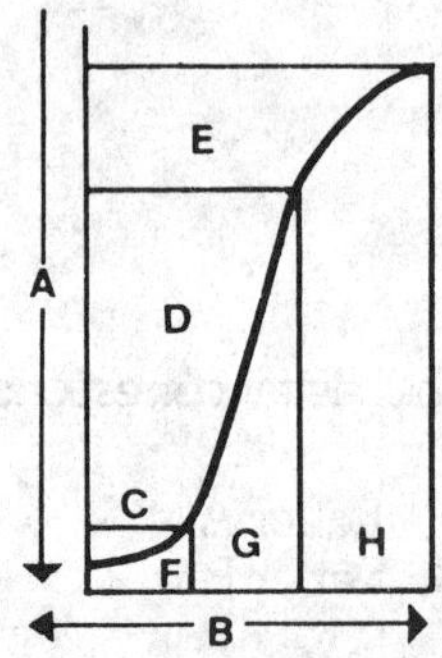

32. Which of the letters above corresponds to the normal exposure region? 2:274

A. Letter A
B. Letter D
C. Letter E
D. Letter F

33. Which of the letters above corresponds to the latitude of the overexposure region? 2:274

A. Letter B
B. Letter C
C. Letter F
D. Letter H

34. Letter A corresponds to the ___________ of this radiographic film. 2:277

A. Exposure axis
B. Contrast axis
C. Density axis
D. Latitude axis

35. The principal silver halogen molecule found in a modern radiographic film is: 2:191

A. Silver bromide
B. Silver iodide
C. Silver sulfate
D. Silver sulfite

Pertaining to the diagram of the characteristic curve, answer questions 36. and 37.

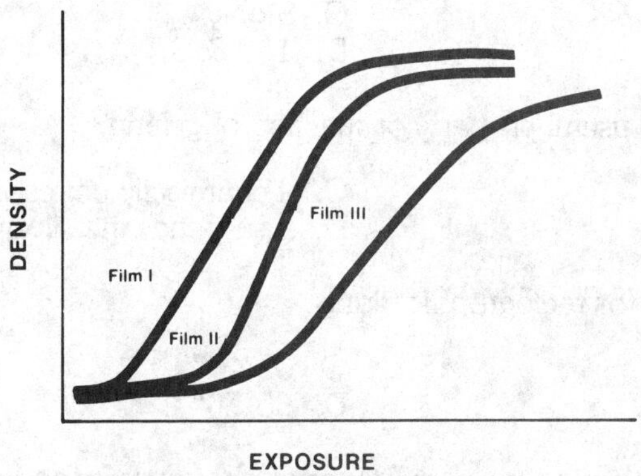

36. Which of the numbered films above will have the greatest exposure latitude? 2:274

A. Film 1
B. Film 2
C. Film 3
D. All are the same

37. Which of the numbered films above will demonstrate the greatest sensitivity? 2:274

A. Film 1
B. Film 2
C. Film 3
D. All are the same

38. The great majority of processed radiographic images that are recorded on film are considered:

A. Positive images
B. Negative images
C. Subtracted images
D. Duplicated images

39. Since the 1960's, the universally accepted film base material is called: 2:190

A. Triacetate
B. Cellulose acetate
C. Chromater
D. Polyester

40. The density of a film is expressed by the formula in which Io is the incident light and It is the transmitted light: 2:271

A. $D = \frac{Io}{It}$
C. $D = \frac{Io^2}{It^2}$
A. $D = \text{Log } \frac{Io}{It}$
C. $D = \text{Log } \frac{Io^2}{It^2}$

41. At the region above the shoulder of an H & D curve, continued exposures may cause the slope of the line begin to curve downward. This area is termed the __________ region. 3:595

A. Solarization
B. Density
C. Polarization
D. Postexposure

42. The property of the film base to resist stretching or shrinking, thus resulting in more consistent radio–graphic images, is termed: 2:190

A. Neutrality
B. Permeability
C. Dimensional stability
D. Retentivity

43. The manufacturing process of a radiographic film that must occur in total darkness is the: 12:271
1. Formation of silver halide crystals *2. Coating of the adhesive layer to base*
3. Production of the film base

A. 1 only
B. 2 only
C. 3 only
D. 1, 2 & 3

44. Which of the following is not a useful property or function of gelatin? 2:191

A. It swells in warm solution
B. It hardens when dried
C. It chemically unites with the silver crystals
D. It suspends the silver crystals

45. The ideal storage temperature for radiographic film is: 2:200

A. 5-10° C
B. 15-20° C
C. 35-40° C
D. 65-70° C

46. The first type of film base used during the early stages of photography and roentgenography was (a): 2:190

A. Barium plate
B. Glass plate
C. Gelatin plate
D. Cellulose nitrate

47. The principle controlling factor for the speed of the film is the: 2:192

A. Size of the lattice network
B. Concentration of the gelatin
C. Location of the sensitivity specks
D. Concentration of the silver halide crystals

48. A protective coating placed on most modern film helps to __________ of the film. 2:190
1. Increase sensitivity *2. Increase the lucency* *3. Prevents scratching*

A. 1 only
B. 2 only
C. 3 only
D. 1, 2 & 3

49. During latent image formation, the concentration of metallic silver in the exposed crystal occurs in the region of the: 2:206

A. Bromide pool
B. Sensitivity speck
C. Silver speck
D. Film base

50. Static buildup on a film most commonly occurs on a: 2:200

A. Hot, humid day
B. Cold, humid day
C. Hot, rainy day
D. Cold, dry day

Automatic Processing

Chapter 29

1. During automatic processing, the proper order of steps are as follows: 2:205

   A. Fixation, development, washing, drying
   B. Development, washing, fixation, drying
   C. Development, fixation, washing, drying
   D. Development, drying, washing, fixation

2. The proper developer action will only occur in a/an _________ solution. 2:208

   A. Acidic
   B. Neutralized
   C. Alkaline
   D. Saline

3. In order for the correct amount of solution replenishment to occur, in an automatic processor, radiographic films are normally processed: 15:257

   A. Lengthwise
   C. Widthwise
   B. Either of the above

4. During automatic processing the time the film is in the processing solutions is primarily controlled by the: 2:213

   A. Replenisher system
   B. Dryer system
   C. Recirculation system
   D. Transport system

5. In most 90 second automatic processors, the developer temperature is critical and is usually maintained at about: 2:212

   A. 80° F (28° C)
   B. 95° F (35° C)
   C. 110° F (68° C)
   D. 125° F (95° C)

6. The addition of restrainers in the developer solutions will tend to prevent the development of unexposed silver bromide crystals, thus reducing: 2:208

   A. Film contrast
   B. Developmental fogging
   C. Aerial oxidation
   D. Radiation fogging

Pertaining to the diagram of the characteristic curves of the three identical films, answer question 7.

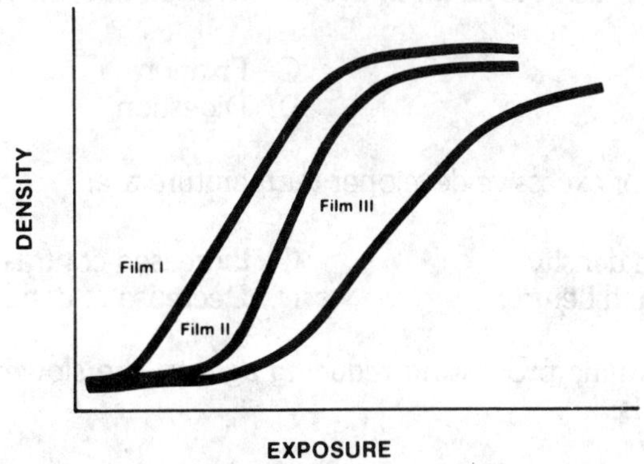

7. The three identical films were processed using the same processor and solutions. Which of the films show the effects of excessive developer temperature? 2:208

   A. Film I
   B. Film II
   C. Film III
   D. Unable to determine

Pertaining to the radiographic images, answer questions 8 & 9.

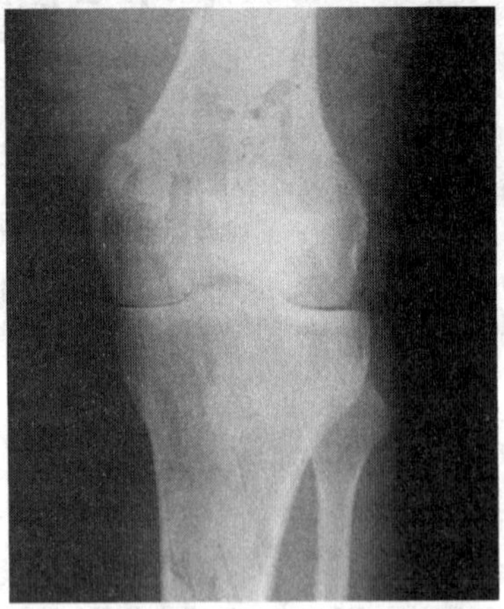

A

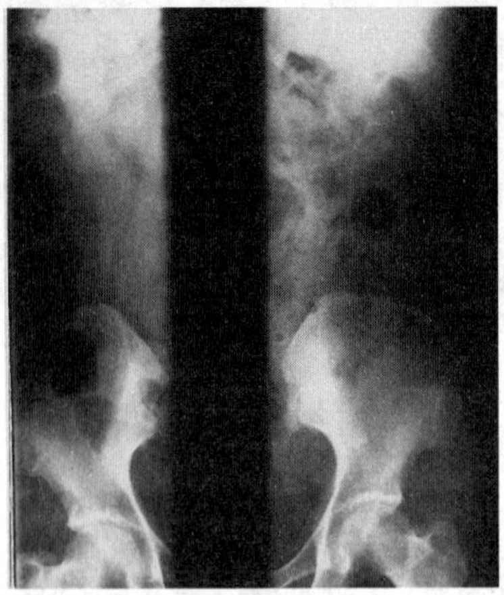

B

8. In the radiographic image in Film A, the increased density marks seen throughout the image were caused by: 2:444

A. Static electricity
B. Solutions on the feed tray
C. Bending of the film
D. Insufficient fixation

9. In the radiographic image in Film B, the large area of increased density in the center of the film was most likely caused by: 2:244

A. Improper safe light filtration
B. Excessive developer temperature
C. Exposure to light during processing
D. Dirty wash water

10. The most commonly employed reducing agents used for automatic processing solutions are: 2:206

A. Water and glutaraldehyde
B. Hydroquinone and phenidone
C. Sodium sulfate and acidic acid
D. Thiosulfate and glutaraldehyde

11. Which of the following is/are components of the transport system? 2:210
1. Drying tubes *2. Transport rollers* *3. Crossover rack*

A. 1 & 2 only
B. 1 & 3 only
C. 2 & 3 only
D. 1, 2 & 3

12. The stage of processing in which the latent image is converted into the manifest image is called: 2:203

A. Development
B. Solarization
C. Fixation
D. Digestion

13. The most common results of excessive developer temperature are: 2:208

A. Increased contrast and density
B. Decrease in contrast and density
C. Increased contrast and decrease in density
D. Decrease in contrast and increase in density

14. Of the two common automatic processing reducing agents, the slower acting agent responsible for the darkest shades of density is: 2:206

A. Sodium carbonate
B. Hydroquinone
C. Sulfuric acid
D. Phenidone

15. A film comes out of a processor which is wet and tacky. A possible cause is: 2:208

A. Excessive development
B. Insufficient fixation
C. Excessive drying temperature
D. Depleted hardener

16. During automatic processing, solution carryover between the developer and fixer tanks is prevented by the 3:298

A. Squeegee rollers
B. Rinse bath
C. Overflow valves
D. Microswitches

17. A solution that is capable of giving up negative ions (electrons) is classified as a/an: 2:205

A. Reducing agent
B. Oxidizing agent
C. Bleaching agent
D. Washing agent

18. Processor monitoring is necessary to insure consistent quality and should be done at least once every: 2:443

A. Day
B. Week
C. Month
D. Year

19. The total amount of solution replenishment for both the fixer and developer solutions is effected by the: 2:213
1. Thickness of the film emulsion 2. Size of the films processed 3. Number of films processed

A. 1 & 2 only
B. 1 & 3 only
C. 2 & 3 only
D. 1, 2 & 3

20. During automatic processing, the activation of the replenisher pump microswitch is accomplished in the: 2:213

A. Crossover assembly
B. Entrance roller assembly
C. Guideshoe assembly
D. Developer squeegee assembly

Pertaining to the radiographic images, answer questions 21. & 22.

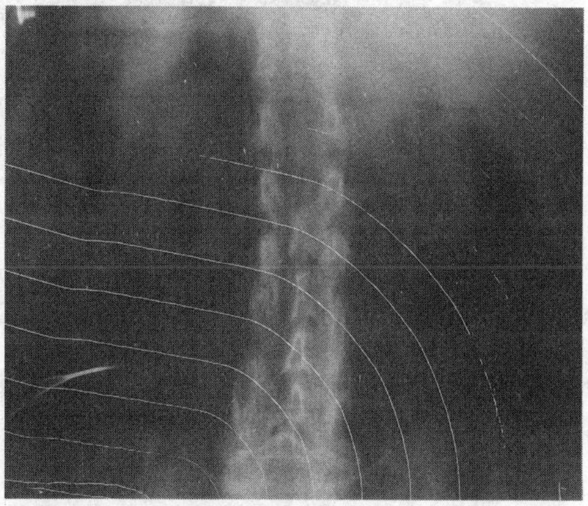

A

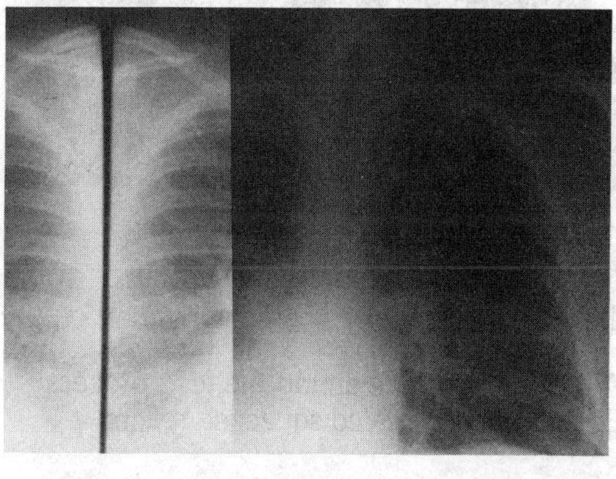

B

21. The decreased density artifacts seen on the radiographic image of Film A were most likely caused by: 2:445

A. Scratches on the x-ray table
B. Processor guide shoes
C. Radiographic parallex
D. Improper film storage

22. The unusual appearance of the radiographic image of Film B most likely resulted from: 2:444

A. An improperly feeded film
B. A double exposure
C. Low developer temperature
D. An improperly loaded cassette

23. The most likely cause of a film's failure to transport in an automatic processor is: 2:208

A. Dirty wash water
B. Over-replenished developer
C. Exhausted developer
D. Low developer temperature

24. The developing agents, when placed together, have a chemical action stronger than the sum of the two working independently. This chemical action is termed: 2:206

A. Syntax
B. Synchronism
C. Synarthrosis
D. Synergism

25. A possible cause of under-replenished solutions in an automatic processor is: 15:278
1. Faulty microswitches *2. Faulty replenisher pump* *3. Air in replenisher lines*

A. 1 only
B. 2 only
C. 3 only
D. 1, 2 & 3

26. The liquid clearing agent used in most automatic fixing solutions is: 2:209

A. Sodium sulfite
B. Ammonium thiosulfate
C. Aluminum chloride
D. Hydroquinone

27. The principal solvent used in all wet automatic processing solutions is: 2:205

A. Hydroquinone
B. Thiosulfate
C. Water
D. Sodium sulfate

Pertaining to the radiographic images answer questions 28. & 29.

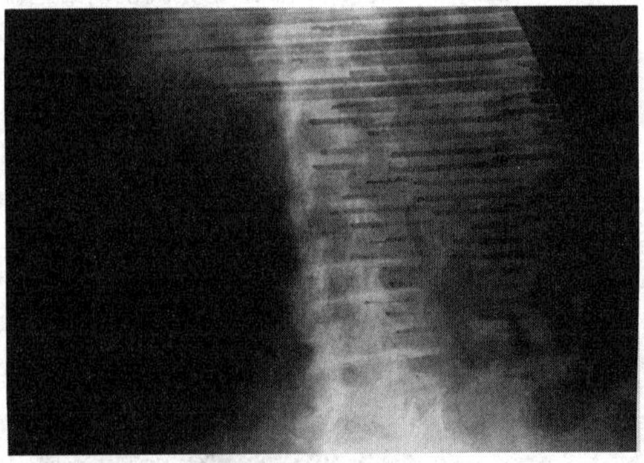

A

B

28. The increased density streaks appearing on the radiographic image of Film A were most likely caused by:

A. Light exposure during the feed process
B. Improperly seated squeegee rollers
C. The processor guide shoes
D. Developer solution on the entrance rollers

29. The decreased density marks seen on the radiographic image of Film B resulted from:

A. Improper film storage
B. Wraped cassette front
C. An exposure to the back of the cassette
D. Upside down focus grid

30. One of the more common causes of damp films and processor jam up is the depletion of the developer's hardener called: 2:208

A. Glutaraldehyde
B. Aluminum chloride
C. Silver iodide
D. Hydroquinone

31. The solution that is maintained at the lowest temperature of approximately 90° F (32° C) is the: 2:210

A. Wash
B. Fixer
C. Developer
D. Stop bath

32. The most common restrainer used in automatic processing solutions is: 2:208

A. Silver iodide
B. Sodium sulfite
C. Aluminum chloride
D. Potassium bromide

33. The temperature stabilization in a processor is most often a function of the: 2:212

A. Circulation system
B. Replenisher system
C. Dryer system
D. Transport system

34. Which of the following conditions will likely cause fog on a radiographic film? 2:209
1. Increased developer time *2. Lowered developer temperature* *3. Low solution activity*

A. 1 only
B. 2 only
C. 3 only
D. 1, 2 & 3

35. A film is processed in dirty solutions and numerous lines appear on the radiograph opposite the direction of film travel. These are referred to as: 2:444

A. Guideshoe marks
B. Wet processing artifacts
C. Crescent marks
D. Pi lines

36. Which of the following factors will effect the time that the film should be developed? 2:206
1. Base thickness *2. Developer activity* *3. Developer temperature*

A. 1 & 2 only
B. 1 & 3 only
C. 2 & 3 only
D. 2, 3 & 4

37. The toe (lower density) portion of the characteristic curve is principally controlled by the reducing agent termed: 2:206

A. Phenidone
B. Sodium thiosulfate
C. Potassium bromide
D. Glutaraldehyde

38. The replenisher rates for the developer solution are generally ________ those for the fixer solution. 2:213

A. Greater than
B. Less than
C. The same as

39. When developer temperature exceeds recommended values, it may lead to: 2:208
1. Excessive amount of chemical fog *2. A marked decrease in radiographic contrast*
3. Rapid deterioration of the solutions

A. 1 only
B. 2 only
C. 3 only
D. 1, 2 & 3

40. The termination of development process is accomplished by placing the film in a/an ______ solution. 2:209

A. Brine
B. Alkaline
C. Acid
D. Acrid

41. The normal developer replenisher rate per 14 inches of film is approximately: 2:213

A. 10-30 ml
B. 30-40 ml
C. 60-70 ml
D. 90-100 ml

42. The conversion of the exposed silver halide crystals into the black metallic silver image is the primary function of the: 2:208

A. Replenisher solution
B. Developer solution
C. Restrainer solution
D. Fixer solution

43. The principal causes of oxidation of the automatic developer solution is the exposure to: 2:208

A. X-rays
B. Heat
C. Air
D. Light

44. A film comes out of an automatic processor with numerous scratches in the direction that the film travels. These are caused by improperly seated: 2:444

A. Entrance rollers
B. Guideshoes
C. Drying tubes
D. Replenisher pumps

Pertaining to the radiographic images, answer questions 44. & 45.

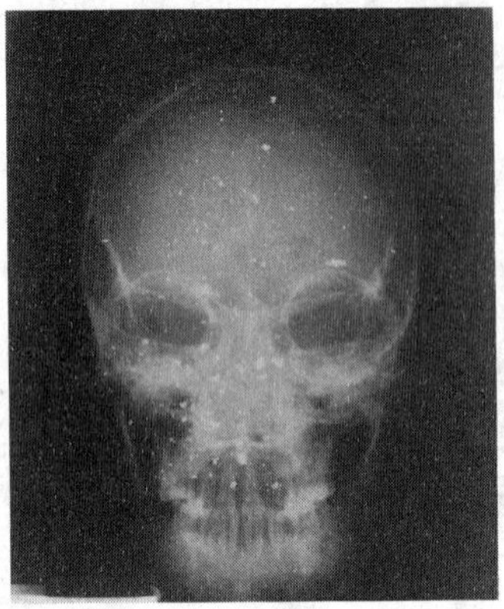

A

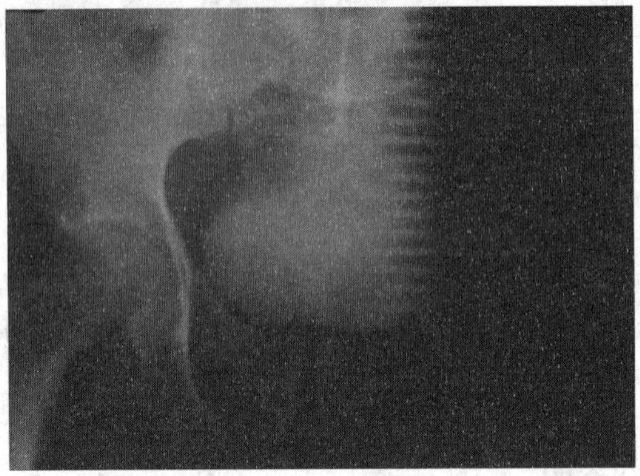

B

45. In the radiographic image of Film A, the decreased density marks appearing throughout the film were most likely caused by: 12:288

A. Dirty processing rollers
B. Dirty intensifying screens
C. Cracks in the safelight filter
D. Light leaks in the cassette

46. The serrated appearance near the center of Film B was caused by a white light exposure during:

A. Drying of the film
B. Fixation of the film
C. Washing of the film
D. Feeding of the film

47. The mixing and agitation of the processing solution during automatic film processing is accomplished by the: 2:212

1. Circulation system *2. Replenisher system* *3. Dryer system*

A. 1 only
B. 2 only
C. 3 only
D. 1, 2 & 3

48. The hardening agents (aldehydes) which are added to the automatic processing solutions also aid in the reduction of: 9:202

A. Rapid oxidation
B. Excessive acidity
C. Fungus growth
D. None of the above

49. A silver reclaimer unit serves to collect the silver that is removed from the: 3:306

1. Developer solution *2. Fixer solution* *3. Wash water*

A. 1 only
B. 2 only
C. 3 only
D. 1, 2 & 3

50. The major function of the clearing agent is to remove ________from the film emulsion during processing. 2:209

A. Unexposed silver bromide
B. Exposed silver bromide
C. Unexposed black metallic silver
D. Exposed black metallic silver

Tomography

Chapter 30

1. The visualization of a single tomographic section involves the motion of the tube and image receptor movement: 2:326

   A. At right angles to each other
   B. In the same direction as each other
   C. In opposite directions of each other
   D. Vertical to each other

2. Which is not a common term that has been employed to describe the visualization of a single tissue plane? 5:242

   A. Body section radiography
   B. Tomography
   C. Zonography
   D. Scanography

3. During tomography, all tissue planes ___________ the focus plane will show a marked degree of blurring. 15:257

   A. Above
   B. Below
   C. Both of the above
   D. Neither of the above

4. Which of the following tomographic movements is generally employed to provide the most uniform blur pattern? 2:328

   A. Circular
   B. Rectilinear
   C. Elliptical
   D. Hypocycloidal

5. The single thickness of tissue that remains in focus during tomography is called the: 4:180

   A. Objective plane
   B. Fulcrum plane
   C. Tomographic angle
   D. Grossman angle

6. The distance the tube travels during the time the tomographic exposure is being made is termed: 5:246

   A. Fulcrum arc
   B. Amplitude
   C. Total travel distance
   D. Source-to-image receptor distance

7. Of the following types of tube movement patterns, the one that requires the longest tube travel time during tomography is: 4:176

   A. Hypocycloidal
   B. Curvilinear
   C. Spiral
   D. Linear

8. In order to maximize the blurring during rectilinear tomography, the tube movement should be at a/an __________ angle to the dominant axis of the part radiographed. 5:245

   A. 30 degree
   B. 60 degree
   C. 90 degree
   D. 180 degree

9. Which type of tomographic movement is generally described as a cloverleaf pattern? 5:244

   A. Spiral
   B. Sinosoidal
   C. Rectilinear
   D. Hypocycloidal

10. If the amplitude remains constant during a tomographic exposure and the source-to-image receptor distance is increased, the tomographic section remaining in focus will: 4:180

    A. Increase
    B. Decrease
    C. Remain the same

11. In general, as the tomographic section becomes thinner: 5:247
1. Less blurring occurs 2. A longer scale of contrast results 3. Image definition is reduced

A. 1 only
B. 2 only
C. 3 only
D. 1, 2 & 3

12. Which of the following factors will have an effect on the thickness of the tomographic section? 2:322
1. Amplitude 2. Fulcrum level 3. Source-to-image receptor distance

A. 1 & 2 only
B. 1 & 3 only
C. 2 & 3 only
D. 1, 2 & 3

13. A cassette that contains a multiple set of screens is sometimes employed to produce multiple films with only one exposure. This is termed: 12:424

A. Stratigraphy
B. Plesiotomography
C. Body section radiography
D. Laminography

14. Which of the following exposure angles will produce a tomographic section of less than one millimeter? 5:246

A. 15 degrees
B. 36 degrees
C. 48 degrees
D. 64 degrees

15. The exposure angle during tomography can be reduced by: 2:326
1. Increasing the SID 2. Increasing the amplitude 3. Increasing the fulcum level

A. 1 only
B. 2 only
C. 3 only
D. 1, 2 & 3

16. Which is not a movement associated with pluridirectional tomography? 5:244

A. Hypocycloidal
B. Rectilinear
C. Sinosoidal
D. Elliptical

17. Which of the following technical settings would create the thinnest tomographic cut? 2:327

A. 100 cm SID, 50 cm amplitude
B. 125 cm SID, 75 cm amplitude
C. 100 cm SID, 75 cm amplitude
D. 125 cm SID, 50 cm amplitude

Using the following information, answer questions 18.-20.
A tumor is to be sectioned which is found 6 cm anterior from the posterior chest wall. This patient measures 30 cm.

18. Pertaining to the original statement, in order to visualize this lesion with the patient in the supine position, the fulcrum should be placed at:

A. 6 cm
B. 17 cm
C. 23 cm
D. 30 cm

19. Pertaining to the original statement, if the fulcrum was set at 6 cm, the point of maximum blurring would be found at the level of __________ in this supine patient. 5:244

A. 23 cm
B. 17 cm
C. 6 cm
D. 3 cm

20. In order to obtain a midline section of this patient, the fulcrum should be placed at: 5:244

A. 30 cm
B. 15 cm
C. 11.5 cm
D. 3 cm

Using the following information, answer questions 21. and 22.
A patient's gallbladder is located 10 cm posterior to the anterior abdominal wall. The patient measures 24 cm in the anterior-posterior diameter. 5:245

21. Pertaining to the original statement, at what level would the fulcrum be set to visualize the gallbladder with the patient supine? 1:40:III

A. 24 cm
B. 14 cm
C. 12 cm
D. 10 cm

22. Pertaining to the original statement, at what level should the fulcrum be set to visualize the cystic duct? 1:40:III

A. 18 cm
B. 16 cm
C. 14 cm
D. 12 cm

23. During a tomographic study, if the fulcrum setting is changed from 5 cm to 9 cm with the exposure angle remaining the same for both exposures, the result will be a: 2:326
1. Different cut level 2. Thicker cut 3. Improved resolution

A. 1 only
B. 2 only
C. 3 only
D. 1, 2 & 3

24. The procedure for producing a thick tomographic section using 5°-10° exposure angle is often called: 4:178

A. Zonography
B. Stratigraphy
C. Addography
D. Scanography

25. The amount of stereo-shift distance is based on the ratio of optimum viewing distance and: 2:333

A. Cornea-pupil distance
B. Interpupillary distance
C. Pupil diameter
D. None of the above

26. A method of producing a subjective three-dimensional representation is termed: 2:333

A. Stereoradiography
B. Scanography
C. Tomography
D. Plesiotomography

27. A stereographic examination is performed at 100 cm SID. The total amount of shift required is: 2:332

A. 5 cm
B. 10 cm
C. 15 cm
D. 20 cm

28. Stereoradiography requires: 2:333

A. One film and one exposure
B. One film and two exposures
C. Two films and one exposure
D. Two films and two exposures

29. The total shift distance during stereoradiography is found by taking ________ of the ________. 2:332

A. 5%, OID
B. 10%, SID
C. 15%, OID
D. 15%, SID

30. A stereographic exam of an erect chest is being performed at an 180 cm distance. The proper total tube shift and the direction is: 2:332

A. 18 cm vertical
B. 18 cm horizontal
C. 9 cm vertical
D. None of the above

Exercise No. 1

In each set of technical factors listed, solve for the technique which give the greatest density.

	mA	Time	kVp	Screen Speed	Grid Ratio	SID in cm
1.	A. 200	150 ms	86	50	6:1	150
	B. 300	150 ms	94	200	6:1	150
	C. 100	200 ms	82	100	6:1	200
	D. 200	50 ms	84	50	6:1	100
2.	A. 300	250 ms	82	100	8:1	100
	B. 200	500 ms	78	100	16:1	100
	C. 100	250 ms	72	100	12:1	100
	D. 400	150 ms	78	100	12:1	200
3.	A. 100	150 ms	86	100	8:1	100
	B. 100	100 ms	88	50	6:1	200
	C. 200	100 ms	84	100	5:1	100
	D. 300	750 ms	76	200	12:1	100
4.	A. 300	330 ms	74	100	12:1	200
	B. 400	50 ms	78	100	No grid	100
	C. 100	500 ms	76	300	8:1	200
	D. 300	150 ms	76	50	12:1	200
5.	A. 200	50 ms	96	400	12:1	100
	B. 200	70 ms	84	100	8:1	200
	C. 200	150 ms	86	200	8:1	200
	D. 200	150 ms	96	400	6:1	100
6.	A. 400	50 ms	90	100	5:1	100
	B. 600	30 ms	88	100	8:1	100
	C. 400	70 ms	94	50	6:1	100
	D. 300	150 ms	96	100	No grid	100
7.	A. 300	25 ms	78	400	8:1	100
	B. 800	25 ms	78	100	8:1	100
	C. 600	33 ms	72	400	12:1	200
	D. 300	15 ms	84	200	16:1	100
8.	A. 1000	1 ms	90	100	6:1	100
	B. 800	1 ms	92	100	12:1	150
	C. 400	7 ms	94	200	8:1	100
	D. 800	2 ms	88	50	16:1	200
9.	A. 300	5 ms	80	50	16:1	200
	B. 200	1.5 sec.	90	No screen	No grid	100
	C. 500	1 ms	82	50	12:1	200
	D. 300	10 ms	87	No screen	6:1	100
10.	A. 50	1.5 sec.	67	50	6:1	200
	B. 25	5.0 sec.	60	100	8:1	100
	C. 100	1.5 sec.	67	200	No grid	100
	D. 50	2.5 sec.	62	320	6:1	100

Density Analysis (All Factors)

Exercise No. 2

In each set of technical factors listed, solve for the technique which give the greatest density.

	mA	Time	kVp	Screen Speed	Grid Ratio	SID in cm
1.	A. 600	50 ms	72	50	8:1	100
	B. 400	50 ms	88	50	6:1	200
	C. 300	50 ms	92	50	6:1	100
	D. 200	50 ms	86	50	8:1	200
2.	A. 200	150 ms	78	100	8:1	100
	B. 300	250 ms	86	400	6:1	200
	C. 200	150 ms	86	200	12:1	100
	D. 300	250 ms	84	160	5:1	200
3.	A. 500	50 ms	82	320	12:1	200
	B. 500	150 ms	84	160	8:1	200
	C. 100	600 ms	80	100	No grid	100
	D. 300	50 ms	82	50	No grid	100
4.	A. 50	35 ms	62	100	6:1	200
	B. 100	15 ms	70	200	8:1	200
	C. 50	25 ms	64	100	6:1	200
	D. 200	15 ms	72	200	6:1	100
5.	A. 100	300 ms	89	160	12:1	100
	B. 300	300 ms	91	320	16:1	100
	C. 400	200 ms	87	320	8:1	100
	D. 400	100 ms	86	100	8:1	100
6.	A. 300	70 ms	105	100	12:1	100
	B. 800	30 ms	103	200	16:1	150
	C. 500	40 ms	100	100	12:1	100
	D. 300	150 ms	103	200	6:1	100
7.	A. 600	70 ms	86	50	12:1	200
	B. 100	150 ms	84	320	12:1	200
	C. 800	1.0 sec.	87	No screen	6:1	200
	D. 100	150 ms	92	100	8:1	100
8.	A. 200	50 ms	80	100	No grid	100
	B. 200	10 ms	82	200	12:1	200
	C. 200	70 ms	80	400	8:1	200
	D. 200	70 ms	82	400	No grid	100
9.	A. 400	250 ms	78	100	6:1	200
	B. 800	150 ms	76	160	8:1	200
	C. 600	300 ms	74	320	8:1	200
	D. 400	150 ms	82	100	6:1	200
10.	A. 300	750 ms	92	300	12:1	200
	B. 200	500 ms	88	400	8:1	150
	C. 50	2.5 sec.	94	320	6:1	100
	D. 600	300 ms	86	100	6:1	200

BIBLIOGRAPHY FOR SECTION II & III & IV

1. Berrett, A., Brunner, S. Valvassori, G.E. Modern thin Section Tomography, 1st Edition, Charles C. Thomas Co., 1975
2. Bushong, S.C., Radiologic Science For Technologists, 5th Edition C.V. Mosby Co., 1993
3. Coulam, C.M., Erickson J.J., Rollo F.D., The Physical Basis of Medical Imaging, 1st Edition, Appleton-Century-Crofts, 1981
4. Cullinan, J.E., Cullinan A.M., Producing Qualityradiographs, 2nd Edition, J.B. Lippincott Co., 1994
5. Curry T.S., Dow Dey J.E., Murry E.C., Christensen's Introduction To The Physics of Diagnostic Radiology, 4th Edition, Lea & Febiger, 1991
6. Donohue D.P., An Analysis of Radiographic Quality, 2nd Edition, University Park Press, 1984
7. Frankel R., Radiation Protection for Radiologic Technologists, 1st Edition, McGraw Hill Book Co., 1976
8. Ehrilch, R.A., Givins, E.M., Patient Care in Radiography, 4th Edition, C.V. Mosby, 1993
9. Jenkins D., Radiographic Photography, 1st Edition, University Park Press, 1980
10. Noz, M.E., Maguire, G.Q., Radiation Protection in the Radiology Health Sciences, 2nd Edition, Lea & Febiger, 1985
11. Prasad K.N., Human Radiation Biology, 1st Edition, Harper Row, 1974
12. Selman J., Fundamentals of X-Ray and Radium Physics, 8th Edition, Charles C. Thomas Publishers, 1994
13. Statkiewicz M.A., Radiation Protection in Medical Radiography, 2nd Edition, Multimedia, 1993
14. Thomas C. Tabers, Cyclopedic Medical Dictionary, 16th Edition, F.A. Davis, 1985
15. Thompson T.T., Cahoons, Formulating X-Ray Techniques, 9th Edition, Duke University Press, 1977
16. Torres L.S., Morrill C., Basic Medical Techniques and Patient Care, 4th Edition, J.B. Lippincott Co., 1993
17. National Council on Radiation Protection, No. 39, 1974
18. National Council on Radiation Protection, No. 49, 1976
19. National Council on Radiation Protection, No. 116, 1993
20. National Council on Radiation Protection, No. 102, 1989
21. National Council on Radiation Protection, No. 105, 1989

To the right of each question is a multi-digit number, eg. 2:342. The first number indicates the reference text; after the colon is the page number for the information.

Section IV
Explanations & Answer Keys
Image Production and Evaluation

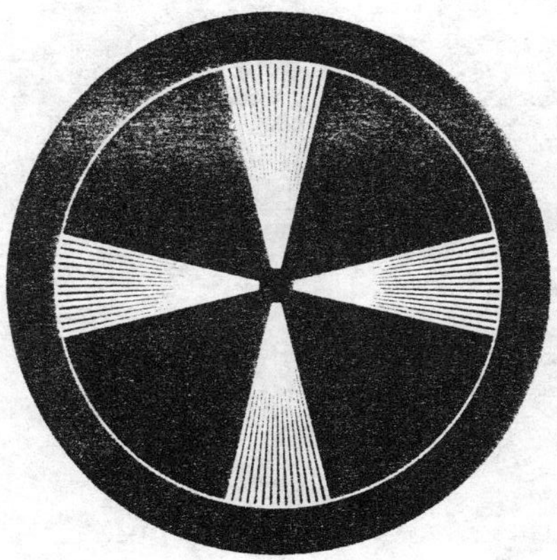

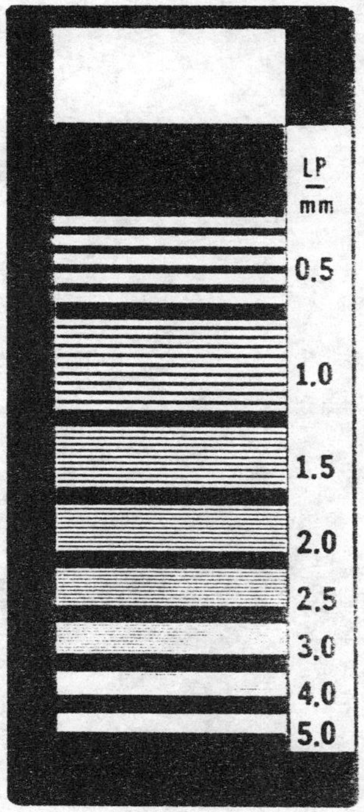

Radiographic Quality

Chapter 23

1. B The radiographic density or the amount of blackness in the image is directly proportional to the exposure of the film's silver bromide crystals to the light emitted by the intensifying screens and the direct action of the x-rays. The action of the development solution speeds the conversion of the exposed silver bromide crystals into the black metallic silver components of the radiographic image.

2. D Technique charts must take into account all the technical and equipment factors which effect radiographic density and contrast of the image. These include the intensifying screen speed, grid construction, collimation, filtration and source-to-image receptor distance.

3. B This problem is based on the principle stated in the 15% kVp rule. The first step in solving the problem is to determine that 20 mAs will be required at 86 kVp to maintain a similar density. The second step uses the formula mA=mAs/time to obtain the appropriate mA station at a 50 ms (.05 sec.) timer setting.

4. D The milliampere - seconds value determines the number or quantity of photons produced in the x-ray beam. This factor has a strong direct relationship to the patient's dose and the radiographic density of the image.

5. C The kilovoltage value indicates the energy and penetrability of the photons in the beam. This factor effects both the radiographic density and the relative difference between the adjacent densities termed contrast.

6. B The problem can be solved by using the formula New mAs = Original mAs x New SID^2/Original SID^2.

7. B A radiograph with excessive density appears darker than desired due to excessive amounts of radiation that have reached the imaging receptor.

8. B Scattered radiation is an unavoidable consequence of x-radiation's interaction with matter. The scattered radiation produced in the tissue can be reduced by using a lower kVp setting, reducing the thickness of the tissue, increasing the degree of beam limitation.

9. C The 15% kVp rule is used to equate the change required in the amount of kVp which results in the same change in radiographic density as a doubling or halving (50%) of the mA's value.

10. A Many pathological conditions have a considerable effect on the overall density of a tissue and the resulting radiographic density. Conditions associated with excessive bone growth (acromegaly) and bone repair (Paget's disease) would require an increase in technique of about 30-50% in the mA's value to maintain the desired density.

11. C The increased number of density steps in the image III indicates this film has a long scale or low contrast caused by the greater penetration of a higher kVp setting. The decreased number of density steps in image I is associated with a less penetration, a lower kVp and a short scale contrast.

12. A See answer to question 11 of this section.

13. D An increase in the thickness of the tissue will have a number of negative imaging consequences, including an increased scattered radiation production, a lowering of the contrast, more primary beam attenuation and the resulting loss of radiographic density, as well as an increase in the amount of image unsharpness and magnification.

14. A The amount of radiation absorbed in a patient (dose) is directly proportional to the number of x-ray photons to which the patient is exposed; i.e., the milliampere - seconds (mAs) value.

15. C Though scattered radiation is directly effected by the patient's size and density, the main technical factor used for the control of contrast is the kilovoltage peak (kVp).

16. D The kilovoltage peak is perhaps the single most important technical factor since its modification effects beam quality, penetrability, attenuation, image contrast and density, as well as the exposure latitude.

17. D A lower radiographic contrast combined with an excessive radiographic density will require a reduction in penetration (kVp), provide a higher contrast and a decrease in mAs value to reduce the density of the image.

18. B Any pathological condition associated with excessive fluid production (pulmonary edema) and its build up in tissues will require a 30 - 50% increase in mA's value to compensate for the increased attenuation of fluid and maintain the appropriate radiographic density.

19. B As the number of x-ray photons passing through the patient and hitting the film increases, more energy is transferred into the film emulsion increasing the amount of darkness or radiographic density of the image.

20. C Tissues with relatively low atomic numbers such as air and gas are more easily penetrated and therefore will have the lower absorption or attenuation coefficients to photons in the diagnostic energy region.

21. D The distance between the crests of a photon, the wavelength, is inversely proportional to the energy of the photon. Therefore, as the wavelength increases, the energy decreases causing less scatter production, less penetration, higher beam attenuation and an image having a higher contrast.

22. A The relative difference in density between adjacent areas of a radiograph is termed radiographic contrast.

23. C Tissues with the greatest differences in their atomic numbers will have the greatest differences in their degree of differential attenuation. The greatest difference in the listed tissues is between soft tissue and bone.

24. A Most osteoblastic disorders are related to a substantial increase of bone growth and a higher degree of attenuation. Therefore, an increase in the mAs value is required to maintain the appropriate image density. Osteolytic, porotic and necrotic conditions are more easily penetrated by the beam.

25. D The excess fluids seen in pleural effusion, empyema, ascites and anthracosis will all require an increase in technique to compensate for the greater radiation attenuation occurring in these fluids.

26. D A photon depending on its energy and nature of the tissue traversed will result in one of three processes; penetration, attenuation, or scattering. By far the most common event in thick tissues possessing a substantial density is the development of scattered radiation. During radiographic studies, it has been estimated that scatter would form about 90% of the image in a non-grid abdomen image and nearly 50% of the image in a non-grid chest radiograph.

27. B The problem is solved using the formula:

$$\text{New SID} = \sqrt{\frac{\text{New mAs x Original SID}^2}{\text{Original mAs}}}$$

28. B Substances with a low atomic number (nitrogen, oxygen, hydrogen) are only poorly attenuated by x-rays and are classified as radiolucent or radiotransparent.

29. B Voluntary motions or those under the conscious control of the patient are most easily accomplished through use of proper, clearly stated instructions and a cooperative patient.

30. A Empyema refers to a condition in which a fluid (pus) is leaking into the pleural cavity resulting in an increase in the tissue density of the chest. Gout, acute osteomyelitis and a pneumothorax all require a dec.'ease in the degree of penetration.

31. B Since bone has the highest tissue density in the body, the radiographic densities corresponding to bone tissue will posses a low radiographic density.

32. C A high contrast image consisting of a small number of density shades (short scale) with a wide density difference between them is most commonly seen when low kVp techniques are employed.

33. C The shortest exposure time setting will be associated with the least involuntary motion appearing on the radiograph.

34. B A higher kilovoltage setting is associated with higher penetration, a lower or longer scale contrast and an increased exposure latitude.

35. D According to the law of reciprocity, the same mAs value, assuming all other variables are maintained, should provide the same radiographic density. On a non-screen system, 100 mA at .8 sec. and 800 mA at .1 sec. will provide the same density. With a cassette type imaging system, the radiograph taken with the longest timer setting will be darker due to the failure of the law of reciprocity.

36. A Excessive penetration will lower contrast of a radiographic image thereby reducing its quality. The use of lower kilovoltage settings for smaller body parts will help prevent over-penetration and maintain the high contrast desired in these images.

37. B A higher kilovoltage (kVp) setting will result in a higher penetration, wider exposure latitude, lower contrast, long scale contrast, increased scattered radiation production and a higher radiographic density.

38. C An x-ray beam with a low quality has a low penetrability and is sometimes referred to as a soft x-ray beam.

39. C Because the substance with the highest atomic number has the highest ability to stop the x-rays, lead with an atomic number of 86 will have the highest degree of radiopacity.

40. A A small difference in the attenuation properties of a tissue will provide only small differences in the radiographic densities and produce a low or long scale contrast.

41. D A significant change in radiographic density will require a change of 25-30% in the mAs value or about 6% change in the kVp.

42. A In order to accomplish a 60% reduction in time, the initial timer setting is multiplied by .6. This value is then subtracted from the original timer value.

43. C The longest timer setting would be associated with the greatest amount of involuntary motion.

44. D The marked expansion in the diameter of the chest cavity in patients with acute kyphosis (humpback) will often necessitate an increase in technique to maintain the appropriate radiographic density. Leprosy, giant cell tumors and osteoporosis are all associated with a decrease in overall density of the tissues.

45. A The principal factors effecting the amount of attenuation of a tissue are its effective atomic number and thickness.

46. A The log value of the light transmitted through a film is called a radiographic density.

47. D The problem is solved using the formula New mAs = Original mAs x New SID^2/Original SID^2.

48. D The mAs can be reduced by 1/2 by a 15% increase in the kVp without a significant effect on radiographic density. This technical modification can be used to reduce the patient dose and the exposure time while increasing the exposure latitude and penetration of the beam.

49. C Since a homogenous object shows no differences in the ability to attenuate radiation, it will show only a single density and no radiographic or subject contrast.

50. B The higher average tube current of a 3-phase unit enables the use of about half of the mAs necessary in a full wave rectified unit to provide the same beam intensity.

51. D The shortest exposure time is associated with the least patient motion.

52. A When a variable kVp technique chart is used, a 2 kVp change is recommended for every 1 cm change in tissue thickness to maintain a consistent radiographic density on the image.

53. B See explanation number 39 of this section.

54. B The damage to the alveolar sacs that occurs in patients with emphysema increases the volume of air in the lungs necessitating a decrease in technique to provide the desired radiographic density. Empyema, pulmonary edema and silicosis are all associated with an increased tissue density of the lung.

55. B The increased thickness of the tissue in the oblique position would result in less radiation reaching the image receptor, and a lower radiographic density. The higher degree of scattered radiation produced by this thick tissue will result in a lower contrast of the radiographic image.

56. C Assuming the law of reciprocity can be applied, a technique providing the same mAs and kVp should produce a similar density.

57. C The added density of the fluid from ascites will tend to attenuate more radiation resulting in a lower than normal radiographic density.

58. A The combination of the milliampere station and the timer value (mAs) is used to give an estimate of the amount of radiation used during the imaging process.

59. A See explanation number 20 of this section.

60. C A small number of widely varying density shades is associated with a high or short scale contrast will result from the use of the lowest kVp value.

61. C An increase in the radiographic density is accomplished by an increase in kVp, mA, time or decreased source to image receptor distance.

62. D Large amounts of intestinal gas which may be present in obstructed patients often causes a decrease in the average tissue density. The resulting image will show excessive penetration and an increase in radiographic density.

63. C The problem is solved by using the formula mAs = mA x time.

64. B The vast majority of the scattered radiation produced in a patient exposed to x radiation in the diagnostic energy region results from the Compton's event. During this interaction the scattered photons produced are normally only slightly less energetic than the primary photons that caused the interaction.

65. A The problem is solved using the formula mA = time/mAs.

66. B The air gap effect causes divergent scattered photons to miss the image receptor. This effect can be maximized by an increase in the object-to-image receptor distance. The effect is <u>not</u> due to the absorption by air as its name implies.

67. D The subject contrast of an individual is related to one's general state of health, age, degree of development, and number of pathological conditions which can modify the density of the tissue.

68. D Scattered radiation is more likely to occur in thick, dense tissues and at higher kilovoltage levels.

69. B See explanation number 9 of this section.

70. C This problem is solved using the formula New Time = Original Time x New SID^2/Original SID^2.

71. C The damage caused by Paget's disease often results in an excessive over-growth of bone which may require additional amounts of radiation to provide the appropriate density. Emphysema, osteoporosis and pneumothorax are all associated with a decrease in tissue density.

72. A Contrast and scale contrast are primarily controlled by changes in the kilovoltage peak (kVP).

73. D This problem is solved using the formula New mAs = Original mAs x New SID^2/Original SID^2.

74. C By definition, short scale contrast is described as having an image consisting of a small number of widely varying density shades.

75. D See explanation number 9 of this section.

76. C This problem is solved by using the formula: $\text{New SID} = \sqrt{\dfrac{\text{New mAs x Original SID}^2}{\text{Original mAs}}}$

77. B See explanation number 35 of this section.

78. D — The addition of a high atomic number contrast agent such as barium or iodine will increase the amount of radiation attenuation of the organ into which it is placed. An increase in the technical setting; i.e., mA, time or kVp would be required to maintain the appropriate density due to increased attenuation of the beam.

79. C — The high subject or tissue contrast of the chest combined with its moderate tissue density will permit acceptable radiographs of this area without the necessity of a grid.

80. D — In order to provide accurate results, technique (exposure) charts will always assume the acceptable calibration and performance of the x-ray machine, processor and all accessory equipment. To ensure the proper technical settings are selected, all patients must be carefully measured using callipers.

81. C — An increase in radiographic density will require an increase in the mAs value. According to the 15% kVp rule, 15 mAs at 90 kVp and 60 mAs at 66 kVp will give the same density as 30 mAs at 78 kVp.

82. C — See explanation number 37 of this section.

83. B — If a reduction in penetration is desired, the kilovoltage setting should be reduced.

84. C — The problem can be solved using the formula: Time = mAs/mA.

85. B — The problem can be solved by using the formula:

$$\text{New SID} = \sqrt{\frac{\text{New mAs} \times \text{Original SID}^2}{\text{Original mAs}}}$$

86. A — A special ruler-like device called a caliper is used to measure tissue thickness.

87. B — The reduction of scattered radiation by a collimator will reduce the radiographic density while producing images possessing a higher contrast.

88. A — According to the mAs formula a decrease in mA setting will require a proportional increase in the timer setting to maintain the radiographic density.

89. D — The chart indicates that the intensity of scatter is directly proportional to the thickness of the patient; i.e., larger patient creates more scattered radiation.

90. B — A compression device will reduce the amount of scattered radiation produced in the tissue by reducing its thickness.

91. 50 ms — All problems can be solved through the use of the 15% kVp rule to determine the amount of mAs required at a given kVp value, and then an application of the appropriate formula: time = mAs/mA or mA=mAs/time.

92 100 ms — See explanation number 91 of this section.

93. 64 kVp — See explanation number 91 of this section.

94. 86 kVp — See explanation number 91 of this section.

95. 600 mAs — See explanation number 91 of this section.

96. 100 mAs — See explanation number 91 of this section.

97. A — The lowest kVp provides the highest contrast.

98. C — The highest mAs results in the greatest dose.

99. A The shortest time provides for the least motion.

100. D The need to penetrate through the barium type contrast agent will require the use of kilovoltage settings in the 90 - 100 kVp range.

Time - Distance

Exercise No. 1

1. 5 ms
2. 10 ms
3. 40 ms
4. 46.2 ms (rounded to 50)
5. 110 ms
6. 970 ms
7. 28 ms
8. 840 ms
9. 27 ms (rounded to 30)
10. 100 mA
11. 100 mA
12. 900 mA
13. 200 mA
14. 800 mA
15. 150 mA
16. 100 mA
17. 22.5 mAs
18. .2 mAs
19. .5 mAs
20. 1.3 mAs
21. 35.5 mAs
22. 6.25 mAs
23. 36 mAs
24. 270 mAs
25. 165.3 mAs

Time - Distance

Exercise No. 2

1. 60 ms
2. 30 ms
3. 80 ms
4. 81 ms
5. 800 ms
6. 400 ms
7. 1.35 sec
8. 375 ms
9. 40 ms
10. 100 mA
11. 800 mA
12. 100 mA
13. 200 mA
14. 400 mA
15. 2400 mA
16. 100 mA
17. 50 mAs
18. 60 mAs
19. 5 mAs
20. 4 mAs
21. 270 mAs
22. 6 mAs
23. 14.2 mAs
24. 222.2 mAs
25. 16.5 mAs

Advanced Time - Distance

Exercise No. 1

1. 400 cm
2. 200 cm
3. 400 cm
4. 90 cm
5. 56.6 in.
6. 100 cm
7. 17.3 in.
8. 173.2 in.
9. 244.9 in.
10. 120 cm
11. 50 cm
12. 100 cm
13. 113. 1 cm
14. 50 inches
15. 56.6 in.
16. 50 cm
17. 50 cm
18. 50 cm
19. 400 cm
20. 300 cm
21. 56.6 inches
22. 142.3 cm
23. 365.1 cm
24. 183.7 cm
25. 346.4 cm

Advanced Time - Distance

Exercise No. 2

1. 400 cm
2. 200 cm
3. 600 cm
4. 75 cm
5. 51.9 in.
6. 100 cm
7. 30.9 in.
8. 141 cm
9. 252.9 cm
10. 120 cm
11. 200 cm
12. 360 cm
13. 200 cm
14. 120 in.
15. 50.9 in.
16. 57.7 cm
17. 100 cm
18. 100 cm
19. 400 cm
20. 108 cm
21. 51.6 inches
22. 74.5 cm
23. 132. 3 cm
24. 75 cm
25. 322.5 cm

Recorded Detail

Chapter 24

1. C The image appearing on a radiograph consists of a clearly defined central region called the umbra surrounded by an area of unsharpness called the geometric or focal-spot blur.

2. A The geometrically recorded detail of an image is under the influence of four main elements: the distance factors (SID, OID, SOD), the size of the focal spot, type and degree of beam restriction, and imaging system design.

3. C — The imaging systems factors which increase geometrically recorded detail are those which decrease the speed; namely, smaller crystal size and thinner crystal layer of the screen. Increasing the amount of collimation reduces the amount off-focus radiation that adds to the geometric (focal-spot) blur or unsharpness.

4. C — The degree of image radiation blur or image unsharpness can be obtained from the unsharpness formula: Us = focal spot size (FSS) x SID)/SID - OID. The exposure factors kVp and mAs do not effect image blur.

5. A — Size distortion or magnification caused by the relationship of the distance factors is determined by the formula: Magnification factor (MF) = SID/SOD or SID/(SID - OID).

6. C — Shape distortion or the visualization of an image which does not conform in shape to the actual object is commonly seen in radiography due to the prevalence of 3 dimensional objects being presented on a 2 dimensional image display and the use of angulated techniques. Depending on the shape and orientation of the original object, a tube angulation may cause a shape distortion called elongation or foreshortening.

7. B — Size distortion (magnification) is increased by employing a long object-to-image receptor distance (OID), shorter source-to-object distance (SOD), and a shorter source-to-image receptor distance.

8. B — Use the formula in explanation number 5 of this section.

9. B — Improper alignment of the tube, object and image receptor may expose the object to the divergent rays off of the central axis causing shape distortion of the image.

10. B — Use the formula used in explanation number 4 of this section.

11. C — All disks or spheres on the central axis will form images with a circular shape When the disk (object) is lateral to the central axis, their image will appear elliptical.

12. B — The long object-to-image receptor distance (OID) employed in all magnification techniques provides an air gap by which much of the divergent scattered radiation misses the image receptor.

13. C — Even though tube angulation will introduce some degree of shape distortion to the image, they are extremely beneficial for the separation of superimposed tissues.

14. D — Use the formula in explanation number 5 of this section.

15. C — The reduction of the radiation received by a given object as the distance varies is directly related to spread or divergence of the beam.

16. B — An increased source-to-image receptor distance (SID) is associated with less magnification, less unsharpness (blur) less radiographic density and greater geometrically recorded detail.

17. D — Use the formula used in explanation number 5 of this section.

18. D — The unsharp border surrounding the true image shadow (umbra) is effected by changes in the distance factors, effective focal spot size, image receptor design and the degree of beam limitation.

19. A — This problem can be solved by the use of the second magnification formula: MF = Image width/Object width.

20. C Magnification (size distortion) can be reduced by increasing the source-to-image receptor distance (SID) and/or source-to-object distance (SOD).

21. B A combination of the smallest focal spot and the slowest imaging system will normally provide for the greatest geometrically recorded detail.

22. B See explanation number 11 of this section.

23. D Shape and size distortion are the two common forms of image distortion. Shape distortion is effected by tube angulation and patient rotations, where size distortion (magnification) is controlled by the distance factors. The size of the focal spot has a great effect on geometrically recorded detail, but little or no effect on the degree of size distortion.

24. B This problem is solved using the combined magnification formula: Image width = (Object width x SID)/(SID - OID).

25. A During any technique requiring a small degree of image unsharpness (blur), the use of an extremely small fractional focal spot of less than .3 mm in size is required.

26. D Use the formula in explanation number 4 of this section.

27. D See the explanation number 18 of this section.

28. A Magnification (size distortion) is principally controlled by the distance factors, the source-to-image receptor (SID), object-to-image receptor (OID), and source-to-object (SOD) distances.

29. C Use the formula in explanation number 4 of this section.

30. A If an image were produced by theoretical point source, no image unsharpness (blur) would be produced regardless of the degree of magnification.

31. D Use the formula in explanation number 24 of this section.

32. D See explanation number 18 of this section.

33. A The anode angulation and the resulting anode heel effect has a direct influence on the degree of coverage by beam.

34. D Since there is no change in the distance, no uneven magnification of the object and their corresponding images is seen.

35. C Magnification (size distortion) is principally effected by changes in the distance factors. The size of the focal spot will not significantly alter the magnification of the object.

Magnification

Exercise No. 1

1. 2.5	11. 2	21. 4
2. 4	12. 1.33	22. 2.5
3. 2	13. 2.5	23. 1.66
4. 1.5	14. 2	24. 1.11
5. 1.33	15. 2	25. 1.25
6. 1.5	16. 3	
7. 2	17. 2	
8. 3	18. 1.66	
9. 5	19. 1.33	
10. 4	20. 2	

Magnification

Exercise No. 2

1. 2	11. 4	21. 1.5
2. 2	12. 2	22. 1.5
3. 3	13. 1.66	23. 2.5
4. 2.66	14. 3	24. 1.25
5. 1.66	15. 2	25. 1.2
6. 2.5	16. 2	
7. 3	17. 4	
8. 3	18. 1.25	
9. 10	19. 4	
10. 3	20. 1.5	

Advanced Magnification

Exercise No. 1

1. 70 cm	13. 50 cm
2. 30 cm	14. 50 cm
3. 60 inches	15. 135 cm
4. 120 in.	16. 66.7 cm
5. 75 cm	17. 100 cm
6. 66.6 cm	18. 60 cm
7. 20 cm	19. 100 %
8. 41.5 cm	20. 33 %
9. 45 cm	21. Four
10. 60 in.	22. 6cm by12cm
11. 50 in.	23. 15 cm
12. 33.3 cm	24. 16cm by 20cm
	25. 45 cm^2

Advanced Magnification

Exercise No. 2

1. 50 cm	13. 100 cm
2. 41.6	14. 33.3 cm
3. 150 in.	15. 150 cm
4. 20 in.	16. 20 cm
5. 400 cm	17. 60 cm
6. 30 cm	18. 33.3 cm
7. 80 cm	19. 300%
8. 12.5 cm	20. 100%
9. 25 cm	21. 16 times
10. 10 in.	22. 15cm by 22.5cm
11. 12.5 in.	23. 18 cm
12. 60 cm	24. 21cm by 45cm
	25. 36 cm^2

Image Unsharpness

Exercise No. 1

1. 1	11. .18
2. .60	12. 1.5
3. .50	13. .02
4. .22	14. .05
5. .13	15. C
6. .86	16. A
7. .10	17. A
8. 1	18. C
9. .66	19. A
10. .1	20. B

Image Unsharpness

Exercise No. 2

1. 2	11. .33
2. .60	12. 1.3
3. .33	13. .38
4. .22	14. .06
5. .13	15. C
6. 1.14	16. B
7. .15	17. A
8. 2	18. C
9. .33	19. A
10. .11	20. B

Beam Limiting Devices

Chapter 25

1. A — Beam restricting devices; i.e., cones and collimators, tend to both reduce the scattered radiation produced in the patient and that reaching the image receptor. This loss of radiation will decrease the radiographic density of the resulting image unless an increase in incident beam intensity (mAs, kVp) is made prior to the exposure.

2. B — The tilting of the anode in the x-ray tube results in a beam intensity that varies from the head to the foot of the x-ray table. This anode heel effect provides a beam with a higher intensity towards the cathode side of the tube and a lower output on the anode side of the tube. In most modern rooms the cathode is on the right of the tube closest to the foot of the table.

3. A — The term collimation has to a large extent replaced the old term called coning. Collimation refers to the restriction of the primary beam to the desired dimensions on the radiograph.

4. B Most tube assemblies are designed to place the anode to the left or head side as a radiographer faces the table. This arrangement will tend to increase the intensity of the beam toward the heavier body parts in the lower trunk and pelvis.

5. C An automated collimator consists of two sets of motor driven shutters, a light localizing mirror, light source, and series of electronic sensors located in or under the bucky tray which automatically collimate the beam to the borders of the image receptor.

6. B Since the greatest radiation intensity is found at the cathode side of the tube, the heaviest body parts should be placed under the cathode toward the foot of the table.

7. B The entrance shutters which extend from the collimator into an opening in the bottom of the tube housing serve to reduce the widely divergent, off-focus radiations produced from the anode disk or stem.

8. D In order to produce light and radiation fields with the same physical size, the distance from the x-ray source to the mirror must be the same as the light source to the mirror distance. This also requires mirror to be oriented to an angle of 45 degrees to the two sources.

9. A All beam limiting devices serve to reduce the size and amount of the tissue exposed to the beam, thus reducing the scattered radiation produced in the patient. This results in an image consisting of improved resolution, decreased radiographic density, and reduced dose to the patient.

10. B Use the formulas: area = length x width and area $=\pi r^2$ to find the smallest field covered by the beam.

11. D In order to provide a higher intensity to the lower thoracic cavity in an erect patient, the anode is placed toward the head, superior to the cathode.

12. B In order to restrict the radiation beam before it reaches the patient, all BID's are located between the source of the radiation and the patient.

13. C The sensing of the cassette size in the bucky tray and the subsequent automatic shutter adjustments by the collimator is termed positive beam limitation (automatic collimation).

14. C The use of tight collimation or use of an accessory cylinder will reduce both the scattered radiation produced in the patient and the percentage of radiation reaching the image receptor. This will produce a higher contrast while decreasing the radiographic density. A 30-50% increase in mAs or a 6-10 increase kVp setting is recommended to compensate for this expected drop in radiographic density.

15. D See explanation number 9 of this section.

16. B The anode heel effect is maximized when the entire length of the anode (large exposure field) is employed. This is most commonly the situation at a shorter (40") SID using the largest image receptor and collimator openings.

17. D This problem is solved by using the field coverage formula: Field Size = (SID x Size of the aperture opening)/Source-to-aperture distance (SAD).

18. C This problem is solved by using the field dimension formula: New field dimension = (Original field dimension x New SID)/Original SID.

19. A — The variation in radiation intensity across the x-ray beam, the anode heel effect, reduces the relative intensity of the beam on anode side (A) and increases the relative intensity on the cathode side (E). The intensity at central ray or axis is designated as 100%.

20. C — See explanation number 19 of this section.

21. B — The maximum reduction in off-focus radiation occurs from a beam limiting device which is placed as close to the patient as possible or as far from the source as possible.

22. B — In order to provide light and radiation field coincidence, the mirror must be in the beam equidistant from the x-ray and light sources at an angle of 45 degrees.

23. C — In order to meet current regulations, the collimator shutters like the tube housing, are considered to be primary barriers and are therefore required to have a lead equivalent thickness of 1.5 mm.

24. C — The four main types of beam limiting devices include the aperture diaphragm, radiographic cone, radiographic cylinder and the variable aperture diaphragm or collimator.

25. C — See explanation number 8 of this section.

26. 314 cm2 — This problem is solved using the area = πr^2 formula.

27. 1661 cm2 — This problem is solved using the area = πr^2 formula.

28. 450cm^2 — This problem is solved using the area = length x width formula.

29. 360 cm2 — This problem is solved using the area = length x width formula.

30. 600 cm2 — This problem is solved using the area = length x width formula.

Radiographic Grids

Chapter 26

1. A — In addition to holding the image receptor in the center of the table for an exposure, the bucky contains a stationary grid and a spring or motor driven device which moves the grid to blur the appearance of the grid lines.

2. D — A reduction in scattered radiation can be accomplished by the restriction of the radiation source prior to its reaching the object through the use of beam limiting devices or after it leaves the object by air gap techniques and radiographic grids.

3. C — The grid ratio (r) which is the single best factor for predicting grid efficiency and latitude is determined by the ratio of the height of the grid strips (h) to the distance between the grid strips (D): r=h/D.

4. C — A focus grid is designed with grid strips that parallel the divergence of the x-ray beam. The angled grid strips converge at a point over the center of the grid; a line drawn perpendicular from this point to the grid defines the optimum distance for the use of the grid.

5. D The amount of scattered radiation a grid eliminates is controlled by the grid ratio, total lead content and the number of lines per inch (frequency). The highest efficiency or amount of scatter reduction occurs with a lower frequency, high lead content and high grid ratio.

6. A The type of grid pattern which most parallels the divergence of the primary beam (focus grid) will also absorb the least amount of primary radiation to which it is exposed.

7. C Hollis Potter developed a device for the movement of a grid during the exposure. This device was later incorporated in the Potter-Bucky diaphragm (Bucky tray) to eliminate the appearance of the grid strips on the radiographic image.

8. C Radiographic contrast is primarily controlled by the kilovoltage peak of the primary beam and the ratio of the radiographic grid. The lowest or longest scale contrast results from the combination of a high kVp technique and low ratio grid.

9. A The amount of off-centering permitted before noticeable grid cutoff is seen is termed centering latitude. Distance latitude defines the range of source-to-image receptor distances that can be used before causing a noticeable amount of cutoff. Both distance and centering latitude ranges are increased as the grid ratio is reduced. High grid ratios are associated with more scatter clean up but narrow latitude ranges.

10. D The appropriate grid for a given situation is based on the degree of clean up required, the centering and distance latitudes that may be necessary and the range of kilovoltages that will be used.

11. B The reduced density seen on the right side of this image is caused by an uneven attenuation or grid cutoff due to failure to direct the central ray to the center of the grid.

12. D When grids are employed during a radiographic examination, they serve to improve contrast by the absorption or attenuation of the scattered radiation produced in the patient. Since this will also reduce the total number of photons reaching the image receptor, an increase in the mAs value selected is necessary to maintain the appropriate image quality. This image was made without the necessary technical compensation and shows a marked reduction in the radiographic density.

13. A The movement of the grid within the bucky tray must begin prior to the exposure and continue until the exposure is terminated to avoid the appearance of grid lines on the image. The movement of the grid in most modern units is started by the activation of the prep or boost stage of the exposure switch.

14. B Any grid which is designed with grid strips which run parallel with each other are called linear parallel grids.

15. B The increase in attenuation (absorption) of the primary beam and a slight increase in the object-to-image receptor distance required when the bucky tray is used are the two main reasons for the increase in technique that is required when a grid is utilized.

16. D The problem is solved using the grid ratio formula: grid ratio $r = h/D$. See explanation number 3 of this section.

17. D This problem is solved using the grid frequency formula: frequency $= 25.4/(D+d)$, when D is the distance between the strips and d is the thickness of the lead strip.

18. A See explanation number 8 of this section.

19. C — A linear grid is designed to have a series of grid strips running lengthwise parallel to the x-ray table. The term focus applies to the titling of this strip to match the divergence of the primary beam. The inability of a linear focus grid to maintain its alignment with the primary beam results in an uneven absorption of the primary radiation called grid cut-off. This occurs when the central ray is improperly centered, using the wrong distance or angling the tube across the grid strips. All linear grids permit tube angulations parallel to the grid strips without resulting in grid cut-off.

20. C — The efficiency or scatter clean up of a grid is directly proportional to grid ratio. Therefore, grids with a higher ratio eliminate more scattered radiation and have a higher contrast improvement factors than grids with lower ratios.

21. A — The design of many high ratio grids (16:1) requires the use of a relatively high 90-120 kilovolt peak range in order to provide acceptable attenuation factors.

22. A — In all grids, it is the high atomic number (82) of the lead strips that makes the attenuation of scattered radiation possible.

23. C — See explanation number 3 of this section.

24. B — Because the design of a focus grid will more readily follow the divergence of the primary beam, it has the lower amount of primary beam cut than a parallel grid.

25. D — See explanation number 4 of this section.

26. B — The amount of scatter clean up from a grid is primarily effected by two factors: grid ratio and the grid frequency. The highest clean up or grid efficiency will result from a combination providing the highest grid ratio and the lowest grid frequency.

27. B — The use of a higher grid ratio is associated with a lower radiographic density, higher radiographic contrast, improved scatter clean up and higher radiation exposure to the patient when the appropriate technical compensations are made to maintain the radiographic density.

28. B — Crossed grids which are comprised of two individual grids with their grid strips directed and right angles to each other have a major disadvantage of not permitting any tube angulated techniques. These grids otherwise have an excellent scatter clean-up, wide distance, and centering latitudes.

29. A — See explanation number 26 of this section.

30. C — Grid cut-off with a focus grid may occur from the use of the wrong source-to-image receptor distance, improper centering, off-leveling of the grid and angulating across the grid strips.

31. A — Though grids are associated with higher patient exposures, the reduction in the amount of scattered radiation reaching the image receptor dramatically improves the contrast of the image. The use of the grid will not decrease the amount of scattered radiation produced in the patient or object, only that which reaches the image receptor.

32. D — See explanation number 3 of this section.

33. B — Grid lines associated with moving (bucky) grid, though rarely seen, may result from failing to place the grid in motion prior to the exposure, having the grid moving too slowly, or having the grid stop during the exposure.

34. C The first practical radiographic grid was developed by Dr. Gustave Bucky in the early 1910's and though many design changes have been made, the basic form of the grid is still close to the original design.

35. B No device has been shown to be better at reducing the amount of scattered radiation reaching the image receptor than the radiographic grid.

36. B The combined (crossed) grid has a latitude which is only as large as its narrowest element or in this case the 8:1 grid.

37. B A cross grid consisting of two individual 6:1 grid at right angels with each other has the latitude ranges of a 6:1 grid and the grid efficiency of a 12:1 grid.

38. A The unusual (moire effect) appearance of this image was due to the placement of two separate grids on top of each other with their grid strips parallel to each other.

39. B A reduction of image density on the lateral edges of the image and a near normal density in the center of the image may occur from the use of an excessive SID, insufficient SID or, as it was in this case, an upside down grid.

40. B Though many different references exist, a general consensus finds that a 3 - 4 time increase in mAs is required to maintain appropriate density when an 8:1 grid is employed. Other conversion values are provided at the end of grid conversion exercises on pages 81 and 82 of this book.

41. D See explanation number 39 of this section.

42. C High ratio grids (16:1) are recommended for use with kilovoltage above 90 kVp. Low ratio grids (5:1 -6:1) are usually limited to kilovoltages below 80 kVp.

43. D See explanation number 8 of this section.

44. A A major advance in the application of grid was made in the 1920's by Dr. Hollis Potter and his design of a diaphragm capable of moving the grid during the exposure. Now commonly called a bucky tray, its proper name is the Potter-Bucky diaphragm.

45. C Though rarely used at the present time, the term canting refers to the tilting of the grid strips to center line in a focus grid.

46. B See explanation number 33 of this section.

47. A See explanation number 3 of this section.

48. B See explanation number 5 of this section.

49. C See explanation number 38 of this section.

50. D The radiographic contrast of an image is dependent upon the patient factors of tissue density and thickness, degree of beam limitation, the kilovoltage peak selected and the use of air gap techniques.

Grid Conversion

Exercise No. 1

1. 20 mAs	11. 20 mAs
2. 63 mAs	12. 75 mAs
3. 27 mAs	13. 10 mAs
4. 30 mAs	14. 25 mAs
5. 240 mAs	15. 12 mAs
6. 125 mAs	16. 14 mAs
7. 75 mAs	17. 10 mAs
8. 140 mAs	18. 12 mAs
9. 25 mAs	19. 4 mAs
10. 24 mAs	20. 45 mAs

All mAs values have been rounded off to the nearest whole number.

Grid Conversion

Exercise No. 2

1. 15 mAs	11. 30 mAs
2. 120 mAs	12. 5 mAs
3. 50 mAs	13. 1 mAs
4. 63 mAs	14. 5 mAs
5. 38 mAs	15. 5 mAs
6. 21 mAs	16. 75 mAs
7. 20 mAs	17. 12 mAs
8. 30 mAs	18. 18 mAs
9. 10 mAs	19. 20 mAs
10. 400 mAs	20. 24 mAs

All mAs values have been rounded off to the nearest whole number.

Density Analysis Grid

Exercise No. 1

1. A	6. C
2. C	7. D
3. D	8. B
4. A	9. C
5. B	10. A

Density Analysis Grid

Exercise No. 2

1. A	6. C
2. B	7. C
3. A	8. C
4. A	9. D
5. B	10. B

Intensifying Screens

Chapter 27

1. D — In a film-screen imaging system, about 95 -98% of the resulting radiographic density is due to the blue or green color light emitted by the intensifying screens. The remaining 2 - 5% is caused by the direct action of the x-rays with the film.

2. C — Though rarely used because the large amount of exposure which is required to achieve the desired radiographic quality, a non-screen or direct exposure image provides exceptional detail which is unobtainable during screen radiography or with any other analog or digital modality.

3. C — All of the light emitted by an intensifying screen is known as luminescence. Two forms of luminescence are fluorescence, the portion of light emitted during the actual exposure to the radiation source and phosphorescence (lag), the light emitted after the exposure has ended. In modern phosphor, the preferred materials for a film-screen imaging systems will have a minimal amount of phosphorescence.

4. C Radiographic parallax is an artifact which occurs from the use of a double emulsion film and the two distinct images that are formed on the two emulsions. If a tube angulation is employed on a film with a thick base, the image on the tube side of the emulsion and the back image are formed in two separate locations on the film. When viewed, these images are not perfectly aligned forming the so-called parallax effect. Films with a thin base are preferred to reduce this effect.

5. B The principal factor which effects x-ray absorption of any substance; (phosphor) is the effective atomic number of the compound. Elements with high atomic numbers are associated with more photoelectric interactions and higher absorption of the incident photon.

6. C During x-ray production, the number of photons emitted by the tube will vary from one area to the next. This unavoidable random distribution of photons called quantum mottle becomes more noticeable and objectionable at the lower mAs values used with higher speed rare earth intensifying screens.

7. A For a given crystal size and layer thickness a rare earth phosphor emits about four times more light than a calcium tungstate phosphor for a given exposure. The higher atomic number of the rare earth elements not only stops more radiation but also converts more of the absorbed radiation into light. The ratio of the incoming radiation to the amount of light produced is termed conversion efficiency.

8. B A larger crystal produces more light and has a larger corona than smaller crystals. The recorded detail of the image is in part related to this factor since no structure smaller than the corona of the crystals light can be imaged. Therefore, larger crystals are associated with a loss of recorded detail and reduced spatial resolution.

9. B Three principal intrinsic factors can be used to increase the speed of the intensifying screen; namely, an increase in the crystal size, an increase in the thickness of the crystal layer and/or use of a reflective backing for the screen. Light absorbing yellow or pink dyes reduce the amount of light emitted and therefor reduce the screen speed and increase the detail of screen type imaging systems.

10. B The higher speed rare earth phosphors which are universally employed in modern intensifying screens will substantially reduce the patient dose while maintaining the same recorded detail of the calcium tungstate (CaW) screens they replaced.

11. B Nearly 2000 phosphors were discovered or categorized by Thomas Edison including the first practical intensifying screen phosphor calcium tungstate (CaW) which was used extensively in radiography from the 1930's through the mid 1970's.

12. A The more energy an x-ray photon possess, the more energy it is able to transfer to an exposed material (screen). The use of higher kVp beams is associated with a greater transfer of energy and a higher screen speed.

13. D The intensification factor or the ratio of exposures required to produce the same radiographic density with or without the use of intensifying screens is most closely related to the relative screen speed numbers. An intensifying screen with the highest screen speed number also has the highest intensification factor.

14. A The principal factor that effects the conversion of x-rays into visible light in an intensifying screen (conversion efficiency) is the chemical nature of the material. The rare earth phosphors such as gadolinium oxysulfide (Lanex), and lanthanum oxybromide (Quanta), have a 15% conversion efficiency or about four times that of calcium tungstate screens.

15. B A combination of a small crystal, a thinner crystal layer and the use of a light absorbing dye will reduce the relative speed of the intensifying screen while improving the recorded detail and resolution of the resulting image.

16. A In order to maximize the speed of a film-screen imaging system, the light emitted by the screen should be matched to the highest sensitivity of the film (spectral matching). A film-screen combination with a poor spectral match may be as much as four times slower than those with appropriate components.

17. A Lanthanum, gadolinium, and yttrium are the three main elements employed in the manufacture of rare earth intensifying screen materials.

18. B See explanation number 9 of this section.

19. D Like all other sources of electromagnetic radiations, x rays are emitted with equal intensity in all directions. The term which best describes this property is termed isotropic propagation.

20. C The use of a high speed imaging system is associated with lower radiation exposure to achieve the desired radiographic density or fluoroscopic brightness level. As the exposures are decreased, the effect of quantum mottle and electronic noise levels rise proportionally.

21. B When a luminescent crystal or phosphor is struck by x rays, gamma radiation or high speed electrons, the outer shell electrons in the crystals are excited, producing a characteristic visible as they return to their ground state.

22. A A high atomic number material such as lead is normally placed in the back lid of a cassette to eliminate the majority of the secondary scatter (back scatter) radiation which is directed toward the back of the cassette from nearby walls or floors.

23. B The size of the corona of light around the crystal can be reduced by the use of light-absorbing dyes. These dyes will reduce the speed of the screen while increasing the recorded detail or resolution of the image.

24. A Though an image produced with an intensifying screen has less recorded detail and a higher noise level than non-screen image, the huge dose reduction to the patient makes their relatively small losses acceptable.

25. B The simplest means for determining the resolution of an imaging system is by taking an exposure on a specially designed test pattern formed by alternating lead and radiolucent strips called a line pair phantom. See page 46 of this text for an image of this device.

26. A With extremely high speed imaging systems, the radiation exposure may be small enough to visualize a random variation in beam intensity called quantum mottle. See explanation number 6 of this section.

27. B The use of light absorbing dyes tends to absorb the light photons which are directed towards the film at steep angles, will tend to improve the recorded detail of the image.

28. D In addition to the intrinsic factors of crystal size and phosphor layer thickness which are directly related to the screen speed, a colder phosphor will emit substantially more light than a warmer screen. This effect is rarely significant because of the opposite effect temperature has on the speed of the film.

29. D The intensification factor is the ratio of exposures that will produce the same density with and without the use of intensifying screens.

30. B The quantum mottle (noise) seen on the radiograph taken at low intensity levels can degrade the appearance of the low contrast structures on the image.

31. A — The process by which a phosphor or crystalline material gives off light by a process other than heating is called luminescence.

32. C — After its exposure to x-rays, a phosphor may re-emit energy in the form of heat or light. In calcium tungstate, this rate of conversion is about 4%, in rare earth phosphors a 15-20% conversion efficiency is expected.

33. A — Before the switch to the rare earth phosphors in the mid 1970's, the blue light emitting calcium tungstate was the main screen material employed between the 1930's and 1970's.

34. C — The main controlling factors for the resolution of the screen are the size of the crystals and the thickness of the phosphor layer. Since most intensifying screens are manufactured to similar size specifications they will provide a similar resolution.

35. C — Though blurring of the radiographic image due to poor screen contact is rare in a modern cassette, an improperly mounted screen or damaged cassette can be evaluated by placing a wire screen or mesh over the cassette and examining the radiographic for aberrations.

36. C — In order to enable x rays to reach the x-ray film with only minimal attenuation, the cassette front should be made of a low atomic number material (carbon fiber) which is homogeneous so that no density differences are recorded.

37. B — Good film screen contact is critical in the maintenance of acceptable recorded detail. Among the factors that are responsible for this are overall cassette design, status of the latching devices, and the compressibility of the foam rubber (felt) backing which helps to apply pressure to the film when it is placed between the screens.

38. A — See explanation number 18 of this section.

39 D — See explanation number 3 of this section.

40. C — See explanation number 29 of this section.

Screen Conversion

Exercise No. 1

1.	5 mAs	11.	20 mAs
2.	2 mAs	12.	48 mAs
3.	10 mAs	13.	24 mAs
4.	4 mAs	14.	60 mAs
5.	15 mAs	15.	40 mAs
6.	50 mAs	16.	280 mAs
7.	17 mAs	17.	200 mAs
8.	35 mAs	18.	270 mAs
9.	1 mAs	19.	45 mAs
10.	6 mAs	20.	40 mAs

All mAs values have been rounded off to the nearest whole number.

Screen Conversion

Exercise No. 2

1.	10 mAs	11.	10 mAs
2.	120 mAs	12.	60 mAs
3.	60 mAs	13.	48 mAs
4.	6 mAs	14.	75 mAs
5.	23 mAs	15.	100 mAs
6.	50 mAs	16.	18 mAs
7.	40 mAs	17.	10 mAs
8.	1 mAs	18.	45 mAs
9.	40 mAs	19.	67 mAs
10.	3 mAs	20.	20 mAs

All mAs values have been rounded off to the nearest whole number.

Density Analysis (Screen)

Exercise No. 1

1. A
2. B
3. D
4. B
5. B
6. C
7. C
8. D
9. A
10. A

Density Analysis (Screen)

Exercise No. 2

1. C
2. D
3. C
4. A
5. A
6. A
7. C
8. A
9. D
10. C

Grid and Screen Conversion

Exercise No. 1

1. 30 mAs
2. 11 mAs
3. 4 mAs
4. 75 mAs
5. 4 mAs
6. 5 mAs
7. 8 mAs
8. 13 mAs
9. 3 mAs
10. 20 mAs
11. 2 mAs
12. 27 mAs
13. 7 mAs
14. 3 mAs
15. 15 mAs
16. 240 mAs
17. 240 mAs
18. 60 mAs
19. 300 mAs
20. 360 mAs

Grid and Screen Conversion

Exercise No. 2

1. 10 mAs
2. 8 mAs
3. 20 mAs
4. 5 mAs
5. 8 mAs
6. 8 mAs
7. 500 mAs
8. 150 mAs
9. 120 mAs
10. 15 mAs
11. 30 mAs
12. 60 mAs
13. 45 mAs
14. 45 mAs
15. 75 mAs
16. 50 mAs
17. 33 mAs
18. 150 mAs
19. 31 mAs
20. 13 mAs

Radiographic Film

Chapter 28

1. B — The most commonly employed radiographic films have a double emulsion which places each of the two surfaces of the intensifying screen in contact with an emulsion surface. This design helps to reduce the exposure to the patient by maximizing the speed of the imaging system.

2. A — The high density artifact appearing in this image is caused by the pressure associated with the bending of the film. These artifacts often appear as small half moon shaped areas and are referred to as crescent marks.

3. C — The branch-like increased density artifacts called tree static on this image were caused by a static discharge will not always have this appearance and may appear as a dark smudge mark or a king's crown. The grounding loading bench and the maintenance of higher humidity levels in the darkroom will help in the prevention of these artifacts.

4. C — The full range of densities recorded by an x-ray film is from about .1 - 4. The human eye being less sensitive can only visualize a smaller range, between .25 -2.5, constituting the useful range of densities.

5. A In the late 1800's, two students, Hurter and Driffeld, developed a method for expressing the properties of a photograph through the plotting of the density as a function of its log relative exposure to light. This characteristic or sensitometric (H+D) curve is used to determine the sensitivity (speed), contrast and exposure latitude of a film.

6. A Similar to intensifying screens, the size of the silver halide crystals and the thickness of the crystal layer in the films emulsion are the main determining factors for the speed of radiographic film.

7. B The two most important relationships of the characteristic (H+D) curve are the distance of the curve from the density axis which indicates the film's sensitivity (speed) and the slope of the useful density region which is used in the determination of the relative contrast and exposure latitude. Comparing the curves, one finds that film I closest to the density axis has the highest sensitivity (speed), Film II is the most vertical and has the highest contrast and film III has the greatest latitude (lowest contrast).

8. A See explanation number 7 of this section.

9. C The term artifact is used in radiography to describe any unwanted mark on a radiographic image. Artifacts can result from improper film handling, processing problems, equipment errors or patient related factors.

10. A See explanation number 5 of this section.

11. C After a film is exposed to light or radiation, a chemical change in the silver halogen crystals of the film occurs. Because this change is imperceptible to the human eye, it is called a latent image. This image is made visible by the film's immersion into the processing (developer) solutions which completes the conversion of the silver halide crystals into the black metallic silver appearing in the dark regions of the film.

12. B All films are unavoidably exposed to heat, moisture and natural background radiation that over time will cause an unwanted fog on the film. For this reason, each box of film is given an expiration date to eliminate the loss of speed and contrast this unwanted fog may cause.

13. D Film density is found to be a logarithmic function of the incident light from the view box to the light transmitted through the film: Density = Log IO/IT where IO is the original intensity of light from the view box and it is the amount of light transmitted through the film. The corresponding density values for a number of common fractions of original to transmitted intensities are provided. 1/10 = 1; 1/100 = 2; 1/1000 = 3; 1/10,000 = 4.

14. A Even under ideal conditions, an unexposed film after it has been through the processor is not perfectly clear but shows a small but measurable density of about .1 - .15. This so-called base plus fog density is caused by the absorption of light by the blue tint of most x-ray film bases and the unavoidable fog produced during film processing.

15. B The bright light from the view boxes has been known to cause eye strain in persons required to look at these for prolonged periods of time. This problem was helped by the incorporation of a blue tint into the base of most modern x-ray films.

16. C In order to provide the maximum speed, the intensifying screens must produce light in the blue-green spectrum to which film is most sensitive.

17. C See explanation number 5 of this section.

18. D The main chemical reaction by which silver bromide crystals are converted into the black metallic silver image involves the neutralization of the silver halide lattice and its chemical reduction by the (developer) solution.

19. B See explanation number 5 of this section.

20. A The ability of film to record and display a series of varying densities is termed film contrast. This inherent factor of the film is different for each type of film and the type of processing solutions in the development of the film.

21. C The active component of all films are the light sensitive silver halide (bromide) crystals. To be effective, these must be evenly coated and disbursed over the entire film. A clean porous substance called gelatin serves as the mechanical support medium for the silver bromide (AgBr) crystals.

22. B Pertaining to the corresponding diagram, A=density axis; B=log exposure axis; C=over-exposure region; D=useful density region; E=under-exposure region; F=latitude of the under-exposed region; G=Latitude of the useful density region; H=latitude of the over-exposure region.

23. D See explanation number 22 of this section.

24. C See explanation number 22 of this section.

25. B See explanation number 3 of this section.

26. A A fog which a film obtains after it has been exposed to radiation is called post-exposure fog. The most common cause of this exposure is improper shielding of the safe light fixtures or improper seals around the processor opening allowing white light leakage into the darkroom. Safe light fixtures must be used sparingly with low wattage bulbs and minimum safe light to film distances observed.

27. B Because a perfect silver bromide crystal will not permit a chemical change to occur, a contaminant such as sulfur (mustard oil) is added during the manufacturing process to create small faults on the crystals called sensitivity specks or developmental centers.

28. B After the light sensitive silver bromide crystals are mixed with gelatin and coated on to the film base, this layer is referred to as the film emulsion.

29. A The slope of the straight line portion of the characteristic curve is used in the determination of the film contrast and latitude. The more vertical the slope, the higher the contrast and the more narrow the film latitude.

30. B See explanation number 28 of this section.

31. D See explanation number 13 of this section.

32. B See explanation number 22 of this section.

33. D See explanation number 22 of this section.

34. C See explanation number 22 of this section.

35. A Any of the compounds formed by silver and the halogen family of elements will form light sensitive compounds suitable for films. At the present time, the vast majority of films' crystals are composed of silver bromide (AgBr).

36. C Film I possesses the greatest sensitivity (speed) because of its location nearest the density axis. Film II being the most vertical has the highest contrast and lowest latitude. Film III has the lowest contrast and speed but possesses the widest exposure latitude.

37. A See explanation number 36 of this section.

38. B A radiograph is most similar to a photographic negative. Unlike a photographic image which is normally viewed only after the image is printed to make it into positive form, a radiographic is viewed in this negative form.

39. D Films have employed a number of transparent base materials. Until the late 1800's, film bases were made of glass plates that were fragile and difficult to work with. In the early 1900's, cellulose nitrate and cellulose triacetate, though improved, failed to meet all of the desirable properties possessed by the present day polyester base.

40. B See explanation number 13 of this section.

41. A The characteristic (H&D) curve has been described as consisting of four basic regions: the low density toe, the useful straight line portion and the high density shoulder region. The fourth and least commonly depicted is the near mirror image of the first three portions called the reversal or solarization region.

42. C The foundation of the film, the base, must resist shrinkage or expansion and be able to maintain shape (dimensional stability) under various conditions.

43. A The process by which silver is formed into silver bromide crystal must take place in total darkness to prevent activation of the crystals prior to their radiographic applications.

44. C The gelatin which suspends the silver halide crystals will swell in a warm solution enabling the processing solution access to all of the silver crystals. When dried the gelatin shrinks and hardens forming the film's familiar texture. A chemical union with the AgBr crystals and the gelatin would deactivate the compound.

45. B Films must be stored in a cool dry area ideally at temperatures in the 50-70°F (10-20°C) range. Exposure to excessive heat may cause the film to be fogged resulting in its loss of contrast.

46. B See explanation number 39 of this section.

47. D Until the advent of the T-grain process for making silver grains, the main controlling factor for film speed was the concentration of the crystals. A new technique for producing T-grain crystals by Kodak enables a higher speed, better contrast with a smaller amount of silver.

48. C A protective super-coating covering the emulsion of the film is employed to help prevent scratches, pressure and fingerprint artifacts.

49. B The secondary electrons produced when x-ray photons are absorbed into the crystal lattice of a silver bromide crystal migrate toward a silver sulfide fault called a sensitivity speck. These free electrons begin the conversion of the AgBr into neutral black metallic silver.

50. D In the northern states, the low humidity conditions of the winter provide the ideal dryness for static build-up to occur.

Automatic Processing

Chapter 29

1. C The normal sequence of the film passage during automatic processing is development, fixation, washing and drying.

2. C The action of the developing (reducing) agents are greatly enhanced in the presence of the low PH, alkali agents such as sodium carbonate or sodium hydroxide.

3. B The replenisher rates in an automatic processor are adjusted for the crosswise placement of the film on the feed tray and the action of the microswitch which determines the length of the replenishment. The placement of the film lengthwise is wasteful and may lead to the over-replenishment of the developer and fixer solutions.

4. D The processing time in an automated unit is directly related to the speed of the transport system. This speed is critical to maintaining image quality and should be maintained to within a ±2% limit.

5. B Strict temperature control in an automatic processor is especially important for the developer solution because of its direct relationship to the resulting density and contrast of the image. In most automated systems, the developer is maintained between 90-95°F (33-35°C). Precise regulation is required to prevent temperature fluctuations more than ± 2°F which can noticeably effect the density and contrast of the image.

6. B During rapid processing the high activity of the developer solution may cause the development of even the unexposed silver bromide crystals. The addition of potassium bromide to this solution helps restrain the action of the reducing agents so they only act on exposed silver bromide crystals.

7. A An excessive rise in the temperature of the developer solution will increase the density of the image while producing film with a lower contrast. Film I shows these effects on the characteristic curve.

8. B Water or developer solutions which are spread over parts of the film before processing may produce this type of positive density artifacts.

9. C The artifact appearing in radiograph B was caused by a momentary opening of the processor lid while the film was in the developer section.

10. B The process by which exposed silver bromide is converted in the black metallic silver is termed reduction. The most common reducing agents are hydroquinone, phenidone and metol.

11. C The transport system which consists of the drive motor, the transport rollers, and transport racks, also contain a series of grooved metal plates called guide shoes. Guide shoes assist the films in changing its direction as it passes into the turn around or crossover racks during normal processing.

12. A The invisible latent image will become visible during the development stage of film processing as the exposed silver bromide crystals obtain free electrons from the reducing agents hydroquinone and phenidone.

13. D Excessive radiographic density combined with a lower radiographic contrast is most commonly associated with an excessive temperature of the developer solution.

14. B The two principal reducing agents, hydroquinone and phenidone, work together in maintaining the desired film density. The faster acting phenidone begins the development process and is most responsible for lighter shades of density seen in the toe portion of the H + D curve. Hydroquinone acts more slowly but provides the darkest shades of density on the film.

15. B Improperly dried films may be caused by directly improperly seated drying tubes or problems with the heater or thermostat. Chemical causes can include excessive processing solution temperatures and depleted fixer solutions.

16. A The neutralizing effect of the developer solution with the fixer can have detrimental effects on the fixation (clearing) process. For this reason, a pair of closely spaced squeegee rollers are placed at the end of the developer solution tank to remove excess developer prior to its immersion in the fixer section of the unit.

17. A The developer (reducing) solution provides an excess of electrons to the film which aid in the reduction of exposed silver bromide crystals. These electrons neutralize the positively charged silver ions forming a compound termed black metallic silver.

18. A In order to maintain consistent image quality, an automatic processor should be checked daily for proper solution temperature (thermometer), activity (hydrometer) and have a sensitometric image made and interpreted.

19. C When a 14" - 17" (35-43 cm) is processed correctly, about 60 cc of developer replenisher and 100 cc of fixer replenisher are required to maintain the solutions activity. The microswitch in the entrance roller assembly controls the appropriate replenishment rate by detecting both the size and the number of film processed.

20. B The microswitch within the entrance roller assembly serves both to regulate the developer and fixer replenishment rates and provide a visual or audible signal which assists with proper film spacing when feeding films into the processor.

21. B The decreased density of radiograph A is due to scratches in the film emulsion caused by the pulling of a jammed film over the guide shoes of the fixer-wash cross over rack.

22. D The image of radiograph B resulted from a bent film that was improperly loaded cassette.

23. C The most common chemical causes for processor jam-ups is exhausted developer or fixer solutions which contain insufficient hardener enabling the excessive swelling of the film emulsion. Mechanical jam ups are nearly all related to problems of the transport system.

24. D The action of the two developer agents independently would be substantially less than their action together. This chemical phenomenon is termed synergism or superaddidity.

25. D Any malfunction of the replenisher system such as air in the replenisher lines, faulty micro-switches, and faulty replenisher pumps, result in a lowering of film's speed and contrast. This becomes progressively worse over a series of images as the developer and fixer solutions are slowly depleted.

26. B The clearing or fixing agent most commonly employed in most automatic processors is ammonium thiosulfate, which serves to remove the unexposed silver bromide crystals from the film's emulsion during the second processing step.

27. C The mixing of the processing solutions has changed substantially over the years with the hand mixing of the chemical in water (solvent) becoming extinct. Most modern solutions are provided in 3 - 5 gallon concentrates which are mixed with the water so to achieve the desired activity level.

28. D This radiograph shows a series of evenly spaced streaks which were caused when an improperly fed film was pulled backwards from the processor. Since portions of the film were in the developer, excess solution was dragged over the entrance rollers and redeposited on the film when processing was attempted the second time.

29. C The low density areas of this radiograph correspond to the straps and latching devices of an older type strap cassette. These are seen because the cassette was inadvertently exposed with its back surface facing the x-ray tube.

30. A A hardening agent called glutaraldehyde is added to the developer solution to prevent excessive swelling of the film emulsion when it is immersed into a warm solution. A depleted or exhausted developer solution is unable to prevent the over expansion of the film emulsion causing it to adhere to the rollers and jamming the processor.

31. A In a 90 second automatic processor, the developer and fixer solution temperatures are in part stabilized by a series of heat exchange tubes from the wash tank. The wash water in most processors is maintained at 85-90°F (30-32° C) or about 5° F (3° C) below the temperature of the developer solution.

32. D See explanation number 6 of this section.

33. A In a 90 second automatic processor, the circulation system serves to agitate and maintain the solution temperature. In most processors, a heat exchanger is provided at the bottom of the developer, fixer and wash tanks to maintain the temperature within the 1°F accuracy desired.

34. A Excessive developer temperature, increased development time, over-replenished developer solution, and exhausted fixer can all lead to an increase in the production of chemical fog on the radiograph.

35. D Evenly spaced lines usually 3.14 inches apart are most often caused by dirt or chemical stains deposited by the revolution of the 1" diameter processor rollers. The term pi lines is derived from the formula for the circumference of a circle $=\pi d$.

36. C The proper development time is dependent on the type of film emulsion employed, its compatibility with the solution, the developer activity and temperature.

37. A The lower density shades which comprise the toe of the characteristic curves are formed in the first part of the development process by the action of the chemical phenidone.

38. B See explanation number 19 of this section.

39. D Excessive developer temperature can lead to numerous processor related problems such as chemical fog formation, rapid chemical oxidation and increased radiographic density and lower contrast of the image. The radiograph coming out of the processor would feel wet and tacky if they did not get jamed within the unit.

40. C Film development which requires a highly alkaline medium is rapidly terminated by its placement into the strongly acid fixer solution.

41. C See explanation number 19 of this section.

42. B See explanation number 10 of this section.

43. C Both the developer and fixer solutions are highly active chemical solutions which are not limited to their action with the film emulsion but also readily combine with the oxygen atoms of the air. The addition of sodium sulfite to the processing solutions slows this process by more readily combining with the oxygen than the balance of the solution, thereby extending its life.

44. B Improperly seated guide shoes in the crossover or turn around racks of an automatic processor may result in evenly spaced scratches on the film in the direction of the film travel, called guide shoe marks.

45. B The most common causes of low density artifacts are related to the inability of light emitted from the intensifying screen to reach the film.

46. D The alternating low and high densities on the center of this image correspond to the light penetrating through the spaces of the entrance roller when the darkroom light was turned on before the film completely entered the developer section of the processor.

47. A See explanation number 33 of this section.

48. C The gluteraldyde hardening agent added to the developer also has antifungal properties which help to restrain the growth of bacteria or fungi in the developer.

49. B In order to prevent the loss of silver and avoid the hazards of its disposal into the environment, all automatic processors should have the unexposed silver bromide removed by placing a silver reclaimer unit to the fixer tank of the processor.

50. A See explanation number 26 of this section.

Tomography

Chapter 30

1. C In order to radiographically demonstrate a single plane of tissue within the body, requires the movement of the tube and image receptor in equal but opposite directions by a technique called tomography.

2. D Over the last few years, a number of terms have been used to describe the visualization of a single tissue plane including body section radiography, tomography, zonography and pleisotomography. Scanography is a technique used for the measurement of the long bones of the legs.

3. C Tomography provides the visualization of a single tissue plane by causing a motion blur in all tissues above and below a desired plane.

4. D The movement pattern called hypocycloidal which looks similar to a clover leaf involves the formation of 3 overlapping ellipses providing a very long amplitude. This pattern is associated with a thin tomographic section and exceptional uniformity of blurring.

5. A The objective plane which is found at the level of the fulcrum is the only region of the object to remain in sharp focus during tomography.

6. B — The linear distance the tube travels during the tomographic exposure is termed the amplitude. In most newer units, the term amplitude is replaced by the term arc which more properly describes the arc-like movement of the tube above the table.

7. A — See explanation number 4 of this section.

8. C — During tomography, there is a direct relationship between the amount of blurring and the orientation of the object. When the tube movement is at a 90 degree or right angle to the major lines of interest (bony structures) a maximum degree of blurring is seen. Correspondingly, as tube moves parallel to the dominate lines of the anatomic region of interest, a minimal blurring is noted.

9. D — See explanation number 4 of this section.

10. A — The thickness of the tomographic section is inversely proportional to the amplitude, exposure angle and tomographic arc. Since an increase in the source-to-image receptor distance at a fixed amplitude will reduce the exposure angle, a thicker tomographic cut would be visualized.

11. B — Wide-angle tomography which produces very thin tomographic sections is used to visualize structures normally obscured by overlaying tissues; i.e., cochlea, semicircular canals. The main disadvantage of these techniques is the low or long scale contrast associated with thin tomographic slices and the need for precise localization of the cuts.

12. B — The tomographic cut thickness is effected by changes in the amplitude, source-to-image receptor distance, the arc and the exposure angle.

13. B — Though rarely used today because of its inability to maintain the desired amount of film-screen contact, a plesiotomographic technique involved the use of multiple screens and films in a sandwich arrangement to produce numerous thin cuts during a single radiographic exposure.

14. D — The principal controlling factor for the cut thickness is the tomographic arc. In order to obtain images in the 1 mm thickness range requires an arc in excess of 60 degrees. This large arc is most easily obtained through the use of pluridirectional movement patterns (polytomography).

15. A — The exposure or tomographic angle is the angle through which the central ray moves during exposure. This angle can be reduced by increasing the source-to-image receptor distance, decreasing the amplitude or decreasing the exposure arc.

16. B — Polytomography is a special tomographic technique which uses pluridirectional movements to improve the blurring characteristics of an image. This technique uses circular, elliptical, hypocycloidal or spiral movement patterns to provide extremely thin tomographic sections.

17. C — The thinnest tomographic section is obtained through a combination possessing the longest amplitude and the shortest source-to-image receptor distance.

18. A — The tissue plane that remains in focus (objective plane) during tomography is determined by the height of the fulcrum. Measured from the top of the table, fulcrum can be set at any level within the patient to be sectioned. In this supine patient measuring 30 cm the tumor is found using a 6 cm fulcrum level since this is its height above the table. The point of maximum blurring is simply the point most distant from the fulcrum level. A midline section is obtained by simply measuring the patient (30 cm) and dividing that number in half.

19. A — See explanation number 18 of this section.

20. B See explanation number 18 of this section.

21. B In order to place the gall bladder at the fulcrum level in this supine patient (24 cm) the fulcrum is set at 14 cm or 14 centimeters above the table top. Since the cystic duct projects posteriorly from the gall bladder it would be best demonstrated in this supine patient by reducing the fulcrum setting by about 2 cm to 12 cm.

22. D See explanation number 21 of this section.

23. A The fulcrum is exclusively related to the height or level of the tomographic image and will have no effect on the cut thickness or image resolution.

24. A Narrow-angle tomography, also known as zonography, is used to produce sharply defined images of the objective plane with little interference from the overlaying structures. A non-linear movement pattern; i.e., circular, is preferred for this technique.

25. B Due to normal human physiology, the eyes are best able to focus on objects which are seen at a viewing distance of 25 inches. The measured distance between eyes (interpupillary) is 2.5" providing a 10:1 viewing to distance ratio. This ratio is used in calculating the appropriate amount of stereographic shift that is required during this imaging procedure. This can be simplified to become 1/10 or 10% of the source-to-image receptor distance.

26. A In order to add depth to the two dimensional radiographic image requires a stereoradiographic technique. During this technique, two radiographs are taken from different locations of a single object. The radiographs are optically merged through the use of a stereoscopic viewing device to form a single image with the added dimension of depth.

27. B See explanation number 25 of this section. The total tube shift at a 100 cm distance is 10 cm. During a stereoradiographic technique the x-ray tube is shifted one half of this distance to each side object prior to exposing each of the two images.

28. D See explanation number 27 of this section.

29. B See explanation number 25 and 27 of this section.

30. A Similar to a tomographic procedure, the major lines of interest (body structures) are used in the determination of the shift direction. In an upright chest exam, the tube should be shifted at right angles to the ribs. A vertical shift of 18 cm is employed for a 180 cm chest radiography.

Density Analysis (All Factors)

Exercise No. 1

1. B
2. A
3. C
4. B
5. D
6. D
7. A
8. C
9. B
10. C

Density Analysis (All Factors)

Exercise No. 2

1. C
2. B
3. C
4. D
5. C
6. D
7. D
8. D
9. C
10. C

Section V
Anatomy and Physiology

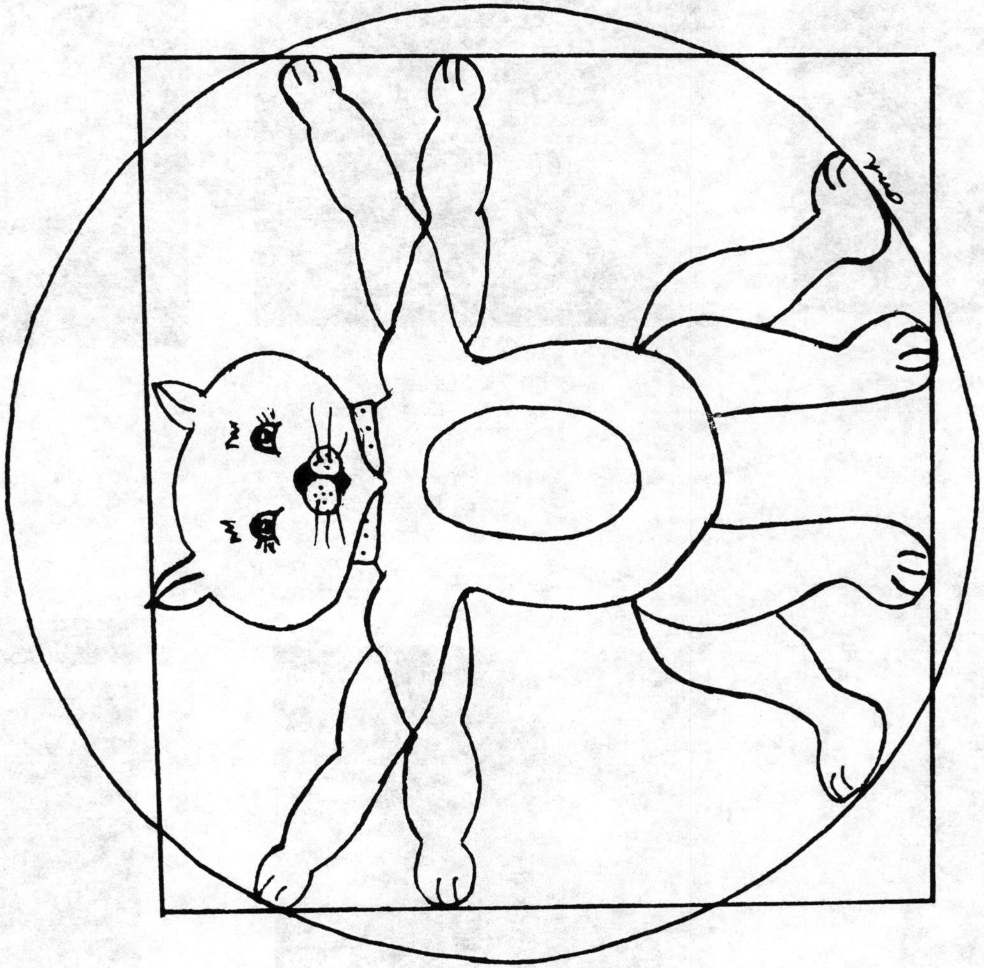

Radiographic Graffiti II

Can you identify these common patient related artifacts? Answers appear at the bottom of the page.

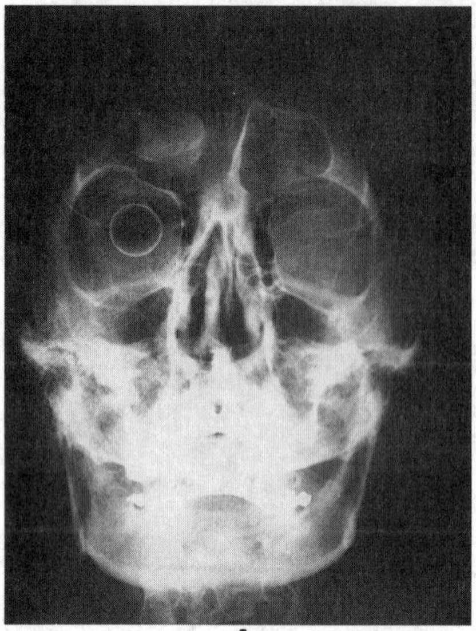

A

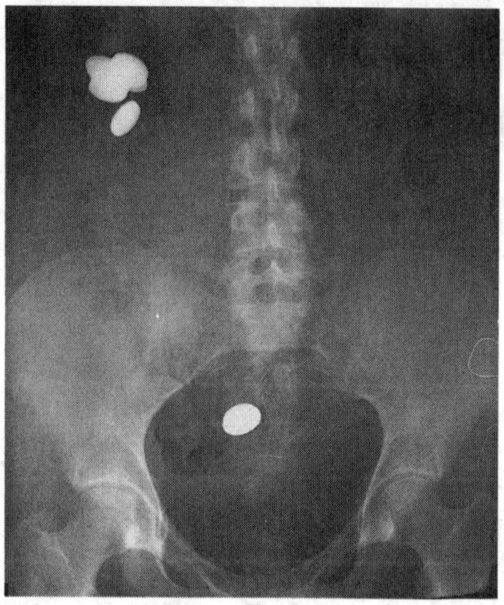

B

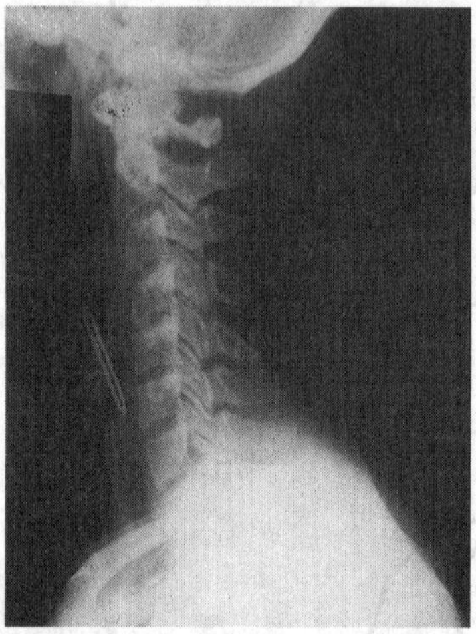

C

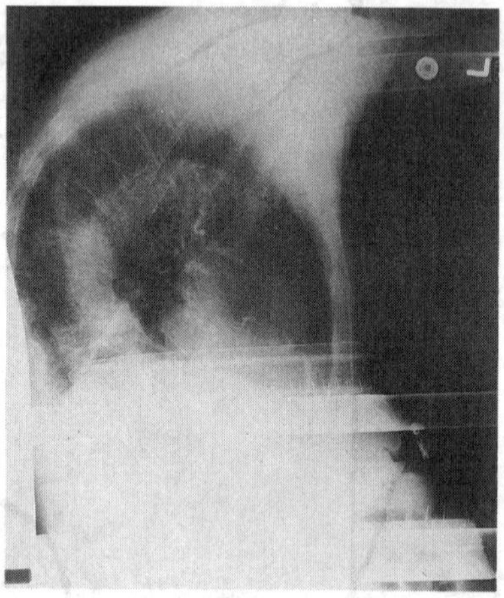

D

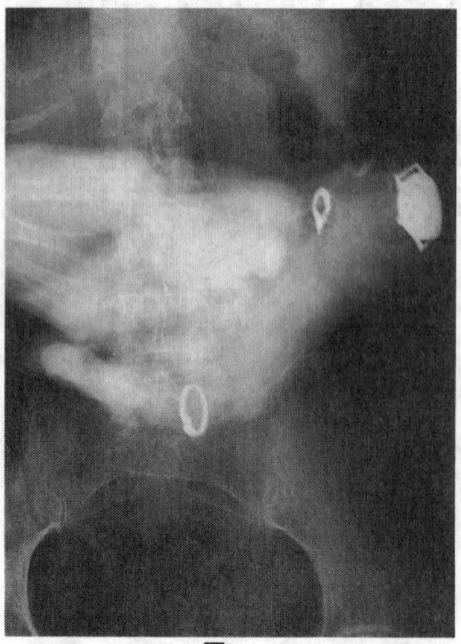

E

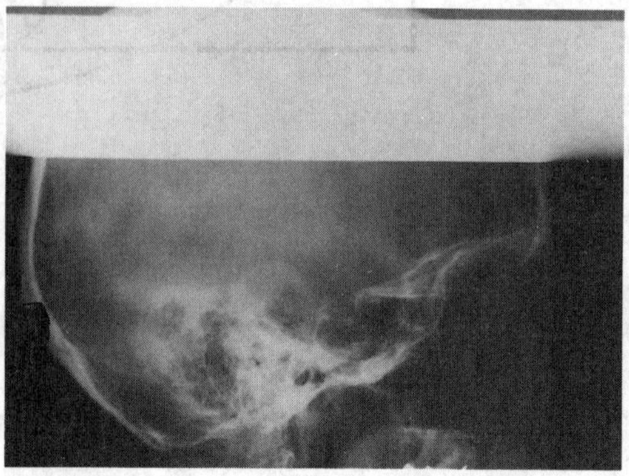

F

A - Patient with a false eye. Film B - Patient who has swallowed coins and other metallic objects. Film C - Patient with a razor le in the upper esophagus. Film D - Side rails of the wheelchair. Film F - Patient with his hands over his abdomen. Film F - port bar located under the top of the x-ray table.

Bone Development and Articulations

Chapter 31

1. The primary ossification center on the shaft of a bone is the point at which bone replacement begins and is called the: 4:176

A. Diaphysis
B. Epiphysis
C. Metaphysis
D. Synapse

2. The tough, fibrous covering of the bones that aids in the repair of bones is called the: 4:176

A. Peridontal membrane
B. Articular cartilage
C. Periosteum
D. Epiphysis

3. Bone cells receive nourishment and eliminate wastes from a series of microscopic blood vessels contained within small bony canals called: 10:144

A. Haversian canals
B. Lacunae
C. Canaliculi
D. Lamella

4. The bone-forming cells that serve to replace the embryonic skeleton by building up the bony matrix are termed: 4:179

A. Osteoclasts
B. Osteoblasts
C. Osteocytes
D. Osteons

5. The roughened areas on a bone that serve for the attachment of tendons are called: 10:161

A. Condyles
B. Foramen
C. Meati
D. Tubercles

6. The flat bones of the adult skull are usually joined together along irregular articulations called: 4:189

A. Sutures
B. Fontanells
C. Sinuses
D. Diaphyses

7. The entire normal human adult skeleton consists of __________ distinct bones. 4:189

A. 238
B. 206
C. 184
D. 126

8. The hollowed out central cavity that is found in the shaft of the long bones is called the: 4:178

A. Nutrient cavity
B. Diploe
C. Medullary cavity
D. Haversian canal

9. The flexibility of bone tissue is primarily due to the action of the chief organic component of bone called: 10:150

A. Collagen
B. Calcium
C. Vitamin D
D. Marrow

10. Bones that form in a joint capsule and serve to reduce the friction created at the joint are termed: 10:198

A. Flabella
B. Fontanells
C. Sesamoid bones
D. Carbonites

11. With the exception of the wrist and foot bones, all the bones of the upper and lower extremities are classified as: 4:176

A. Short bones
B. Irregular bones
C. Flat bones
D. Long bones

12. The gland that has the greatest single effect on bone development and growth is the: 4:183

A. Pituitary gland
B. Thyroid gland
C. Reproductive gland
D. Adrenal gland

13. The cranial sutures and other immovable joints are classified as: 4:226

A. Diarthrodal
B. Synarthrodal
C. Amphiarthrodal
D. Inarthrodal

14. The movement involving the turning of a part toward the midline of the body is termed: 5:36

A. Circumduction
B. Flexion
C. Abduction
D. Inversion

15. The stability of synovial joints is dependent, in large part, upon the action of: 4:112

1. Ligaments *2. Tendons* *3. Muscles*

A. 1 & 2 only
B. 1 & 3 only
C. 2 & 3 only
D. 1, 2 & 3

16. Most of the bones of the body are pre-formed by a special type of cartilage. This embryonic skeleton is made of: 4:113

A. Articular cartilage
B. Hyaline cartilage
C. Cancellous cartilage
D. Sesamoid cartilage

17. Which of the following is considered a function of the skeletal system? 5:33

1. Protection for internal organs *2. Framework for the body* *3. Levers to permit movement*

A. 1 only
B. 2 only
C. 3 only
D. 1, 2 & 3

18. The action in which the angle between two bones increases is termed: 5:36

A. Adduction
B. Luxation
C. Extension
D. Rotation

19. The outer layer (cortical layer) of bone is made up of tightly packed bone cells and is given the name: 5:34

A. Spongy bone
B. Compact bone
C. Cartilaginous bone
D. Epiphyseal bone

20. The most common type of bone marrow found within the bones of the adult skeletal system is: 5:34

A. Yellow bone marrow
B. Red bone marrow
C. White bone marrow
D. Reticular bone marrow

21. The connection of two bones at a joint is accomplished by a specialized fibrous connective tissue called: 4:112

A. Tendons
B. Cartilage
C. Plasma
D. Ligaments

22.-26. In the following, match the individual bone to its classification. 5:33

22. ________ Tenth dorsal vertebra
23. ________ Metacarpal bone
24. ________ Sternum
25. ________ Patella
26. ________ Navicular bone

A. Long bone
B. Short bone
C. Flat bone
D. Irregular bone
E. Sesamoid bone

27. Bone is the most rigid of the connective tissues. Its hardness is principally due to the presence of mineral salts such as: 10:142

A. Collagen sulfate
B. Calcium phosphate
C. Lead phosphate
D. Radium carbonate

28. In a newborn infant, the soft spot (fontanel) located at the junction of the frontal and parietal bones is termed the: 5:104

A. Anterior fontanel
B. Posterior fontanel
C. Lateral fontanel
D. Medial fontanel

29. The true bone cell (osteocyte) can be distinguished from an osteoblast by the presence of the surrounding: 4:115

A. Bone marrow
B. Hyaline cartilage
C. Lacunae
D. Skeletal muscle

30. The ends of many long bones contain several branching bony plates which increase the strength of the bone. This type of bone is classified as: 4:178

A. Compact bone
B. Osteoporitic bone
C. Volkman's bone
D. Cancellous bone

31. In freely movable joints, the ends of the bone are covered by ____________ which prevents direct bone to bone contact during movement. 4:227

A. Ligaments
B. Articular cartilage
C. Yellow bone marrow
D. Periosteum

32. The secondary growth centers that appear in the ends of the long bones in the first few years after birth are called: 4:176

A. Metaphyses
B. Diaphyses
C. Epiphyses
D. None of the above

33. In the developing infant, large amounts of red bone marrow are found in both the flat bones and at the ends of the long bones. This is important for the production of: 5:34
1. Growth hormones *2. Blood cells* *3. Calcium salts*

A. 1 only
B. 2 only
C. 3 only
D. 1, 2 & 3

34.-38. In the following, match the given joint to its proper classification. 10:217

34. ______ Pivot joint	A. First carpal metacarpal joint
35. ______ Hinge joint	B. Intervertebral (apophyseal) joints
36. ______ Ball and socket joint	C. First and second cervical vertebrae
37. ______ Saddle joint	D. Metacarpal phalangeal joint
38. ______ Gliding joint	E. Hip joint

39. Many bones of the adult skull possess a layer of cancellous bone between the inner and outer layers of bone. This is called the: 10:160

A. Diploe
B. Canaliculi
C. Suture
D. Haversian canal

40. Most freely movable joints have a special lining membrane which secretes a lubricating fluid called: 4:227

A. Meniscus
B. Mucus
C. Synovial fluid
D. Cerumen

41. The broad, flat bones of the skull develop from a special connective tissue and are termed: 4:179

A. Endochondral bones
B. Intramembranous bones
C. Trabecular bones
D. None of the above

42. The medullary cavity is in large part formed by specialized cells which reabsorb bone tissue called: 4:181

A. Osteoblasts
B. Osteomites
C. Osteocytes
D. Osteoclasts

43. Individual bone cells are able to communicate and pass nutrients between each other by a series of: 4:178

A. Porocytes
B. Endosteum
C. Canaliculi
D. Hemotomas

44. Muscle tissue is attached to bone by a specialized, connective tissue known as (a): 4:112

A. Tendon
B. Ligament
C. Cartilage
D. Epiconium

45. Aging, especially in women, may lead to an abnormally large amount of calcium loss in the bones causing a condition termed: 4:188

A. Osteoporosis
B. Osteopetrosis
C. Osteomyelitis
D. Osteitis

46. The improper intake of calcium and Vitamin D can lead to a softening of developing bones called: 4:183

A. Rickets
B. Osteoderma
C. Acromegaly
D. Gout

47. Chronic rheumatoid arthritis often damages the joint membranes and may lead to an abnormal joint stiffening called: 4:234

A. Achondroplasmia
B. Ankylosis
C. Hemoarthrosis
D. Osteonecrosis

48. The shoulder joint is the most common of the larger joints to be injured. The term employed for a dislocation is: 8:839

A. Hyperextension
B. Subluxation
C. Luxation
D. Sprain

49. The overproduction of growth hormones may result in a condition called: 4:234

1. Gigantism *2. Dwarfism* *3. Acromegaly*

A. 1 only
B. 2 only
C. 1 & 3 only
D. 1, 2 & 3

50. Tennis elbow and housemaid's knee are two common disorders involving a specialized sac near the joint capsule called the: 4:235

A. Meniscus
B. Cursate ligament
C. Tendon
D. Bursa

Upper Extremity

Chapter 32

1. The bony framework for the body of the hand is formed by the: 4:216
   1. Proximal phalanges 2. Wrist bones 3. Metacarpal bones

   A. 1 only
   B. 2 only
   C. 3 only
   D. 1, 2 & 3

2. The tip of the shoulder joint is formed by a large, flat, roughened area of the lateralmost part of the scapular spine called the: 5:44

   A. Glenoid process
   B. Acromion
   C. Coracoid process
   D. Coronoid process

3. The second digit of the normal adult hand contains ____________ interphalangeal joint(s). 5:49

   A. One
   B. Two
   C. Three
   D. Four

4. The pointed extensions on the distal ends of the ulna and radius, which serve for ligament attachments with the wrist bones, are called the: 4:216

   A. Styloid processes
   B. Tuberosities
   C. Coronoid processes
   D. Epicondyles

5. Many of the muscles for lateral rotation of the humerus attach to a tubercle found on the lateral surface of the proximal humerus called the: 4:214

   A. Greater tubercle
   B. Lesser tubercle
   C. Radial tubercle
   D. Deltoid tubercle

6. The hook is a characteristic extension found on the wrist bone called the: 5:49

   A. Navicular
   B. Capitate
   C. Trapezoid
   D. Hamate

7. The epiphysis of the proximal humerus after its closure near the time of puberty forms a bony ridge called the: 5:45

   A. Bicipital groove
   B. Humeral head
   C. Anatomic neck
   D. Trochlea

8. The large, cupshaped structure on the lateral border of the scapula that articulates with the humeral head is termed the: 5:45

   A. Acromion
   B. Glenoid fossa
   C. Semilunar notch
   D. Subscapular fossa

9. The head of the first proximal phalanx articulates with the: 5:49

   A. Head of the metacarpal bone
   B. Base of the metacarpal bone
   C. Base of the distal phalanx
   D. Greater multangular bone

10. The shoulder girdle is a structure formed by the: 5:45

1. Clavicle *2. Scapula* *3. Sternum*

A. 1 & 2 only
B. 1 & 3 only
C. 2 & 3 only
D. 1, 2 & 3

11.-18. Pertaining the diagram of the hand, place the letter that corresponds to the following structures: 4:216

11. ______ Head of the proximal phalanx
12. ______ Distal phalanx
13. ______ Ulnar styloid process
14. ______ Navicular bone
15. ______ Radial styloid process
16. ______ Base of the first metacarpal
17. ______ Head of the fifth metacarpal
18. ______ Middle phalanx

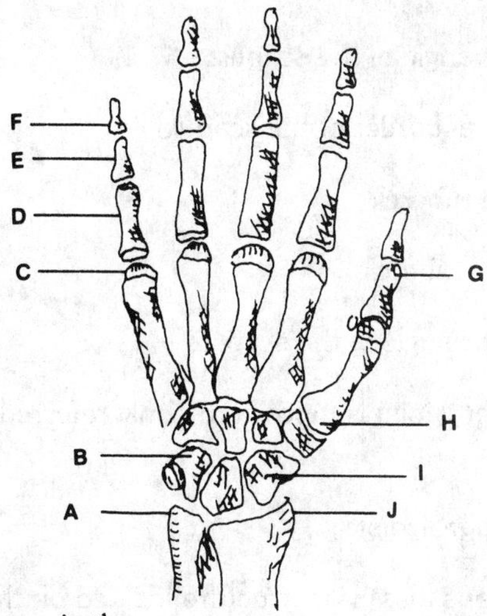

19. In anatomical position, the most lateral bone of the forearm is the: 4:47

A. Tibia
B. Ulna
C. Radius
D. Humerus

20. Projecting anteriorly from the upper scapula is a fingerlike process that serves for numerous muscle attachments called the: 5:44

A. Coracoid process
B. Acromial process
C. Coronoid process
D. Lesser tubercle

21. The wrist is formed by __________ distinct bones which are collectively known as the __________ bones. 5:49

A. 6/tarsal
B. 8/carpal
C. 6/carpal
D. 8/tarsal

22. The rotation of the radius around the ulna during act of pronation occurs between the radial head and the: 5:47

A. Capitulum
B. Acromion
C. Trochlea
D. Semilunar notch

23. The depression on the proximal humerus separating the greater and lesser tubercles is called the: 5:45

A. Intercondylar groove
B. Nutrient fossa
C. Bicipital groove
D. Glenoid fossa

24. During acute flexion of the elbow, the coronoid process of the ulna articulates with the coronoid fossa which is located on the: 5:47

A. Anterior humerus
B. Posterior humerus
C. Posterior radius
D. Medial radius

25. The common abbreviation used for the articulation formed by the medial end of the clavicle and the sternum is: 5:45

A. PIP joint
B. AC joint
C. SC joint
D. TM joint

26.-34. Pertaining to the diagram of the shoulder, place the letter that corresponds to the following structures: 4:212

26. ______ Coracoid process

27. ______ Humeral head

28. ______ Axillary border of the scapula

29. ______ Acromion

30. ______ Inferior angle of the scapula

31. ______ Vertebral border of the scapula

32. ______ Deltoid tubercle

33. ______ Greater tubercle

34. ______ Superior angle of the scapula

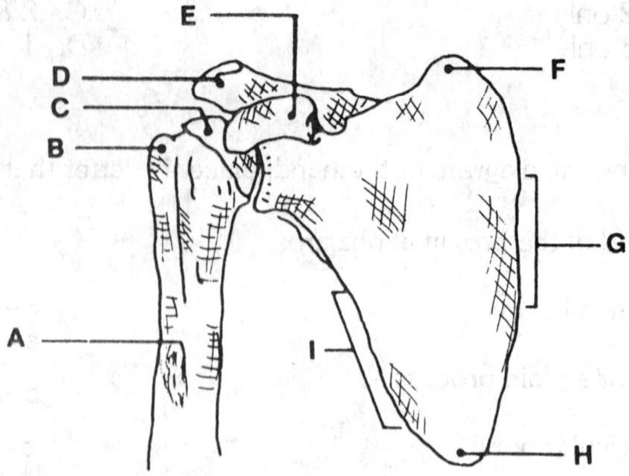

35. The glenohumeral joint is more commonly referred to as the: 5:45

A. Elbow
B. Interphalangeal joint
C. Shoulder
D. Acromioclavicular joint

36. The superspinous fossa is a structure located on the ____________ aspect of the scapula. 5:44

A. Anterior
B. Posterior
C. Lateral
D. Inferior

37. The largest and most lateral bone in the proximal row of wrist bones is the: 5:47

A. Os magnum
B. Hamate
C. Greater multangular
D. Navicular

38. The normal adult hand contains a total of ____________ phalanges. 5:49

A. Fourteen
B. Fifteen
C. Twenty-four
D. Twenty-six

39. During the act of pronation, the moves ____________ from lateral to medial. 5:47

1. Humerus *2. Ulna* *3. Radius*

A. 1 only
B. 2 only
C. 3 only

40. The greater multangular bone of the wrist articulates with the: 5:49

A. Head of the first metacarpal
B. Head of the fifth metacarpal
C. Base of the first metacarpal
D. Base of the fifth metacarpal

41. The proximal end of the ulna articulates with a large fossa on the posterior humerus called the: 5:47

A. Olecranon fossa
B. Condyloid fossa
C. Glenoid fossa
D. Acromial fossa

42. The large, bony prominence of the distal humerus sitting medial to the trochlea is called the: 10:194

A. Anterior condyle
B. Posterior condyle
C. Medial epicondyle
D. Lateral epicondyle

43. The bicep muscles attach at the proximal end of the radius to a structure called the: 4:216

A. Bicipital groove
B. Coronoid process
C. Radial tuberosity
D. Styloid process

44. The semilunar notch of the proximal ulna articulates with a smooth, rounded condyle called the: 5:47

A. Trochlea
B. Styloid process
C. Capitulum
D. Olecranon process

45. Each normal human hand contains ____________ interphalangeal joints. 5:49

A. Four
B. Five
C. Nine
D. Ten

46. On the anterolateral surface of the mid-humerus is a tubercle that serves for the insertion of the: 4:212

A. Deltoid tendon
B. Pectoralis tendon
C. Bicep tendon
D. Teres minor tendon

47. The medial border of the scapula derives its name from its proximity to the: 5:44

A. Vertebral column
B. Axilla
C. Clavicle
D. Diaphragm

48. The intertubercular groove of the humerus is formed by the action of the: 5:42

A. Deltoid tendon
B. Biceps tendon
C. Supraspinous tendon
D. Subscapular tendon

49. The largest wrist bone in the normal adult is the: 5:49

A. Capitate
B. Lunate
C. Pisiform
D. Hamate

50. The clavicle articulates with which two bones? 5:45

A. Humerus and sternum
B. First rib and scapula
C. Humerus and scapula
D. Sternum and scapula

Chapter 33

1. The total number of phalanges in the normal adult foot is: 5:65

A. Five
B. Seven
C. Fourteen
D. Fifteen

2. The expanded portion of the proximal end of the fibula (fibular head) articulates with the tibia at the: 5:63

A. Proximal medial surface
B. Distal medial surface
C. Proximal lateral surface
D. Distal lateral surface

3. When compared to the male pelvis, the female pelvis is characterized by a: 5:59
1. Wider transverse diameter *2. Large pelvic outlet* *3. Larger pubic arch*

A. 1 only
B. 2 only
C. 3 only
D. 1, 2 & 3

4. Located on the posterior surface of the femur is a prominent ridge of bone serving for the attachment of many of the muscles of the lower extremity. This process is called the: 5:61

A. Intercondylar eminence
B. Epicondylar ridge
C. Linea aspera
D. Intertrochanteric crest

5. The most superior portion of the tibia (tibial eminence) projects upwards from the plateau into the: 5:61

A. Intercondylar space
B. Acetabulum
C. Mortal space
D. Obturator foramen

6. The two large bones that help to form the majority of the adult pelvis are termed the: 5:56

A. Sesamoid bones
B. Innominate bones
C. Styloid bones
D. Crusate bones

7. The two small, rounded bones that are found on the plantar surface of the first metatarsophalangeal joint to help reduce friction are termed: 5:65

A. Sesamoid bones
B. Pisiform bones
C. Premetatarsal bones
D. Premetacarpal bones

8. The largest bone in the normal adult foot is called the: 5:63

A. Talus
B. Navicular
C. Calcaneus
D. Cuboid

9. The base of the patella is found on the ____________ surface of the bone. 5:61

A. Inferior
B. Superior
C. Medial
D. Lateral

10. The junction of the two innominate bones at the anteroinferior margin is termed the: 5:58

A. Symphysis menti
B. Acetabulum
C. Symphysis pubis
D. Sacroiliac joint

11. The smooth, upper surfaces of the proximal tibia (plateau) serves as the articulating surface for the: 5:67

A. Talus bones
B. Lateral malleoli
C. Fibular condyles
D. Femoral condyles

12.-20. Pertaining to the diagram of the foot, place the letter that corresponds to the following structures: 5:65

12. ______ Cuboid bone

13. ______ Head of the metatarsal

14. ______ Distal phalanx

15. ______ Calcaneus

16. ______ Base of the metatarsal

17. ______ Navicular

18. ______ Medial cuneiform

19. ______ Proximal phalanx

20. ______ Talus

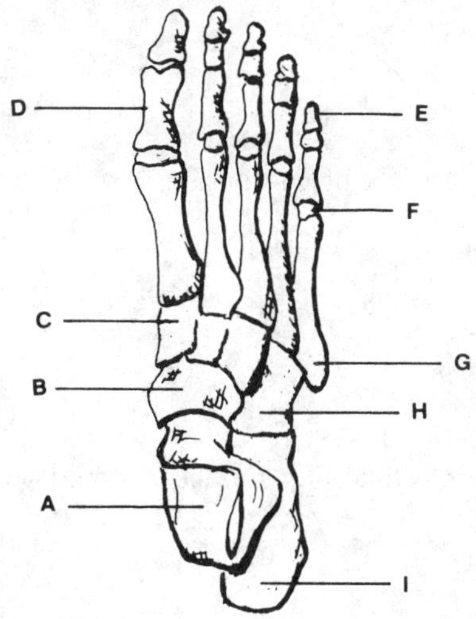

21. The larger and more medial bone of the lower leg is the: 5:63

A. Tibia
B. Fibula
C. Femur
D. Tibula

22. The rounded condyles of the femur are separated by a deep depression termed the: 5:61

A. Symphysis femorus
B. Intercondylar fossa
C. Bicipital groove
D. Ischial notch

23. The head of the femur articulates with a cupshaped socket on the lateral surface of the pelvis called the: 5:58

A. Obturator foramen
B. Pelvic brim
C. Olecranon fossa
D. Acetabulum

24. In the normal sitting position, most of the weight of the trunk is supported by the ____________ of the pelvis. 5:58

A. Iliac crest
B. Ischial spine
C. Ischial tuberosity
D. Pubic ramus

25. The tendon of the quadraceps femoris attaches on the anterior surface of a raised process of the proximal tibia called the: 5:61

A. Tibial spine
B. Tibial eminence
C. Tibial malleolus
D. Tibial tuberosity

26. The majority of the greater or false pelvis is formed by all or part of the: 5:58

A. Sacrum
B. Ischium
C. Ilium
D. Pubis

27. The largest and strongest bone of the appendicular skeletal system is the: 1:63:I

A. Tibia
B. Femur
C. Fibula
D. Humerus

28.-31. Pertaining to the diagram of the ankle, place the letter that corresponds to the following structures: 5:63

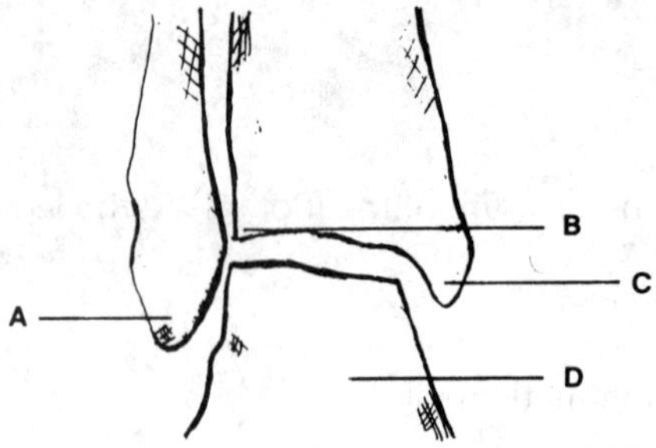

28. ______ Lateral malleolus

29. ______ Medial lateral malleolus

30. ______ Medial malleolus

31. ______ Talus bone

32. During development, the pelvis consists of six separate bones that fuse during childhood. These paired bones are the: 5:58
1. Ilium 2. Ischium 3. Pubis

A. 1 & 2 only
B. 1 & 3 only
C. 2 & 3 only
D. 1, 2 & 3

33. The large, smooth articulating surfaces located on the distal femur are termed the: 5:61

A. Femoral condyles
B. Greater trochanters
C. Linea asperae
D. Epicondyles

34. The astragulus (talus) bone of the foot is normally located ____________ to the calcaneus (os calsis). 5:65

A. Anterior
B. Posterior
C. Superior
D. Inferior

35. The protuberance that is felt on the lateral surface of the lower leg distal to the knee joint is the: 5:63

A. Medial epicondyle
B. Lateral epicondyle
C. Fibular head
D. Tibial tuberosity

36. The acetabulum is composed by part of the: 5:58
1. Ilium 2. Ischium 3. Pubis

A. 1 & 2 only
B. 1 & 3 only
C. 2 & 3 only
D. 1, 2 & 3

37. The cuneiform bones articulate anteriorly with the: 5:65

A. Calcaneus
B. Base of the metatarsals
C. Head of the metatarsals
D. Navicular bone

38. Fallen arches (flat feet) is often due to weakening of the ligaments in the feet and results in the loss of the ____________ arch. 1:35:I

A. Longitudinal
B. Vertical
C. Transverse
D. Neural

39. The pointed, bony prominence located on the distal fibula is called the: 5:63

A. Head of the fibula
B. Medial malleolus
C. Lateral malleolus
D. Fibula eminence

40.-49. Pertaining to the diagram of the hip joint, place the letter that corresponds to the following structures: 4:218

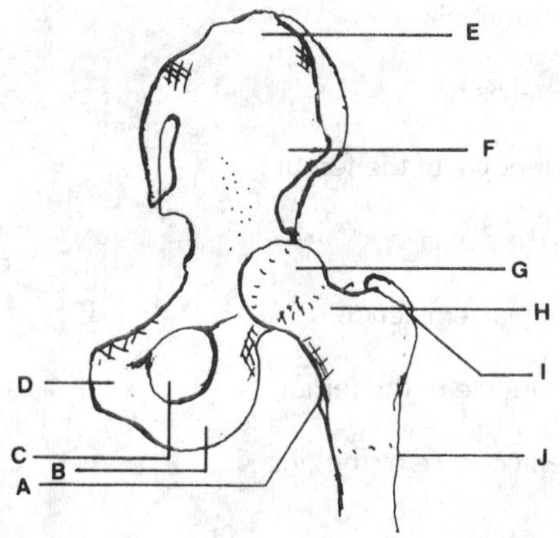

40. ______ Ischial tuberosity

41. ______ Anatomic neck

42. ______ Lesser trochanter

43. ______ Surgical neck

44. ______ Iliac crest

45. ______ Anterior superior iliac spine

46. ______ Obturator foramen

47. ______ Pubic bone

48. ______ Head of the femur

49. ______ Greater trochanter

50. The navicular (scaphoid) bone of the foot is located: 5:65
1. Inferior to the cuboid *2. Posterior to the cuneiforms* *3. Anterior to the talus*

A. 1 & 2 only
B. 1 & 3 only
C. 2 & 3 only
D. 1, 2 & 3

51. The large bony ridges of the proximal femur that serve to attach many of the muscles for moving of the hip are called the: 5:61

A. Greater and lesser malleoli
B. Greater and lesser trochanters
C. Greater and lesser condyles
D. Greater and lesser styloids

52. In the normal adult, the angle formed by the neck and shaft of the femur is about: 1:231:I

A. 90 degrees
B. 105 degrees
C. 120 degrees
D. 145 degrees

53. The knee joint is formed by the articulation of the ____________ bones. 5:63

A. Femur and tibia
B. Tibia and fibula
C. Femur, tibia and fibula
D. Femur and fibula

54. The cuboid bone of the foot articulates posteriorly with the: 5:65

A. Navicular
B. Talus
C. Calcaneus
D. Cuneiform

55.-62. Pertaining to the diagram of the knee, place the letter that corresponds to the following structures: 4:234

55. ______ Lateral epicondyle of the femur

56. ______ Shaft of the fibula

57. ______ Tibial tuberosity

58. ______ Lateral condyle of the femur

59. ______ Head of the fibula

60. ______ Intercondylar eminence

61. ______ Medial condyle of the femur

62. ______ Medial epicondyle of the tibia

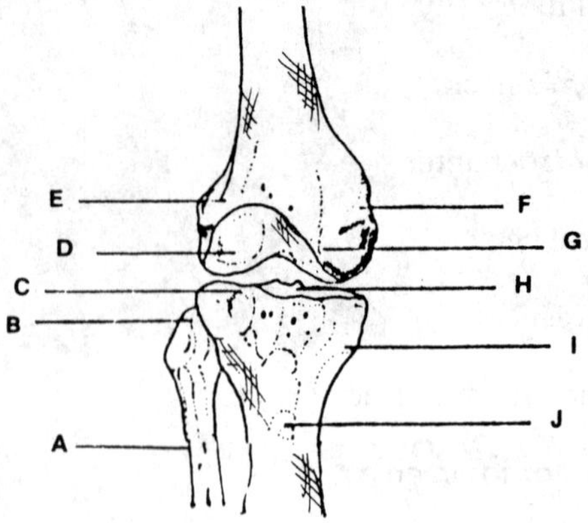

63. The obturator foramen of the pelvis is formed by the junction of the following bones(s): 5:58
1. Ilium   2. Ischium   3. Pubis

A. 1 & 2 only
B. 1 & 3 only
C. 2 & 3 only
D. 1, 2 & 3

64. The ankle joint is formed by all or part of the: 5:63
1. Navicular   2. Tibia   3. Talus bone

A. 1 & 2 only
B. 1 & 3 only
C. 2 & 3 only
D. 1, 2 & 3

65. The deep notch located on the posterior margin of the innominate bone between the posterior inferior iliac spine and the ischial spine is called the: 5:58

A. Greater sciatic notch
B. Pubic angle
C. Symphysis pubis
D. Ischial tuberosity

Axial Skeleton

Chapter 34

1. The largest number of vertebrae are found in the ________ portion of the vertebral column. 5:72

A. Cervical
B. Thoracic
C. Lumbar
D. Sacrum

2. The segment of the vertebral column that is characterized by foramina in the transverse process is termed the: 5:73

A. Sacral vertebrae
B. Lumbar vertebrae
C. Thoracic vertebrae
D. Cervical vertebrae

3. The fibrous cartilage wedge that separates the vertebral bodies of the spinal column is termed the: 5:75

A. Meningeal spacer
B. Intervertebral disc
C. Laminal wedge
D. Transverse disc

4. In the normal adult, the ribs are subdivided into three types based upon their: 5:82

A. Sternal attachments
B. Type of costal cartilage
C. Vertebral attachments
D. Direction of articulation

5. The odontoid process is a toothlike extension arising from the anterior aspect of the: 5:73

A. First cervical vertebrae
B. Second cervical vertebrae
C. Seventh cervical vertebrae
D. Fifth lumbar vertebrae

6. The intervertebral foramina, which are found between all the upper segments of the vertebral column, serve to allow passage of the: 5:72

A. Vertebral arteries
B. Spinal cord
C. Spinal nerves
D. Cauda equina

7. The first cervical vertebra that serves to support the weight of the skull is often referred to as the: 5:73

A. Odontoid
B. Hyoid
C. Axis
D. Atlas

8. The greatest amount of flexibility is permitted by the vertebrae in the: 3:102

A. Cervical region
B. Thoracic region
C. Lumbar region
D. Sacral region

9. The posterior rib articulation with the body of a thoracic vertebra is called a/an: 5:74

A. Costotransverse joint
B. Costovertebral joint
C. Apophyseal joint
D. Intervertebral joint

10. In the normal adult, the space for the spinal cord extends to the vertebral level of: 5:226

A. First lumbar vertebra
B. Fourth lumbar vertebra
C. Tenth thoracic vertebra
D. Fifth sacral vertebra

11. The terminal segments of the vertebral column are formed by the fusion of four rudimentary vertebrae and, in the adult, are known collectively as the: 5:74

A. Ala
B. Lamina
C. Coccyx
D. Facets

12.-18. Pertaining to the diagram of the cervical vertebrae, place the letter that corresponds to the following structures: 4:208

12. ______ Articulating condyle

13. ______ Spinous process

14. ______ Vertebral foramen

15. ______ Vertebral body

16. ______ Transverse process

17. ______ Transverse foramen

18. ______ Lamina

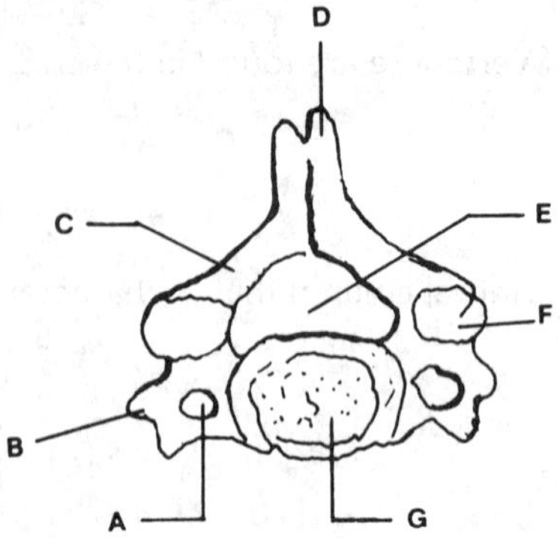

19. The exaggerated lip on the body of the first sacral vertebra is often termed the: 5:75

A. Sacral promontory
B. Sacral spine
C. Lateral mass
D. Ala

20. The sacrum articulates with the innominate bones on the sacrum's: 10:198

A. Anterior medial surface
B. Posterior medial surface
C. Inferior lateral surface
D. Superior lateral surface

21. The sternum has a small, pointed process on its inferior surface. This is a useful landmark in the location of the level of the: 5:83

A. Seventh cervical vertebra
B. Second thoracic vertebra
C. Tenth thoracic vertebra
D. Third lumbar vertebra

22. The body of a vertebra serves to support the majority of weight impressed upon it and is normally located: 5:74

A. Posterior to the spinous process
B. Anterior to the vertebral foramen
C. Posterior to the pedicle
D. All of the above

23. The spinal cord is protected along its course by passing within a large foramen of the vertebrae called the: 5:72

A. Vertebral foramen
B. Intervertebral foramen
C. Spinal foramen
D. Transverse foramen

24. In the normal adult, the lumbar vertebrae form a convex forward or ________ curve. 5:77

A. Lordotic
B. Kyphotic
C. Scoliotic
D. Spinal

25. The dorsal vertebrae are distinguished from the other vertebrae by the presence of the: 5:94

A. Lamina
B. Rib facets
C. Articular processes
D. Transverse foramina

26. The vertebral prominence that is found on the seventh cervical vertebra derives its name from its: 5:73

A. Long transverse process
B. Large vertebral body
C. Long spinous process
D. Large articular process

27. The spinal cord is protected along its course by a bony foramen of the vertebrae. The spinal cord proper usually ends at the vertebral level of the: 5:226

A. Tenth dorsal vertebra
B. Second lumbar vertebra
C. Fifth lumbar vertebra
D. Second sacral vertebra

28. The joint that allows the rotation or turning of the skull around a pivot point occurs between the: 5:73

A. Occipital bone and the first cervical vertebra
B. Occipital bone and the second cervical vertebra
C. First and second cervical vertebrae
D. None of the above

29. The last two pairs of ribs are classified as floating ribs because they have: 5:84

A. No vertebral attachment
B. No sternal attachment
C. Both of the above
D. Neither of the above

30. The first cervical vertebra is characterized by its lack of a/an: 5:73

A. Neural arch
B. Pedicle
C. Vertebral body
D. Articulating process

31. The sternal or manubrial notch is often employed for the localization of the: 5:82

A. Fifth cervical vertebra
B. Seventh cervical vertebra
C. Second thoracic vertebra
D. Tenth thoracic vertebra

32. The most anterior portion of a normal vertebra is the: 5:72

A. Arch
B. Pedicle
C. Spinous process
D. Body

33. The vertebral foramen is formed by all or part of the: 5:72

1. Vertebral body *2. Cornu* *3. Pedicles*

A. 1 & 2 only
B. 1 & 3 only
C. 2 & 3 only
D. 1, 2 & 3

34. The manubrium serves for the point of attachment for the costal cartilage of the first ________ pairs of ribs. 5 :82

A. Two
B. Three
C. Five
D. Seven

35. The semi-fluid central portion of the intervertebral disk is called the: 5:75

A. Nucleus pulposus
B. Nucleus fibrous
C. Annulus fibrous
D. Annulus pulposus

36. The first seven pairs of ribs are classified as: 5:82

A. False ribs
B. Ribbon ribs
C. Cartilaginous ribs
D. True ribs

37. An example of an apophyseal joint is the articulation between the: 1:207:I

A. Bones of the skull
B. Ribs and the vertebrae
C. Superior and inferior articular processes
D. First vertebra and the occipital bone

38.-42. Pertaining to the diagram of the sternum, place the letter that corresponds to the following structures: 4:210

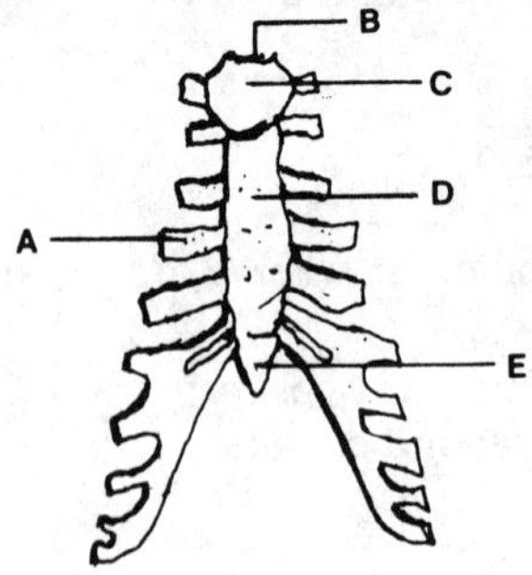

38. ______ Xiphoid

39. ______ Sternal notch

40. ______ Gladiolus

41. ______ Costal cartilage

42. ______ Manubrium

43. Which foramina of the vertebral column are formed by portions of more than one individual vertebra? 5:72

A. Transverse foramina
B. Vertebral foramina
C. Optic foramina
D. Intervertebral foramina

44. The atlanto-occipital joint of the upper cervical region allows for: 5:76
1. Rotation of the head *2. Flexion of the head* *3. Extension of the head*

A. 1 & 2 only
B. 1 & 3 only
C. 2 & 3 only
D. 1, 2 & 3

45. The two laminae project posteriorly from the transverse process to their junction in the midline. They help to form the: 5:72

A. Spinous process
B. Superior articulating process
C. Inferior articulating process
D. Vertebral body

46. The usual points of artiuclation between the ribs and a thoracic vertebra are the: 5:74
1. Bodyand spinous process *2. Lamina and pedicle* *3. Transverse process and the body*

A. 1 only
B. 2 only
C. 3 only
D. 1, 2 & 3

47. The bony ridge formed by the junction of the two upper portions of the sternum is called the: 5:83

A. Sternal notch
B. Sternal angle
C. Xiphoid ridge
D. Costal ridge

48.-53. Pertaining to the diagram of the lateral aspect of the lumbar vertebra, place the letter that corresponds to the following structures: 4:206

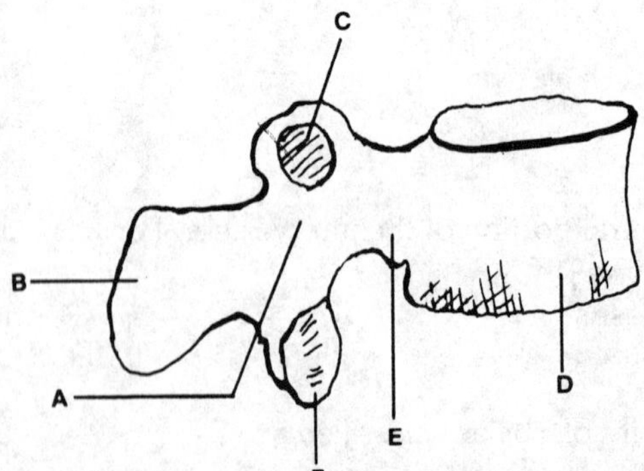

48. ______ Vertebral body

49. ______ Lamina

50. ______ Superior articulating process

51. ______ Pedicle

52. ______ Spinous process

53. ______ Inferior articulating process

54. The articulation between the first and second cervical vertebrae is called the: 5:76

A. Axio-occipital joint
B. Atlanto-epistrophyseal joint
C. Denso-occipital joint
D. Axio-epistrophyseal joint

55. The destruction of a vertebra, which can lead to spinal cord and nerve involvement, is termed: 8:1731

A. Lordosis
B. Spondylolisthesis
C. Spondylitis
D. Spondylolysis

56. The abnormal lateral curvature of the spine results in a condition called: 10:187

A. Scoliosis
B. Lordosis
C. Subluxation
D. Ankylosis

57. Protrusion of the central portion of an intervertebral disc on the spinal cord leads to a painful condition known as (a): 8:882

A. Hunchback
B. Thoracic spine
C. Sciatica
D. Herniated disc

58. Kyphosis is a relatively common condition of the elderly that results in a pronounced forward curvature of the: 10:187

A. Cervical spine
B. Thoracic spine
C. Lumbar spine
D. Sacrum

59. A congenital deformity resulting in the incomplete formation of the spinal canal is called: 10:187

A. Spondylolisthesis
B. Lordosis
C. Spina bifida
D. Spondylitis

60. The total number of sacral foramina found in the adult human sacrum is: 10:184

A. Four
B. Eight
C. Twelve
D. Sixteen

Chapter 35

1. The adult cranial vault or calveria that houses the brain is formed by all or part of ________ cranial bones. 5:93

A. Four
B. Eight
C. Fourteen
D. Twenty-eight

2. The large prominent ridges on the anterior surface of the frontal bone (superciliary ridges) are closely related to the surface structure termed the: 5:93

A. Acanthion
B. Bregma
C. Inner canthus
D. Eyebrows

3. The pterygoid processes that form part of the posterior nasal cavity are the inferior extensions of the: 1:213:II

A. Sphenoid bone
B. Temporal bone
C. Maxillary bone
D. Mandible

4. The roof of the mouth (hard palate) is formed by the inferior surfaces of the: 1:220:II
1. Maxillary bone *2. Palatine bones* *3. Malor bones*

A. 1 & 2 only
B. 1 & 3 only
C. 2 & 3 only
D. 1, 2 & 3

5. The meninges, the membranes that help to protect the brain, have an attachment on a small triangular process of the ethmoid bone called the: 4:197

A. Falx cerebri
B. Clinoid process
C. Crista galli
D. Styloid hamulus

6. The mastoid air cells are located posterior to the ear within a large, bony prominence of the: 5:94

A. Temporal bone
B. Sphenoid bone
C. Malor bone
D. Occipital bone

7. The majority of the sides and top of the cranial vault is formed by a pair of bones called the: 4:197

A. Parietal bones
B. Frontal bones
C. Ethmoid bones
D. Palatine bones

8. The large foramen that passes through the occipital bone which allows for the passage of the spinal cord and brain stem into the calveria is called: 4:202

A. Foramen ovale
B. Foramen magnum
C. Foramen rotundum
D. Foramen lacerum

9. The majority of the middle cranial fossa is formed by the body of the: 5:97

A. Occipital bone
B. Maxillary bone
C. Temporal bone
D. Sphenoid bone

10. The nasal cavities have a continuous membrane with a series of cavities within the cranial bones called the: 5:101

A. Paranasal sinuses
B. Infraorbital fissure
C. Mastoid air cells
D. Olfactory foramina

11. The junction of the nasal bones and the frontal bones corresponds to a useful surface landmark called the: 5:103

A. Inion
B. Gonion
C. Nasion
D. Acanthion

12.-24. Pertaining to the diagram of the skull, place the letter that corresponds to the following structures: 5:92

12. ______ Mandible

13. ______ Lambdoidal suture

14. ______ Frontal bone

15. ______ Occipital bone

16. ______ Zygomatic (malor) bone

17. ______ Parietal bone

18. ______ Styloid process

19. ______ Nasal bones

20. ______ Mastoid process

21. ______ Anterior nasal spine

22. ______ Frontal (coronal) suture

23. ______ Maxillary bone

24. ______ Temporal bone

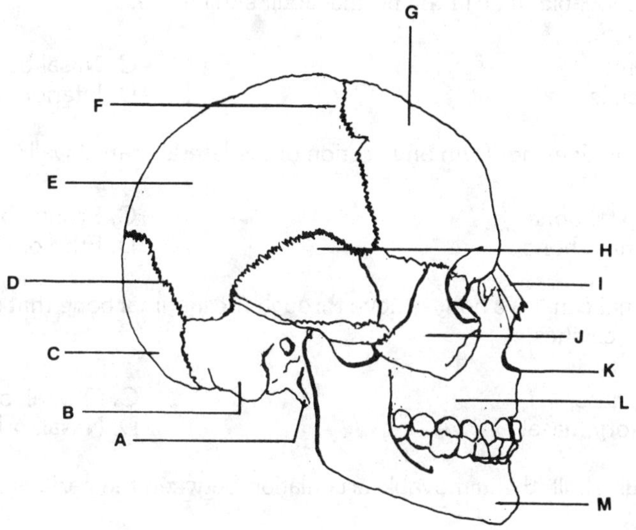

25. The flattened, horizontal portion on the superior aspect of the ethmoid bone is called the: 1:211:II

A. Crista galli
B. Perpendicular plate
C. Superior conchae
D. Cribriform plate

26. In the adult skull, the immovable articulation between the parietal and temporal bones is called the: 5:90

A. Lambdoidal suture
B. Squamosal suture
C. Sagittal suture
D. Coronal suture

27. Which of the following bones does not contribute to the formation of the bony orbit? 5:94

A. Lacrimal bone
B. Maxillary bone
C. Temporal bone
D. Ethmoid bone

28. The lateral border of the eyelids meet at a point termed the: 5:91

A. Acanthion
B. Nasion
C. Outer canthus
D. Optic fissure

29. The widest transverse measurement of the skull can be found between the two: 5:94

A. Parietal bones
B. Temporal bones
C. Malor bones
D. Lacrimal bones

30. The condyles for the articulation with the first cervical vertebra are located on the inferior surface of the: 5:94

A. Malar bone
B. Occipital bone
C. Vomer bone
D. Ethmoid bone

31. The most superior paranasal sinus is the: 5:101

A. Ethmoidal sinus
B. Maxillary sinus
C. Frontal sinus
D. Zygomatic sinus

32. The thick, spongy ridge of bone for the reception of the roots of the teeth is termed the: 1:220:II

A. Antrum of Highmore
B. Condyloid process
C. Parietal eminences
D. Alveolar process

33. The only movable bone in the normal adult skull is the: 5:100

A. Zygoma
B. Mandible
C. Nasal bone
D. Inferior conchae

34. Which bone does not form any portion of the lateral cranial wall? 5:97

A. Occipital bone
B. Sphenoid bone
C. Frontal bone
D. Ethmoid bone

35. The lacrimal canal is a bony groove through the lacrimal bone that serves to connect the ________ and the ________ cavities. 5:99

A. Oral, nasal
B. Auditory, nasal
C. Orbital, oral
D. Nasal, orbital

36. In the adult skull, the immovable articulation between the parietal and occipital bones is called the: 5:103

A. Sagittal suture
B. Coronal suture
C. Lambdoidal suture
D. Squamosal suture

37. The maxillary sinus is often referred to as the: 1:220:II

A. Pyloric antrum
B. Venous sinus
C. Antrum of Highmore
D. Hypoglossal antrum

38. The surface areas of the nasal cavities are divided into two more or less equal sides by a bony separation called the: 1:211:II

A. Alveolar process
B. Nasal septum
C. Horizontal sinus
D. Charismatic groove

39. The hard, dense bony process of the temporal bone that houses the structures of the middle and inner ear is called the: 5:95

A. Mastoid process
B. Carotid canal
C. Petrous pyramid
D. Tympanic ridge

40. The coronal suture is located between the ________ and the ________ bones. 5:103

A. Frontal, parietal
B. Parietal, temporal
C. Frontal, temporal
D. Occipital, parietal

41. The localizing point that is found between the eyebrows over the frontal bone is called the: 5:93

A. Nuchel point
B. Articular point
C. Glabella
D. Lambda

42.-48. Pertaining to the diagram of the mandible, place the letter that corresponds to the following structures: 1:222:II

42. ______ Mental foramen

43. ______ Gonion (angle)

44. ______ Coronoid process

45. ______ Body

46. ______ Ramus

47. ______ Mental protuberance

48. ______ Condyloid process

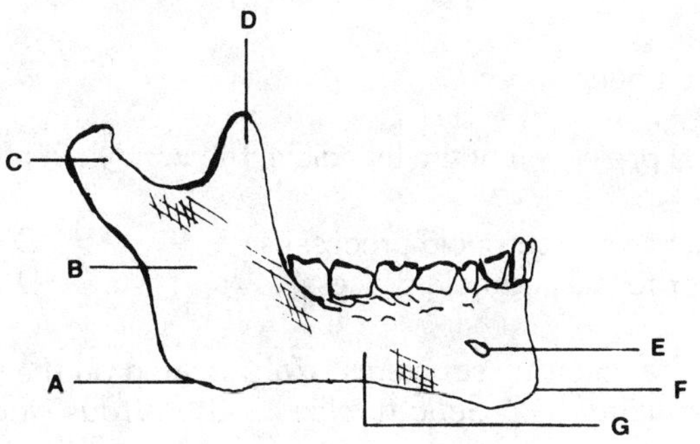

49. The majority of the lateral orbital wall and the cheek is formed by the: 5:98

A. Malor bone
B. Maxillary bone
C. Parietal bone
D. Temporal bone

50. Which of the following cranial bones contains a paranasal sinus? 5:101
1. Frontal bone 2. Ethmoid bone 3. Sphenoid bone

A. 1 & 2 only
B. 1 & 3 only
C. 2 & 3 only
D. 1, 2 & 3

51. The rounded, articulating condyles of the mandible that help form the temporomandibular joint are called the: 1:222:II

A. Coronoid processes
B. Coracoid processes
C. Condyloid processes
D. Alveolar processes

52. The greatest number of cranial foramina are located on the: 4:202

A. Temporal bone
B. Sphenoid bone
C. Occipital bone
D. Frontal bone

53. The external occipital protuberance is often referred to as the: 5:94

A. Inion
B. Gonion
C. Pterion
D. Asterion

54. Extending anteriorly from the dorsum sella are two bony processes which help in the formation of the lesser wing of the sphenoid. These are termed the: 1:212:II

A. Petrous ridges
B. Pterygoid processes
C. Anterior clinoid processes
D. Anterior condyloid processes

55. In the adult, the junction between the sagittal and lambdoidal sutures is known as the: 5:103

A. Bregma
B. Pterion
C. Inion
D. Lambda

56. The superior and middle nasal conchae that serve to increase the surface area of the nasal cavities are found on the: 5:97

A. Sphenoid bone
B. Ethmoid bone
C. Frontal bone
D. Nasal bones

57. What is the name of the two L-shaped facial bones that form part of the posterior hard palate and nasal septum? 5:99

A. Maxillae
B. Palatine bones
C. Vomer bones
D. Inferior conchae

58. In the lateral projection of the mandible, the condyloid process is: 5:100

A. Superior to the coronoid process
B. Anterior to the coronoid process
C. Posterior to the coronoid process
D. More than one but not all the above

59. Which of the following structures is/are located on the sphenoid bone? 5:95
1. Greater wing *2. Sella turcica* *3. Petrous ridges*

A. 1 & 2 only
B. 1 & 3 only
C. 2 & 3 only
D. 1, 2 & 3

60. The long, sliver-like processes that extend inferiorly from each temporal bone serving for the attachment of the muscles of speech are called: 5:95

A. Styloid processes
B. Zygomatic processes
C. Sphenoid struts
D. Coronoid processes

61. The small, U-shaped hyoid bone that sits below the floor of the mouth aids the act of: 10:175

A. Phonation
B. Swallowing
C. Both of the above
D. Neither of the above

62. The floor of the cranial vault is formed by all or part of the: 5:95
1. Parietal bones *2. Occipital bone* *3. Frontal bone*

A. 1 & 2 only
B. 1 & 3 only
C. 2 & 3 only
D. 1, 2 & 3

63. The majority of the upper portion of the nasal septum is formed by the ______ of the ethmoid bone. 5:97

A. Perpendicular plate
B. Cribriform plate
C. Lateral masses
D. Crista galli

64. The sphenoid sinus is located anterior and inferior to what other structure of the sphenoid bone? 5:95

A. Optic foramen
B. Styloid process
C. Greater wing
D. Sella turcica

65. The surface landmark associated with the anterior nasal spine at the junction of the upper lip and nose is called the: 5:91

A. Glabella
B. Acanthion
C. Gonion
D. External occipital protuberance

66. Which bone does not form any portion of the nasal cavity or septum? 5:101

A. Vomer bone
B. Palatine bones
C. Ethmoid bone
D. Malor bone

67. The most posterior paranasal sinus is the: 5:101

A. Frontal sinus
B. Sphenoid sinus
C. Ethmoid sinus
D. Maxillary sinus

68. The sagittal suture is located between the: 5:103

A. Two parietal bones
B. Frontal and parietal bones
C. Occipital and temporal bones
D. Occipital and parietal bones

69. The zygomatic arch is formed by portions of the: 5:98

A. Zygomatic and temporal bones
B. Temporal and occipital bones
C. Sphenoid and temporal bones
D. Frontal and malor bones

70. Nerves, arteries and veins pass in and out of the floor of the cranium by way of the: 5:103

A. Auditory canals
B. Eustachian tubes
C. Basal foramina
D. Paranasal sinuses

Respiratory System

Chapter 36

1. The entire process by which the body exchanges gasses with the atmosphere is called: 4:568

A. Osmosis
B. Hypoxia
C. Respiration
D. Transferral

2. Some of the dust and other particles in the air that enter the nose are filtered by a series of: 4:568

A. Internal nasal hairs
B. Mucus membranes
C. Lymphatic ducts
D. Nasal foramina

3. The separation of the nasal cavity into right and left portions is accomplished by a vertical partition called the: 5:157

A. Turbinates
B. Conchae
C. Nostrils
D. Septum

4. Which of the following is the correct sequence of air passage in the upper respiratory tract? 5:158

A. Nose, larynx, pharynx, trachea
B. Mouth, pharynx, larynx, trachea
C. Nose, trachea, pharynx, larynx
D. Mouth, larynx, trachea, pharynx

5. The vocal cords are housed in a structure formed by a number of singular and paired cartilages called the: 4:572

A. Pharynx
B. Trachea
C. Larynx
D. Bronchi

6. Due to its large diameter and more vertical placement, the most common location of foreign body blockages is the: 5:159

A. Right primary bronchus
B. Right pleural cavity
C. Left primary bronchus
D. Left pleural cavity

7. The bifurcation of the trachea into the mainstem bronchi occurs at the vertebral level of the ________ thoracic vertebra. 5:159

A. First or second
B. Fourth or fifth
C. Seventh or eighth
D. Ninth or tenth

8. The principal muscle involved with pulmonary ventilation is a dome-shaped muscle called the: 4:581

A. Intercostal muscle
B. Psoas muscle
C. Diaphragm
D. Pleura

9. The upper, bluntly pointed portion of the lung sitting just above the clavicles is called the: 5:159

A. Lingula
B. Hilum
C. Superior fissure
D. Apex

10. The numerous U-shaped cartilages which form the anterior and lateral borders of the trachea serve to: 4:575

A. Separate the esophagus and trachea
B. Aid in the swallowing of food
C. Prevent collapse during respiration
D. Produce the sounds of speech

11. The small sac at the end of a respiratory bronchiole that serves in the exchange of respiratory gasses is called the: 5:161

A. Alveolus
B. Glottis
C. Bronchiolus
D. Larynx

12. The left lung of the normal adult contains ________ distinct lobe(s). 5:161

A. A single
B. Two
C. Three
D. Five

13. The large, shield-shaped cartilage that forms the anterior portion of the larynx is called the: 8:154

A. Hyoid cartilage
B. Thyroid cartilage
C. Cricoid cartilage
D. Arytenoid cartilage

14. The inner layer of pleura that attaches directly to the lung tissue is termed the: 5:161

A. Visceral pleura
B. Parietal pleura
C. Axillary pleura
D. Alveolar pleura

15. The nose functions to: 4:568

1. Moisten incoming air *2. Warm incoming air* *3. Filter incoming air*

A. 1 & 2 only
B. 1 & 3 only
C. 2 & 3 only
D. 1, 2 & 3

16.-21. Pertaining to the diagram of the bronchial tree, place the letter that corresponds to the following structures: 5:158

16. ______ Cricoid cartilage

17. ______ Right primary bronchus

18. ______ Thyroid cartilage

19. ______ Left primary bronchus

20. ______ Secondary bronchus

21. ______ Trachea

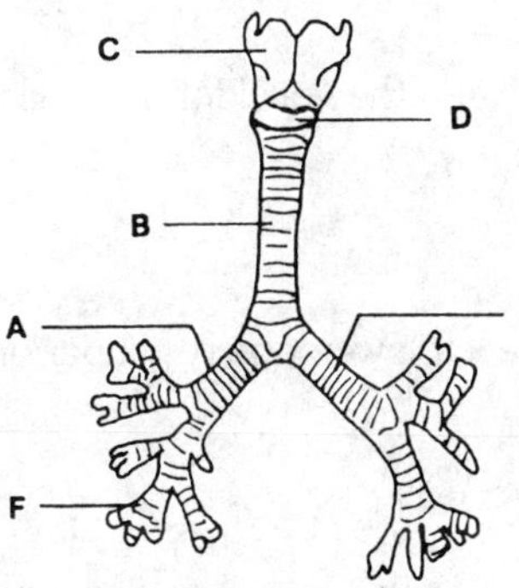

22. During normal respiration, the diaphragm will reach the lowest point of its excursion during: 5:162

A. Supine inspiration
B. Erect inspiration
C. Supine exhalation
D. Erect exhalation

23. The openings of the mainstem bronchi at their bifurcation are separated by a ridge of cartilage called the: 4:575

A. Bronchial fissure
B. Bronchial septum
C. Carina
D. Vocal fold

24. Each individual alveolar sac is surrounded by a: 4:577

A. Mucus membrane
B. Pleural membrane
C. Cartilaginous ring
D. Capillary network

25. The medial border of each lung has a deep depression for entry of the pulmonary vessels and bronchi termed the: 5:159

A. Mediastinum
B. Apex
C. Hilus
D. Diaphragmatic recess

26. In the lower respiratory tract, dust is filtered and moved by the action of mucus and small, hairlike structures called: 4:574

A. Cilia
B. Villi
C. Flagella
D. Cornu

27. The lungs are covered by a transparent, double-layered serous membrane called the: 5:161

A. Dura
B. Pleura
C. Meninges
D. Lingula

28. The mediastinum, the potential space between the lungs, contains all of the following except the: 5:163

A. Heart
B. Thyroid gland
C. Esophagus
D. Thymus gland

29. The thin, serous fluid that exists between the layers of the pleura serves to: 4:579

A. Remove dust particles
B. Prevent collapse
C. Reduce friction
D. Prevent infection

30. The blood supply to the lungs is accomplished by the: 5:162

A. Systemic arteries
B. Portal arteries
C. Aorta
D. Pulmonary arteries

31. The normal amount of air that enters the lungs during quiet inspiration is about 500cc and is called the: 4:584

A. Tidal volume
B. Vital capacity
C. Inspiratory reserve
D. Residual volume

32. The slitlike opening between the vocal folds is called the: 5:159

A. Glossus
B. Glottis
C. Carina
D. Septum

33. The upper portion of the left lower lobe lies superior to the lower portion of the left upper lobe of the lung. 10:699

A. True
B. False

34. The separation of the lobes of the lungs is accomplished by long, narrow grooves called: 10:699

A. Faci
B. Lobules
C. Fissures
D. Vagi

35. Contractions of the diaphragm will result in an action of the lung tissue called: 5:162

A. Exhalation
B. Inhalation
C. Segmentation
D. Coartation

36. The collection of fluids or gasses in the pleural space may lead to a lung collapse. This condition may be referred to as a: *1. Pneumothorax* *2. Hydrothorax* *3. Hemothorax*

A. 1 only
B. 2 only
C. 3 only
D. 1, 2 & 3

37. If any portion of the upper respiratory tract becomes swollen or blocked, an artificial opening can be made to improve breathing which is called a: 4:575

A. Laryngotomy
B. Rhinoplasty
C. Bronchography
D. Tracheostomy

38. Many air movements not related to breathing are termed non-respiratory movements, which are reflex in nature. A common reflex for clearing the lower respiratory tract is called a: 4:587

A. Sneeze
B. Cough
C. Hiccup
D. Laugh

39. A common degenerative disease characterized by the destruction of the alveolar walls is termed: 4:589

A. Asthma
B. Pneumonia
C. Emphysema
D. Dyspnea

40. The collapse of the part or all of the lung, which often results from bronchial obstruction, is termed: 5:163

A. Bronchiectasis
B. Atelectasis
C. Pneumitis
D. Pleurisy

41. The insertion of a needle into the pleural cavity to remove air or fluid is called a: 5:163

A. Tracheotomy
B. Lobectomy
C. Thoracoplasty
D. Thoracocentesis

42. The allergic reaction to a foreign substance, such as pollen, may cause the constriction of the bronchial passages. This condition is termed: 4:588

A. Bronchial asthma
B. Bronchiectasis
C. Bronchopneumonia
D. Bronchoesophageal fistula

43. The production of free fluid in the chest cavity is often seen on an erect chest radiograph in the lower portion of the lung called the: 5:163

A. Mediastinum
B. Apical region
C. Costaphrenic angle
D. Axillary border

44. A condition resulting from a marked deficiency of oxygen and an excess of carbon dioxide in the blood from a respiratory obstruction is termed: 4:603

A. Bradypnea
B. Asphyxia
C. Eupnea
D. Pulmonary edema

45. In the normal adult, the resting inspiration rate is about: 5:162

A. 70 respirations/minute
B. 40 respirations/minute
C. 15 respirations/minute
D. 10 respirations/minute

46.-50. Match the following: 4:603

46. ______ Apnea
47. ______ Dyspnea
48. ______ Eupnea
49. ______ Anoxia
50. ______ Bradypnea

A. Abnormally fast breathing
B. Normal breathing
C. Difficult breathing
D. Temporary absence of breathing
E. Low oxygen in tissues
F. Abnormally slow breathing

Urinary System

Chapter 37

1. The principal function of the urinary system is the elimination of the body's major metabolic byproduct which is: 5:189

A. Carbon dioxide
B. Urea
C. Hydrocholoric acid
D. Thrombin

2. The medial border of each kidney has a deep longitudinal fissure where vessels and nerves enter and leave the kidney called the: 4:750

A. Calyx
B. Medulla
C. Hilus
D. Cortex

3. The kidneys lie on either side of the vertebral column resting on the psoas and quadratus lumborum muscle's ________________ the peritoneum. 4:750

A. In front of
B. Inside of
C. Between the layers of
D. Behind

4. The upper expanded end of each ureter enters the kidney at an area termed the: 4:750

A. Renal pelvis
B. Renal pyramid
C. Renal column
D. Renal cortex

5. The kidneys are held in the upper abdominal position by a connective tissue called: 4:750

A. Renal columns
B. Renal papillae
C. Renal fascia
D. Convoluted tubules

6. The principal function of the ureters is to actively aid the passage of urine from the: 4:770

A. Renal pelvis to the urinary bladder
B. Urinary bladder to the urethra
C. Urethra to the urinary bladder
D. Urinary bladder to the kidney

7. The term applied to the expelling of urine from the urinary bladder is: 4:772

A. Mastication
B. Catheterization
C. Tubular secretion
D. Micturition

8. In the normal adult female, the urethra functions as a/an: 10:848
1. Reproductive organ *2. Excretory organ* *3. Endocrine organ*

A. 1 only
B. 2 only
C. 3 only
D. 1, 2 & 3

9. The muscular coat of the urinary bladder that plays an important part in expelling urine is called the: 10:847

A. Trigone
B. Psoas muscle
C. Detrusor muscle
D. External urinary sphincter

10. The renal pyramids have a striated appearance due to a number of __________ which are directed into the renal papillae. *1. Collecting ducts* *2. Calyces* *3. Renal arteries* 4:751

A. 1 only
B. 2 only
C. 3 only
D. 1, 2 & 3

11. The glandular appearance of the cortex of the kidney is due to the arrangement of the functional units of the kidney called the: 4:751

A. Renal sinuses
B. Nephrons
C. Peritubular capillaries
D. Afferent arterioles

12. In the normal adult's urinary system, the ureters will attach to the urinary bladder at its: 5:188

A. Anterior lateral surface
B. Posterior lateral surface
C. Anterior inferior surface
D. Posterior inferior surface

13. Due to the effect of other abdominal structures, the right kidney is normally located above the left kidney. 5:187

A. True
B. False

14. The renal corpuscle consists of a thin-walled, saclike structure surrounding a cluster of capillary beds called the: 4:754

A. Proximal convoluted tubule
B. Glomerulus
C. Afferent arteriole
D. Collecting duct

15.-20. Pertaining to the diagram of the kidney, place the letter that corresponds to the following structures: 4:752

15. ______ Renal pyramid

16. ______ Major calyx

17. ______ Ureter

18. ______ Renal pelvis

19. ______ Minor calyx

20. ______ Cortex

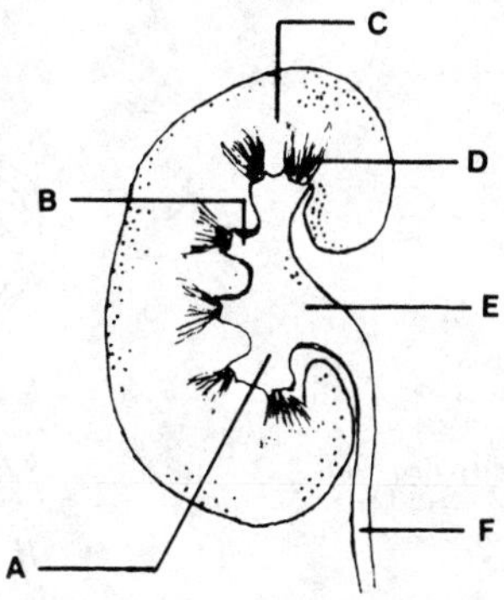

21. In the normal adult female, urine passes from the urinary bladder to the outside in a short muscular tube called the: 4:773

A. Urethra
B. Ureter
C. Trigone
D. Renin

22. The filtration of the organic salts, fluids and nitrogenous compounds from the blood occurs across a membrane called the: 5:189

A. Collecting ducts
B. Hypothalamus
C. Bowman's capsule
D. Major calyx

23. In the male adult, a mechanical urinary bladder obstruction may result from a/an: 10:890

A. Ureteral reflux
B. Renal colic
C. Enlarged prostate gland
D. Paralyzed external urethra sphincter

24. Blood is carried to the glomerulus by which of the following: 4:758

A. Vasa recta
B. Loop of Henle
C. Afferent arteriole
D. Efferent arteriole

25. The urinary bladder is situated: 4:771
1. Below the peritoneum *2. Anterior to the rectum* *3. Anterior to the uterus*

A. 1 only
B. 2 only
C. 3 only
D. 1, 2 & 3

26. The concentration of urine and tubular reabsorption occurs within the: 4:763
1. Proximal convoluted tubules *2. Distal convoluted tubules* *3. Loop of Henle*

A. 1 only
B. 2 only
C. 3 only
D. 1, 2 & 3

27. The normal adult's urinary bladder can hold a maximum of about: 4:772

A. 100cc of fluid
B. 300cc of fluid
C. 600cc of fluid
D. 1500cc of fluid

28. In the normal adult male, which is not one of the divisions of the male urethra? 4:773

A. Prostatic urethra
B. Membranous urethra
C. Erectile urethra
D. Penile urethra

29.-36. Pertaining to the diagram of the nephron, place the letter that corresponds to the following structures: 4:758

29. ______ Proximal convoluted tubule

30. ______ Afferent arteriole

31. ______ Loop of Henle

32. ______ Bowman's capsule

33. ______ Distal convoluted tubule

34. ______ Efferent arteriole

35. ______ Collecting duct

36. ______ Glomerulus

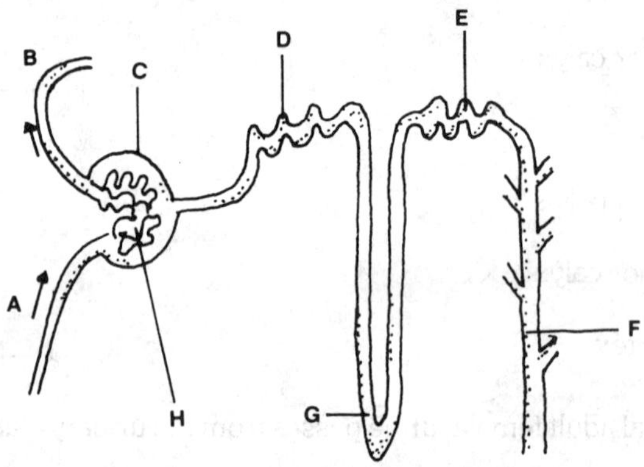

37. An individual who has lost all kidney function must have the blood cleaned out. This is normally accomplished by some type of renal: 10:59

A. Micturition
B. Diuresis
C. Urinalysis
D. Dialysis

38. Which of the following is not normally a function of the urinary system? 5:192

A. Elimination of urea
B. Maintenance of electrolite balance
C. Regulation of fluid balance
D. Production of erythrocytes

39. Many substances, such as alcohol, caffeine and digitalis, may cause an excessive urine output termed: 10:649

A. Pyuria
B. Diuresis
C. Uremia
D. Hemoturia

40. The obstruction of the blood flow to the kidney is one of the common causes of: 10:836

A. Hypertension
B. Hypotension
C. Diabetes
D. Renal colic

41. A blockage of the ureter by a calculus or other mechanical obstruction will often lead to the dilatation of the renal pelvis and calyces called: 5:192

A. Ptosis
B. Hydronephrosis
C. Polyuria
D. Agenesis

42. Water and fluid waste products can be eliminated by way of the: 10:826
1. Skin 2. Kidneys 3. Liver

A. 1 only
B. 2 only
C. 3 only
D. 1, 2 & 3

43. The term employed for a kidney not found in its normal location is: 5:192

A. Ectopic kidney
B. Staghorn kidney
C. Horseshoe kidney
D. Micro-kidney

44. The buildup of poisonous waste products in blood is commonly caused by poor kidney function which is called: 5:192

A. Ptosis
B. Anuria
C. Uremia
D. Renal reflux

45. Many types of trauma may lead to a kidney falling from its original location. This is termed (a): 5:192

A. Pyelitis
B. Horseshoe kidney
C. Extravasation
D. Ptosis

Digestive System

Chapter 38

1. The process by which food particles are mechanically reduced into smaller particles in the mouth is termed: 4:482

A. Absorption
B. Peristalsis
C. Mastication
D. Deglutition

2. Projecting from the middle arch of the soft palate is a small soft tissue structure concerned with the gag reflex called the: 4:484

A. Vulva
B. Uvula
C. Macula
D. Sebum

3. Small circular bands of smooth muscle fibers that serve to restrict the flow of liquids and solids within the digestive tract are termed: 4:494

A. Sphincter muscles
B. Cardiac muscles
C. Diaphragmatic muscles
D. Erectile muscles

4. The small raised structures of the tongue, papillae, contain the receptors for the sensation of: 4:485
1. Smell 2. Speech 3. Taste

A. 1 only
B. 2 only
C. 3 only
D. 1, 2 & 3

5. The underside of the tongue is attached to the floor of the mouth by a membranous fold called the: 4:483

A. Omentum
B. Uvula
C. Gingiva
D. Frenulum

6. The neural intervention that controls much of the digestive function results primarily from the action of the 10th cranial nerve called the : 4:482

A. Vagus nerve
B. Trigeminal nerve
C. Hypoglossal nerve
D. Buccal nerve

7. The act of swallowing involves both voluntary and involuntary actions by the muscles of the: 4:489
1. Tongue 2. Pharynx 3. Upper esophagus

A. 1 only
B. 2 only
C. 3 only
D. 1, 2 & 3

8. The wavelike muscular contractions that move the food bolus through the digestive system is termed: 4:482

A. Fauces
B. Micturition
C. Peristalsis
D. Mastication

9. The teeth that fit into the alveolar processes of the jaw bones are held in place by the substance: 4:486

A. Dentin
B. Cementum
C. Enamel
D. Pulp

10. The first liquid secretion (saliva) encountered by food during the digestive process is formed by the: 4:488

A. Tonsils
B. Salivary glands
C Glottis
D. Eustachian glands

11. The process by which food is actively transported from the mouth into the esophagus is termed: 4:491

A. Pepination
B. Salivation
C. Deglutition
D. Phonation

12. The different types of teeth are concerned with specific functions. The grinding of food is normally accomplished by the larger, flatter teeth called: 4:484

A. Molars
B. Incisors
C. Canines
D. Milk teeth

13. The nerves and blood vessels of the teeth are located in the centralized cavity called the: 4:486

A. Crown
B. Dentin
C. Enamel
D. Pulp cavity

14. The Wharton's ducts that open under the tongue is concerned with the passage of saliva from the: 4:488

A. Sublingual gland
B. Submandibular gland
C. Parotid gland
D. Thyroid gland

15. The first set of temporary teeth, which appear in early childhood, are referred to as: 4:485

A. Permanent teeth
B. Deciduous teeth
C. Wisdom teeth
D. Partition teeth

16. The bulk of a mature tooth lying beneath the outer enamel covering is composed of: 4:486

A. Cementum
B. Periosteum
C. Caries
D. Dentin

17. The largest set of salivary glands sitting in front and below each ear are the: 4:488

A. Parotid glands
B. Sublingual glands
C. Submaxillary glands
D. Submandibular glands

18. The portion of the tooth that extends above the gum line is termed the: 4:486

A. Root
B. Capitellum
C. Crown
D. Stalk

19. Saliva from the parotid gland passes within the ____________ to its opening onto the inner surface of the cheek. 4:488

A. Wharton's duct
B. Stensen's duct
C. Ducts of Wirsung
D. Ducts of Rinivus

20. In the normal adult, the esophagus is located ____________ to the trachea. 5:171

A. Posterior
B. Anterior
C. Medial
D. Lateral

21. Which of the following structures does not contribute to the mechanical or chemical digestive process? 4:489

1. Larynx 2. Esophagus 3. Mouth

A. 1 & 2 only
B. 1 & 3 only
C. 2 & 3 only
D. 1, 2 & 3

22. The protection of the intestinal tract from the action of digestive acids and enzymes is primarily accomplished by the secretion of: 4:494

A. Maltese
B. Cholecystokinin
C. Sodium bicarbonate
D. Mucus

23. The highest portion of the normal adult stomach sitting just below the left diaphragm is the: 5:173

A. Pylorus
B. Cardiac orifice
C. Fundus
D. Greater curvature

24. The stomach is involved with the ____________ digestion of food. 4:494

A. Chemical
B. Mechanical
C. Both of the above
D. Neither of the above

25. The semifluid, partially digested food found in the upper gastrointestinal tract is referred to as: 4:497

A. Chyme
B. Meconium
C. Flora
D. Bolus

26. The terminal segment of the normal human stomach is the: 5:172

A. Duodenum
B. Pylorus
C. Fundus
D. Trigone

27. The periodontal membrane is a close fitting protective membrane associated with the structure of the: 4:486

A. Stomach
B. Intestines
C. Teeth
D. Peritoneum

28. In the stomach, the production of hydrochloric acid aids the digestive process. This is produced by specialized cells called: 4:494

A. Acini cells
B. Goblet cells
C. Parietal cells
D. Zymogenic cells

29. Most of the capillary beds concerned with the absorption of food are located within numerous finger-like soft tissue extensions of the small intestines called the: 4:509

A. Cilia
B. Villi
C. Papillae
D. Taenia coli

30. The empty stomach is thrown into numerous mucosal folds that give its characteristic appearance called: 10:746

A. Rugae
B. Haustra
C. Epiplocicae
D. Voluli

31. The name of the last portion of the small intestines is the: 5:173

A. Ilium
B. Jejunum
C. Duodenum
D. Ileum

32. The junction of the transverse and descending colons takes place at the: 5:174

A. Angular notch
B. Splenic flexure
C. Hepatic flexure
D. Rectosigmoid junction

33.-41. Pertaining to the diagram of the stomach, place the letter that correspons to the following structures: 4:493

33. ______ Pyloric sphincter

34. ______ Cardiac orifice

35. ______ Duodenum

36. ______ Esophagus

37. ______ Fundus of the stomach

38. ______ Lesser curvature of the stomach

39. ______ Body of the stomach

40. ______ Greater curvature of the stomach

41. ______ Pylorus

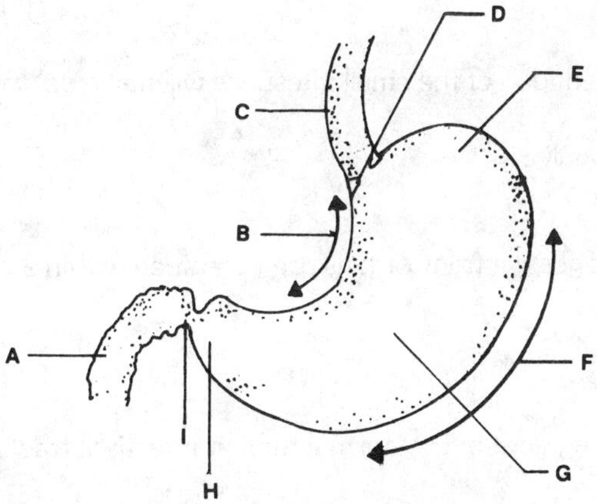

42. Food entering the stomach is prevented from exiting into the small intestines by the action ring of smooth muscles called the: 5:172

A. Duodenal sweep
B. Ileocecal valve
C. Cardiac sphincter
D. Pyloric sphincter

43. The majority of blood vessels and nerves, Meissner's plexus, are located in the ______________ of the intestinal tract walls. 4:480

A. Serosa layer
B. Submucosa layer
C. Muscular layer
D. Mucosa layer

44. The digestive tract organ most responsible for the absorption of water is the: 4:515

A. Large intestine
B. Stomach
C. Esophagus
D. Duodenum

45. The most distal portion of the large intestine is called the: 4:516

A. Rectum
B. Sigmoid colon
C. Ileum
D. Cecum

46. The folds of mucosa that help to increase the surface area of the small intestines are termed: 4:510

A. Rugae
B. Circular folds
C. Angulations
D. Maculi

47. Which of the following organs is/are concerned with the chemical digestion of food? 4:494
1. Colon 2. Small intestines 3. Salivary glands

A. 1 & 2 only
B. 1 & 3 only
C. 2 & 3 only
D. 1, 2 & 3

48. The small intestines receive large amounts of sodium bicarbonate from pancreatic secretions to help neutralize the ______________ present in the stomach. 4:500

A. Fatty acids
B. Digestive enzymes
C. Hydrogen choloride
D. Bile salts

49. The walls of the intestinal tract are composed of ______________ distinct layers of tissue. 5:169

A. Two
B. Four
C. Six
D. Eight

50. The undigested food of the small intestines will make its way into the colon through the: 10:759

A. Ileocecal valve
B. Pyloric canal
C. Cardiac orifice
D. Sphincter of Oddi

51. Most of the digestive tract organs are contained within a double-walled membrane called the: 4:508

A. Omentum
B. Mesentery
C. Peritoneum
D. Pleural cavity

52. The longest segment of the normal human intestinal tract is the: 4:508

A. Esophagus
B. Jejunum
C. Ileum
D. Cecum

53. The external layer of the digestive organs are composed of fibrous membranes called the:

A. Serosa
B. Mucosa
C. Endothelium
D. Mesentery

54. The first portion of the small intestines, which is about 25cm in length, is called the: 4:507

A. Pylorus
B. Duodenum
C. Ileum
D. Cecum

55. The veriform appendix is a small, blind pouch which extends from the inferior surface of the: 5:174

A. Jejunum
B. Duodenum
C. Cecum
D. Rectum

56. Which of the following organs is/are concerned with the absorption of digested foods? 4:507
1. Small bowel *2. Colon* *3. Esophagus*

A. 1 only
B. 2 only
C. 3 only
D. 1, 2 & 3

57. The fibrous structure that serves to attach the intestines to the posterior abdominal wall is termed the:
1. Omentum *2. Peritoneum* *3. Mesentery* 5:168

A. 1 only
B. 2 only
C. 3 only
D. 1, 2 & 3

58. A small flap of cartilage on the pharynx that helps to prevent solid or liquid foods from entering the trachea is called the: 5:157

A. Frenulum
B. Uvula
C. Epiglottis
D. Adenoids

59. The digestive enzymes of the small intestines are produced in specialized cells called: 4:510

A. Chief cells
B. Crypts of Lieberkuhn
C. Islets of Langerhans
D. Brunner's gland

60. The haustra are the folds that give the ______________ its characteristic appearance. 5:174

A. Esophagus
B. Small intestines
C. Stomach
D. Colon

61.-72. Pertaining to the diagram of the large intestine, place the letter that corresponds to the following structures: 5:175

61. ______ Rectum

62. ______ Splenic flexure

63. ______ Sigmoid colon

64. ______ Ileum

65. ______ Transverse colon

66. ______ Cecum

67. ______ Taeniae coli

68. ______ Descending colon

69. ______ Hepatic flexure

70. ______ Appendix

71. ______ Ileocecal valve

72. ______ Ascending colon

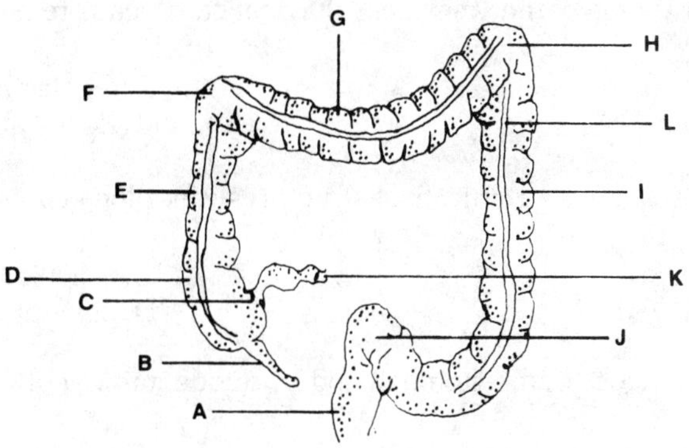

73. The ascending and transverse colon join at an area under the right lobe of the liver called the: 5:174

A. Splenic flexure
B. Gastric flexure
C. Hepatic flexure
D. Duodenal flexure

74. Digestive enzymes are manufactured by the body in which of the following organs? 5:174
1. Stomach 2. Pancreas 3. Salivary glands

A. 1 & 2only
B. 1 & 3 only
C. 2 & 3 only
D. 1, 2 & 3

75. The appearance of haustral folds is due in large part to the action of three flat bands of muscle fibers on the exterior of the colon called the: 1:79:II

A. E-coli
B. Taeneia coli
C. Epiploicae
D. Gyri

76. The formation of a stalked tumor that protrudes into the intestinal tract is called a: 8:1135

A. Volvulus
B. Fistula
C. Hemorroid
D. Polyp

77. The telescoping of one part of the bowel into another is termed: 5:180

A. Herniation
B. Varices
C. Intussusception
D. Atresia

78. An outpouching or weaking of the intestinal wall most commonly seen in the large intestine is termed a/an: 5:180

A. Diverticulum
B. Aneurism
C. Calculus
D. Fistula

79. Bright red blood in the stool is a common indication of bleeding in the: 3:175
1. Colon 2. Stomach 3. Small intestines

A. 1 only
B. 2 only
C. 3 only
D. 1, 2 & 3

80. The valsalva maneuver is used to increase abdominal cavity pressure and is useful for the demonstration of: 1:19:II

A. Diverticuli
B. Polyps
C. Esophageal varices
D. Ascites

81. An abnormal dilatation of the veins near the anal canal leads to a condition termed: 4:519

A. Colitis
B. Hemorroids
C. Cholelithiasis
D. Rectal abscess

82. Which of the following gastrointestinal structures is not located within the peritoneum? 4:508

A. Stomach
B. Duodenum
C. Splenic flexure
D. Jejunum

83. The loss of the mucosa of the stomach and/or duodenum will often lead to development of (an): 4:496

A. Appendicitis
B. Jaundice
C. Intussusception
D. Ulceration

84. The only major secretion of the large intestines is: 5:179

A. Digestive enzymes
B. Hormones
C. Mucus
D. Water

85. The abnormal protrusion of the stomach through the diaphragm is called: 5:179

A. Hiatal hernia
B. Pyloric stenosis
C. Esophageal atresia
D. Meckel's diverticulum

86. In the normal adult male, the rectum is located: 5:175

A. Anterior to the urinary bladder
B. Posterior to the urinary bladder
C. Superior to the diaphram
D. None of the above

87. In the normal adult, the normal sequence for the passage of food is: 4:507

A. Esophagus, Stomach, SI, Colon
B. Stomach, Colon, Esophagus, SI
C. Colon, SI, Stomach, Esophagus
D. SI, Esophagus, Stomach, Colon

88. A common abnormality occurring more frequently in male infants resulting from a persistent spasm of the terminal portion of the stomach is called: 5:179

A. Pyloric stenosis
B. Umbilical hernia
C. Gastritis
D. Hypoglycemia

89. An abnormal collection of serous fluids in the peritoneal cavity is termed: 8:129

A. Peritonitis
B. Ascites
C. Varices
D. Dehydration

90. The abnormal twisting of a loop of intestine around itself can lead to an abnormal condition called (a/an): 5:180

A. Imperforate colon
B. Ileitis
C. Volvulus
D. Thrombosis

Accessory Digestive System

Chapter 39

1. The liver is known to have over 100 functions. Among these is/are: 4:502

   1. Formation of blood cells *2. Storage of glycogen* *3. Detoxification of toxins*

   A. 1 & 2 only
   B. 1 & 3 only
   C. 2 & 3 only
   D. 1, 2 & 3

2. The bulk of the pancreas is made up of pancreatic acinar cells that serve to secrete a number of enzymes which aid in the: 4:499

   A. Emulsification process
   B. Digestion of food products
   C. Hemopoietic process
   D. Detoxification process

3. The spleen is concerned with both the formation and breakdown of blood cells and is normally located under the: 5:147

   A. Right lobe of the liver
   B. Left lobe of the liver
   C. Right diaphragm
   D. Left diaphragm

4. The two major lobes of the liver are separated by folds of visceral peritoneum called the: 4:502

   A. Caudate ligament
   B. Broad ligament
   C. Quadrate ligament
   D. Falciform ligament

5. A musculomembranous sac that serves to store and concentrate bile between meals is called the: 4:504

   A. Gallbladder
   B. Alveolar sac
   C. Biliary node
   D. Urinary bladder

6. The biliary ducts are principally concerned with the transport of bile and digestive enzymes into the: 5:177

   A. Colon
   B. Stomach
   C. Small intestines
   D. Lymphatic system

7. The gallbladder is normally found by its attachment to the ventral surface of the: 4:504

   A. Right lobe of the liver
   B. Left lobe of the liver
   C. Spleen
   D. Duodenum

8. The duct formed at the junction of the duct from the gallbladder and the common hepatic ducts is termed the: 5:177

   A. Ampulla of Vater
   B. Common bile duct
   C. Duct of Santorini
   D. Common cystic duct

9. The head of the pancreas will normally be located within the: 4:499

   A. Lesser curvature of the stomach
   B. Splenic flexure
   C. Sweep of the duodenum
   D. Falciform ligament

10. The largest single organ contained within the abdominal cavity is the: 5:177

A. Spleen
B. Pancreas
C. Small intestines
D. Liver

11. The breakdown of the red blood cells' hemoglobin into bilirubin is principally the function of the: 10:757

A. Spleen
B. Pancreas
C. Liver
D. Kidney

12. The exocrine secretions of the pancreas are carried to the common bile duct by way of the: 4:499

A. Duct of Santorini
B. Stenson's duct
C. Pancreatic duct
D. Wharton's duct

13. The absorption of water is an important factor in the maintenance of homeostasis. Which of the following organs is/are concerned with this function? 4:504
1. Esophagus *2. Gallbladder* *3. Large intestine*

A. 1 & 2 only
B. 1 & 3 only
C. 2 & 3 only
D. 1, 2 & 3

14. The control of glucose utilization is directly associated with a hormonal secretion of the pancreas called: 5:177

A. Urea
B. Insulin
C. Bile
D. Bilirubin

15.-25. Pertaining to the diagram of the biliary system, place the letter that corresponds to the following structures: 10:753

15. ______ Gallbladder
16. ______ Sphincter of Oddi
17. ______ Duodenum
18. ______ Left hepatic duct
19. ______ Pancreatic duct
20. ______ Common hepatic duct
21. ______ Duct of Santorini
22. ______ Common bile duct
23. ______ Right hepatic duct
24. ______ Cystic duct
25. ______ Sphincter of Boyden

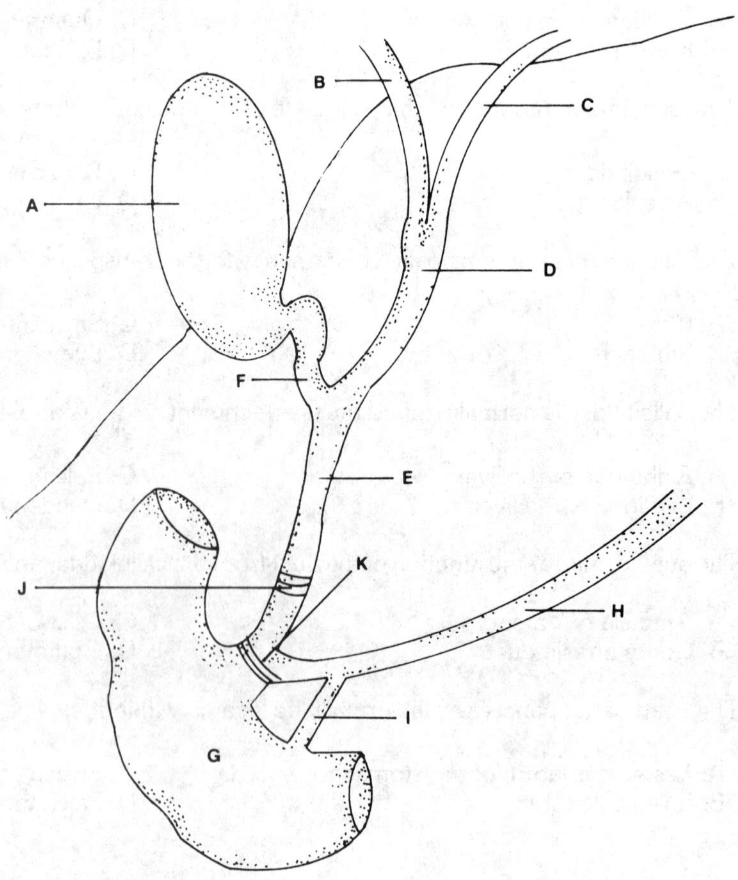

26. The release of bile by the gallbladder is controlled by the hormonal secretion of the small intestines called: 4:504

A. Cholegraphin
B. Cholecystokinin
C. Ferritin
D. Bilirubin

27. The tail of the pancreas is attached to what other abdominal cavity structure? 5:175

A. Kidney
B. Duodenum
C. Liver
D. Spleen

28. The gallbladder is filled by the reverse flow of bile caused by this strong sphincter muscle located at the distal end of the common bile duct called the: 10:759

A. Sphincter of Boyden
B. Sphincter of Wirsung
C. Sphincter of Rivinus
D. Sphincter of Kupffer

29. The formation of bile from bile salts and bilirubin is one of the functions associated with the: 5:177
1. Spleen *2. Liver* *3. Kidney*

A. 1 only
B. 2 only
C. 3 only
D. 1, 2 & 3

30. The stored bile from the gallbladder passes into the biliary duct system by way of the: 4:505

A. Cystic duct
B. Ampulla of Vater
C. Hepatic duct
D. Papillary duct

31. Which of the following is not a common function of the liver? 4:502

A. Storage of blood
B. Storage of Vitamins A and D
C. Deamination of amino acids
D. Secretion of digestive enzymes

32. The junction of the biliary collecting system and the duodenum is guarded by a sphincter muscle called the: 4:505

A. Sphincter of Lieberkuhn
B. Sphincter of Oddi
C. Sphincter of Beta
D. Sphincter of Pauli

33. The major function of bile in the gastrointestinal tract is the: 4:505

A. Emulsification of fats
B. Production of amino acids
C. Emulsification of carbohydrates
D. Absorption of vitamins

34. The process by which the liver manufacturers sugars from non-carbohydrate materials is termed: 4:529

A. Glycolysis
B. Metabolism
C. Gluconeogenesis
D. Maturation

35. The release of bile by the gallbladder is triggered by the presence of ____________ in the stomach. 4:497
1. Carbohydrates *2. Lear Liquids* *3. Fats*

A. 1 only
B. 2 only
C. 3 only
D. 1, 2 & 3

36. The intake of fecal contaminated food or water may lead to an inflammation of the liver called: 4:503

A. Hepatitis
B. Cirrhosis
C. Choleitis
D. Jaundice

37. Cholelithiasis is a common condition of the biliary tract and is commonly referred to as: 4:505

A. Duodenal ulcers
B. Arteriosclerosis
C. Gallstones
D. Jaundice

38. A serious condition that results from the hypofunction of the islets of Langerhans of the pancreas is termed: 4:462

A. Hemophilia
B. Diabetes mellitus
C. Hydrophobia
D. Gastroenteritis

39. The failure of the liver to absorb bilirubin will often lead to a yellowish discoloration of the skin and eyes called: 4:504

A. Cyanosis
B. Yanthosis
C. Jaundice
D. Chlorosis

40. Cirrhosis of the liver is often associated with the excessive intake of toxic material known as: 4:519

A. Cholesterol
B. Heparin
C. Renin
D. Alcohol

Special Senses

Chapter 40

1. The ability to detect the external environment is accomplished by a series of specialized structures called: 4:390

A. Motor end plates
B. Adaptors
C. Sensory receptors
D. Accommodators

2. The sense of smell is related to the first cranial nerve, which passes through the ethmoid bone into the nasal cavity, called the: 5:245

A. Acoustic nerve
B. Olfactory nerve
C. Optic nerve
D. Cochlear nerve

3. The normal adult possesses a number of cutaneous (exteroceptive) senses. Among these is/are: 4:391
1. Pain 2. Pressure 3. Heat

A. 1 & 2 only
B. 1 & 3 only
C. 2 & 3 only
D. 1, 2 & 3

4. The detection of certain low energy forms of electromagnetic radiation can be accomplished by photoreceptors located in the: 4:390

A. Eye
B. Ear
C. Nose
D. Mouth

5. The degree of distention in some muscular organs, such as the rectum and urinary bladder, is accomplished by: 4:395

A. Thermoreceptors
B. Chemoreceptors
C. Acoustoreceptors
D. Stretch receptors

6. The transparent outer portion of the eye that allows passage of light into the eye is called the: 5:239

A. Cornea
B. Pupil
C. Optic disc
D. Retina

7. The tympanic cavity is a small irregular space within the temporal bone that serves to house the structures associated with: 5:243

A. Taste
B. Hearing
C. Smell
D. Sight

8. The specialized organs on the tongue for the detection of taste are called: 4:396

A. Gustatory glands
B. Auditory receptors
C. Taste buds
D. Ossicles

9. The external funnel-like structure of the ear that helps to direct sound vibrations to the ear is called the: 4:401

A. Tragus
B. Auricle
C. Ceruminous
D. Eustachian cartilage

10. The tympanic membrane is a thin membrane which separates the structures of the: 4:401

A. Chambers of the eye
B. Middle and inner ear
C. External and middle ear
D. None of the above

11. Pressure differences in the middle ear must be maintained to prevent hearing impairment. The structure associated with this function is the: 4:403

A. Eustachian tube
B. Nasolacrimal duct
C. Semicircular canal
D. Fallopian tube

12. The small bones of the middle ear that serve to transmit sound vibrations to the inner ear are the: 5:243

1. Cochlea *2. Stapes* *3. Malleus*

A. 1 & 2 only
B. 1 & 3 only
C. 2 & 3 only
D. 1, 2 & 3

13. The sense of equilibrium (balance) is controlled by a series of small fluid canals called the: 4:404

A. Cochlear canals
B. Semicircular canals
C. Canaliculi
D. Lacrimal canal

14. The specialized receptor for hearing located on the basilar membrane of the cochlea is called the: 4:407

A. Utricle
B. Conjunctiva
C. Organ of Corti
D. Ciliary body

15.-20. Pertaining to the diagram of the bony labyrinth, place the letter that corresponds to the following structures: 4:405

15. ______ Stapes at oval window

16. ______ Cochlear nerve

17. ______ Vestibule

18. ______ Semicircular canals

19. ______ Cochlea

20. ______ Vestibular portion of the acoustic nerve

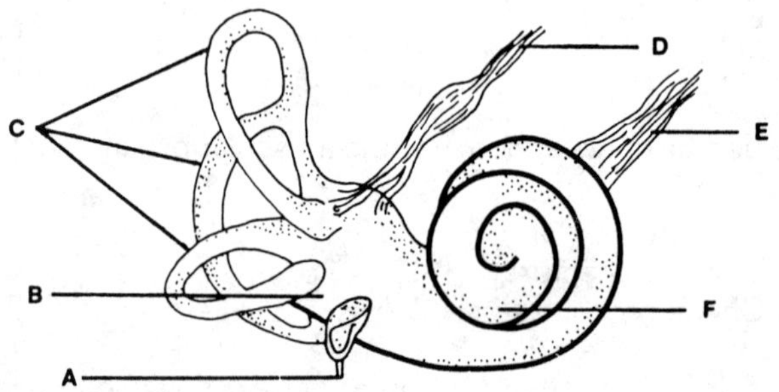

21. The membranous labyrinth of the inner ear is surrounded by a protective fluid called: 4:404

A. Perilymph
B. Plasma
C. Vitreous humor
D. Aqueous humor

22. The action of otoliths on the hairlike extensions of the macula in response to gravity is interpreted by the brain to help control: 4:408

A. Positional orientation
B. Visual reflexes
C. Tympanic reflex
D. Sensory adaptation

23. The tongue is able to detect four primary taste sensations. Among these is: 4:399
1. Sweet *2. Sour* *3. Salt*

A. 1 & 2 only
B. 1 & 3 only
C. 2 & 3 only
D. 1, 2 & 3

24. Visceral pain impulses may be felt at body parts other than where the stimulus is occurring. This phenomenon is called: 4:393

A. Adaptation
B. Discrimination
C. Referred pain
D. Anosmia

25. The upper eyelid that forms a protective covering for the eye also houses a small gland that produces tears called the: 4:411

A. Lacrimal gland
B. Conjunctiva
C. Rectus gland
D. Basilar gland

26. The production of tears helps to moisten and protect the eye from bacterial infection and will drain into the nasal cavity through the: 5:241

A. Iris
B. Vestibule
C. Eustachian tube
D. Nasolacrimal duct

27. The visual receptor cells are located on the back of the eye in a cellular layer called the: 5:239

A. Sclera
B. Vitreous humor
C. Retina
D. Ciliary body

28. The amount of light passing through the pupil of the eye is controlled by a set of circular muscles called the: 5:239

A. Lens
B. Iris
C. Retus muscle
D. Obicularis oculi

29. The majority of the eyeball is composed of a white, fibrous protective covering called the: 5:239

A. Sclera
B. Cornea
C. Retina
D. Macula

30. The impulses for visual sensation will pass into the occipital lobe of the brain by way of the second cranial nerve, which is also called the: 4:414

A. Olfactory nerve
B. Oculomotor nerve
C. Optic nerve
D. Ophthalmic nerve

31. The action of the ciliary bodies control the shape of a transparent oval body called the ____________ which serves to focus light on the back of the eye. 4:415

A. Lens
B. Cornea
C. Retina
D. Iris

32. Eye color is due principally to the appearance of a circular muscle called the: 4:416

A. Lens
B. Iris
C. Suspensory ligament
D. Optic disc

33. The retina is composed of two distinct types of cells called: 4:425

A. Actin and myocin
B. Fovea and stroma
C. Cones and rod
D. Stroma and incus

34. The point on the retinal surface with the greatest amount of visual acuity is the: 4:419

A. Fovea centralis
B. Fovea femoris
C. Fovea lateralis
D. Fovea rectus

35.-44. Pertaining to the diagram of the eye, place the letter that corresponds to the following structures: 4:414

35. ______ Vitreous humor

36. ______ Cornea

37. ______ Lens

38. ______ Optic nerve

39. ______ Retina

40. ______ Sclera

41. ______ Iris

42. ______ Ciliary body

43. ______ Pupil opening

44. ______ Optic disc

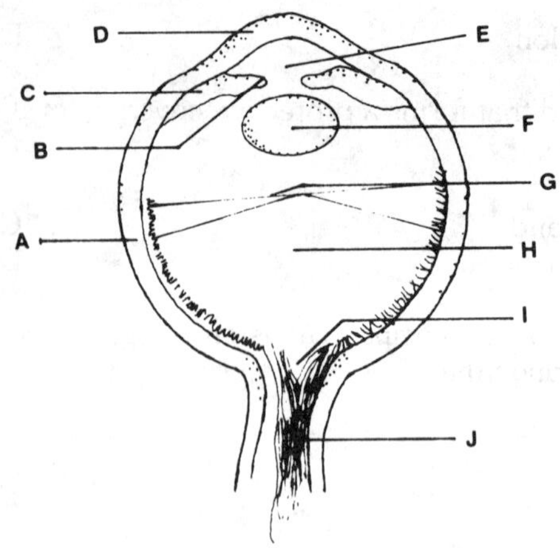

45. The normal value for the interpupillary measurement is: 4:426

A. 1.5" (3-4cm)
B. 2.5" (6-7cm)
C. 3.5" (9-10cm)
D. None of the above

46. High interocular pressure most commonly results from excessive amounts of ______________ buildup in the anterior chamber. 4:417

A. Scotopin
B. Phodopin
C. Vitreous humor
D. Aqueous humor

47. There are no photoreceptors in the area where the optic nerve joins the posterior portion of the eye leaving a blind spot called the: 5:239

A. Fovea centralis
B. Choroid plexus
C. Optic disc
D. Focal point

48. The clouding of the lens of the eye is a common cause of blurred vision. This condition is called (a): 4:416

A. Glaucoma
B. Night blindness
C. Conjunctivitis
D. Cataract

49. The specialized cell of the retina that is primarily responsible for color vision is the: 4:423

A. Fovea
B. Stroma
C. Cone
D. Myocin

50. The principal muscle group responsible for eye movements are the: 4:413

A. Rectus muscles
B. Palpebrae muscles
C. Oculi muscles
D. Ciliary muscles

Male Reproductive System

Chapter 41

1. The principal hormone that is responsible for both proper function and development of the male reproductive organs is: 4:822

   A. Acrosome
   B. Progesterone
   C. Estrogen
   D. Testosterone

2. In the male fetus, the testes originate and develop behind the peritoneum near the: 4:810

   A. Prostate gland
   B. Kidneys
   C. Urinary bladder
   D. Symphysis pubis

3. The male sex cell (spermatozoa) is produced by a specialized lining in tiny coil passages of the testes called the: 4:812

   A. Epididymis
   B. Vas deferens
   C. Prostate gland
   D. Seminiferous tubules

4. Before birth, the testes descend into a pouchlike sac below the abdominal cavity called the: 5:203

   A. Prepuce
   B. Acrosome
   C. Vas deferens
   D. Scrotum

5. Which of the following structures is/are concerned with the formation of seminal fluids? 10:890

   1. Prostate gland *2. Prepuce* *3. Seminal vesicle*

   A. 1 & 2 only
   B. 1 & 3 only
   C. 2 & 3 only
   D. 1, 2 & 3

6. The male germ cells undergo a special type of cell division that reduces the number of chromosomes by half. This is termed: 4:813

   A. Osmosis
   B. Mitosis
   C. Meiosis
   D. Chromosis

7. In the adult male, urinary tract problems often result from enlargement of the ________ gland which surrounds the urethra. 5:203

   A. Seminiferous
   B. Scrotal
   C. Inguinal
   D. Prostate

8. After spermatogenesis has occurred, the immature sperm passes into the ________ for final maturation. 4:815

   A. Epididymis
   B. Urethra canal
   C. Scrotal sac
   D. Ejaculatory duct

9. The proper temperature required in the formation of the male sex cell is: 4:812

   A. Above normal body temperature
   B. Below normal body temperature
   C. The same as normal body temperature

10.-19. Pertaining to the diagram of the male reproductive system, place the letter that corresponds to the following structures: 4:810

10. ______ Epididymis

11. ______ Urinary bladder

12. ______ Ejaculatory duct

13. ______ Seminiferous tubules

14. ______ Scrotal sac

15. ______ Prostate gland

16. ______ Urethra

17. ______ Vas deferens

18. ______ Seminal vesicle

19. ______ Head of the penis

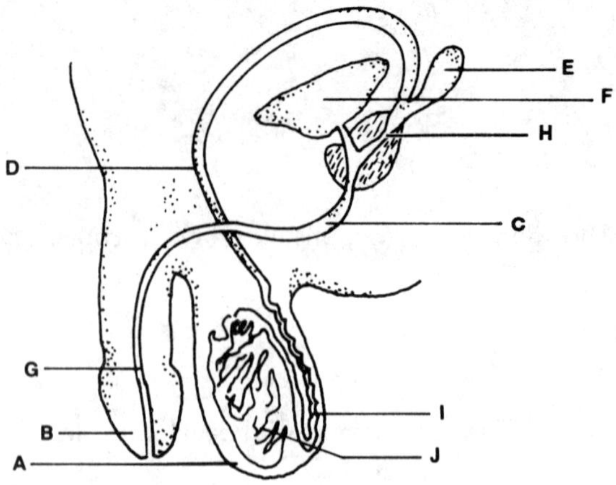

20. The normal mature spermatozoa will contain ____________ autosomes and ____________ sex chromosome(s). 10:883

A. 44, two
B. 45, one
C. 21, one
D. 22, one

21. The body of the penis is composed of three columns of specialized tissue that allows for the stiffening necessary during intercourse, which is called: 4:819

A. Seminal tissue
B. Erectile tissue
C. Nocturnal tissue
D. Pubescent tissue

22. A loose fold of skin covering the glans penis (prepuce) is often removed shortly after birth by a procedure called: 5:206

A. Caversionization
B. Circumcision
C. Cryptorectomy
D. Vasectomy

23. The average male ejaculation will normally contain about spermatozoa. 4:818

A. Ten thousand
B. Fifty thousand
C. One million
D. One hundred million

24. A vasectomy is a relatively simple procedure for male sterilization. This involves the cutting or tying off of a portion of the: 5:206

A. Urethra
B. Seminal duct
C. Vas deferens
D. Ejaculatory duct

25. A possible cause of male infertility is: 4:818

1. Cryptorchidism *2. Low sperm count* *3. Insufficient sperm motility*

A. 1 & 2 only
B. 1 & 3 only
C. 2 & 3 only
D. 1, 2 & 3

Female Reproductive System

Chapter 42

1. The process by which a secondary oocyte is released from a mature ovarian follicle is termed: 4:827

A. Oogenesis
B. Implantation
C. Ovulation
D. Fertilization

2. After its release from the ovary, the egg is directed into the uterine tube by a series of fingerlike extensions of the infundibulum called the: 4:827

A. Fimbriae
B. Villi
C. Cilia
D. Fornices

3. The normal site for implantation of a fertilized egg is the wall of the: 4:828

A. Vagina
B. Uterus
C. Oviduct
D. Ovary

4. During intercourse, the fibromuscular tube that serves as the depository for the spermatozoa is called the: 4:829

A. Clitoris
B. Mons pubis
C. Labia major
D. Vagina

5. If fertilization of the egg cell does not occur, menstruation will begin. Which of the following layers of the uterine wall disintigrate and slough away? 4:835

1. Endometrium *2. Myometrium* *3. Perimetrium*

A. 1 only
B. 2 only
C. 3 only
D. 1, 2 & 3

6. Shortly after giving birth, the alveolar glands within the mammary tissue begin producing a special nutrient fluid called: 4:845

A. Muconium
B. Colostrum
C. Androgen
D. Chorion

7. In the normal female, the development of the mature oocyte (egg cell) is accomplished within the: 4:823

A. Vagina
B. Ovary
C. Uterus
D. Oviducts

8. The lower constricted portion of the thick-walled muscular sac (uterus) is called the: 4:828

A. Myometrium
B. Labia major
C. Clitoris
D. Cervix

9. The maintenance of the uterus, ovaries and oviducts position within the abdominal cavity is accomplished by the: 4:824

1. Ligament of Treitz *2. Broad ligament* *3. Ovarian ligament*

A. 1 & 2 only
B. 1 & 3 only
C. 2 & 3 only
D. 1, 2 & 3

10. The penetration of a secondary oocyte by a spermatozoan is termed: 10:928

A. Implantation
B. Parturition
C. Fertilization
D. Ovulation

11. The small collection of erectile tissue in the female that corresponds in structure and origin to the male penis is called the: 4:830

A. Hymen
B. Areola
C. Clitoris
D. Prepuce

12. The secondary oocyte passes from its origin in the ovary into the uterus by way of the: 5:197

A. Cervix
B. Vaginal orifice
C. Urethral tubes
D. Fallopian tubes

13.-20. Pertaining to the diagram of the female productive system, place the letter that corresponds to the following structures: 5:199

13. ______ Fundus of the uterus

14. ______ Ovary

15. ______ Suspensory ligament

16. ______ Cervix

17. ______ Myometrium

18. ______ Vagina

19. ______ Fimbriae

20. ______ Oviduct

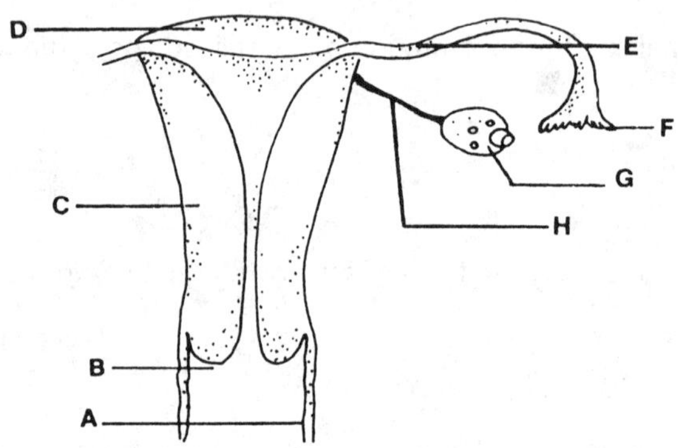

21. During fetal development, the highly vascular structure that serves as respiratory, nutritive and excretory organs of the fetus is called the: 10:936

A. Placenta
B. Zygote
C. Amion
D. Mammary glands

22. In the female, the hormone principally responsible for the development of the secondary sexual characteristics is: 4:846

A. Testosterone
B. Oxytocin
C. Estrogen
D. Lactogen

23. In the female, puberty begins around the thirteenth year and is marked by the first menstrual flow called: 4:833

A. Menarche
B. Phimosis
C. Menopause
D. Parturition

24. Many routine radiographic procedures should be performed during the first third of the menstrual cycle. This cycle begins on the: 4:833

A. First day of menses
B. Last day of menses
C. Day of ovulation

25. After fertilization of the female egg cell by a single male spermatozoa, a single cell containing 46 chromosomes is formed called a: 4:838

A. Zona
B. Zygote
C. Polar body
D. Primary follicle

26. Upon entering the oviducts, the egg cell is actively transported toward the uterus by: 4:828

1. Wipelike motions of the egg's tail 2. Ciliary action 3. Peristaltic contractions

A. 1 & 2 only
B. 1 & 3 only
C. 2 & 3 only
D. 1, 2 & 3

27. Implantation of the blastocyst into the endometrium will generally occur within ____________ about after fertilization. 4:838

A. 2 hours
B. 24 hours
C. 2 days
D. 1 week

28. At the present time, the most effective form of temporary contraception for females is: 4:847

A. Coitus interruptus
B. Rhythm method
C. Intrauterine device
D. Oral contraceptives

29. Pregnancy normally continues for about 40 weeks or 9 calendar months. This development stage is referred to as: 4:849

A. Conization
B. Gestation
C. Menopause
D. Partuition

30. A surgical method for permanent sterilization of the female in which the oviducts are cut and tied is called a/an: 4:849

A. Curettage
B. Eclampsia
C. Abortion
D. Tubal ligation

Nervous System

Chapter 43

1. The normal adult nervous system is subdivided into two major divisions called the: 4:363

A. Central and peripheral nervous systems
B. Meningeal and cutaneous nervous systems
C. Thalamic and spinal nervous systems
D. Skeletal and vascular nervous systems

2. The sensory areas of the brain associated with hearing, sight and smell are located in the: 4:349

A. Cerebrum
B. Cerebellum
C. Midbrain
D. Brain stem

3. In a cross-section cut of the spinal cord, the gray appearance of the central portion is due in large part to the: 5:220

A. Neuro-fibrils
B. Myelinated fibers
C. Nissl bodies
D. Nerve cell bodies

4. The activity of the pituitary gland is controlled by the action of the: 4:358

A. Telencephalon
B. Cerebellum
C. Hypothalamus
D. Midbrain

5. The coordination of fine muscular movement of the extremities is primarily accomplished by the: 4:362

A. Cerebrum
B. Cerebellum
C. Limbic system
D. Diencephalon

6. The folds of the brain giving the appearance of ridges and valleys are called the: 4:347

A. Gyri and sulci
B. Villi and cilia
C. Ganglia and fissures
D. Axons and dendrites

7. The lateral ventricles are the largest of the fluid-filled cavities of the brain and are located within the mass of the: 4:354

A. Brain stem
B. Cisterna magnum
C. Cerebral hemispheres
D. Falx cerebri

8. The brain and spinal cord are protected by three membranous layers collectively known as the: 4:338

A. Diploe
B. Meninges
C. Sinuses
D. Fissures

9. The cranial venous sinuses that drain the blood from the brain are formed by the two layers of the: 10:635

A. Periosteum
B. Dura mater
C. Lateral ventricles
D. Arachnoid villi

10. The brain and spinal cord develop from a tubelike nerve bundle called the: 4:345

A. Neural tube
B. Cerebral nuclei
C. Commissural fibers
D. Neural synapse

11. The dorsal root of the spinal nerve is composed of: 10:360
1. Motor fibers *2. Sensory fibers* *3. Afferent fibers*

A. 1 & 2 only
B. 1 & 3 only
C. 2 & 3 only
D. 1, 2 & 3

12. The medulla oblongata exerts a great deal of control over: 4:359
1. Respiration rate *2. Heart rate* *3. Blood pressure*

A. 1 only
B. 2 only
C. 3 only
D. 1, 2 & 3

13. The largest portion of the normal adult human brain is the: 4:347

A. Cerebellum
B. Pons
C. Cerebrum
D. Diencephalon

14. The normal connection between the third and fourth ventrals is accomplished by a narrow canal called the: 4:354

A. Cerebral aqueduct
B. Chorid plexus
C. Foramen of Monro
D. Falx cerebri

15. The innermost layer of the membrane covering the brain is termed the: 4:339

A. Arachnoid mater
B. Dura mater
C. Subarachnoid mater
D. Pia mater

16. The cerebrospinal fluid that is secreted by masses of specialized capillary beds in the walls of the ventricles serves for the: 4:355

A. Regulation of hormones
B. Protection of the brain
C. Filtration of wastes
D. Production of antibodies

17. How many distinct pairs of spinal nerves originate from the spinal cord? 4:369

A. Twelve
B. Twenty-one
C. Twenty-four
D. Thirty-one

18. The subarachnoid space that contains the cerebrospinal fluid is formed by the: 4:339

A. Periosteum and dura mater
B. Dura and pia mater
C. Arachnoid and pia mater
D. None of the above

19. The junction of two or more nerve cell extensions occurs at a conduction pathway called the: 5:220

A. Syntaxis
B. Ganglion
C. Synapse
D. Reflex arc

20. The conduction of electrical impulses from the central nervous system to the peripheral nerves occurs over the: 5:229
1. Efferent fibers *2. Sensory fibers* *3. Motor fibers*

A. 1 & 2 only
B. 1 & 3 only
C. 2 & 3 only
D. 1, 2 & 3

21. The interventricular foramen permits the passage of fluids between the __________ ventricles. 4:355
1. Third and fourth *2. Fourth and lateral* *3. Lateral and third*

A. 1 only
B. 2 only
C. 3 only
D. 1, 2 & 3

22. The phrenic nerve that innervates the primary muscle of respiration, the diaphragm, is a branch off of the: 10:371

A. Cervical plexus
B. Brachial plexus
C. Lumbar plexus
D. Sacral plexus

23. A neuron consists of a cell's body and two thin, branching extensions called the: 5:220

A. Actin and myosin
B. Axon and dendrite
C. Synapse and ganglion
D. Myelin and neurolemma

24. The fluid-filled cavities of the brain and spinal canal communicate by way of the foramina of: 5:225

A. Sylvius and Moore
B. Schwan and Ranvier
C. Monro and Rinivus
D. Luschka and Magendie

25. Severe pain in the facial region, called tic douloureux, is caused by mechanical irritation of the: 10:411

A. Oculomotor nerve
B. Vagus nerve
C. Facial nerve
D. Trigeminal nerve

26. Drop foot, due to the loss of the dorsiflexor muscles of the foot, often results from damage to the: 10:377

A. Musculocutaneous nerve
B. Obturator nerve
C. Median nerve
D. Peroneal nerve

27. The cutaneous areas supplied by a single dorsal root is called a: 4:370

A. Dermatome
B. Ganglia
C. Spinothalamic tract
D. Nerve plexus

28. The regeneration of peripheral nerve fibers is in large part due to a special nerve covering called the: 10:336

A. Nodes of Ranvier
B. Myelin sheath
C. Neurolemma
D. Synaptic cleft

29. The conduction pathway between the right and left sides of the cerebrum is accomplished by a large band of white fibers called the: 4:347

A. Corpus callosum
B. Basal ganglia
C. Cerebral aqueduct
D. Islets of Reil

30.-34. Pertaining to the diagram of the superior aspect ventricles, place the letter that corresponds to the following structures: 5:224

30. ______ Anterior horn (lateral ventricle)

31. ______ Fourth ventricle

32. ______ Cerebral aqueduct

33. ______ Posterior horn (lateral ventricle)

34. ______ Third ventricle

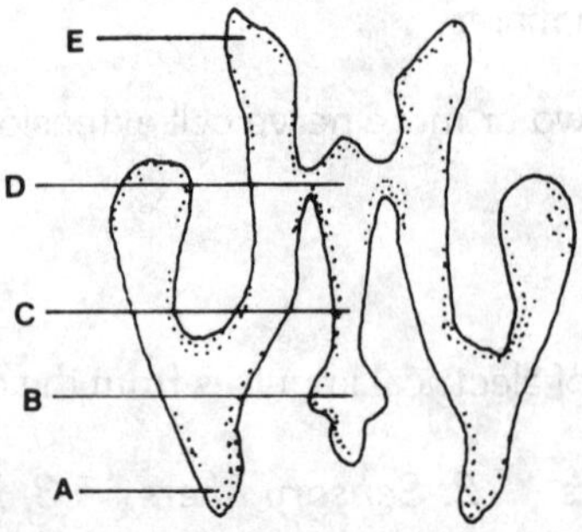

35. The sensory and motor impulses to and from the brain for the head and neck region are carried by the ______________ pairs of cranial nerves. 10:408

A. Six
B. Twelve
C. Eighteen
D. Thirty-one

Cardiovascular System

Chapter 44

1. The movement of blood to the lungs and other tissues of the body is accomplished by a specialized, muscular organ called the heart. The normal human heart consists of: 5:129

A. Two chambers
B. Four chambers
C. Six chambers
D. Eight chambers

2. The movement of blood away from the heart occurs in thick-walled elastic muscular tubes called: 4:671

A. Arteries
B. Veins
C. Capillaries
D. Lymphatic vessels

3. The right and left sides of the heart are separated by a fibrous membrane called the: 4:653

A. Chordae tendineae
B. Bundle of His
C. Septum
D. Mitral valve

4. Blood is returned to the heart by way of two major trunk veins that empty into the: 5:136

A. Right atrium
B. Right ventricle
C. Left atrium
D. Left ventricle

5. The majority of the heart is composed of a specialized type of cardiac muscle called the: 4:652

A. Pericardium
B. Endocardium
C. Epicardium
D. Myocardium

6. The prevention of the back flow of blood as it passes through the heart is the principal function of the: 5:134

A. Septum
B. Valves
C. Trunk vessels
D. Coronary sinus

7. The heart is covered by a serous, double-walled fibrous membrane called the: 4:650

A. Pericardium
B. Papillary muscle
C. Purkinge fibers
D. Sino-atrial nodes

8. The largest, most muscular chamber of the heart concerned with the movement of blood to the tissues of the body is the: 5:132

A. Right atrium
B. Right ventricle
C. Left atrium
D. Left ventricle

9. The right side of the heart is primarily concerned with the passage of deoxygenated blood into the tissues of the: 4:657

A. Liver
B. Kidney
C. Lungs
D. Intestines

10. The muscle of the heart derives its blood supply by way of the: 5:134

A. Pulmonary vessels
B. Coronary vessels
C. Pericardial vessels
D. Atrioventricular vessels

11.-17. Pertaining to the diagram of the heart, place the letter that corresponds to the following structures: 4:654

11. ______ Tricuspid valve

12. ______ Superior vena cava

13. ______ Left atrium

14. ______ Septum

15. ______ Right ventricle

16. ______ Right atria

17. ______ Left ventricle

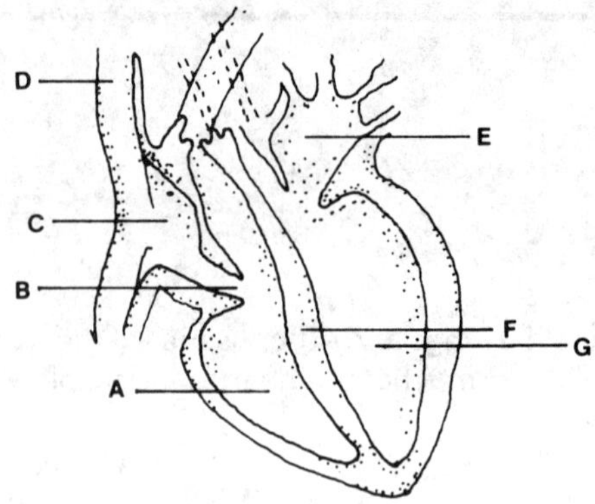

18. The cardiac conduction system that both initiates and distributes impulses to the heart muscles consists of: 4:663

1. Sino-atrial node *2. Atrioventricular node* *3. Purkinge fibers* *4. Coronary vessels*

A. 1 & 2 only
B. 1 & 3 only
C. 2 & 3 only
D. 1, 2 & 3

19. The exchange of oxygen, carbon dioxide and other nutrients and wastes to the tissues of the body occurs across the walls of the: 5:130

A. Arteries
B. Veins
C. Lymphatic vessels
D. Capillaries

20. The highest blood pressure measurements are most often recorded in the: 5:129

1. Arteries *2. Veins* *3. Capillaries*

A. 1 only
B. 2 only
C. 3 only
D. 1, 2 & 3

21. The celiac axis is the first branch of the abdominal aorta which serves to supply arterial blood to the: 4:692

1. Spleen *2. Liver* *3. Pancreas* *4. Stomach*

A. 1 & 2 only
B. 1 & 3 only
C. 2 & 3 only
D. 1, 2 & 3

22. The external division of the common carotid artery supplies many of the structures located in the: 4:691

A. Anterior cranium
B. Posterior cranium
C. Face and neck
D. Shoulder and clavicle

23. The diaphragm is supplied with arterial blood by branches of both the thoracic and abdominal aorta called the inferior and superior: 5:137

A. Phrenic arteries
B. Intercostal arteries
C. Apical arteries
D. Coronary arteries

24. The femoral artery is the continuation of the ____________ below the inguinal ring. 5:138

A. Popliteal artery
B. Subclavian artery
C. Internal iliac artery
D. External iliac artery

25. Which is not a paired branch of the abdominal aorta? 4:692

A. Renal arteries
B. Mesenteric arteries
C. Spermatic arteries
D. Lumbar arteries

26. The arterial blood supply to the brain is accomplished by the: 4:693
1. Internal carotid arteries *2. Vertebral arteries* *3. Jugular arteries*

A. 1 & 2 only
B. 1 & 3 only
C. 2 & 3 only
D. 1, 2 & 3

27. The first major division of the aorta is termed the: 5:137

A. Ascending aorta
B. Aortic arch
C. Descending aorta
D. Pulmonary aorta

28. The vessels principally involved with the circulation of blood through the lungs are the: 4:689

A. Cardiac vessels
B. Portal vessels
C. Pulmonary vessels
D. Azygos vessels

29.-34. Pertaining to the diagram of the aortic arch, place the letter that corresponds to the following arteries: 4:691

29. ______ Right vertebral

30. ______ Left common carotid

31. ______ Left subclavian

32. ______ Innominate

33. ______ Right common carotid

34. ______ Right subclavian

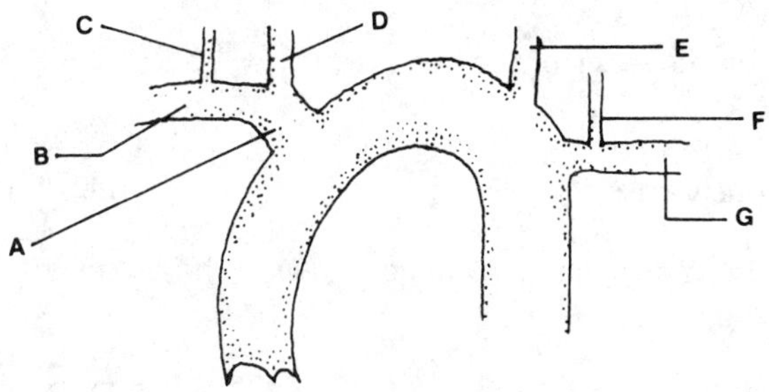

35. The bifurcation of the abdominal aorta normally occurs at the level of the: 5:138

A. 10th thoracic vertebra
B. 1st lumbar vertebra
C. 4th lumbar vertebra
D. 2nd sacral vertebra

36. The largest portion of the intestinal tract is supplied with arterial blood by way of the: 4:692

A. Celiac axis
B. Superior mesenteric artery
C. Inferior mesenteric artery
D. Hepatic artery

37. The popliteal artery is named for its location in the posterior fossa of the: 4:699

A. Ankle
B. Elbow
C. Armpit
D. Knee

38. The ovarian arteries are a branch off of the: 5:137

A. Internal iliac artery
B. External iliac artery
C. Abdominal aorta
D. Obturator artery

39. A specialized arterial structure in the brain that helps in the prevention of irreparable damage in cases of arterial blockage is called the: 4:693

A. Circle of Willis
B. Loop of Henle
C. Plantar arch
D. Cerebral sinuses

40. The urinary bladder, reproductive organs and gluteal muscles derive the majority of blood from the: 4:698

A. External iliac arteries
B. Internal iliac arteries
C. Mesenteric arteries
D. Femoral arteries

41. From the aortic arch to the hand, describe the correct sequence of arterial branches. 5:142

A. Axillary, subclavian, brachial, radial
B. Subclavian, brachial, axillary, ulnar
C. Brachial, subclavian, axillary, radial
D. Subclavian, axillary, brachial, ulnar

42. The bifurcation of the common carotid arteries occurs at the level of the: 5:137

A. Manubrial notch
B. External auditory meatus
C. Thyroid cartilage
D. Inion

43. The largest artery in the normal human body is the: 5:136

A. Aorta
B. Vena cava
C. Portal artery
D. Pulmonary artery

44. Arterial blood is delivered to the liver primarily by the: 5:137

A. Hepatic artery
B. Portal artery
C. Basilic artery
D. Renal artery

45. The longest blood vessel of the body is the: 5:143

A. Femoral artery
B. Saphenous vein
C. Aorta
D. Vena cava

46. The jugular veins are primarily concerned with the drainage of blood from the: 4:702

A. Feet
B. Arms
C. Lungs
D. Brain

47. The most important deep vein of the upper arm is the: 5:142

A. Cephalic
B. Brachial
C. Antebranchial
D. Basilar

48.-52. Pertaining to the diagram of the arteries of the lower abdomen and pelvis, place the letter that corresponds to the following vessels: 4:698

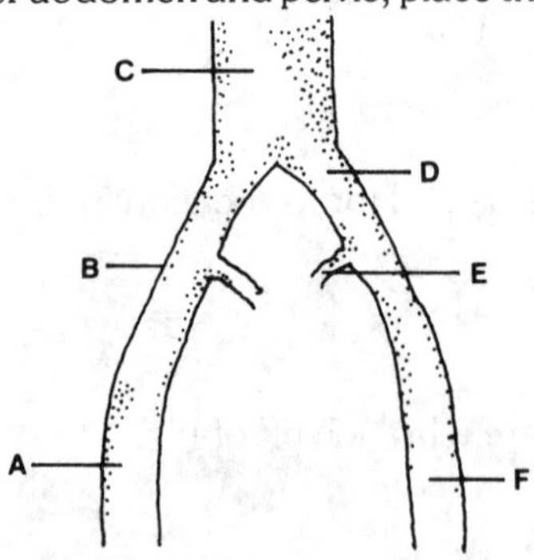

48. ______ Left common iliac artery

49. ______ Right external iliac artery

50. ______ Abdominal aorta

51. ______ Left internal iliac artery

52. ______ Right femoral artery

53. The largest veins of the normal human that are primarily concerned with the return of blood to the heart are the superior and inferior: 5:139

A. Portal veins
B. Azygos veins
C. Vena cava
D. Saphenous veins

54. The majority of the blood from the small intestines will be drained and returned to the portal circulation by the: 4:704

A. Celiac vein
B. Axillary vein
C. Brachiocephalic veins
D. Superior mesenteric vein

55. The individual layer of the walls of the arteries will receive their own nutrients by way of vessels termed: 10:606

A. Sinusoids
B. Varicosities
C. Vaso vasorum
D. Capillaries

56. Which characteristic(s) is/are generally associated with the venous circulation? 4:701

1. Low blood pressure *2. Low oxygen content blood* *3. High waste content blood*

A. 1 & 2 only
B. 1 & 3 only
C. 2 & 3 only
D. 1, 2 & 3

57. In the fetus, shunting of blood between the two atrias is accomplished by an opening called the: 5:148

A. Ductus arteriosus
B. Ductus venosus
C. Foramen lacerum
D. Foramen ovale

58. The umbilical arteries of the fetus which return deoxygenated blood to the mother arise from the: 5:148

A. Abdominal aorta
B. Pulmonary arteries
C. Celiac axis
D. Internal iliac arteries

59. The highly vascular capillary membrane that permits the exchange of nutrients and oxygen between the fetus and mother is termed the: 5:147

A. Placenta
B. Lingula
C. Ovary
D. Macula

60. The white blood cells of the lymphatic system enter the systemic circulation at the: 5:147

A. Cisterna chyli
B. Subclavian veins
C. Cisterna magna
D. Splenic artery

61. In many large veins, back flow of blood is prevented by the presence of: 10:610

A. Sphincter muscles
B. Sinusoids
C. Venules
D. Valves

62. Because of its high oxygen content, arterial blood is characterized by its __________ appearance. 10:548

A. Bright red
B. Dark red
C. Straw-colored
D. Blue-colored

63. The fluid medium in which the blood cells circulate is called: 4:612

A. Serum
B. Plasma
C. Hemocrit
D. Lymph

64. The most common type of blood cell has no nucleus and has a characteristic biconcaval shape. This is called a/an: 4:613

A. Leukocyte
B. Lymphocyte
C. Erythrocyte
D. Thombocyte

65. The major function of platelets (thrombocytes) is: 4:622

A. Destruction of bacteria
B. Transport of oxygen
C. Production of antibodies
D. Clotting of blood

66. A condition in which there is a pronounced deficiency of red blood cells is called: 4:662

A. Cyanosis
B. Hemophilia
C. Anemia
D. Leukemia

67. Insufficiency or regurgitation of the heart valves, or septal defects, are often diagnosed by an abnormal sound called a: 4:662

A. Branch block
B. Heart murmur
C. Fibrillation
D. Hypertrophy

68. A partial loss of blood flow to the heart muscle with subsequent tissue damage is termed: 4:659

A. Mitral stenosis
B. Pericarditis
C. Myocardial infarction
D. Pericardial effusion

69. A localized dilation of the walls of an artery due to a pre-existing weakening leads to a defect called a/an: 5:139

A. Aneurysm
B. Anastomosis
C. Atherosclerosis
D. Phlebitis

70. Vericose veins that sometimes occur in the rectum are termed: 5:151

A. Emboli
B. Thrombi
C. Hemorrhoids
D. Infarcts

Section V

Anatomy & Physiology

Explanations

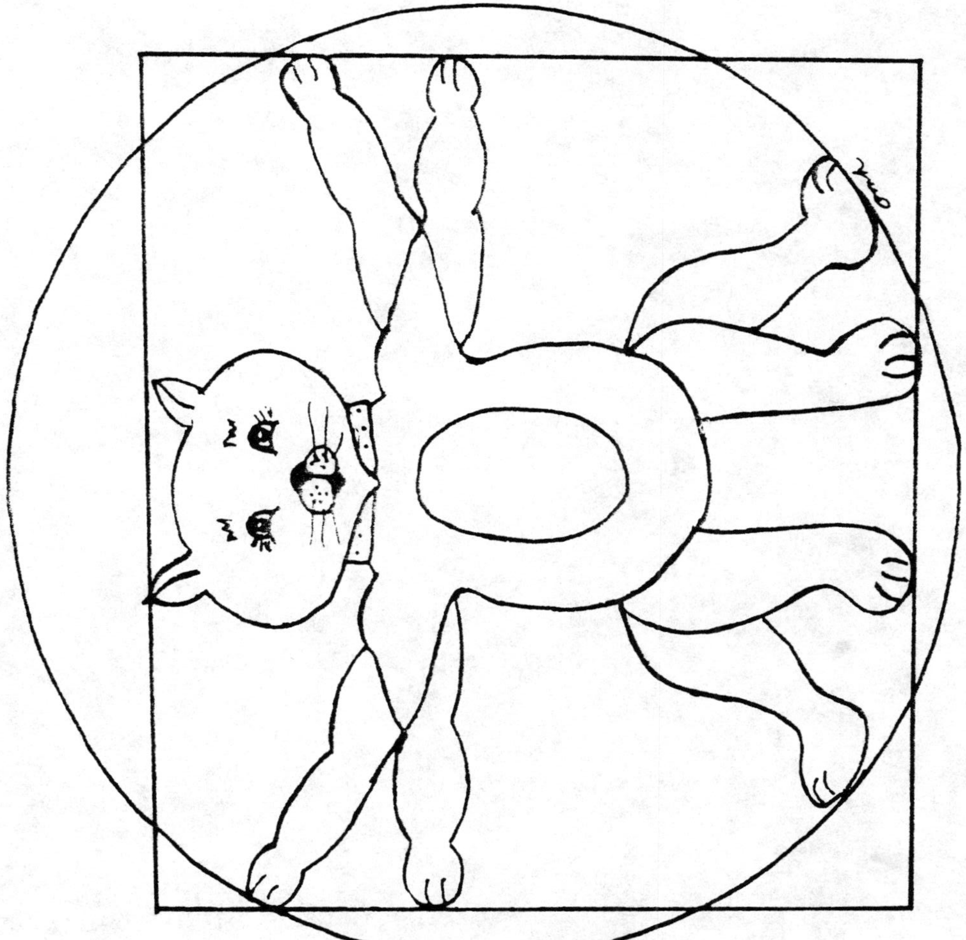

Bone Development

Chapter 31

1. A — Osteogenesis, the conversion of hyaline cartilage to bone, begins in primary ossification centers of the bone called the diaphysis.

2. C — The blood supply to bone which is required for normal growth and repair is augmented by a fibrous membrane covering called the periosteum.

3. A — In order to provide nourishment for the bone cell, a dense matrix of blood vessels forms within tiny vascular cavities of the bone. These Haversain and Volkmans canals extend horizontally and vertically through all of the dense bone tissue of the skeletal system.

4. B — The hyaline cartilage of the embryonic skeleton contains numerous immature bone cells called osteoblasts. The osteoblast develops into a true bone cell in small open spaces called lacunae by surrounding itself with numerous layers of calcium.

5. D — From the latin for "little swelling" a tubercle provides a roughened surface for the attachment of ligaments and tendons.

6 A — The synarthrodial or immovable joints of the adult skull which provide the fixed connections between the skull bones are called sutures.

7. B — The normal adult skeleton, consisting of the axial skeleton (skull and spine), contains 80 distinct bones and the appendicular skeleton containing 126 bones. These numbers exclude the teeth, sesamoid bones and most of those bones which are not completely fused before puberty.

8. C — The hollowed out central (medullary) cavity which serves to reduce the weight of the bone houses the blood forming (hemopoietic) red bone marrow. In an adult most of the red bone marrow is replaced with the fat rich yellow bone marrow.

9. A — Bone consists of two primary substances: the inorganic calcium salts, which provide the brittle strength and the organic rubber-like collagen giving the bone its resiliency.

10. C — Where considerable pressure is exerted on an area such as the ball of the foot, small (sesmoid) bones may develop within the tendons which help to dissipate energy and reduce friction at the joint.

11. D — The long bones characterized by a shaft and two epiphyseal ends make up the majority of the bones of the arm, legs, hands and feet.

12. A — Though a number of glands and their hormonal secretions play a secondary role in bone growth none is as direct or as important as the growth hormones produced in the pituitary gland.

13. B — The small irregular (wormian) bone which develops in the sutures of the skull aids in the fixation of these immovable or synarthrodial joints.

14. D — Inversion and its counterpart eversion describe the turning of the foot or hand toward or away from the midline of the body.

15. D — Freely movable joints contain an articular joint capsule and its associated synovial membrane which bathes the articulating bones with an oil-like fluid. This capsule is penetrated by the numerous tendons, ligaments, cartilages, and muscles which provide additional support for the joint.

16. B — The replacement of cartilage by bone (endochondral ossification) in the fetus primarily involves a special type of cartilage (hyaline). The flat bones of the skull and mandible, which are made of fibrous tissue, undergo a different process termed intramembranous ossification.

17. D — The bony framework known as the internal skeleton protects the critical organs of the body, provides the attachments for muscles, stores mineral salts and houses the hemopoietic or blood forming tissues.

18. C — Flexion, which decreases the angle between the bones and its counterpart, extension, are the two main motions occurring at a hinge joint. In the axial skeleton (neck) a forward bending is called flexion with a backward bending being referred to as extension.

19. B — The dense or compact bony layer covering the outer layer of most bone provides the support and strength to resist the mechanical stresses placed on them during normal activities.

20. A — Red bone marrow which is found in nearly all bones of a child is to a large extent replaced with the fatty yellow bone marrow by adulthood.

21. D — The principal support between two bones at an articulation is provided by a dense connective tissues called ligaments which help to stabilize all amphiarthrodial and diarthrodial joints.

22. D — Long bones: phalanges, metacarpals, tibia, fibula, femur, and humerus, ulna and radius.

23. A — Short bones: wrist and foot bones.

24. C — Irregular bones: all vertebra and most facial bones.

25. E — Flat bones: Ribs, sternum, scapula, ilium and bones of the cranial vault.

26. B — Sesamoid bones: patella, metatarsophalangeal.

27. B — The majority of the intercellular substance of a bone is composed of the mineral salts, calcium phosphate and calcium carbonate.

28. A — At birth, the skull bones are incompletely fused to enable overlapping of the sutures during the birth process. Until their final closure at about 1-2 years of age, four membranous filled spaces called fontanels can be felt on the neonatal head. The anterior fontanel found between the frontal and parietal bones is called the bregma in adulthood. The posterior fontanel found between the occipital and parietal bones is called the lambda in adulthood. The anterolateral and posterolateral fontanels are the two soft spots located on the side of the skull.

29. C — An osteoblast or bone building cell develops in the lake-like lacunae by depositing layers of calcium salts. Now unable to move, the remaining cell is now called an osteocyte or true bone cell.

30. D — The lattice type network seen at the end of the long bones which helps to both maintain its strength and reduce its weight, is called cancellous or spongy bone.

31. B — In order to prevent the damage of bone to bone contact, the articulating surfaces of the bone are covered by a tough fibrous membrane called articular cartilage.

32. C — In most long bones the secondary ossification centers are found at the proximal and distal ends of the bone. These epiphyses or growth plates will normally close at or shortly after puberty.

33. B — The precursor cell for all type of blood cells erythrocytes (RBC), leukocytes (WBC), and thrombocytes (platelets), is the hemocytoblast which resides in the red bone marrow.

34. C — Pivot joint (monoaxial); e.g., atlantoaxial, and the joint at the proximal end of radius and ulna.

35. D — Hinge joint (monoaxial); e.g., elbow, knee, ankle and interphalangeal joint.

36. E — Ball and socket joint (triaxial); e.g., shoulder and hip joints.

37. A — Saddle joint (biaxial); e.g., trapezium and first metacarpal bone.

38. B — Gliding joint (biaxial); e.g., carpal and tarsal articulations, costotransverse, SC, AC, and apophyseal joints.

39. A — The sandwich like layer of spongy bone between the two layers of dense cranial (diploe) bone provides considerable protection to the brain from a blunt trauma.

40. C — The lubrication and nourishment of the joint capsule is largely derived from the synovial membrane and its oil like secretion, synovial fluid.

41. B — The most direct form of ossification involving osteoblasts formed within a fibrous membrane (intramembranous), principally occurs in the flat bones of the skull.

42. D — The formation of the hollow medullary cavity and the spaces within cancellous bone is a function of the bone eating cells called osteoclasts.

43. C — As osteoblasts develop their surrounding calcium walls, small connecting extensions called canaliculi will enable adjacent osteocytes to exchange nutrients and chemical signals with each other.

44. A — The special dense fibrous connective tissue which attaches muscle tissue to the periosteum of the bone is called a tendon.

45. A — A decrease in osteoblastic activity with advancing age may lead to the reduced bone mass seen in osteoporosis. Recent studies indicate that even light daily exercise can delay or even reverse this common condition more often seen in females.

46. A — Rickets though rare in this country, is caused by a severe vitamin D deficiency. This condition is associated with the over-development of the cartilage, reduced ossification which will lead characteristic bowing of the leg bones in patients with rickets.

47. B — Severe arthritic changes which destroy bone and joint tissues will often lead to the formation of scar tissue and eventual loss of movement. This condition involving the stiffening of the joints is called ankylosis.

48. C — Luxation is the term for a full dislocation, subluxation for partial dislocation and hyper-extension for the over-extension or spraining of a joint. These are the three common types of soft tissue injuries.

49. C — Human growth hormone (somatotropin) which is excreted from the pituitary gland or hypophysis is largely responsible for the growth of tissues; i.e., bone. Over production of this hormone leads to gigantism in children and acromegaly in adults. Dwarfism is caused by a reduction in this critical hormonal secretion.

50. D Trauma to the fluid-filled bursal sacs of the knee, elbow, hip or first metatarsal bone may lead to a local inflammation called bursitis. This painful condition is normally treated by rest and anti-inflammatory drugs.

Upper Extremity

Chapter 32

1. C The majority of the hand is made up of the metacarpal bones and their soft tissue and skin coverings. The phalanges are bones of the fingers.

2. B The roughened area on the lateral portion of the scapular spine (acromion) serves for the attachment of the deltoid muscle tendon and coracoacromial ligaments which help to stabilize the shoulder joint.

3. B Each finger (numbers 2-5) consist of 3 phalanges articulating over two interphalangeal (IP) joints.

4. A The concave surfaces formed by the ulnar and radial styloids conform well to the convex border of the proximal row of wrist bones and aid in the stability of this joint.

5. A The supra, infraspinous, and teres minor muscles which enable lateral rotation of the humerus, attach on the posterior surface of the greater tubercle of the humerus.

6. D The hook-like hamulus on the inferior surface of the hamate bone serves for the attachment and origin for 4 muscles and tendons of the wrist and fingers.

7. C The proximal epiphysis between the head and shaft of the humerus closes shortly after puberty at about the age of 20.

8. B The oval articular surface of the scapula (glenoid fossa) is directed forward and laterally toward its junction with the humeral head.

9. C Since the head of the phalanx is always distal and the base is proximal, the head of a proximal phalanx will articulate with the base of the more distal phalanx.

10. A The support structures for the upper extremity are the scapula and clavicle which collectively form the shoulder girdle. The shoulder joint consists of all three bones: the clavicle, scapula and humerus.

See accompanying diagram.

11. G 12. F 13. A 14. I

15. J 16. H 17. C 18. E

19. C The supination of the hand and forearm in the anatomical position places the ulna more medial and the radius more lateral.

20. A The beak-like coracoid process projecting anterolaterally from the neck of the scapula provides ligamentous attachments for the humerus, acromion, clavicle and tendinous attachments for the pectoralis minor, coracobrachialis and biceps muscles.

21. B The normal adult human has 8 carpal bones arranged in two rows which are held together by dozens of strong ligaments.

22. A The humeroradial articulation between the head of the radius, the capitulum and the radio-ulnar articulation, consists of a radial head and radial notch of the ulna. This enables the rotation of the forearm.

23. C The longer of the two heads of the biceps muscle creates a deep groove separating the tubercles of the humerus called the bicipital or intertubercular groove.

24. A The coronoid fossa located on the anterior humerus is a large depression which accepts the coronoid process of the ulna during acute flexion of the elbow.

25. C The sternoclavicular (SC) joint, the only bony connection between the trunk and the upper extremity, is contained within a joint capsule that is continuous with the cartilage of the first rib.

See accompanying diagram.

26. E	27. C	28. I	29. D
30. H	31. G	32. A	33. B
34. F			

35. C The articulation between the glenoid cavity of the scapula and the head of the humerus is commonly called the shoulder joint.

36. B The posterior surface of the scapula is separated into upper and lower sections by the large scapular spine. The smaller upper section, the supraspinous fossa, serves as the origin for the muscle of the same name. The infraspinous fossa forms the majority of the lower scapula and houses the infraspinous muscle.

37. D The scaphoid or navicular bone is the largest and lateral most bone in the proximal row of the wrist. It articulates with the radius, trapezium, trapezoid, capitate and lunate bones.

38. A The fingers of the hand consist of 14 phalanges: 5 distal, 5 proximal, and 4 middle. The first finger (thumb) has no middle phalanx.

39. C The act of turning the hand to display its posterior surface (pronation) involves the medial rotation of the radius across the ulna.

40. C The greater multangular or trapezium which is located on the lateral aspect of the distal row of wrist bones articulates with the base of the first metacarpal bone.

41. A The thick projection on the proximal ulna which forms the point of the elbow, the olecranon process, articulates with the olecranon fossa on the posterior humerus.

42. C The distal humerus contains two articular surfaces, the larger and more medial condyle, the trochlea and the more lateral rounded condyle called the capitulum. A prominent epicondyle extends from the medial surface of the trochlea and the lateral surface of the capitulum.

43. C The roughened surface on the proximal radius, the radial tuberosity, serves for the insertion of the distal end of the biceps muscle.

44. A The larger of the two articulating surfaces of the distal humerus, the trochlea, joins with the semilunar notch of the proximal ulna to form the hinge joint of the elbow.

45. C Each finger (numbers 2-5) contains 3 phalanges separated by 2 interphalangeal joints (IP). The first finger (thumb) contains only 2 phalanges and a single (IP) joint.

46. A The deltoid muscle which originates from the clavicle and scapula passes laterally over the shoulder to its insertion on the deltoid tuberosity of the lateral humerus.

47. A The more medial (vertebral) border of the scapula serves for the attachment of the rhomboid and levator scapulae muscles.

48. B One of the tendinous heads of the biceps muscle of the anterior arm extends between the greater and lesser tubercles forming the bicipital or intertubercular groove.

49. A The largest wrist bone, the capitate, due to its central location, articulates with four of the other wrist bones and three middle metacarpal bones.

50. D The clavicle articulates with the acromion process of the scapula laterally and the manubrium of the sternum medially.

Lower Extremity

Chapter 33

1. C Each foot contains five distal, five proximal and four middle phalanges. The first (big) toe has no middle phalanx.

2. C The shorter and thinner of the two lower leg bones, the fibula, articulates with the tibia on the lateral surface at both its proximal and distal ends. Unlike the larger tibia, the fibula is not a weight bearing bone.

3. D The female pelvis has a larger inlet, larger outlet diameter and wider pubic arc than the male pelvis which aids in the ability to carry and bear children.

4. C The thickened ridge (linea aspera) which extends down the middle two thirds of the posterior femur provides for the attachment of the adductor muscles of the thigh.

5. A Extending upwards from the superior articulating surfaces (plateaus) of the tibia are two eminences which provide the attachments for the anterior and posterior cruciate ligaments of the knee.

6. B The two bones which form the adult pelvis the innominate or coxa bones were formed by three individual bones that fuse in the mid teens called the ilium, pubis, and ischial bones.

7. A The two most commonly found sesamoid bones of the foot which help to reduce friction are located on the inferior surface of the first metatarsal phalangeal joint.

8. C The largest and strongest bone of the foot, the calcaneus or os calcis serves for the origin or insertion for numerous muscles of the foot and toes.

9. B The triangular patella found on the anterior surface of the femur is situated with its apex inferior to the broad base on the superior aspect of the bone.

10. C The two halves of the pelvis are united anteriorly across the superior rami of the pubic bones by the fibrocartilage interpubic disc.

11. D Between the upper articulating condyles of the tibia (plateaus) and the femoral condyles are the two C-shaped knee cartilages called the menisci.

See accompanying diagram.

12. H	13. F	14. E	15. I
16. G	17. B	18. C	19. D
20. A			

21. A The larger bone of the lower leg, the tibia, is the only weight bearing bone of the lower leg.

22. B The deep depression found between the inferoposterior portion of the femoral condyles is called the intercondylar fossa.

23. D The deep hemispheric cavity on the lateral surface of the innominate bone which receives the head of the femur is called the acetabulum.

24. C When seated, the pelvis is tilted so that the majority of weight is transferred onto the large roughened ischial tuberosities.

25. D The roughened area on the upper anterior surface of the tibia (tuberosity) serves for the insertion of the broad tendon of the quadriceps femoris muscle which encapsulates the patella.

26. C The pelvis is subdivided into a true pelvis comprising the inlet and outlet and the ala (wing) of the ilium which forms the lower portion of the abdominal cavity or false pelvis.

27. B The arch shape of the neck and large diameter bone of the femur enables it to support the weight of the entire torso and upper extremities.

See accompanying diagram.

28. A	29. B	30. C	31. D

32. D Before its final fusion in the middle teen years, the pelvis consists of 3 pairs of bones: the ilium, ischium and pubis.

33. A The knee joint is formed by the articulation of the tibial plateaus with the medial and lateral condyles of the femur.

34. C The most superior bone of the foot, the talus (astragalus) sits above the calcaneus and just posterior to the navicular bone. It forms the lower portion of the ankle joint and provides for numerous tendon and ligaments attachments to the tibia, fibula and other bones of the foot.

35. C The expanded upper end of the fibula (head) articulates with the lateral surface of the tibia just distal to the knee joint.

36. C The more or less central location of the acetabulum on the lateral surface of the pelvis enables its formation by all three of the bones of the immature pelvis: the ilium, ischium and pubis.

37. B The three wedge shaped cuneiform bones of the foot articulate posteriorly with the navicular and anteriorly with the base of the first three metatarsal bones.

38. A Pes planus or flat feet often results from excessive stress on the longitudinal arch of the foot and is common in occupations requiring long periods of standing or walking.

39. C The longer of the two distal processes of the ankle joint extending about 1 cm below the superior aspect of the talus bone is the lateral malleolus of the fibula.

See accompanying diagram.

40. B	41. H	42. A	43. J
44. E	45. F	46. C	47. D
48. G	49. I		

50. C The flattened oval navicular bone is centrally located in the foot lying anterior to the talus, posterior to the cuneiforms, and medial to the cuboid bone.

51. B The large bony prominence felt on the lateral aspect of the hip, the greater trochanter provides the attachments for the gluteus maximus and minimus muscles. The greater trochanter is also a useful landmark for finding the level of the symphysis pubis which lies on the same axial plane.

52. C The normal angle between the neck and shaft of the femur is between 115-130 degrees. Deformities causing an excessive angle, coxa volga or insufficient angle, coxa vera will seriously effect the gait of an individual.

53. A The knee joint consists of two true bones, the femur and tibia, and a sesamoid bone, the patella. The fibula, which articulates with the lateral surface of the tibia, does not contribute to the knee joint.

54. C The cuboid bone articulates anteriorly with the fourth and fifth metatarsals, posteriorly with the calcaneus, and laterally with the navicular and lateral cuneiform.

See accompanying diagram.

55. E	56. A	57. J	58. D
59. B	60. H	61. G	62. I

63. C The junction of the ischium and pubic bones of the lower pelvis forms the large obturator foramen found on the lower aspect of the innominate bone.

64. C The medial malleolus of the tibia, the lateral malleolus of the fibula, and the superior portion of the talus all contribute in the formation of the ankle joint.

65. A Just above the ischial spine is a large depression formed by the passage of the pelvic vessels and nerves as they pass in the lower limb called the greater sciatic notch.

Axial Skeleton

Chapter 34

1. B The vertebral column of the adult is separated into five divisions containing 26 bones: 7 cervical, 12 thoracic, 5 lumbar, a single sacrum and coccyx bone. The adult sacrum developes from the fussion of 5 rudimentary vertebrae. The terminal segment of the spine, the coccyx, is formed by the fussion of the last 4 vertebral segments.

2. D A unique characteristic of the cervical vertebra is a large oval foramen in each transverse process which protects the vertebral arteries and veins in their journey toward the brain.

3. B Each intervertebral disc is composed of an outer fibrous ring, the annulus fibrosus, and the pulpy elastic center called nucleus pulposus.

4. A The first 7 pairs are classified as true ribs because of their direct costal cartilage attachment to the sternum. The 3 pairs of false ribs have only an indirect sternal connection and the last 2 pairs are termed floating because they have no sternal attachment.

5. B The tooth-like odontoid process found on the second cervical vertebra most likely represents the remnants of the body of first cervical vertebra which is no longer present in modern man.

6. C The notches found on the inferior surface of the upper vertebra and superior surface of an inferior vertebra form a new structure called the intervertebral foramen. This foramen allows the passage of spinal nerves in the cervical, thoracic, and lumbar regions to exit from the spinal canal.

7. D Because it supports the head in a similar manner to the Greek god Atlas holding the world, first cervical vertebra shares this name. The junction between first cervicle vertebra and the occipital bone is called the atlantooccipital joint.

8. A The greatest flexibility is found in the cervical region of the spine followed by that of the lumbar region. The thoracic spine permits only a small amount of motion while the sacrum and coccyx permit virtually no movement.

9. B The first 10 pairs of ribs articulate with a thoracic vertebra at two separate locations: the head of the rib joins with the body across the costovertebral joint while the tubercle of the rib attaches to the transverse process at the costotransverse joint.

10. D The vertebral or sacral canal extends through the five sacral vertebrae and ends in a small opening at the level of fourth sacral vertebra called the sacral hiatus.

11. C See explanation number 1 of this section.

See accompanying diagram.

12. F 13. D 14. E 15. G

16. B 17. A 18. C

19. A The large prominence on the anteriosuperior surface of the first sacral vertebrae is termed the sacral promontory.

20. B The ear-shaped auricular surfaces on the upper lateral border of the sacrum articulate with the auricular surface of the ilium.

21. C The most inferior portion of the sternum, the xiphoid process is normally located at the same level as the 10th thoracic vertebra and shaft of the 6th rib.

22. B The body of a normal vertebra extends anteriorly from the structures of the neural arch.

23. A The vertebral foramen which protects the spinal cord is bounded anteriorly by the body, laterally by the pedicles, and posteriorly by the lamina.

24. A At birth, the vertebral column contains only the two kyphotic curves, those in the thoracic and sacral regions. With the beginning of the weight bearing erect posture, the lordotic curves in the cervical and lumbar region develop.

25. B The principal characteristic of the thoracic or dorsal vertebra is the presence of the articulating processes for the ribs called rib facets.

26. C The transitional vertebra between the cervical and thoracic spines, the seventh cervical vertebra is known for its long horizontal spine called the vertebral prominence.

27. B The 18" or 45 cm long spinal cord terminates at the lower border of the first lumbar vertebra to a tapered structure called the conus medullaris.

28. C The atlantoaxial joint between the first and second cervical vertebra is responsible for the lateral rotation of the head.

29. B The small, slender 11th and 12th ribs, which have no sternal connection, are called the floating ribs.

30. C The first cervical vertebra is characterized by a lack of a spinous process and vertebral body. In addition, it has 2 specialized superior and inferior articulating processes that join the condyles of the occipital bone. This joint called the atlantooccipital joint permits the nodding of the head.

31. C The large sternal or jugular notch on the superior surface of the manubrium is normally found at the same vertebral level as the body of first or second thoracic vertebra.

32. D The body of a vertebra extends anteriorly from the structures of the neural arch.

33. C The vertebral foramen or canal is formed anteriorly by the vertebral body, laterally by the pedicles, and posteriorly by the lamina.

34. A The upper segment of the sternum, the manubrium, articulates with the cartilages from the first 2 pairs of ribs.

35. A The highly resilient semi-liquid center of an intervertebral disk, the nucleus pulposus, aids in its action as a shock absorber for the spinal column.

36. D The direct sternal attachments to the costal cartilages of the first 7 pairs of ribs leads to their classification as true ribs.

37. C By definition, an apophysis is a bony outgrowth lacking its own ossification center. Examples include the superior and inferior articulating processes of the vertebra.

See accompanying diagram.

38. E	39. B	40. D	41. A
42. C			

43. D — The deep notches above and below each pedicle creates an oval foramen formed when two vertebrae are articulated called the intervertebral foramen. These structures surround and protect the pair of spinal nerves which exit at each vertebral level.

44. C — The junction between the atlas (CI) and the occipital bone permits the forward and backward bending of the head called flexion and extension.

45. A — The posterior portion of the vertebral foramen called the neural arch is formed by the two lamina and the spinous process which projects posteriorly from their junction.

46. C — To increase the stability of the posterior connection of the ribs with a thoracic vertebra two points of articulation are present. The head of the rib connects with the body of the vertebrae and the tubercles connect with the transverse process.

47. B — The manubrium, the upper portion of the sternum, is fused to the body or gladiolus at a raised ridge (sternal angle) at the level of the third thoracic vertebra.

See accompanying diagram.

48. D	49. A	50. C	51. E
52. B	53. F		

54. B — The atlantoaxial or atlantoepistrophyseal joint refers to the articulation between CI (atlas) and C2 (axis).

55. D — Spondylolysis is defined as the destruction of a vertebra. Spondylolisthesis is the forward slippage of one vertebra over another. Spondylitis is inflammation of a vertebra.

56. A — If the spine has an abnormal curve (scoliosis), the body will attempt to keep the head centered over the spine. In most individuals these compensatory actions worsen the condition leading to the progressive nature of the untreated scoliosis.

57. D — The rupture (herniation) of the nucleus pulposus and subsequent leakage and pressure on the spinal cord causes pain and muscle weakness on the effected side.

58. B — The exaggerated curve of the dorsal spine gives rise to a condition called humpback, hunchback, or kyphosis.

59. C — Rachischisis, or spina bifida, is a serious condition in which the lamina of the neural arch fail to fuse. The spinal cord and its membranes are pushed through this opening leading to brain and/or spinal cord damage.

60. D — The sacrum has 8 ventral and 8 dorsal foramen which enable the passage of the lower spinal nerves from the spinal canal. These nerves innervate the anterior and posterior structures within the pelvis.

Skull

Chapter 35

1. B The cranium or cranial vault is formed by 8 separate bones; namely, the frontal, 2 parietal, 2 temporal, occipital, sphenoid and ethmoid bones.

2. D The raised bony prominence above the orbit, the superciliary ridges, form the under-structure for the eyebrows.

3. A The slender feet-like extensions on either side of the posterior nasal cavity are called the pterygoid processes of the sphenoid bone. The appearance of these structures is often compared to the feet of a bat.

4. A The arched roof of the mouth (hard palate) is composed of the palatine processes of the maxillae and the horizontal portions of the palatine bones.

5. C The falx cerebri, a specialized layer of dura mater occupying the cerebral fissure is attached to the crista galli of the ethmoid bone to increase stability of the brain.

6. A The air cells which provide a resonance chamber for sound waves are located within a bony (mastoid) process at the inferolateral portion of the temporal bone.

7. A A majority of the superior and lateral portions of the cranium are formed by a pair of flat parietal bones.

8. B The lower central portion of the skull contains a large foramen (magnum) which enables the passage of the brainstem, vertebral and spinal arteries into and out of the skull.

9. D The deep depressions (greater wings) found on the base and side of the sphenoid bone comprise the majority of the floor of the middle cranial fossa.

10. A The mucus membranes that line the four paranasal sinuses of the frontal, ethmoid, sphenoid and maxillary bones are continuous with those of the nasal cavity.

11. C The midpoint of the frontonasal suture at the roof of the nose is termed the nasion.

See accompanying diagram.

12. M	13. D	14. G	15. C
16. J	17. E	18. A	19. I
20. B	21. K	22. F	23. L
24. H			

25. D The horizontal or cribriform plate of the ethmoid bone contains numerous perforations which allow the passage of the olfactory nerves for the sensation of smell to pass into the roof nasal passages.

26. B The cranial sutures, the fixed junctions between the bones of the skull, are normally named for the articulating bones. One exception is the connection between the temporal and parietal bones, called the squamosal suture.

27. C The bony orbit consists of seven bones including the frontal, malar, maxillary, sphenoid, lacrimal, ethmoid and palatine bones.

28. C The two corners of the eye where the eyelids meet are called the inner and outer canthi.

29. A The widest point in the adult skull is between the two parietal eminences of the parietal bones.

30. B Just lateral to the foramen magnum are the 2 oval shaped occipital condyles which serve to articulate with the superior articular process of the first cervical vertebra.

31. C The four paranasal sinuses are arranged with the frontal being most superior, the maxillary most inferior, and the sphenoid being most posterior.

32. D From the latin for cavity, the alveolar process containing the tooth sockets are found on the inferior border of the maxillae and superior surface of the body of the mandible.

33. B The only movable bone of the skull is the lower jaw bone or mandible.

34. D The lateral cranial wall is formed by the frontal, temporal, sphenoid, parietal, and occipital bones.

35. D The nasolacrimal duct passes through the lacrimal bone on the middle surface of the orbits conveying tears from the eye into the nasal cavity.

36. C The most posterior suture formed at the junction of the parietal and occipital bones is called the lambdoidal suture.

37. C The largest and most inferior paranasal sinus located within the maxillary bone is often referred to as the antrum of Highmore.

38. B The two halves of the nasal cavity are separated by a bony septum formed by the ethmoid, palatine, maxillae, and vomer bones.

39. C The air-filled middle ear and the fluid-filled bony and membranous labyrinths are housed in hollowed cavities within the petrous ridge of the temporal bone.

40. A The fixed articulation which passes laterally across the skull between the frontal and parietal bone is called the frontal or coronal suture.

41. C A useful landmark in skull positioning is a flattened area between the superciliary ridges called the glabella.

See accompanying diagram.

42. E 43. A 44. D 45. G

46. B 47. F 48. C

49. A The lateral orbit, cheek and anterior portion of the zygomatic arch is formed by a superficial facial bone called the zygoma or malar bone.

50. D The mucus membrane, the lining of the nasal passages are continuous with those of the paranasal sinuses of the frontal, sphenoid, ethmoid and maxillary bones.

51. C The junction between the capitulum on the superior surface of the condyloid process of the mandible and the mandibular fossa on the temporal bone form the temporomandibular (TMJ) joint.

52. B In addition to the large foramen magnum of the occipital bone, the floor of the cranium contains five paired foramina, 4 are associated with the sphenoid bone: foramen ovale, lacerum, spinosum and rotundum; while the fifth, jugular foramen, is found between the temporal and occipital bones.

53. A The most prominent structure on the posterior aspect of the occipital bone is the external occipital protuberance (EOP) or the inion.

54. C The deep saddle shaped depression on the superior surface of the sphenoid bone, the sella turcica, is bounded anteriorly by the anterior clinoid process of the lesser wing of the sphenoid and dorsally by the posterior clinoid process.

55. D The sagittal and lambdoidal sutures converge at the back of the head in an area named for its similarity to the greek letter λ (lambda).

56. B The two upper scroll-like processes of the ethmoid bone are called the nasal conchae. The third set or inferior conchae are extensions of a deep facial bones called the turbinates.

57. B The posterior one-third of the hard palate is formed by the horizontal portions of the palatine bones, with the remaining two-thirds consisting of the palatine portion of the maxillae.

58. C The ramus of the mandible divides superiorly into an anterior coronoid process and the more posterior condylar process which articulates with the temporal bone.

59. A The structures of the sphenoid bone include the sella turcica, a pair of clinoid processes, two wings, two pterygoid processes and a sphenoid sinus.

60. A The styloid processes of the petrous portion of the temporal bone project infero-anteriorly and serve to attach the stylohyoid, stylopharyngeus and styloglossus muscles used to aid in speaking.

61. C The hyoid bone has numerous muscular attachments to the mandible, styloid process, and tongue which help in phonation and deglutition.

62. C The floor of the cranial vault is formed by portions of the frontal, ethmoid, temporal, sphenoid, and occipital bones.

63. A The majority of the superior aspect of the nasal septum is formed by a thin bony process called the cribriform or horizontal plate of the ethmoid bone.

64. D The air filled sphenoid sinus is normally found under the sella turcica just posterior to the back of the nasal passages.

65. B The fusion of the maxillae shortly before birth leaves a small ridge and a sharp spine at the base of the nasal aperture called the anterior nasal spine.

66. D The nasal septum is formed by portions of the ethmoid, palatine, maxillae and vomer. Other bones contributing to the nasal cavity are the inferior conchae, sphenoid, nasal, lacrimal, and frontal bones.

67. B The most posterior of the paranasal sinuses is the sphenoid sinus which extends to the back of the sella turcica.

68. A The sagittal plane which divides the body into right and left sides derives its name from the cranial suture between the two parietal bones.

69. A The prominence of the cheek is due in large part to the underlying zygomatic arch formed by the temporal process of the malar bone and the zygomatic process of the temporal bone.

70. C — The numerous pairs of foramen found on the floor of the cranium are collectively termed the basal foramina.

Respiratory System

Chapter 36

1. C — Respiration or the exchange of gases may take one of two forms; internal, which involves the gaseous transfer within the tissues of the body, and external respiration, which moves air in and out of the lungs and the pulmonary exchange of gases.

2. A — The build-up of foreign particles in the lungs is prevented by the filtering effects of coarse nasal hairs and ciliated membrane of the bronchial passages. Any trapped particles are carried to the oropharynx for their elimination by the gastrointestinal tract.

3. D — The two sides of the nasal cavity are separated by a bony partition called the nasal septum, which is formed by portions of the ethmoid, palatine and vomer bones.

4. B — Outside air passes from the mouth or nose into the pharynx, larynx, trachea, main stem bronchi, primary bronchiole, terminal bronchiole and terminates at the alveolar sacs.

5. C — The voice box or larynx that houses the true and false vocal cords and the glottis is formed by nine individual cartilages, namely the thyroid, epiglottic, cricoid and the 3 paired cartilages: the arytenoid, corniculate and cuneiform.

6. A — The right primary bronchus is shorter, wider and more vertical than the left and is therefore more readily accessed by foreign objects or infectious agents.

7. B — The last tracheal cartilage is expanded inferiorly forming a hook-like process at the 4th or 5th thoracic level called the carina.

8. C — The large dome shaped diaphragm which has extensive muscular attachments to the xiphoid process, ribs, costal cartilages and vertebrae is the principal muscle of respiration.

9. D — The narrow superior portion of the lung located behind the first three pairs of ribs and clavicle is called the apex of the lung.

10. C — The solid ring of C-shaped cartilages which form the larynx, trachea and bronchiole provide the rigid support to prevent the inward collapse of these structures during respiration.

11. A — At the terminal end of the respiratory passages are a series of thin-walled membranous sacs called alveoli. These are surrounded by a capillary network that enables the free exchange of oxygen and carbon dioxide between the lungs and the red blood cells.

12. B — The left lung has two distinct lobes called the upper and lower lobes. The right lung contains an upper, middle, and lower lobe. Surprisingly both lungs have approximately the same surface area and capacity.

13. B — The characteristic shape of the larynx is due in a large part to the fused triangular shaped thyroid cartilage which forms the anterior prominence or adam's apple.

14. A The lungs are protected and separated by a double layered serous membrane called the pleura. The parietal or outer layer of pleura is attached to the wall of the thoracic cavity while the inner visceral layer forms a direct covering of the lung tissue. Between these is a thin fluid filled cavity called the pleural space.

15. D The nose is lined with coarse nasal hairs and mucus membranes to filter out dust particles. The scroll shaped nasal conchae create the turbulence which circulates and warms the air. Tears from the orbital cavity pass into the nasal cavity to humidify the air prior to its passage into the lungs.

See accompanying diagram.

16. D	17. A	18. C	19. E
20. F	21. B		

22. B On inspiration, the intrathoracic pressure is lowered by the contraction of the diaphragm and the resulting expansion of the lung. The further dropping of the abdominal organs in the erect position aids in maximizing the volume lung tissue which can be visualized radiographically.

23. C The prominent last tracheal cartilage at the bifurcation of the trachea forms a ridge called the carina.

24. D The exchange of respiratory gasses in the lungs occurs across the thin walls of the alveolus and the surrounding capillary vessels.

25. C The centrally located bronchi, pulmonary arteries, veins and lymphatics extend laterally from their origins causing an indentation as they enter the lung called the hilum.

26. A The trachea and bronchi are covered with a specialized membrane containing goblet cells for mucus production, ciliated columnar cells whose hair-like cilia traps many of the larger inhaled particles before they reach the lungs.

27. B The parietal or outer layer of pleura is attached to the wall of the thoracic cavity while the inner visceral layer is attached to the lung. Between is a fluid-filled space called the pleural cavity or space.

28. B The two lungs are separated by a space called the mediastinum which contains the heart, great vessels (aorta and vena cava), esophagus, pulmonary vessels, trachea, thymus gland and the vagus, phrenic and cardiac nerves.

29. C The potential space between the two layers of pleura is filled with an oil-like lubricating fluid that helps to reduce the friction occurring during respiration.

30. D The blood supply to the lung is provided by the pulmonary artery originating at the right ventricle of the heart.

31. A The normal volume of air taken into the lung from quiet breathing is about 500 cc and is called the tidal volume. A forced deep breath (inspiratory reserve volume) can pull in about 3600 cc of air.

32. B The glottis or opening between the vocal cords is covered during swallowing by a cartilaginous tissue called the epiglottis.

33. A The lobes of the lung are not separated segmentally by a straight axial plane but instead are divided by a superior to inferior oblique fissure which provide a great deal of lobular overlap when viewed radiographically.

34. C The lobes of each lung are divided by fissures. The oblique fissure of the left lung separates the upper and lower lobes. In the right lung an additional horizontal fissure separates the middle and inferior lobes.

35. B The contraction of the diaphragm increases the volume of the chest cavity and lowers the intrathoracic pressure enabling air to flow into the lungs during the act of inspiration or inhalation.

36. D The presence of excessive fluids in the pleural cavity increase the intrathoracic pressure which collapses the lung. The resulting conditions are often named for the type of agent present in the pleural spaces: blood (hemothorax), fluid (hydrothorax), or air ((pneumothorax).

37. D A blockage of the upper respiratory tract can be alleviated by an artificial surgical opening into the trachea just below the cricoid cartilage called a tracheostomy.

38. B There are a number of modified respiratory movements effecting lung and respiratory passages, namely, the cough, which expels unwanted substances from the lower respiratory passages, sneezing, for the clearing of the upper respiratory passages and the spasmodic contraction of the diaphragm; the hiccup, whose function is still undetermined.

39. C The partial or near complete destruction of the alveolar sacs results in a chronic degenerative process effecting the degree of air exchange between the lungs and blood called emphysema.

40. B The inward recoil of the alveoli and subsequent collapse of the sacs may result in segmental collapse of the lung called atelectasis.

41. D The relief of a pneumothorax or hydrothorax will often involve the removal of these substances from the pleural cavity by a technique called thoracocentesis. This involves the placement of a chest drainage tube into the cavity and the use of a low negative air pressure (suction) device.

42. A The spasmodic contraction of the smooth muscles in the bronchial walls in response to an allergic reaction to pollen, fumes, or physiological stress, may result in an asthmatic attack.

43. C The domed shape of the diaphragm creates 2 recesses in the lower lung near the junction with the ribs. In the erect position free fluids in the chest cavity will tend to collect in these costophrenic angles at the bottom of the lungs.

44. B Insufficient intake of oxygen (asphyxia) may lead to symptoms including cyanosis, tachycardia, mental impairment and in severe cases, convulsions and death.

45. C A complete respiration, including one inspiration and one expiration normally occur at a rate of 12-16 respirations per minute.

46. D The prefix a, meaning without, combined with pnea for breathing, apnea (without breathing).

47. C The prefix dys, meaning difficult, dyspnea (difficult breathing).

48. B The prefix eu, meaning average or normal, eupnea (normal breathing).

49. E The prefix an, meaning without, and oxia for oxygen, anoxia (without oxygen).

50. F The prefix brady, meaning slow, bradypnea (slow breathing).

Urinary System

Chapter 37

1. B — The breakdown of food (proteins) by the body is associated with a number of toxic waste products such as ammonia and urea which are normally removed by the urinary system.

2. C — The notch on the concave medial border of the kidney for the passage of the ureters, blood and lymphatic vessels into the kidney is called the hilum.

3. D — The kidney rests between the level of T12-L3 behind the peritoneum on the posterior abdominal wall.

4. A — The expanded proximal end of the ureter (pelvis) is located in the renal sinus and receives urine from the major and minor calyces of the kidney.

5. C — Both the surrounding adipose capsule and the fibrous layer called the renal fascia of the kidney help in anchoring this organ to the adjacent structures on the posterior abdominal wall.

6. A — The muscular walls of the ureters which contain both longitudinal and circular muscle fibers enable the peristaltic motion of this tubular structure and encourage the movement of urine toward the bladder.

7. D — Also called urination or voiding, micturition is the act by which urine is passed from the urinary bladder into the urethra for its expulsion from the body.

8. B — Unlike the male urethra, which has dual urinary and reproduction functions, the female urethra, which is attached to the anterior bladder wall between the clitoris and vaginal opening, serves only to discharge urine.

9. C — The outer coat of the bladder consists of a muscular layer, the detrusor, which can contract to actively force urine from the urinary bladder.

10. A — The appearance of the collecting ducts and the associated blood vessels give the pyramid its striated appearance when the kidney is seen in cross-section.

11. B — The tubular components, with their associated vasculature, make up the essential components or functional units of the kidney, the nephrons.

12. B — The ureters descend from the renal pelvis through the posterior abdomen to their final resting point near the lateromedial portion of the posterior bladder.

13. B — The large space occupied by the liver on the right side of the abdomen displaces the right kidney inferiorly and the right diaphragm superiorly.

14. B — The renal (Bowman's) capsule and the vascular glomerulus provide the membrane over which wastes are filtered from the blood stream and directed into the collecting system of the kidney.

See accompanying diagram.

15. D 16. A 17. F 18. E

19. B 20. C

21. A The female urethra with its inner mucosa, middle venous coat and outer muscular coat directs urine from the urinary bladder to the outside of the body.

22. C The first step in urine production glomerular filtration occurs as the blood pressure forces water and dissolved blood particles through the capillary wall and basement membrane of the Bowman's capsule.

23. C Since the male urethra descends vertically through the prostate gland before its junction with the penis, any enlargements of this gland may be associated with a lower urinary tract obstruction.

24. C The 1200 ml of blood that passes through the kidney each minute is distributed to the renal cortex by the afferent arterioles and the interlobular arterial branches.

25. D The midline urinary bladder is bordered anteriorly by the vagina and uterus in the female, posteriorly by the rectum and peritoneum in both males and females.

26. D The glomerular filtration rate which provides about 180 liters of filtrate each day is largely reabsorbed in the epithelial layers of the proximal and distal convoluted tubules and the loop of Henle.

27. C Even though the maximum capacity of the urinary bladder is about 6-800 ml, specialized stretch receptors in the wall will trigger a desire to urinate when the bladder reaches one half of its maximum capacity.

28. C The three main divisions of the male urethra are the prostatic, membranous and penile or spongy portions.

See accompanying diagram.

29. D 30. A 31. G 32. C

33. E 34. B 35. F 36. H

37. D Kidney failure results in a dangerous build up in toxic wastes in the blood. Hemodialysis (artificial kidney) is used in these patients to clean blood and maintain the appropriate electrolyte balance.

38. D The kidney aids in the removal of the nitrogenous wastes, ammonia and urea from the blood stream, and maintains the electrolyte balance by its control of sodium, chloride, sulfate and phosphate ions in addition to regulating the volume of fluid in the tissues and blood.

39. B Any drug which stimulates the flow of urine is classified as a diuretic (Lasix) and may be prescribed to assist patients in the control of edema, congestive heart failure and hypertension.

40. A The restriction to the blood flow in the renal artery causes the production of renin that raises the blood pressure.

41. B The failure of urine to pass from the kidney into the bladder increases the pressure in the calyces causing their distension (hydronephrosis) and eventual atrophy if the condition is chronic in nature.

42. D Though the urinary tract is the principal system involved in the maintenance of the fluid balance, the lungs, skin, biliary and GI tract all contribute to normal fluid homeostasis.

43. A Ectopic kidneys or those located in other than the usual areas are most commonly found in the pelvis and only rarely cause significant problems for the patient.

44. C Renal failure and the resulting toxic build-up causes uremia, which is diagnosed by increased levels in a blood urea nitrogen (BUN) or creatinine test.

45. D Damage to the supporting tissues of the kidney may result in its dropping to an abnormal abdominal location called nephroptosis.

Digestive System

Chapter 38

1. C Of the various movements that aid in the digestive process the most important for the pulverization of food in the mouth is chewing or mastication.

2. B The uvula and surrounding fauces when irritated trigger gagging and vomiting which serves to clear objects obstructing the pharynx.

3. A In its normal contracted state, a circular sphincter muscle serves to close an orifice. The relaxation of a sphincter muscle enables the passage of materials into more distal sections of the organ or tract.

4. C The four primary tastes, sweet, sour, salt and bitter are detected on the receptors (taste buds) located on the fungiform papillae of the tongue.

5. D In order to prevent the accidental swallowing of the tongue and the resulting asphyxiation, the tongue is attached to the floor of the mouth by the short flap of tissue called the frenulum.

6. A The wide distribution of the vagus nerve to the muscles of the pharynx, larynx, respiratory passages, lungs, heart, esophagus, stomach, small intestine and colon make it critical for normal respiratory and digestive functions.

7. D The act of swallowing which can be initiated voluntarily ends in a reflex action involving the muscles of the tongue, pharynx, larynx and upper esophagus which push the food into the proximal esophagus.

8. C In organs possessing both longitudinal and circular layers of smooth muscle fibers, a wavelike constriction called peristalsis assists in the movement of substances within the tract.

9. B A bone-like substance which covers the dentin on the roots of the tooth and attaches it to the periodontal membrane of the alveolar process is called cementum.

10. B Saliva is a watery substance which assists in the swallowing of food, and triggers the digestive process. This slightly acidic fluid contains both bacteriolytic and digestive enzymes produced in the three sets of salivary glands.

11. C The act in which the tongue, pharynx and esophagus move the food toward the stomach is termed deglutition or swallowing.

12. A A study of teeth indicates a relationship exists between their shape and function. The chisel shaped incisors are adapted to cutting food, the pointed canines tear and shred, while the larger molars crush and grind the food.

13. D The outer dentin encloses the hollowed pulp cavity and its slender root canals. This cavity houses the blood vessels, nerves and lymphatic vessels needed to nourish the tooth.

14. B Saliva passes from the submandibular gland beneath the base of the tongue to its openings just behind the central incisors through the Wharton's ducts.

15. B The normal human goes through two dentitions during their lives: the first ending between 6 and 12 years of age consists of 20 deciduous or baby teeth, the second is the permanent dentition consisting of the 32 adult teeth.

16. D The majority of a normal tooth is composed of a bone-like substance called dentin which gives the tooth its characteristic shape.

17. A The largest and most posterior set of salivary glands, the parotids, are located in front of the ears on top of the masseter muscle.

18. C The portion of the tooth extending above the gum (crown) is covered with the hardest material of the body, enamel.

19. B A small opening in the cheek near the second molar provides the entry point for Stensen's ducts which pass anteriorly from the parotid gland to carry saliva into the mouth.

20. A The esophagus is a midline structure which connects the oropharynx and the stomach . It is located anterior to the cervical and dorsal spines and posterior to the trachea within the mediastinum.

21. A Chemical digestion begins in the mouth with the secretion of saliva and continues in the stomach and small intestine. The mechanical digestive process begins in the mouth with chewing and is completed by the churning action of the stomach and small intestine. The esophagus has no active chemical or mechanical digestive function.

22. D The acidic conditions which aid the digestive process in the stomach would, without protection, digest the stomach itself. The gastric mucosa and its associated mucus producing glands secrete a protective mucus for the stomach. Mucus producing cells are also present in the esophagus, small intestine, and large intestine.

23. C The esophagus descends within the mediastinum of the thoracic cavity to its junction with the stomach (cardia). The fundus, the highest portion of the stomach, extends superiorly to a space under the left hemidiaphragm.

24. C The digestive enzymes and acids produced in the gastric walls and the churning motion provided by the muscles of the stomach assist in both the chemical and mechanical digestive process.

25. A The food entering the stomach is subjected to rippling peristaltic movement (mixing waves) which macerate the food into a thin partially digested liquid called chyme.

26. B The body of the stomach narrows at its distal end in an area called the pylorus just proximal to its junction with the duodenum.

27. C The roots of the teeth and their articulating surfaces of the alveolar processes or tooth sockets are lined with a periodontal membrane or ligament.

28. C The gastric mucosa contains three types of secreting cells: the goblet cell that produces mucus, the chief or zymogenic cells for the digestive enzyme pepsinogen, and the parietal cells which secrete hydrochloric acid.

29. B The mucosa of the small intestine contains 4-5 million, 1 mm high projections called villi which greatly increase the surface area of the epithelium over which the absorption of the digested food takes place.

30. A The interior of the empty stomach is not smooth and contains a number of large deep folds of mucosa called rugae. When fully distended these folds stretch and are no longer visible.

31. D The small intestine is divided into three segments based on the appearance of the mucosa. The first and shortest portion, the duodenum, is only about a foot long. The eight foot long jejunum extends to the 12 foot long final portion called the ileum.

32. B The ascending and descending portions of the colon have abrupt turns at their junction with the transverse colon. The hepatic (right colic) flexure is named for its proximity to the liver. The higher splenic (left colic) flexure derives its name from its association to the spleen.

See accompanying diagram.

33. I	34. D	35. A	36. C
37. E	38. B	39. G	40. F
41. H			

42. D The narrowed end of the stomach, the pylorus, is separated from the duodenum by a thick circular sphincter muscle called the pyloric sphincter.

43. B The intestine consist of four distinct tissue layers. The innermost layer of the mucosa contains the crypts of Lieberkuhn which produce digestive enzymes. The vascular submucosa possesses the nerves and vessels of the Meissner's plexus. The third layer consists of smooth muscle tissue which aids in peristalsis and the outer covering membrane called the serosa or adventitia.

44. A The absorption of water is an important function of both the small intestine and large intestine.

45. A From proximal to distal, the large intestine consists of a cecum, ascending colon, hepatic flexure, transverse colon, splenic flexure, descending colon, sigmoid colon, rectum, and anus.

46. B The plicae circularis or circular folds are the true mucosa of the walls in the small intestine. They act to agitate the chyme and provide additional surface area for the absorption of food.

47. C The chemical digestion of food begins in the mouth by the action of saliva which is augmented by the stomach and small intestine.

48. C The hormonal secretions from the intestinal mucosa trigger the release of the pancreatic juice which is rich in bicarbonates to assist in the neutralization of the acidic chyme entering the small intestine.

49. B The 4 layers of the digestive tract are the mucosa, submucosa, muscular, and serosa or adventitia.

50. A The small intestine and large intestine are joined across a muscular sphincter called the ileocecal valve.

51. C The largest serous membrane of the body that covers most of the organs in the gastrointestinal tract is called the peritoneum.

52. C At a length of nearly 25 feet, the small intestine is by far the longest segment of the GI tract. The colon has a length of about 8 feet.

53. A The serosa or outer layer for most structures in the gastrointestinal tract is a double layered membrane known as the visceral peritoneum.

54. B The first and shortest 10-12 inch long segment of the small intestine is called the duodenum.

55. C Derived from the word vermis, for worm, the coiled tube called the vermiform appendix attaches to the lower margin of the cecum. The appendix, once thought to be functional, no longer plays any detectable role in the digestive process.

56. A The structural variation including the circular fold, villi, microvilli, and the extended length of the small intestine, make it well adapted for the absorption of food and water.

57. C The peritoneum has two major extensions: the mesentery, which attaches the small intestine to the posterior abdomen wall, and the omentum (fatty apron) which surrounds the transverse colon and aids in the spread of infections.

58. C In response to deglutition, the larynx moves upward shifting the epiglottis into a position which seals the glottis and larynx.

59. B The intestinal glands or crypts of Lieberkuhn of the small intestine produce sucrose, lactase and maltase which break down the named sugars into glucose.

60. D Unlike other portions of the gastrointestinal tract, the longitudinal muscle layer of the colon is gathered into 3 flat bands called the taeniae coli that rest on the external surface. Under the normal tonic conditions a series of pouches called haustral are formed.

See accompanying diagram.

61. A	62. H	63. J	64. K
65. G	66. D	67. L	68. I
69. F	70. B	71. C	72. E

73. C The colon ascends from the ileocecal valve in the right lower quadrant into the right upper quadrant before turning just under the liver to become the right colic or hepatic flexure.

74. D Digestive enzymes are produced in many areas including the salivary glands, stomach, small intestine, pancreas and liver (bile).

75. B The contraction of the taenia coli gives the large intestine its characteristic puckered (haustral) appearance.

76. D Both benign and malignant tumors may occur in the lower GI tract. A class of stalked tumors called polyps are among the most common, though histological examinations are required for a definite diagnosis, short stemmed polyps with irregular borders are generally cancerous while a longer stemmed smooth bordered polyp is most often benign.

77. C Most commonly seen in the ileocecal region in children, an intussusception involves the slippage of one part of the intestine into another. The result is often an obstruction which is alleviated by abdominal surgery or in some cases by a therapeutic barium enema.

78. A A weakening in the lining of the walls of the colon may result in the formation of small blind pouches (diverticulum). When inflamed (diverticulitis), these may bleed causing cramp-like pains due to the muscle spasms of the colon.

79. A The examination of the stool can be helpful in distinguishing the region of gastrointestinal bleeding. Dark tar-colored stools indicate bleeding in the stomach or small intestine. Fresh blood mixed with the stool is associated with lesions in the large intestine, rectum, and anus.

80. C The increased intra-abdominal pressure associated with the bearing down action of the valsalva maneuver is helpful in the evaluation of esophageal varices, stress incontinence, laryngopharynography and nasopharynography.

81. B Tortuous veins that have developed in the anorectal region can result in a painful swelling and inflammation called hemorrhoids or piles.

82. B The double serous membrane, the peritoneum, encapsulates all or part of the stomach, jejunum, ilium and the majority of the colon. The duodenum, pancreas, liver, spleen, and urinary tract organ are located outside of this covering.

83. D The body's own protective mucosa is not a perfect barrier to the effects of gastric and intestinal fluids. Its failure may result in the erosion of the mucosal lining and the formation of an ulcer.

84. C The large intestine's main function is the absorption of water. Its only secretion is the protective mucus needed to avoid irritation of its walls.

85. A The protrusion of all or part of an organ through an adjoining membrane or cavity is called herniation. Three types of hernias are commonly seen in the abdominal cavity. The diaphragmatic or hiatal hernia involves the protrusion of the stomach or intestine into the thoracic cavity. The umbilical hernia in which the abdominal organs pass through a weakening near the umbilicus and the inguinal hernia in which a loop of intestine passes through a weakness of the inguinal ring of the groin.

86. B In a normal male, the rectum is located anterior to the sacrum and coccyx and posterior to the urinary bladder and peritoneum.

87. A The correct sequence in the gastrointestinal tract is the mouth, pharynx, esophagus, stomach, small intestine, and colon.

88. A The narrowing of the pyloric valve (pyloric stenosis) by an enlargement of the circular muscle fibers and persistent spasm of the pylorus (pylorospasm) are two common conditions which more frequently occur in the stomach of male infants.

89. B Interference with the venous return from the abdomen, can cause lymphatic obstructions or cirrhosis of the liver, and may result in an excess fluid accumulation in the peritoneum called ascites.

90. C Obstruction in the intestinal tract caused by a twisting of the bowel (volvulus) is most commonly seen in the sigmoid and ileocecal regions.

Accessory Digestive System

Chapter 39

1. C — The liver is one of only a few organs that is essential to life. Among its important functions are the manufacture of heparin, prothrombin, fibrinogen, and albumin. It also detoxifies toxins such as nitrogen wastes and alcohol, stores whole blood, glycogen and many other vitamins and essential metals as well as producing bile from bilirubin.

2. B — The exocrine secretion of the pancreas (pancreatic juice) is a mixture of digestive enzymes including trypsin, chymotrypsin, pancreatic amylase and lipase which are produced in the specialized acinar cells of the gland.

3. D — The largest mass of lymphatic tissue is contained within a 5 inch oval organ called the spleen located on the posterolateral surface of the stomach just below the left hemidiaphragm.

4. D — The liver maintains its position in the large part due to a fold of the peritoneum, the falciform ligament, which attaches this organ to the anterior abdominal wall.

5. A — The gall bladder serves as a reservoir for bile excreted into the biliary collecting system by the liver. In response to hormonal signals from the intestinal mucosa in the presence of fats its muscular wall contracts to expel the bile due to the duct network.

6. C — The digestive juices of the pancreas, bile from the liver and gall bladder, empty into the duodenum by way of the duodenal sphincter or sphincter of Oddi.

7. A — The gall bladder is a muscular sac which normally contains bile and is located along the under surface of the right lobe of the liver.

8. B — The cystic duct provides a passageway for the flow of bile from the gall bladder into the common hepatic duct just above its junction with the larger common bile duct.

9. C — The 6" oblong pancreas has a broad head that is situated within the C-shaped curve of the duodenum. Its tapered tail extends laterally to its attachment on the medial surface of the spleen.

10. D — Weighing in at about 4 pounds, the liver is the largest single organ in the body. It occupies most of the right hypochondrium and part of the epigastric regions.

11. A — The phagocytic bacteria present in the spleen, break down worn-out red blood cells and platelets converting the released hemoglobin into a green colored fluid called bilirubin.

12. C — The pancreatic juices pass into the biliary collecting system by way of a long duct running the length of the pancreas known as the pancreatic duct or duct of Wirsung.

13. C — To maintain the proper fluid balance in the body, a number of systems aid in this process including the kidneys, small intestine and large intestine, lungs and gall bladder.

14. B — The islets of Langerhans in the pancreas contain two types of cells, the alpha cells which secrete glucagon that tend to increase blood glucose levels, and the insulin producing beta cells which serve to lower blood glucose levels.

See accompanying diagram.

15. A	16. K	17. G	18. C
19. H	20. D	21. I	22. E
23. B	24. F	25. J	

26. B — Stimulated by fats in the stomach or duodenum, the intestinal mucosa secretes cholecystokinin which triggers the release of bile and pancreatic juices.

27. D — The upper end of the pancreas extends laterally across the abdomen to its attachment on the outer surface of the spleen.

28. A — The filling of the gall bladder with bile is principally due to the action of the sphincter of Boyden, located on the common bile duct just above to its junction with the pancreatic duct.

29. B — The principal secretion of the liver, bile, is composed of a number of compounds including bilirubin, cholesterol, fatty acids, and bile salts.

30. A — The gall bladder is connected to the common hepatic duct by way of a thin hollow tube called the cystic duct.

31. D — The functions of the liver as outlined in explanation number 1 include all of the following except the production of digestive enzymes.

32. B — The weak sphincter at the end of the ampulla vater which guards the opening into the duodenum is called the sphincter of Oddi.

33. A — Bile and more specifically the bile salt component of bile formed during the breakdown of cholesterol is the active constituent which serves to emulsify fats and aid in the absorption of fatty acids from the intestine.

34. C — During periods of low carbohydrate intake (starvation) the liver can form glycogen from amino and fatty acids.

35. C — The release of cholecystokinin by the intestinal mucosa is triggered by the presence of fats or fatty acids in the stomach and small intestine.

36. A — A liver inflammation called hepatitis may have many possible causes including exposure to viruses, bacteria, protozoa or other chemical agents such as carbon tetrachloride, anesthetics or alcohol.

37. C — The presence of stones in the biliary tract sometimes associated with pain and obstruction develops in two main forms: the low density cholesterol stones which tend to float in the bile and the more dense calculi made of calcium salts.

38. B — The main clinical symptoms of insufficient insulin production resulting from diabetes mellitus are hyperglycemia and excretion of glucose in the urine. The inability to reabsorb water, dehydration and electrolytic imbalance are the cause of the life threatening consequences seen with diabetes.

39. C — The yellow-green fluid bilirubin produced in the spleen following the breakdown of hemoglobin can not be removed by a poorly functioning liver therefore it remains in circulation causing the yellow staining of the skin and eyes called jaundice.

40. D — Cirrhosis is a chronic disease of the liver in which fibrous connective tissue replaces the normal cells. This condition can occur from hepatitis, parasitic infection, or chronic exposure to toxic chemicals such as alcohol.

Special Senses

Chapter 40

1. C — The body receives information about the external environment through a number of specialized sensory receptors or exteroceptors.

2. B — The olfactory receptors, for the sense of smell, are found on the membranous lining of the upper nasal cavity. Sensory signals pass along the olfactory nerves into the temporal lobe of the brain for interpretation.

3. D — Each of five cutaneous (skin) sensations, light touch, pressure, pain, cold, and heat are associated with a distinct extroceptor.

4. A — The exteroceptor for vision lies on the retina, the nervous coat, covering the posterior and lateral surfaces of the eyeball.

5. D — The visceroceptors are a class of sensory receptors that provide the body with information concerning the status of the internal organs including hunger, pain, pressure, and distention. These are collectively referred to as stretch receptors.

6. A — The anterior portion of the eye is composed of a clear connective tissue, the cornea, which enables the unimpeded flow of light into the eye.

7. B — The middle ear, or tympanic cavity, is an air filled space housing the auditory ossicles or middle ear bones.

8. C — The gustatory, or taste receptors, called the taste buds, are located on the raised elevations of the tongue, the papillae.

9. B — The skin covered cartilage commonly called the ear is also known as the pinna or auricle.

10. C — At the medial end of the external auditory canal sits the tympanic membrane or eardrum which separates external and middle ears.

11. A — The pressure in the air-filled middle ear is equalized with that of the outside environment to prevent its rupture. The eustachian or auditory tubes which connect the middle ear and nasopharynx serve to balance this pressure.

12. C — The auditory ossicle from external to internal are the malleus (hammer), incus (anvil), and stapes (stirrup).

13. B — The inner ear houses both the organs of hearing and balance. The organs associated with equilibrium and the positional senses are the saccule, utricle, and semicircular canals.

14. C — The true exteroceptors for hearing, the organs of corti, contain hair cells and support structures which generate an electronic signal that passes into the temporal lobe of the brain.

See accompanying diagram.

15. A

16. E

17. B

18. C

19. F

20. D

21. A — A series of sacs and tubes having the same general form as the bony labyrinth, the membranous labyrinth, are protected by a fluid called perilymph.

22. A — Static and positional equilibrium are provided in the utricle, saccule, and semicircular canals. These fluid-filled cavities are lined with sensory hairs that respond to small grains of calcium carbonate (otolith) which trigger the hairs during the changing orientation of the head.

23. D — The 4 primary taste sensations are salt and sweet, detected on the tip and sides of the tongue, and sour on the side, and bitter which is detected on the back of the tongue.

24. C — Because many afferent nerve (pain) fibers have a wide cutaneous distribution, the brain is often unable to pinpoint the exact location of a stimulus. The pain that is felt may be referred to an area other than the effected site; e.g., heart pain may radiate down the left arm, kidney pain may be felt on the sides of the upper leg.

25. A — The superolateral portion of each upper eye lid contains an almond sized lacrimal gland which bathes the eyes with tears through 6-12 small lacrimal duct openings under the lid.

26. D — Tears, which contain both mucus and bactericidal enzymes, spread over the eye before passing to the nasolacrimal duct near the inner canthus. Tears eventually are dispersed within the nasal cavity to help humidify the air passing into the lungs.

27. C — The exteroceptors for vision, the cone and rod cells, are dispersed on the back and sides of the nervous coat or retina of the eye.

28. B — Surrounding the pupil is a doughnut shaped muscle, the iris, which contains the pigments for eye color.

29. A — The tough outer fibrous covering of the eye which both protects and gives shape to the eyeball is formed by the clear cornea and white sclera.

30. C — The retina on which the visual image is formed is continuous with the optic nerve which carries the visual impulses to the occipital lobe of the brain.

31. A — Immediately behind the pupil and iris, the clear oval lens is suspended by the ciliary muscles. Changes in the shape of the lens focus the incoming light providing for both near and far vision.

32. B — The circular muscle, the iris, surrounding the pupil, contains the protective pigment, melanin. The distribution and scattering of the incoming light by the iris provides the various shades of eye color. Blue eyes which lack this pigment tend to be more sensitive to the potential damage caused by ultraviolet radiation.

33. C — The nervous coat of the eye, the retina, contains two separate cell types. The cone cells, which respond to bright and colored light and the more sensitive rod cells, used in dim light conditions which are only capable of black and white vision.

34. A — The distribution of the cells for vision is uneven with the cones being densely packed on the center of the eye (fovea centralis) and the rods predominating on the sides of the retina.

See accompanying diagram.

35. H
36. D
37. F
38. J
39. G
40. A
41. B
42. C
43. E
44. I

45. B — The stereoscopic vision that humans possess is largely due to the 2.5" separation of the eyes and the independent images each sends to the brain.

46. D — The interior of the eye is divided into anterior and posterior chambers by the lens of the eye. The anterior chamber contains a watery fluid called aqueous humor which maintains the desired intraocular pressure. Excessive pressure in this chamber may cause a condition called glaucoma resulting in circulatory disturbances, retinal damage, and eventual blindness. In the posterior chamber, a jelly-like vitreous humor helps maintain the characteristic shape of the eyeball.

47. C — The junction of the optic nerve ganglions at the retina forms a small circular area which lacks visual receptors known as the optic disc or blind spot.

48. D — The loss of transparency of the lens of the eye as is seen with cataracts may be accelerated by exposure to x rays and ultraviolet (UV) radiation.

49. C — The larger of the two cells for vision is the color sensitive cone cell. The cones, which function only at high light levels (daytime) possess a higher level of visual acuity than the smaller rod cells.

50. A — The movements of the eye are controlled by 6 extrinsic eye muscles namely the superior, inferior, medial and lateral rectus muscles and the superior and inferior oblique muscles.

Male Reproductive System

Chapter 41

1. D — In response to the hormones of the anterior pituitary gland, the testes secrete testosterone which stimulates bone and muscle growth, aids in the maturation of sperm and stimulates the development of the secondary sex characteristics.

2. B — In fetal life, the testes develop near the kidneys in the upper abdomen. Their descent out of the abdominal cavity and into the scrotum is normally completed about 2 months prior to birth.

3. D — The immature sperm cells (spermatogonia) are produced in a set of tightly coiled tubules of the testes called the seminiferous tubules.

4. D — The testes are supported in an external pouch, the scrotum, which helps to maintain their temperature at about 3 degrees below that of the body's temperature.

5. B — The slightly alkaline semen or seminal fluid not only provides the sperm with a transportation and nutrient medium it also serves to neutralize the acid environment of the vagina. It is produced in the bulbourethral (Cowper's gland), prostate gland and a pair of seminal vesicles.

6. C The formation of the male sperm and female ova involves a type of haploid type division called meiosis.

7. D The doughnut shaped prostate gland sitting just inferior to the urinary bladder surrounds the first portion of the urethra. In older men the enlargement of the prostate may compress the urethra and obstruct the flow of urine from the bladder.

8. A The immature sperm produced in the seminiferous tubules pass into the epididymis where maturation is completed. During ejaculation, peristaltic contractions of the smooth muscular walls of the ejaculatory ducts system propel sperm toward the urethra.

9. B To increase the production and survival of the sperm, they are produced at a temperature below that of the body. The external location of the testes within the scrotal sac help provide this desired difference in temperature.

See accompanying diagram.

10. I	11. F	12. G	13. J
14. A	15. H	16. C	17. D
18. E	19. B		

20. A Unlike the majority of human cells that contain 44 autosomes and 2 sex chromosomes, the spermatozoa is haploid, possessing only 22 autosomes and a single sex chromosome. Since males have both x and y sex chromosomes, a sperm has either an X (female sperm) or Y (male sperm) genetic potential.

21. B The sponge-like (erectile) tissue of the penis possesses blood sinuses that dilate during sexual excitement or stimulation.

22. B The covering of the glands by the foreskin helps protect this sensitive structure from irritation. The removal of this tissue by circumcision soon after birth is common in some religious faiths and in many medical systems such as our own.

23. D The average volume of ejaculate of about 2-4 cc of semen may contain as many as 100 million sperm per cc of fluid.

24. C The main permanent form of male sterilization is a procedure involving the cutting or tieing of the vas deferens. The interruption of the ducts closer to the urethra are not employed because of the potential negative effects to the urinary tract functions.

25. D There are numerous physiologic causes of male infertility with both physiological and psychological roots. Among the more common are inflammation of the testes, undistended testicles and low sperm count or sperm morbidity.

Female Reproductive System

Chapter 42

1. C The ovum is released from the graafian or ovarian follicle of the ovary. The maturation process which takes approximately 56 days alternates between the two ovaries resulting in the more familiar 28 day cycle.

2. A The ovum is released into the pelvic cavity but is directed by the finger-like fimbriae into the uterine (Fallopian) tube.

3. B Fertilization which normally takes place in the fallopian (uterine) tubes causes a small change in the electrical potential of the egg which encourages its implantation into the uterine wall.

4. D The muscular vagina serves as the passage way for the menstrual flow to leave the body. It also forms the lower portion of the birth canal and the receptacle for the penis during coitus.

5. A During a non-reproductive menstrual cycle, the outer layer of the endometrium, the stratum functionalis, is shed during menses. This layer is reformed during middle portion of the cycle each month.

6. B During pregnancy, the mammary glands reach their final development stage under the influence of estrogen and progesterone. These specialized tissues began with the production of the immune boosting colostrum for 2-5 days after birth. After this time, true mother's milk replaces this fluid. Recent evidence shows that breast fed babies may have better physical and mental advantages compared to formula fed babies.

7. B The ovarian cycle consists of development stages by which the primary follicle matures into the secondary follicle and graafian follicle.

8. D The uterus is shaped like an inverted pear with the fundus at the expanded upper end and the narrowed cervical portion on the lower aspect opening into the vagina.

9. C The majority of the female reproductive organs are held in place by a series of supporting ligaments including the broad, uterosacral, cardinal, round ovarian, and suspensory ligaments.

10. C Fertilization refers to the union of the nuclear material of the sperm and ovum into a new single cell called the zygote.

11. C The small cylindrical mass of erectile tissue, the clitoris, is located anterior to the labia minor just above the vaginal orifice.

12. D After its release from the ovary, the ovum or oocyte enters the fallopian tube and is carried by ciliated cells and muscular waves towards the body of the uterus.

See Accompanying diagram.

13. D 14. G 15. H 16. B

17. C 18. A 19. F 20. E

21. A The fetus is physically connected to the mother by way of a highly vascular organ called the placenta. In the developing infant, the lack of fully functional lungs, liver, GI tract and kidneys necessitates the exchange of food, oxygen and wastes. The placenta provides this essential exchange enabling proper growth and development of the fetus.

22. C Unlike males, who have only one major sex hormone, females possess two. Estrogen which is responsible for the development of the primary and secondary sex characteristics such as hair distribution and breast development, and progesterone which works in conjunction with estrogen to prepare the uterus and egg for implantation and development of mother's milk.

23. A The normal monthly menstrual cycle provides the desired conditions for the development of a viable ovum and the preparation for implantation in the event that fertilization occurs. Beginning with menarche at about 11 years of age, the cycle continues until menopause in the fifth decade of life.

24. A The menstrual cycle progresses in 3 distinct phases. The first, lasting for the first 5 days, is the period of menstruation characterized by the loss of the outer layer of endometrium. In the middle 7-10 days, or pre-ovulatory phase, the endometrium thickens with increased vascularity. During the last week, or ovulatory phase, the egg is released and upon fertilization the egg may attach to the fully developed endometrium.

25. B During fertilization, the 23 chromosomes of the male and female gametes (egg and sperm) are fused to form a single zygote containing 44 autosomes and 2 sex chromosomes. A male zygote possesses XY pattern and the female zygote XX pattern of sex chromosomes.

26. C The ciliary action and muscular contraction of the fallopian tubes provide the necessary force to move the egg cell toward the uterus.

27. D Implantation or the actual attachment of the dividing zygote (blastocyst) and the uterine wall occurs about a week after fertilization takes place.

28. D Temporary contraceptive methods for females take many forms, the diaphragm, sponge and female condom which are designed to prevent the passage of sperm into the uterus. The IUD (intra uterine device) prevents implantation of the egg and the pill acts to prevent ovulation by the hormonal disruption of the menstrual cycle. All have their potential benefits and disadvantages. The most effective at preventing pregnancy at this time is the pill.

29. B The length of time the embryo-fetus is carried by the mother is termed gestation and normally lasts about 40 weeks. The last step in this process is labor which results in the actual birth of the fetus called parturition.

30. D The permanent form of female birth control is termed a tubal ligation, in which the fallopian tubes are cut or tied. Though sometimes reversible with micro-surgery it should only be considered when no other pregnancies are contemplated.

Nervous System

Chapter 43

1. A The brain and spinal cord are collectively referred to as the central nervous system. The remaining nerves which carry signals between the receptors, muscles, and glands are called the peripheral nerves.

2. A The vast majority of the senses: sight, sound, taste, smell, and feeling are interpreted in the area known as the cerebrum or cerebral cortex of the brain.

3. D Neural tissues are grouped by their relative gray or white coloration. White matter's appearance is derived from the covered axons extending from the nerve cell body. The gray matter contains a large number of nerve cell bodies and its short associated dendrites.

4. C The pituitary gland consists of two divisions: the adenohypophysis and neurohypophysis which produces the 9 major hormones. Its physical and functional connection to the hypothalamus of the brain are unique to this gland.

5. B Though gross muscle movements are initiated by the motor areas of the cerebrum, the ability to co-ordinate fine muscle movement is accomplished in the cerebellum.

6. A The convolutions of the brain are divided into raised areas called gyri and depressions known as sulci.

7. C In addition to the protection provided by the cranial bones, the cerebrospinal fluid circulating around the brain and spinal cord provides additional protection from injury. This fluid is contained within the subarachnoid spaces and within the cavities of the cerebral hemispheres called ventricles.

8. B Cerebral spinal fluid is contained between two layers of the meningeal coverings in the subarachnoid space; the arachnoid and pia mater.

9. B The vascular structures of the brain including arteries, capillaries, veins, and specialized cavities formed by the layers of dura mater called venous sinuses.

10. A In early embryonic life, the central nervous system begins to form from a bundle of nerves called the neural tube. Tissues of the brain and spinal cord continue to grow until about the age of 18.

11. C Each spinal nerve arises by 2 roots, the anterior (ventral) root contains the motor or efferent fibers that conduct impulses away from the central nervous system, the afferent or sensory fibers which conduct impulses toward the central nervous system are found in the posterior (dorsal) root.

12. D The upper portion of the spinal column and lower portion of the brain is called the medulla oblongata. The medulla is responsible for cardiac and respiratory functions and constitutes the majority of the primitive brain found in most animals.

13. C The largest portion of the brain is the cerebrum, which is responsible for most of the sensory, gross motor, and all of the creative thought functions.

14. A Cerebral spinal fluid passes between the 3rd and 4th ventricles by way of the sylvian aqueduct or cerebral aqueduct.

15. D The pia mater is the name given to the close fitting membrane which covers the brain and spinal cord.

16. B The clear watery fluid filling the subarachnoid space, the cerebral spinal fluid, contains salts, sugars, and proteins for the exchange of nutrients as well as it's physical protection of the brain and spinal cord.

17. D As the spinal cord descends from the brain it gives rise to 31 pairs of spinal nerves which exit the vertebral canal by way of the intervertebral foramina.

18. C The subarachnoid space which contains the cerebral spinal fluid is formed between pia and arachnoid mater layers.

19. C The nerve cell has 2 cytoplasmic extensions called the axon and dendrites. These connect with the other nerve extensions across a specialized junctions called a synapses.

20. B Electrical signals generated in the brain and spinal cord are directed along efferent or motor fibers to the peripheral structures such as muscles or glands.

21. C — The 2 lateral ventricles are connected by the right and left intervertebral foramina that enables the passage of cerebrospinal fluid between the lateral ventricles.

22. A — The motor fibers which innervate the diaphragm are formed by the ventral rami of C3, C4 and C5 of the cervical plexus. Spinal cord injuries above this point are extremely dangerous because they may result in paralysis of the diaphragm.

23. B — The two cytoplasmic extensions of a nerve fiber are the axon and dendrite.

24. D — The subarachnoid space of the spinal cord and ventricles of the brain communicate at an area at the back of the head called the cisterna magna. Fluids pass from one area to another through the two foramina of Luscka and the foramen of Magendie.

25. D — A painful spasm of the face due to irritation or pressure on the trigeminal nerve is called trigeminal neuralgia or tic douloureux.

26. D — The principal muscles of the anterior lower leg responsible for the lifting of the foot are innovated by the peroneal nerve.

27. A — The extensive pattern of superficial nerve fiber distribution for a single cutaneous nerve is called a dermatome.

28. C — The regeneration of a peripheral nerve is possible if its covering sheath, the neurilemma formed by the schwann cells, remains viable. The nerve cells of the brain and spinal cord lacking this membrane are incapable of repair.

29. A — Although both sides of brain function independently, they must communicate and share information through a pathway of fibers called the corpus callosum.

See accompanying diagram.

30. E   31. B   32. C   33. A

34. D

35. B — Most of the structures of the face and neck area are enervated by branches from the 12 pairs of cranial nerves. The 12 nerves and their general functions are summarized below:

I. Olfactory - sense of smell
II. Optic - vision pathway
III. Oculomotor - movements of the eye and eyelid
IV. Trochlear - movement of the eye
V. Trigeminal - jaw movement and facial sensations
VI. Abducens - eye movement
VII. Facial - muscles of the face
VIII. Vestibulocochlear - hearing and balance
IX. Glossopharyngeal - muscles of swallowing
X. Vagus - muscles of digestion
XI. Accessory - muscles of neck and shoulder
XII. Hypoglossal - muscles of the tongue

Cardiovascular System

Chapter 44

1. B — The interior of the heart contains four distinct chambers, separated into two sides by the septum, and into upper (atria) and lower ventricles by the atrioventricular valves.

2. A — Blood is contained within 3 types of blood vessels: the arteries, capillaries, and veins. The arteries, which transport blood away from the heart, possess thick, elastic walls which aid in the maintenance of the blood flow.

3. C — Between the two sides of the heart is a fibrous partition called the interatrial and interventricular septum.

4. A — Blood for the general circulation is returned to the right atria by way of 2 large veins, the superior and inferior vena cava.

5. D — The 3 types of muscle tissue differ in both their appearance and function. Though smooth muscle can provide only weak contractions, they can be sustained for indefinite periods of time. Skeletal muscles are extremely strong but tire easily and cannot provide the repetitive contraction required of the heart. The cardiac muscles possess moderate strength and short recovery time required for repetitive function.

6. B — Blood which normally travels through the heart in a single direction is assisted by the two atrioventricular valves between the upper and lower chambers and the semilunar valves which sit on top of each ventricle. All of these one way valves help to prevent the backward flow of blood through the heart.

7. A — The heart is encapsulated by a layer of fibrous pericardium which is connected to the mediastinum and helps prevent over distension. Two additional tissue layers called the serous pericardium consisting of visceral and parietal layers and the pericardial space found between the layers reduce the friction associated with the beating of the heart.

8. D — The ventricles are the pumping chambers of the heart. The smaller right ventricle pumps blood into the lungs while the stronger left ventricle provides the force required to distribute the blood to all of the other tissues of the body.

9. C — The poorly oxygenated blood from the vena cava is emptied into the right aorta. As this chamber contracts blood flows into the right ventricle for its transit into the pulmonary arteries and lungs.

10. B — The myocardium or heart muscle is supplied with oxygen rich blood by way of the first branch of the ascending aorta, the coronary artery.

See accompanying diagram.

11. B 12. D 13. E 14. F

15. A 16. C 17. G

18. D The efficient contraction of the chambers of the heart requires a precise timing and the efficient spread of the electrical impulses. Specialized muscle tissue found in the sinoatrial (SA) nodes, atrioventricular (AV) nodes, and bundle of His, generate the electrical signals that pass through the Purkinje fibers to the desired area of the heart muscle.

19. D The thin single cell thick walls of the capillaries are ideally suited for the exchange of respiratory gasses and nutrients between the blood cells and the individual cells of the body.

20. A The proximity of arteries to the heart will result in a high blood pressure measurements not apparent in the capillaries or veins.

21. D After the abdominal aorta descends through the diaphragm its first main branch is the celiac axis which provides oxygen rich blood to the liver (hepatic artery), spleen (splenic artery), and the stomach, gall bladder, and pancreas (gastric artery).

22. C The external and internal divisions of the common carotid artery divide at the level of the angle of the mandible (gonion). The external branches distribute oxygenated blood to the tongue, face, pharynx, scalp, and thyroid gland.

23. A Due to its importance in respiration, the diaphragm derives its blood supply from phrenic arteries originating from both the thoracic and abdominal aorta.

24. D After the external iliac artery descends through the pelvic cavity, at the inguinal ring it is renamed the femoral artery.

25. B The five paired branches of the abdominal aorta are the renal, suprarenal, testicular or ovarian, lumbar, and inferior phrenic arteries.

26. A The four main vessels which supply oxygenated blood to the brain are the right and left internal carotids and the right and left vertebral arteries.

27. A The major divisions of the aorta are the ascending, arch, thoracic, and abdominal. The ascending aorta comes from the left ventricle and ends in the curved aortic arch. As it descends through the chest cavity it becomes the thoracic aorta, and it is renamed the abdominal aorta below the diaphragm.

28. C The pulmonary trunk which arises from the right ventricle bifurcates into the right and left pulmonary arteries which penetrate the lung at its hilus. The pulmonary veins return blood to the left atrium of the heart.

See accompanying diagram.

29. C 30. E 31. G 32. A

33. D 34. B

35. C The abdominal aorta passes along the lumbar spine to an area just in front of the 4th lumbar vertebrae where it bifurcates into the right and left common iliac arteries.

36. B The second major branch off of the abdominal aorta, the superior mesenteric artery, supplies oxygenated blood to the pancreas, duodenum, jejunum, ileum, ascending, and transverse colon.

37. D The femoral artery descends down the medial and posterior side of the femur to the back of the knee where it is renamed the popliteal artery.

38. C Because both the ovaries and testes develop near the kidney, the branches of the abdominal aorta which supply these structures will also follow their descent into the pelvic cavity.

39. A The brain derives its blood from an anastomosis near the base of the skull called the circle of Willis. This structure is formed by right and left internal carotids and basilar artery which give rise to the anterior and posterior cerebral and communicating arteries.

40. B The majority of structures and muscle groups in the mid and lower pelvis receive much of their blood supply by way of the internal iliac arteries.

41. D Blood travels to the hand in the following sequence of arteries: subclavian, axillary, brachial, ulnar, and radial, palmer arch.

42. C The bifurcation of the common carotid artery into its internal and external branches occurs at the same level as the thyroid cartilage of the larynx.

43. A The large vessel originating at the left ventricle of the heart is the ascending aorta.

44. A The majority of the arterial blood entering the liver is supplied by the common hepatic artery, one of the main branches off of the celiac axis.

45. B The longest blood vessel of the body which begins in the foot and extends into the lower pelvis is called the great saphenous vein.

46. D The internal jugular vein drains venous blood from all of the cranial venous sinuses namely the superior and inferior sagittal, the straight, transverse, and sigmoid sinuses.

47. B The largest deep vein of the upper arm is the brachial vein. The 2 other major veins of the arm, the cephalic and basilic veins, are classified as superficial.

See accompanying diagram.

48. D 49. B 50. C 51. E

52. A

53. C The largest veins of the body, the vena cava, collect blood from the upper and lower portions of the body and return it to the right atria.

54. D To prevent high levels of sugars or proteins from gaining access to the general circulation, venous blood from the intestines (mesenteric vein), drain directly into the portal circulation associated with the liver.

55. C The blood vessels which provide oxygen and nutrients to their own walls are called the vaso vasorum.

56. D Venous blood which is being returned to the heart is normally associated with low blood pressure and low oxygen content and a high level of metabolic waste products.

57. D Because the fetal lungs are not functional until birth, blood is directed away from the lungs through an opening between the atria called the foramen ovale.

58. D The umbilical arteries located in the umbilical cord which direct fetal blood to the placenta arise from the internal iliac arteries of the pelvis.

59. A The exchange of gasses, nutrients and wastes between the fetal and maternal circulatory system occur across the membranes of the placenta.

60. B The low blood pressure in the subclavian veins provides a safe entry point for the thoracic ducts of the lymphatic system.

61. D The majority of the veins in the lower and upper limbs contain one-way valvular structures to help ensure that blood continues to flow back towards the heart.

62. A Venous and arterial blood differs in both content and appearance. The presence of oxygen in arterial blood gives this fluid a bright red tint. The lower oxygen content of venous blood is characterized by a deep red color.

63. B The intercellular matrix for the blood cells is a straw colored fluid containing many of the blood proteins called plasma.

64. C The most numerous type of blood cells with a count of between 3-5 million per cc are the erythrocytes or red blood cells.

65. D The platelets or thrombocytes play an essential role in repairing damaged vessels and preventing excessive blood loss through clotting and have a normal count of 300,000 per/cc.

66. C The reduction in the number of erythrocytes or hemoglobin may lead to a condition called anemia. Among the more common causes are poor nutrition, low vitamin B12 levels, bleeding disorders, and deficient bone marrow production.

67. B Many types of heart problems associated with the heart valves or defects in the septum create an abnormal sound called a heart murmur.

68. C One of the most common causes of a heart attack is a blocked coronary artery which reduces the blood supply to the heart muscle called a myocardial infarction.

69. A The abnormal dilatation of an artery due to a weakening in the wall of the vessel is termed an aneurysm.

70. C Tortuous veins resulting from excessive pressure, which are commonly found in the legs, are called varicose veins. Related condition occurring in the gastrointestinal tract are esophageal varices and hemorrhoids or piles of the rectum and anus.

Section VI
Radiographic Procedures

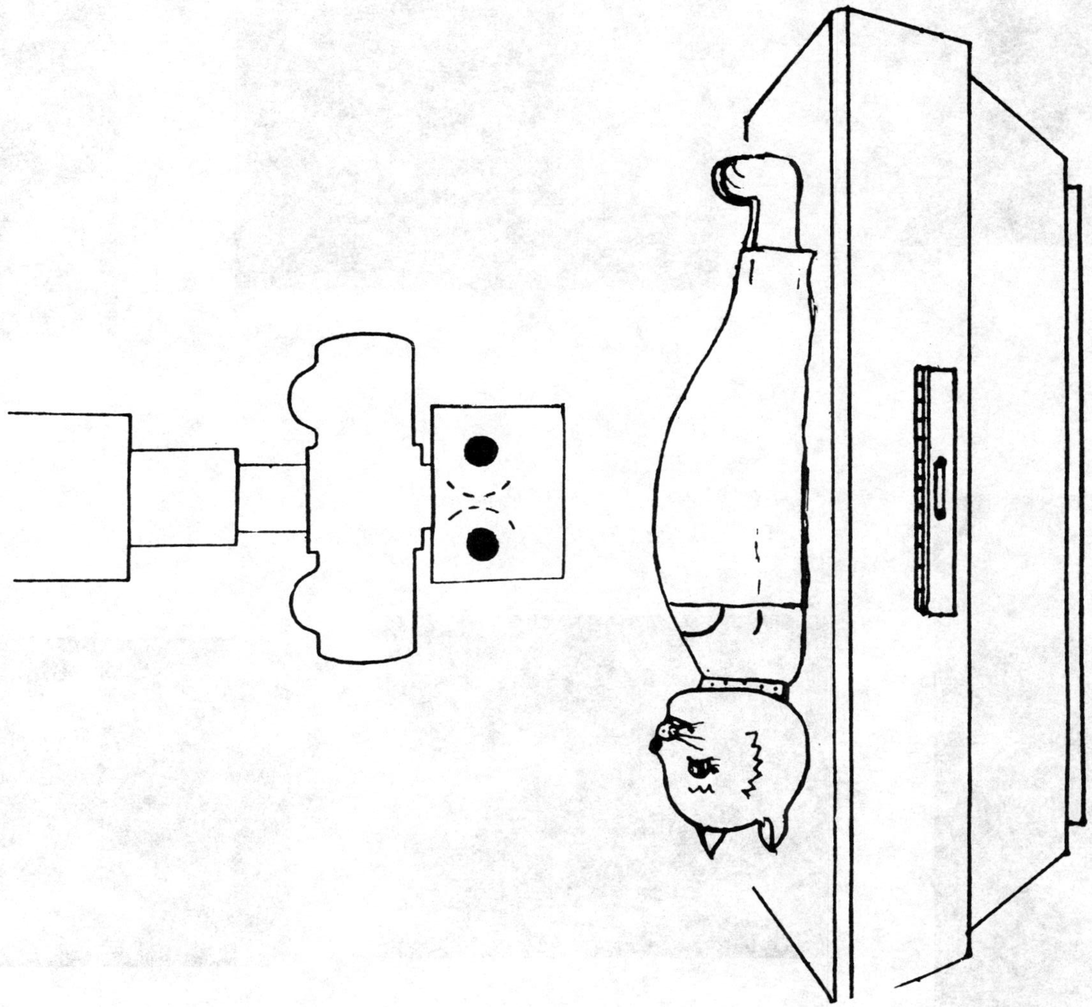

Radiographic Graffiti III

Can you identify what caused these common mistakes. Answers appear at the bottom of the page.

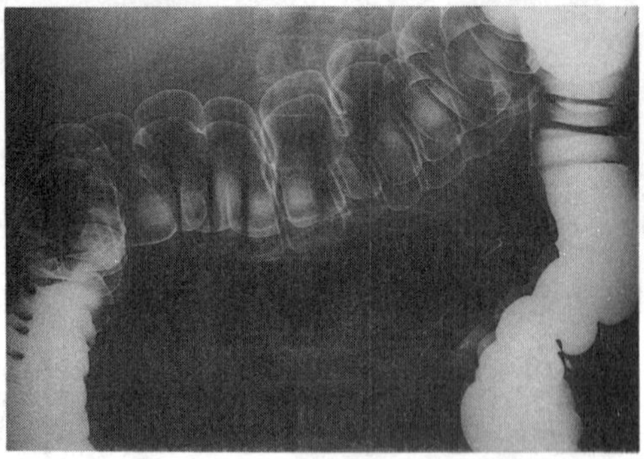

A

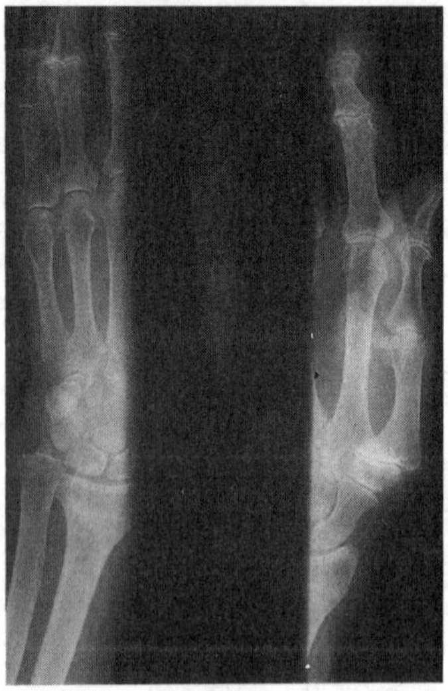

B

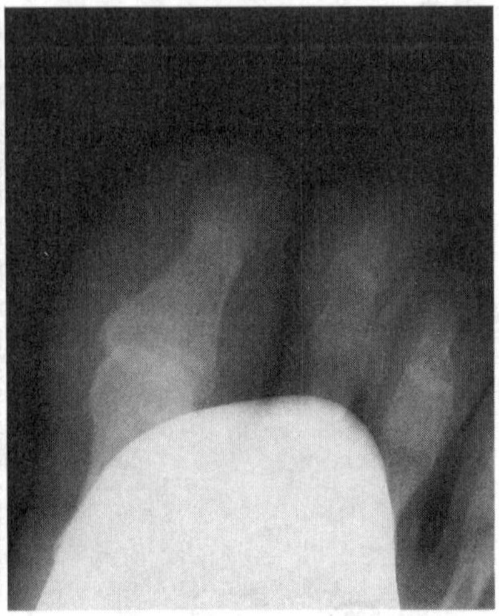

C

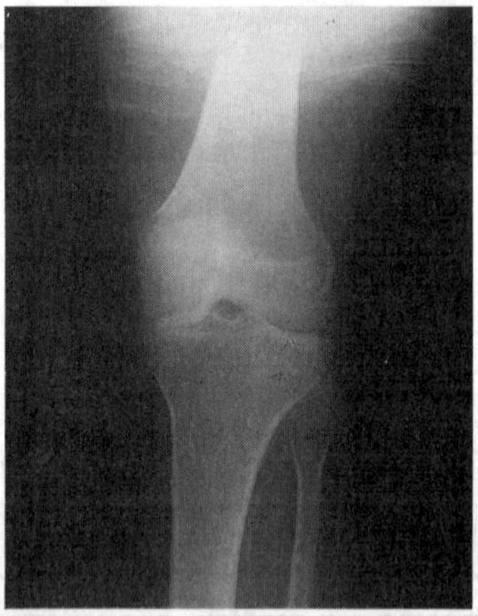

D

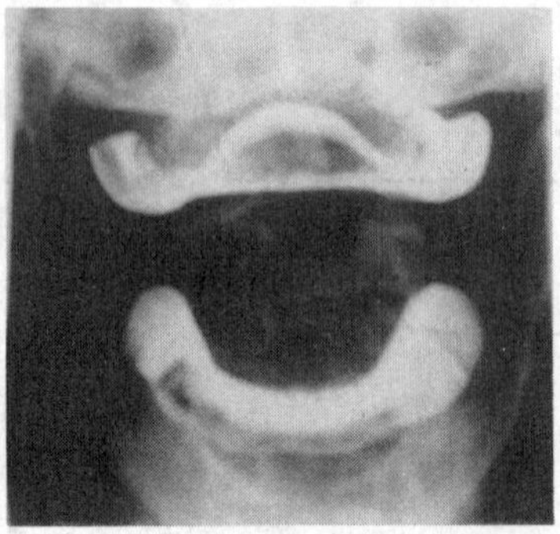

E

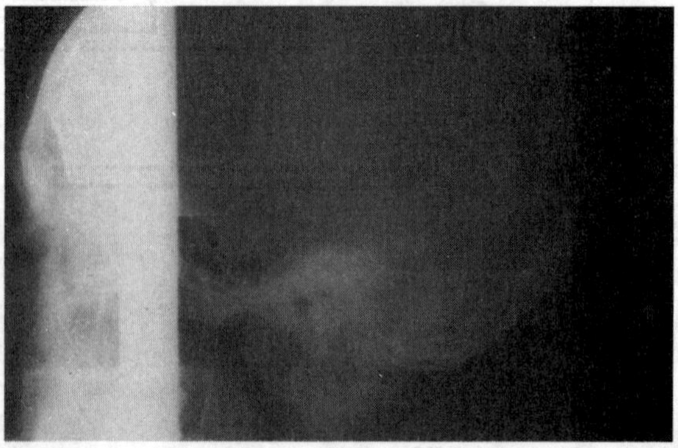

F

ilm A - Double exposed air contrast barium enema radiograph. Film B - Improper film matting. Film C - Improper positioning for sunrise patella: note magnified big toe. Film D - rolled up pants leg on knee radiograph. Film E - Dentures on open mouth view of he C-spine. Film F - Failure to push in bucky tray.

Positioning Terminology

Chapter 45

1. A plane that sections the body into upper and lower portions is called: 1:38:I

A. Sagittal plane
B. Decubitus plane
C. Coronal plane
D. Horizontal plane

2. The most common body habitus consisting of nearly 50% of the population is called the: 1:41:I

A. Hyposthenic type
B. Sthenic type
C. Hypersthenic type
D. Asthenic type

3. Which of the following terms refer to a body part or a central ray angulation directed toward the head? 1:49:I

1. Cranial *2. Cephalic* *3. Caudad*

A. 1 & 2 only
B. 1 & 3 only
C. 2 & 3 only
D. 1, 2 & 3

4. The body position defined as a person standing erect with palms facing forward is termed: 1:38:I

A. Fetal position
B. Recumbent position
C. Habital position
D. Anatomical position

5. A longitudinal angulation of the central ray with the long axis of the body describes a/an: 1:51:I

A. Axial projection
B. Transverse projection
C. Tangential projection
D. Horizontal projection

6. The ventral decubitus position is most similar to the ____________ position. 1:55:I

A. Prone
B. Supine
C. Oblique
D. Lateral

7. The general size, shape, and position of the internal organs can be, to a large degree, predicted by the classification of a person's: 1:41:I

A. Addison's types
B. Bodily habitus
C. Body cavities
D. Primary lobules

8. A vertical plane that divides the body into anterior and posterior portions is called a: 1:38:I

A. Horizontal plane
B. Sagittal plane
C. Coronal plane
D. Squamosal plane

9. The term used to designate a structure away from it's source is: 1:49:I

A. Visceral
B. Medial
C. Proximal
D. Distal

10. The turning of a body part inward is termed: 1:56:I

A. Inversion
B. Eversion
C. Extension
D. Flexion

11. Which of the following terms refers to the front surface of the body? 1:49:I
1. Anterior 2. Posterior 3. Ventral

A. 1 only
B. 3 only
C. 1 & 2 only
D. 1 & 3 only

12. A movement of a joint that increases the angle between the articulating bones is called: 1:56:I

A. Adduction
B. Abduction
C. Flexion
D. Extension

13. Which of the following organs is not located in the pelvic cavity: 1:38:I

A. Spleen
B. Rectum
C. Vagina
D. Urinary bladder

14. A projection in which visualization of a structure is improved by directing the central ray to skim the profiled body part is termed a/an: 1:51:I

A. Tangential position
B. Axial position
C. Decubitus position
D. Oblique position

15. The gonion corresponds to the level of the ____________ cervical vertebra. 1:40:I

A. First
B. Second
C. Fourth
D. Sixth

16. The portions of the body that are located away from the midline of the body are referred to as: 1:49:I

A. Caudal
B. Lateral
C. Cranial
D. Ventral

17. A very thin person will often have a low, vertical stomach and gallbladder. This person would be classified as having the ____________ body type. 1:41:I

A. Sthenic
B. Asthenic
C. Hypersthenic
D. Obesthenic

18. A physician has requested a tightly collimated image of the tenth thoracic vertebra. The external landmark that would be most helpful is the: 1:40:I

A. Costal margin
B. Xiphoid tip
C. Iliac crest
D. Sternal angle

19. During a ventral decubitus examination of the abdomen using a horizontal beam, free fluid will accumulate on the patient's: 1:55:I

A. Right side
B. Left side
C. Anterior surface
D. Posterior surface

20. The term 'projection' describes: 1:50:I

A. The image as seen from the film through the patient.
B. The path of the radiation from the x-ray tube.
C. Both of the above
D. Neither of the above

21. The movement of a structure toward the central axis of the body is termed: 1:56:I

A. Abduction
B. Adduction
C. Pronation
D. Supination

22. A right lateral decubitus radiograph is taken of a patient. This is an example of a/an: 1:55:I

A. Lateral position
B. Oblique position
C. Frontal projection
D. None of the above

23. The dorsum of the foot refers to the: 1:49:I

A. Superior aspect
B. Inferior aspect
C. Medial aspect
D. Lateral aspect

24. The most superior aspect of the iliac crest corresponds to the level of: 1:40:I

A. T-12
B. L-2
C. L-4
D. Sacral promontory

25. When the hand is rotated from anatomic position for a posteroanterior projection, this is termed: 1:56:I

A. Extension
B. Flexion
C. Pronation
D. Supination

26. A person with a large, high peripherally located colon, and high transverse gallbladder and stomach would be classified as: 1:41:I

A. Hypersthenic
B. Hyposthenic
C. Asthenic
D. Sthenic

27. The right anterior oblique position (RAO) is described as a/an ____________ position 1:50:I

A. Anteroposterior (AP) oblique
B. Lateral
C. Decubitus
D. Posteroanterior (PA) oblique

28. The body plane that passes vertically through the midline of the body and divides it into equal right and left portions is termed the: 1:38:I

A. Midcoronal plane
B. Midsagittal plane
C. Midtransverse plane
D. Midlambdoidal plane

29. The turning of the foot outward at the ankle would be an example of: 1:56:I

A. Pronation
B. Extension
C. Eversion
D. Inversion

30. Which of the following structures are found in the same transverse plane? 1:40:I
1. Ischial tuberosity *2. Symphysis pubis* *3. Greater trochanters*

A. 1 & 2 only
B. 1 & 3 only
C. 2 & 3 only
D. 1, 2 & 3

31. The anterior superior iliac spine (ASIS) is found at the same level as: 1:40:I

A. L-4
B. L-5
C. S-1
D. S-3

32. The division of the abdomen into four major quadrants involves a transverse and midsagittal plane which intersects at the: 1:39:I

A. Umbilicus
B. Sternal notch
C. Xiphoid tip
D. Inguinal fold

33. Which of the following can be used to describe a patient lying on his or her back? 1:50:I
1. Supine *2. Dorsal decubitus* *3. Ventral decubitus* *4. Prone*

A. 1 & 2 only
B. 1 & 3 only
C. 2 & 3 only
D. 1, 2 & 3

34. The torso is divided into two major subdivisions called the: 1:38:I

A. Epigastric and hypogastric regions
B. Hypochondral and ilac regions
C. Thoracic and abdominal cavities
D. Pleural and pelvic cavities

35. In the posteroanterior (PA) projection of the abdomen, the gallbladder of a sthenic patient is normally located at a level near the: 1:41:I

A. Sacral promontory
B. Dome of the diaphragm
C. Ninth coastal cartilage
D. Iliac crest

36. The term applied to a position in which the body is rotated so that it is not in a frontal or lateral position is called a/an: 1:50:I

A. Axial position
B. Tangential position
C. Decubitus position
D. Oblique position

37. A club-shaped process usually located at the end of a bone is termed a: 1:49:I

A. Fissure
B. Facet
C. Malleolus
D. Sulcus

38. Which of the following is not located within the pelvic cavity? 1:38:I

A. Bladder
B. Small intestine
C. Rectum
D. Reproductive organs

39. The right PA oblique position will correspond to a patient in the: 1:50:I

A. Right anterior oblique position (RAO)
B. Left anterior oblique position (LAO)
C. Right posterior oblique position (RPO)
D. Left posterior oblique position (LPO)

40. A hole in a bone through which vessels and nerves pass is called a/an: 1:44:I

A. Fovea
B. Fossa
C. Fistula
D. Foramen

Upper Extremity

Chapter 46

1. The degree of obliquity recommended for the hand is ____________ to the plane of the image receptor. 1:76:I

   A. 15 degrees
   B. 25 degrees
   C. 45 degrees
   D. 60 degrees

2. The best demonstration of the clavicle is obtained using a/an ____________ projection/position. 1:158:I

   A. Erect posteroanterior (PA)
   B. Erect oblique
   C. Recumbent lateral
   D. Recumbent oblique

3. The projection of the shoulder which will demonstrate it in true anatomic position is: 1:126:I

   A. Transthoracic lateral
   B. Anteroposterior (AP) internal rotation
   C. Anteroposterior (AP) external rotation
   D. Anteroposterior (AP) neutral rotation

4. To insure proper position of the elbow in the lateral projection, the: 1:102:I

   1. Epicondyles are perpendicular to the film. *2. Radial and ulnar styloids must be superimposed.*
   3. Forearm, elbow, and humerus are in the same plane.

   A. 1 & 2 only
   B. 1 & 3 only
   C. 2 & 3 only
   D. 1, 2 & 3

5. In the posteroanterior (PA) projection of the carpals, the fingers are flexed in order to: 1:83:I

   A. Decrease carpal rotation
   B. Decrease structure to image receptor distance
   C. Open the carpal interspaces
   D. All of the above

6. When bilateral hands or wrists are requested, it is important to radiograph each side separately to prevent:
   1. Lowering patient exposure *2. Distortion of the phalanges* *3. Distortion of the joint spaces* 1:57:I

   A. 1 only
   B. 2 only
   C. 3 only
   D. 1, 2 & 3

7. Radiographs of the acromioclavicular articulations should be performed: 1:152:I
   1. Decubitus *2. Weight-bearing* *3. Bilaterally*

   A. 1 & 2 only
   B. 1 & 3 only
   C. 2 & 3 only
   D. 1, 2 & 3

8. The anteroposterior (AP) internal rotation of the shoulder places the humerus in the: 1:127:I

   A. Anatomic position
   B. Anteroposterior (AP) projection
   C. Lateral position
   D. 45° oblique position

9. The anteroposterior (AP) oblique position of the wrist is used to demonstrate: 1:88:I

   A. Greater + lesser multangular
   B. Hamate + pisiform
   C. Navicular + lunate
   D. More than one but not all the above

10. To avoid crossing the radius and ulna during an anteroposterior (AP) projection of the forearm, the hand should be: 1:98:I

A. Hyperextended
B. Pronated
C. Supinated
D. True lateral

11. The best demonstration of the carpal interspaces is seen in the ____________ projection (position). 1:84:I

A. Anteroposterior (AP)
B. Medial oblique
C. Posteroanterior (PA)
D. Lateral oblique

12.-20. Pertaining to the radiograph of the hand, place the letter that corresponds to the following structures: 1:63:I

12. ______ Navicular bone
13. ______ Proximal interphalangeal joint
14. ______ Head of the metacarpal bone
15. ______ Styloid process of the ulna
16. ______ Base of the proximal phalanx
17. ______ Metacarpophalangeal joint
18. ______ Middle phalanx
19. ______ Distal interphalangeal joint
20. ______ Head of the proximal phalanx

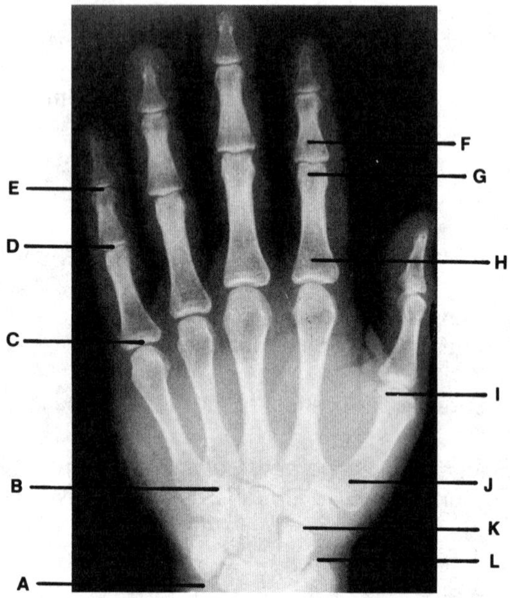

21. The proper centering point for a finger is the: 1:66:I

A. Third metacarpophalangeal joint
B. Middle interphalangeal joint
C. Proximal interphalangeal joint
D. None of the above

22. In evaluating a radiograph of a true lateral scapula, the two structures that should be exactly superimposed are: 1:167:I

A. Vertebral and axillary borders
B. Acromion and coracoid process
C. Scapular body and axillary ribs
D. Humeral head and glenoid fossa

23. When radiographing the proximal humerus using the transthoracic (Lawrence) method, the plane of the epicondyles should be ____________ to the plane of the image receptor. 1:118:I

A. Parallel
B. Perpendicular
C. 45 degrees
D. None of the above

24. A bony outgrowth at the base of the third metacarpocarpal joint (carpe bossu) is best demonstrated using: 1:86:I

A. Acute dorsiflexion
B. Ulnar deviation
C. Acute palmar flexion
D. Radial deviation

25. Which of the following positions will best demonstrate the olecranon process? 1:102:I

A. Anteroposterior (AP)
B. Internal oblique
C. External oblique
D. Lateral

26. When performing a posteroanterior (PA) projection of the hand, the ____________ surface should be in contact with the image receptor. 1:74:I

1. Dorsal *2. Palmar* *3. Ventral*

A. 1 & 2 only
B. 1 & 3 only
C. 2 & 3 only
D. 1, 2 & 3

27. The patient is placed supine with the arm abducted to a right angle with the body. The elbow is flexed. This describes the ____________ of the scapula. 1:164:I

A. Anteroposterior(AP) projection
B. Oblique (Lilienfeld) position
C. Posteroanterior (PA) projection
D. Lateral position

28. The position most commonly recommended for foreign body localization of the hand is a/an: 1:79:I

A. Lateral in flexion
B. Lateral in extension
C. Oblique in flexion
D. Oblique in extension

29. The recommended breathing instruction for the anteroposterior (AP) projection of the scapula is: 1:148:I

A. Full inspiration
B. Full expiration
C. Suspended respiration
D. Quiet breathing

30. In the lateral position for the humerus (non-trauma), the arm is: 1:115:I

A. Neutrally rotated
B. Counter rotated
C. Internally rotated
D. Externally rotated

31. A patient's hand is placed on a cassette with the palmar surface down. It is then dorsiflexed with a slight radial shift. The central ray is directed 25° to a point 1" distal to the base of the third metacarpal. This would demonstrate: 1:94:I

A. Carpal canal
B. Carpe bossu
C. Carpal bridge
D. The navicular bone

32.-37. Pertaining to the radiographs of the elbow place the letter that corresponds to the following structures: 1:102:I

32. ______ Radial head
33. ______ Olecranon process
34. ______ Medial epiconclyle
35. ______ Capitulum
36. ______ Semilunar notch
37. ______ Ulna

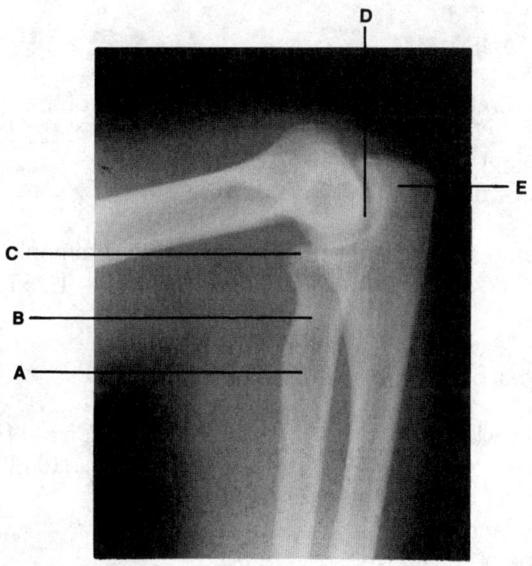

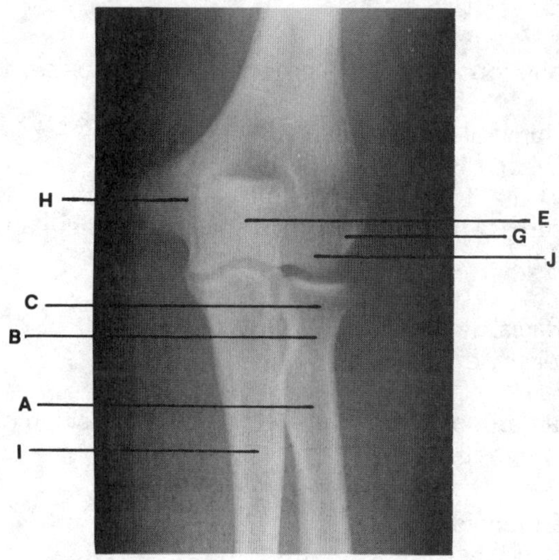

38. When radiographing an elbow in the lateral position for the demonstration of the soft tissue structures, the elbow should be flexed: 1:103:I

A. 45 degrees
B. 90 degrees
C. 60 degrees
D. 35 degrees

39. In the anteroposterior (AP) projection of the first finger, the hand is rotated to a/an: 1:72:I

A. Anteroposterior (AP) projection
B. Posteroanterior (PA) projection
C. 45° oblique position
D. Lateral position

40. For the lateral position of the scapula, the upper part of the scapula is demonstrated as the letter Y. The superior aspect of the Y is formed by: 1:166:I

A. Glenoid and humeral head
B. Scapular spine and vertebral border
C. Acromion and coracoid process
D. Vertebral border and axillary border

41. The humerus is positioned for a true anteroposterior (AP) projection when the epicondyles are ________ to the plane of the image receptor, and the ________ is seen in profile. 1:115:I

A. Perpendicular, greater tuberosity
B. Parallel, lesser tuberosity
C. Perpendicular, lesser tuberosity
D. Parallel, greater tuberosity

42. The carpal interspaces on the medial side of the wrist can be best demonstrated using: 1:90:I
1. Ulnar flexion *2. Radial deviation* *3. Radial flexion*

A. 1 only
B. 2 only
C. 3 only
D. 1, 2 & 3

43. The degree of obliquity utilized for the anteroposterior external oblique position for the elbow is: 1:105:I

A. 15 degrees
B. 25 degrees
C. 40 degrees
D. None of the above

44. The proper centering point for a posteroanterior (PA) projection of the hand is the: 1:74:I

A. Third metacarpocarpal joint
B. Third proximal interphalangeal joint
C. Third metacarpophalangeal joint
D. Midcarpal region

45. For the posteroanterior axial position of the clavicle, the central ray is directed: 1:158:I

A. 25° caudad
B. 25° cephalic
C. 10° caudad
D. 10° cephalic

46. The anteroposterior (AP) projection of the shoulder in internal rotation will demonstrate a profile view of the: 1:127:I

A. Humeral epicondyles
B. Greater tuberosity
C. Glenoid fossa
D. Lesser tuberosity

47. Fractures of the dorsal aspect of the carpal bones can be well demonstrated in the tangential position commonly called the: 1:95:I

A. Radial flexion
B. Ulnar flexion
C. Carpal canal
D. Carpal bridge

48. The anteroposterior (AP) internal oblique position of the elbow will demonstrate the ________ free of superimposition. 1:104:I

A. Coranoid process
B. Radial head
C. Olecranon process
D. None of the above

49. The radial shift of the carpal canal position is useful for the demonstration of the: 1:82:I

1. Pisiform   2. Greater multangular   3. Hamate

A. 1 only
B. 2 only
C. 3 only
D. 1, 2 & 3

50. The patient is supine and the central ray is directed through the posterosuperior region of the shoulder at a caudal angle of 45°. This will demonstrate: 1:174:I

A. Acromioclavicular articulation
B. Glenoid fossa
C. Coracoid process
D. Scapular spine

51.-60. Pertaining to the radiograph of the shoulder place the letter that corresponds to the following structures: 1:165:I

51. ______ Humeral head
52. ______ Inferior angle of the scapula
53. ______ Lesser tuberosity
54. ______ Acromion process
55. ______ Surgical neck of the humerus
56. ______ Axillary border of the scapula
57. ______ Clavicle
58. ______ Greater tuberosity
59. ______ Glenoid fossa
60. ______ Coracoid process

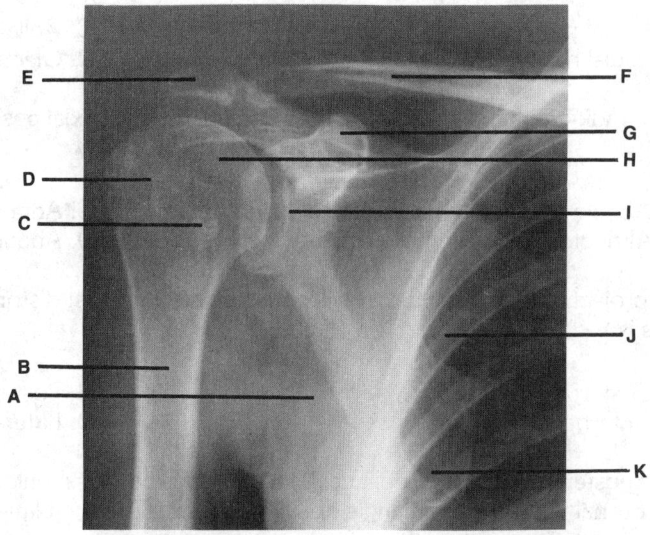

61. To demonstrate the coracoid process using the anteroposterior axial position, the central ray must be directed: 1:172:I

A. Perpendicular to the shoulder
B. 15° - 45° cephalic
C. 15° - 45° caudad
D. 55° caudad

62. In the inferosuperior position (Carpal Canal) of the wrist (Gaynor Hart), the only carpal which is demonstrated free of any superimposition is the: 1:94:I

A. Navicular
B. Capitate
C. Hamate
D. Pisiform

63. For the demonstration of the entire circumference of the radial head, how many lateral positions are necessary? 1:110:I

A. Four
B. Three
C. Two
D. One

64. The early radiologic changes of rheumatoid arthritis can best be demonstrated by radiographing the hands using the: 1:81:I

A. Anteroposterior (AP) oblique position
B. Posteroanterior (PA) projection
C. Anteroposterior (AP) projection
D. Lateral in flexion

65. The Lorenz and Lilienfeld methods are used to demonstrate: 1:169:I

1. Scapular spine   2. Posteroanterior (PA) oblique scapula   3. Acromion

A. 1 only
B. 2 only
C. 3 only
D. 1, 2 & 3

66. In order to visualize the glenoid fossa (Grashey method), the humerus is internally rotated and the patient is obliqued to form an angle of ____________ to the plane of the image receptor. 1:144:I

A. 15 degrees
B. 25 degrees
C. 45 degrees
D. 60 degrees

67. All of the following methods may be used to demonstrate the navicular, except: 1:91:I

A. Wrist PA, fist clenched
B. PA wrist in ulnar flexion
C. PA wrist, central ray 20° to the elbow
D. PA wrist in radial flexion

68. The forearm is placed on the film in acute flexion with the central ray directed perpendicular to the forearm. This position is used to demonstrate the: 1:113:I

A. Radial notch
B. Radial head
C. Collateral ligament tears
D. Olecranon process

69. When radiographing the shoulder for the inferosuperior axial position (Lawrence method), the humerus should be: 1:135:I

A. Abducted 90°, internally rotated
B. Abducted 180°, internally rotated
C. Adducted 90°, externally rotated
D. Abducted 90°, externally rotated

70. The projection of the hand that will demonstrate the most distortion of the phalanges and the interphalangeal joints is: 1:76:I

A. Posteroanterior (PA)
B. Anteroposterior(AP)
C. Posteroanterior (PA) oblique
D. Lateral

71. In the posteroanterior (PA) oblique (Y) scapula position for shoulder dislocations, the patient is rotated so that the midcoronal plane forms an angle of ____________ to the plane of the image receptor. 1:143:I

A. 5-10 degrees
B. 15-20 degrees
C. 25-35 degrees
D. 45-60 degrees

72. The maximum relaxation of the fat pads of the elbow is seen with the joint flexed: 1:102:I

A. 45 degrees
B. 30 degrees
C. 90 degrees
D. None of the above

73. For tangential positions of the bicipital groove, the central ray is directed to skim the ____________ aspect of the shoulder. 1:148:I

A. Medial
B. Anterior
C. Lateral
D. Posterior

74. A posteroanterior (PA) projection of the hand will demonstrate a/an ____________ of the thumb. 1:74:I

A. Lateral position
B. Frontal position
C. Anteroposterior position
D. PA oblique

75. For the inferosuperior axial (West Point) projection of the shoulder, the central ray is directed: 1:134:I

A. 25° anteriorly, 25° medially
B. 10° anteriorly, 25° medially
C. 25° posteriorly, 25° laterally
D. None of the above

Pertaining to the radiographic image, answer questions 76 and 77.

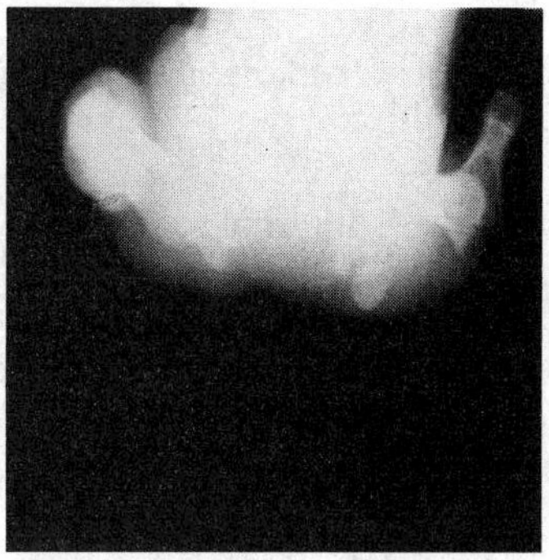

76. This tangential position of the wrist is most commonly referred to as the: 1:94:I

A. Carpal canal
B. Carpal bridge
C. Radial flexion
D. Ulnar flexion

77. The structures which are best shown in this position are the: 1:94:I

1. Pisiform *2. Styloid process* *3. Hook of the hamate*

A. 1 & 2 only
B. 1 & 3 only
C. 2 & 3 only
D. 1, 2 & 3

Pertaining to the radiographic image, answer questions 78 and 79.

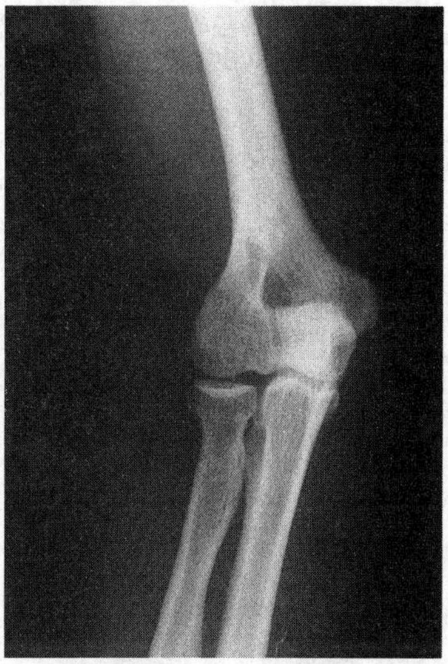

78. This position was obtained with the elbow placed in (the): 1:105:I

A. Medial oblique position
B. Lateral oblique position
C. Acute flexion
D. None of the above

79. The structure that is best demonstrated free of superimposition, is the: 1:105:I

A. Semilunar notch
B. Olecranon process
C. Trochlea
D. Radial head

Pertaining to the radiographic image, answer questions 80 and 81.

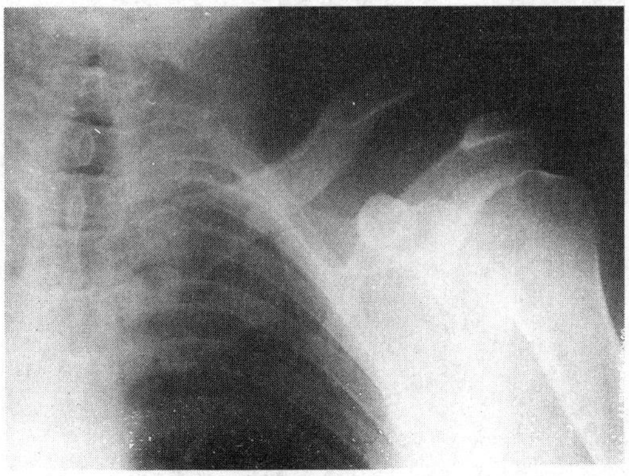

80. The projection shown is used for the demonstration of the: 1:152:I

1. Acromioclavicular articulations *2. Sternoclavicular articulations* *3. Costovertebral articulations*

A. 1 only
B. 2 only
C. 3 only
D. 1, 2 & 3

81. In order to maximize visualization of the dislocation shown, the radiograph is taken with the patient: 1:152:I

A. Supine without stress
B. Erect without stress
C. Supine with weights
D. Erect with weights

Pertaining to the radiographic image, answer questions 82 and 83.

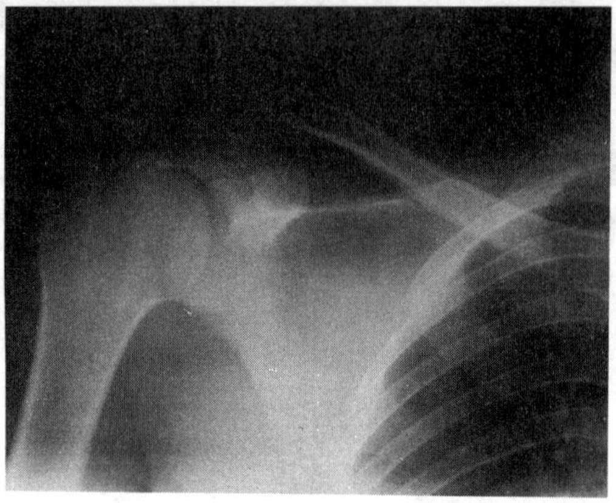

82. The radiographic image shows an AP projection of the shoulder in the: 1:126:I

A. Internal rotation
B. Neutral rotation
C. External rotation
D. Abduction

83. This projection is most commonly employed to show the ______ in profile. 1:126:I

A. Scapular spine
B. Coracoid process
C. Coronoid process
D. Greater tuberosity

Pertaining to the radiographic image, answer questions 84 and 85.

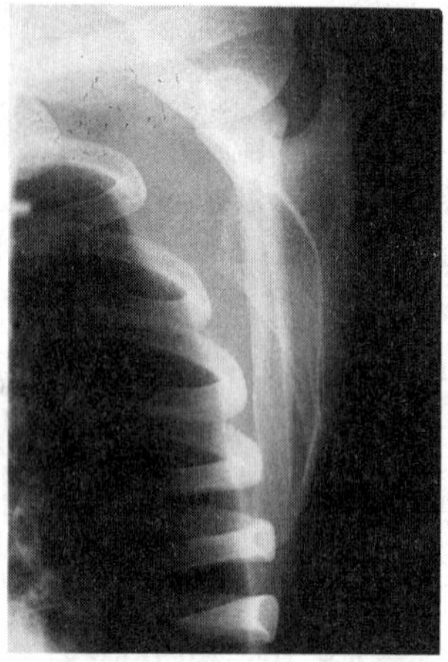

84. This position is principally used in the evaluation of the: 1:166:I

A. Axillary border of the ribs
B. Glenohumeral joint
C. Coracoid process
D. Scapula

85. This position demonstrates the scapula in the: 1:166:I

A. AP position
B. Lateral position
C. Oblique position
D. None of the above

Lower Extremities

Chapter 47

1. A true lateral position for the foot is more consistently obtained with the __________ surface in contact with the image receptor. 1:205:I

   A. Medial
   B. Lateral
   C. Plantar
   D. Dorsal

2. In order to relax the muscles and demonstrate the maximum volume of the knee joint for the lateral position, __________ flexion is used. 1:242:I

   A. 20° - 30°
   B. 45°
   C. 60°
   D. 70°- 90°

3. In order to reduce the normal anteversion of the femoral neck when performing the anteroposterior (AP) projection of the pelvis, the: 1:276:I

   A. Central ray is directed 15° caudad
   B. Central ray is directed 15° cephalic
   C. Toes are inverted 15°
   D. Toes are everted 15°

4. In order to prevent the joints of the lower leg from being projected off the film by the divergence of the central ray, the image receptor should extend __________ beyond each joint. 1:234:I

   A. 13cm (5")
   B. 8cm (3")
   C. 4cm (1 1/2")
   D. None of the above

5. For the lateral position of the calcaneus, the central ray is directed: 1:218:I

   A. 1" proximal to the medial malleolus
   B. 1" distal to the medial malleolus
   C. 1" posterior to the medial malleolus
   D. 1" anterior to the medial malleolus

6. Which of the following would not be a suitable method for examining a possible hip with a suspected fracture? 1:287:I

   1. Axiolateral Danelius-Miller *2. AP axial Cleaves* *3. Lateral Lauenstein-Hickey*

   A. 1 & 2 only
   B. 1 & 3 only
   C. 2 & 3 only
   D. 1, 2 & 3

7. In the oblique plantodorsal position for the foot, a 30° medial rotation is used to demonstrate: 1:204:I

   1. First and second metatarsal bases *2. Fourth and fifth metatarsal bases* *3. Cuboid bone*

   A. 1 & 2 only
   B. 1 & 3 only
   C. 2 & 3 only
   D. 1, 2 & 3

8. When radiographing the knee for the posteroanterior (PA) projection, the central ray is directed __________ due to the slightly inclined position of the lower leg. 1:242:I

   A. 5° caudad
   B. 7° cephalic
   C. 15° caudad
   D. Perpendicular to the film

9. For the anteroposterior (AP) axial position for the femoral neck (Cleaves), the optimum visualization will be obtained if the patient's thighs are abducted ____________ off the vertical 1:283:I

A. 15 degrees
B. 40 degrees
C. 60 degrees
D. 90 degrees

10. If the patient's foot is not dorsiflexed for the mediolateral lateral position of the ankle, the: 1:228:I

A. Toes will obscure the joint
B. Cuboid will obscure the joint
C. Ankle will be rotated
D. None of the above

11. The Settegast method tangential position of the patella is used for the evaluation of: 1:265:I

A. Vertical fractures of the patella
B. Femoropatellar articulation
C. Both of the above
D. Neither of the above

12. The axial position of the pelvis (Chassard-Lapine) is used to demonstrate: 1:280:I
1. Transverse pelvic outlet *2. Urinary bladder* *3. Rectosigmoid area*

A. 1 & 2 only
B. 1 & 3 only
C. 2 & 3 only
D. 1, 2 & 3

13-15 Pertaining to the radiograph of the ankle, place the letter that corrisponds to the following structures:

13. ________ Lateral malleoulus

14. ________ Talus (Astragulus) bone

15. ________ Tibia

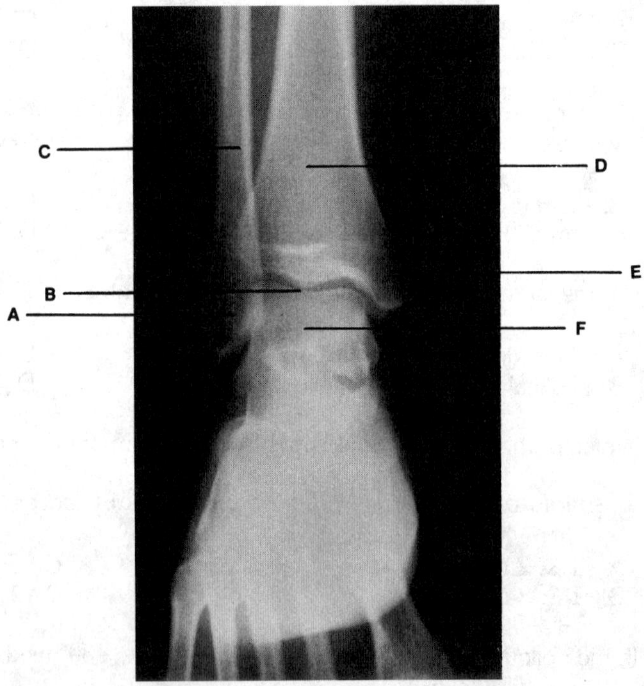

16. The lateral oblique position for the foot will best demonstrate: 1:201:I

A. Medial and intermediate cuneiforms
B. Astragulus
C. Cuboid
D. Calcaneus

17. Which of the following will prevent the medial femoral condyle from obscuring the joint in the lateral position for the knee? 1:242:I
1. Proper degree of flexion *2. True lateral positioning* *3. Angulation of the central ray*

A. 1 only
B. 2 only
C. 3 only
D. 1, 2 & 3

18. In the axiolateral position of the hip (shoot-through), it is important to place the grid strips: 1:291:I
1. Parallel to the femoral neck *2. Perpendicular to the femoral neck* *3. Perpendicular to the table*

A. 1 only
B. 2 only
C. 3 only
D. 1, 2 & 3

19. In the medial oblique position the opening of the joint space is maximized by rotating the ankle to place the malleoli ____________ to the plane of the image receptor: 1:231:I

A. Parallel
B. Perpendicular
C. 45°
D. 25°

20. The axial plantodorsal position for the calcaneus utilizes an angle of ____________ to the long axis of the foot. 1:215:I

A. 15°
B. 25°
C. 40°
D. 60°

21. The anteroposterior (AP) axial position (Cleaves) can be used to demonstrate: 1:282:I
1. Pelvic inlet and outlet *2. Symphysis pubis* *3. Femoral head and neck*

A. 1 only
B. 2 only
C. 3 only
D. 1, 2 & 3

22. The best visualization of the cuboid and it's articulations is obtained in the ____________ of the foot. 1:201:I

A. Anteroposterior (AP) projection
B. Plantodorsal position
C. AP medial oblique position
D. AP lateral oblique position

23. The tangential positions of the patella should only be taken when ____________ has/have been ruled out. 1:264:I

A. Meniscus damage
B. Transverse fractures
C. Arthritis
D. Ligament damage

24. The proper centering for the anteroposterior (AP) projection of the hip requires the centering of the sagittal plane passing 5cm medial to the: 1:286:I

A. Acetabulum
B. Iliac crest
C. Anterior superior iliac spine (ASIS)
D. Greater trochanter

25. The position of the ankle which will place the joint closest to the image receptor requires that the ____________ surface be in contact with the image receptor. 1:228:I

A. Medial
B. Anterior
C. Lateral
D. Posterior

26. For a dorsoplantar position for the third toe, the central ray should be directed to the: 1:187:I

A. Second tarsometatarsal joint
B. Third tarsometatarsal joint
C. Third metatarsophalangeal joint
D. Third proximal interphalangeal joint

27. A patient is brought into the radiography department to rule out a trimalleolar fracture. This will require the routine positions of the: 1:231:I

A. Proximal femur
B. Ankle joint
C. Distal femur
D. Calcaneous

28. In the dorsoplantar oblique positions for the foot, the sole of the foot is rotated to form an angle of ____________ to the plane of the image receptor. 1:201:I

A. 15°
B. 30°
C. 45°
D. 60°

29. The axial posteroanterior (PA) position (Camp-Coventry) is used in the visualization of the: 1:252:I

A. Intercondyloid fossa
B. Popliteal fossa
C. Proximal talofibular articulation
D. Femoropatellar articulation

30.-36. Pertaining to the radiograph of the foot, place the letter that corresponds to the following structures:

30. ________ Navicular (scaphoid) bone
31. ________ Metatarsophalangeal joint
32. ________ Head of the metatarsal bone
33. ________ Calcaneus
34. ________ Interphalangeal joint
35. ________ Cuboid bone
36. ________ Talus (astragulus)

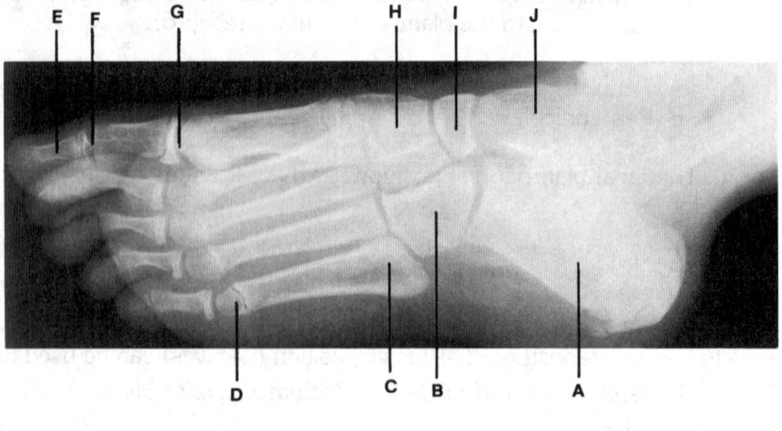

37. The dome of the acetabular cavity can be localized by finding the mid-point of a line drawn between the symphysis pubis and the: 1:274:I

A. Greater trochanter
B. Sacral promontory
C. Ischial tuberosity
D. Anterior superior iliac spine (ASIS)

38. In the axial plantodorsal position of the os calcis, the central ray is directed to enter at the level of: 1:215:I

A. Base of the fifth metatarsal
B. Base of the third metatarsal
C. Head of the fifth metatarsal
D. Head of the third metatarsal

39. In the medial oblique position for the ankle for an open mortise joint, the ankle is rotated: 1:231:I

A. 12 degrees
B. 20 degrees
C. 35 degrees
D. 45 degrees

40. During radiography of the anterior pelvic bones magnification is reduced by employing the: 1:304:I

A. AP projection
B. PA projection
C. Oblique position
D. Lateral position

41. For the AP axial position (Taylor) method of evaluating the anterior pelvic bones, the patient is supine and the central ray is directed: 1:305:I

A. 25° - 45° cephalic
B. 15° caudad
C. 25° caudad
D. Perpendicular to the symphysis

42. For an anteroposterior (AP) projection of the foot, the central ray is directed to pass through the: 1:198:I

A. Base of the third metatarsal
B. Third metatarsophalangeal joint
C. Head of the third metatarsal
D. Third tarsometatarsal joint

43. In the anteroposterior (AP) projection of the knee, the joint space is located 1.5cm distal to the ____________ of the patella. 1:240:I

A. Medial border
B. Lateral border
C. Base
D. Apex

44. In order to obtain a true anteroposterior (AP) projection of the femur, the: 1:286:I

A. Central ray is directed 15° caudad
B. Central ray is directed 15° cephalic
C. Toes are everted 15°
D. Toes are inverted 15°

45. A frontal projection of the ankle with inversion stress is employed for the evaluation of the: 1:233:I

A. Medial collateral ligament
B. Lateral collateral ligament
C. Medial meniscus
D. Lateral meniscus

46. When performing a dorsoplantar position of the toes, the best delineation of the joint spaces is accomplished by directing the central ray: 1:187:I

A. 15° cephalic
B. 15° caudad
C. 45° cephalic
D. 45° caudad

47. When a patient has bilateral hip fractures and the shoot-through lateral method cannot be used, Clements-Nakayama recommends a modification utilizing a central ray angulation of: 1:292:I

A. 15° posteriorly
B. 35° posteriorly
C. 25° anteriorly
D. 45° anteriorly

48. To demonstrate the tarsal articulations in the dorsoplantar projection of the foot, the central ray is directed: 1:199:I

A. 10° Posteriorly
B. 10° Anteriorly
C. Perpendicular to the image receptor
D. None of the above

49. The tangential position of the patella (Hughston) employs a 50° -60° flexion of the affected knee, and the central ray is directed: 1:261:I

A. 5° caudad
B. 15° cephalic
C. 30° caudad
D. 45° cephalic

50. In the inferosuperior position of the hip (shoot-through), the bony structure which will be seen directly posterior to the femoral neck and acetabulum is the: 1:291:I

A. Ischial tuberosity
B. Anterior superior iliac spine (ASIS)
C. Symphysis pubis
D. Lesser trochanter

51. Tangential positions (Lewis and Causton) of the foot are employed for the evaluation of: 1:194:I

A. Sesamoid bones
B. Subtalar joint
C. Calcaneus
D. Longitudinal arch

52. The patient is supine or prone; the foot is dorsiflexed and the central ray is directed vertically to the second metatarsophalangeal joint. This would demonstrate the: 1:194:I

A. Ankle mortise
B. Calcaneus
C. Lateral arch
D. Sesamoids

53. The oblique positions for the knee that will demonstrate the proximal tibiofibular articulation free from overlap is: 1:247:I

A. AP lateral oblique
B. AP medial oblique
C. PA medial oblique
D. More than one but not all of the above

54. For a follow-up examination for a postoperative fractured hip, which of the following would be true? 1:287:I

1. The AP axial (Cleaves may be used) *2. The entire prosthesis must be included*
3. Only a frontal projection is required

A. 1 only
B. 2 only
C. 3 only
D. 1, 2 & 3

55.-64. Pertaining to the radiography of the knee, place the letter that corresponds to the following structures:

55. ________ Head of the fibula

56. ________ Medial condyle of the femur

57. ________ Tibial tuberosity

58. ________ Intercondyloid spaces

59. ________ Lateral condyle of the femur

60. ________ Intercondyloid eminences (tibial spines)

61. ________ Medial tibial plateau

62. ________ Lateral epicondyle of femur

63. ________ Lateral tibial plateau

64. ________ Medial epicondyle of the femur

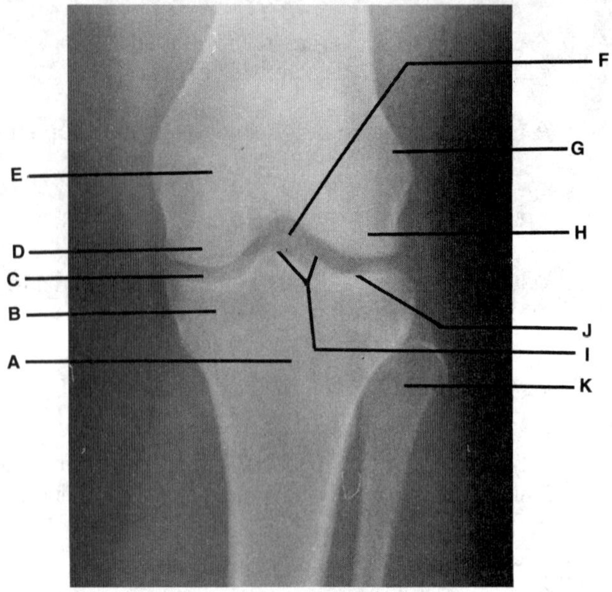

65. In the weight-bearing anteroposterior (AP) projection of arthritic knees, the central ray is directed: 1:244:I

A. 5°-7° cephalic
B. 5°-7° caudad
C. Perpendicular to the joint
D. None of the above

66. The Kite positions are a series of projections recommended for the evaluation of: 1:212:I

A. Arthritis
B. Talipes equinavarous
C. Coxa valga
D. Osgood Slatter's syndrome

67. For adequate demonstration of congenital dislocations of the hip, two anteroposterior (AP) projections are recommended. The first utilizes a vertical central ray, and the second utilizes a ____________ central ray. 1:278:I

A. 25° cephalic
B. 30° caudad
C. 15° caudad
D. 45° cephalic

68. In the anteroposterior (AP) projection of the ankle, the central ray is directed: 1:227:I

A. 15° cephalic - midmalleoli
B. Vertically - midmalleoli
C. 15° cepahlic - 1" distal lateral malleolus
D. Vertically - 1" distal lateral malleolus

69. The Broden, Isherwood, and Feist-Mankin oblique positions are useful for the demonstration of: 1:221:I

A. The subtalar joint
B. Congenital clubfoot
C. Intercondyloid fossa
D. Pelvic outlet

70. In the semi-axial anteroposterior (AP) position (Beclerè) for the knee, the central ray is directed at a right angle to the: 1:256:I

A. Femoropatellar articulation
B. Long axis of the femur
C. Base of the patella
D. None of the above

71. In order to demonstrate the ilium with its broad surface parallel to the image receptor, the unaffected side is elevated: 1:308:I

A. 15° degrees
B. 25° degrees
C. 40° degrees
D. 70° degrees

72. In the posteroanterior (PA) axial position (Holmblad) of the intercondyloid fossa, if the long axis of the femur forms an angle of 70° to the plane of the image receptor, the central ray is directed: 1:250:I

A. 20° caudad
B. 20° cephalic
C. 45° cephalic
D. Vertical to the film

73. Weight-bearing lateral positions for the feet are used to evaluate the status of the: 1:208:I

A. Longitudinal arch
B. Transverse arch
C. Subtalar articulation
D. Sesamoid bones

74. The method in which two exposures demonstrate a frontal projection of the entire foot using the masking effect of the leg is called the: 1:210:I

A. Lewis method
B. Grashey method
C. Weight-bearing lateral method
D. Composite dorsoplantar

75. For the orthoroentgenographic measurement of both lower extremities, the ankles should be: 1:482:I

A. Everted 15°
B. As close together as possible
C. 5"-6" apart
D. None of the above

76.-85. Pertaining to the radiograph of the pelvis, place the letter that corresponds to the following structures:

76. ________ Femoral head

77. ________ Sacrum

78. ________ Ischial tuberosity

79. ________ Anterior superior iliac spine (ASIS)

80. ________ Greater trochanter

81. ________ Sacroiliac joint

82. ________ Pubic symphsis

83. ________ Iliac crest

84. ________ Lesser trochanter

85. ________ Obturator foramen

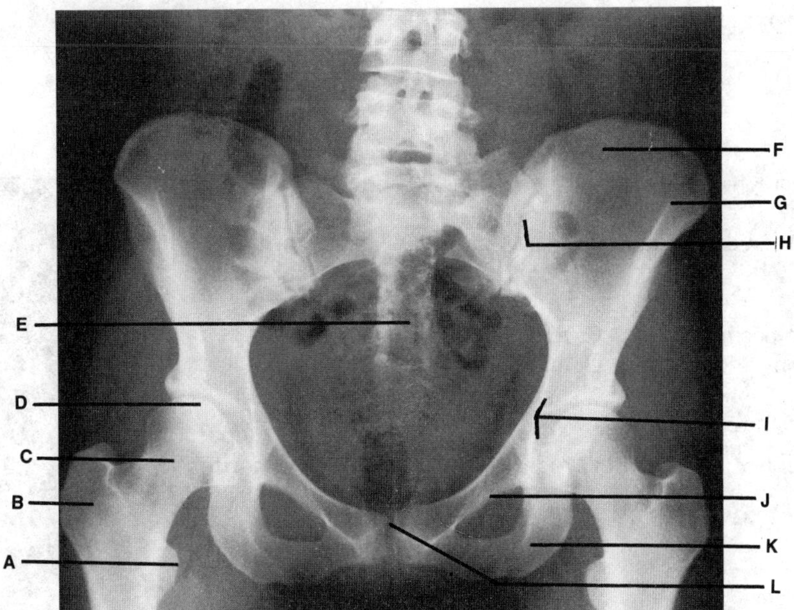

Pertaining to the radiographic image, answer questions 86 and 87.

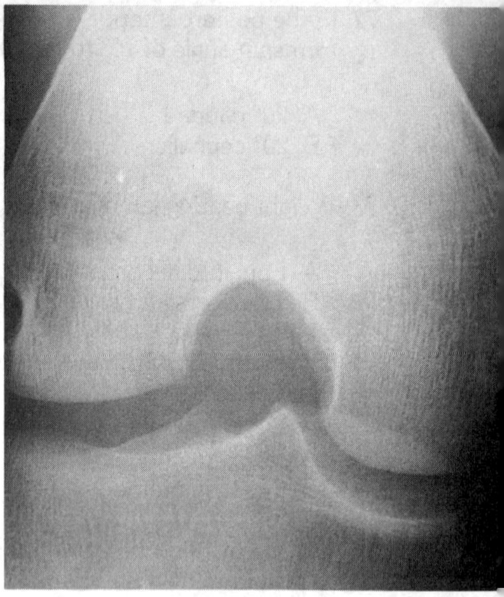

86. This position of the knee is best described as a: 1:250:I

A. PA axial position
B. Tangential position
C. AP oblique position
D. AP position with stress

87. The principle structure of interest in this position is/are the: 1:250:I
1. Intercondylar fossa *2. Fibular head* *3. Tibial eminences*

A. 1 only
B. 2 only
C. 1 & 2 only
D. 1 & 3 only

Pertaining to the radiographic image, answer questions 88 and 89.

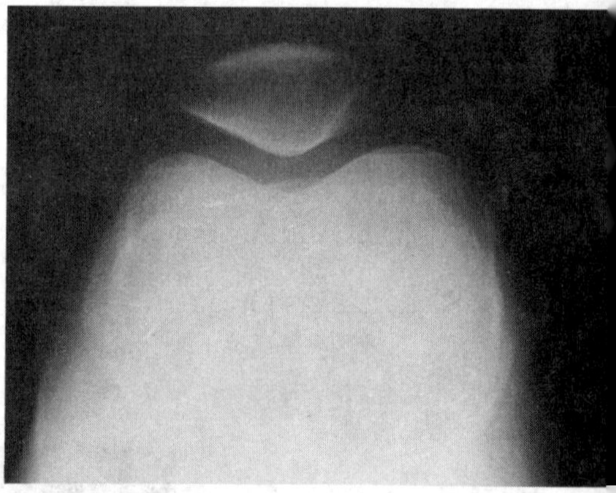

88. In order to accomplish this tangential position of the patella the central ray is directed perpendicular to the: 1:264:I

A. Femur
B. Joint space
C. Tibia
D. Fibula

89. In addition to providing a profile view of the patella this position is often used to evaluate the: 1:264:I

A. Intercondylar fossa
B. Crusate ligaments
C. Femoropatellar articulation
D. Menisci

Pertaining to the radiographic image, answer questions 90 and 91.

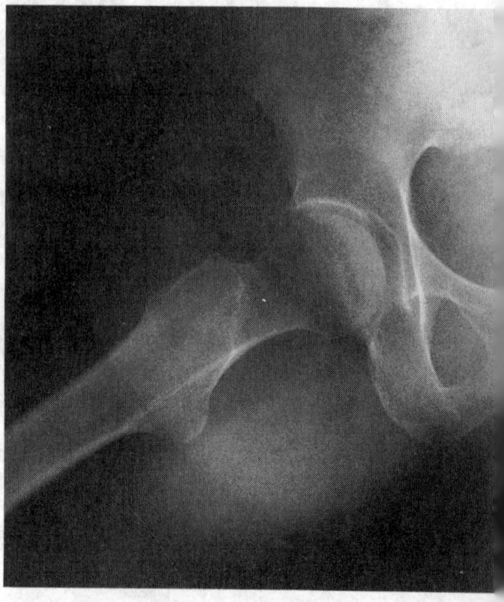

90. The image shown is best described as a/an ____________ of the femoral neck. 1:283:I

A. Superiorinfero position
B. Lateral position
C. AP axial position
D. Inferosuperior position

91. This image demonstrates a profile view of the: 1:283:I
1. Symphysis pubis ***2. Greater trochnter*** ***3. Lesser trochanter***

A. 1 only
B. 2 only
C. 3 only
D. 1, 2 & 3

Pertaining to the radiographic image, answer questions 92 and 93.

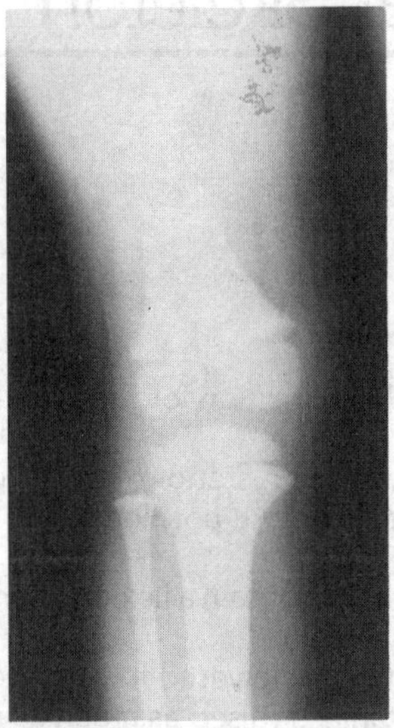

92. In an AP projection of the knee, improved visualization of the joint space can be accomplished by: 1:240:I

A. Acute flexion of the knee
B. Partial flexion of the knee
C. Medial rotation of the knee
D. Increasing the mAs

93. The radiolucent lines seen proximal and distal to the knee joint are termed: 1:240:I

A. Condral folds
B. Epiphyses
C. Collateral ligaments
D. Psuedo joints

Pertaining to the radiographic image, answer questions 94 and 95.

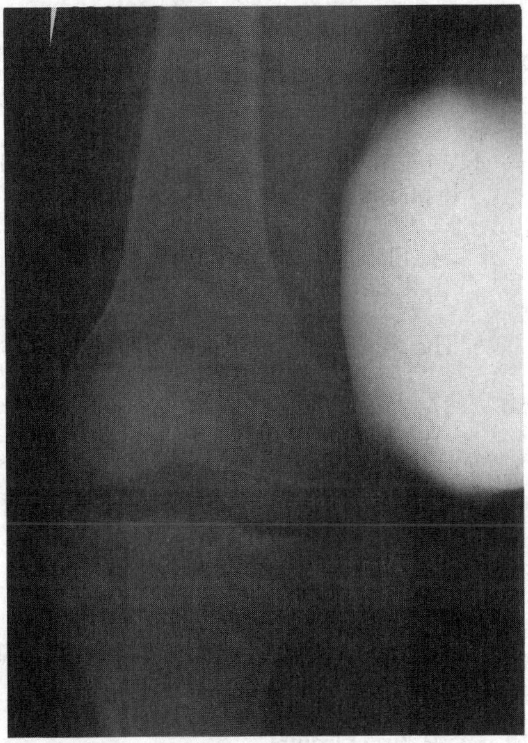

94. This AP projection of the knee was obtained while employing:

A. Weight bearing stress
B. Inversion stress
C. Eversion stress
D. None of the above

95. The radiographic image indicates that the patient has a damaged:

A. Lateral collateral ligament
B. Patellar bursa
C. Crusate ligament
D. Fovea capitus

Chapter 48

1. The best position for the evaluation of the lumbar vertebral bodies and the intervertebral disc spaces is the: 1:369:I

A. Anteroposterior (AP) projection
B. AP oblique position
C. PA oblique position
D. Lateral position

2. In order to obtain a flexion lateral projection of the cervical spine, the patient is positioned true lateral and the: 1:339:I

A. Chin is lowered toward the chest
B. Chin is raised as much as possible
C. Body of the mandible is parallel to the floor
D. More than one but not all the above

3. A lead strip which is placed just behind the posterior surface of a patient in the lateral position serves to: 1:369:I

A. Absorb scattered radiation
B. Filter the x-ray beam
C. Increase the divergence of the beam
D. Eliminate the need for a grid

4. When performing an anteroposterior (AP) projection of the thoracic spine with the patient supine, the normal kyphosis is reduced by: 1:356:I

A. Flexing the knees
B. Extending the legs
C. Inverting the toes
D. Raising the arms

5. The transverse processes of the lumbar spine are best demonstrated in the: 1:365:I

A. Anteroposterior (AP) projections
B. Oblique positions
C. Axial positions
D. Lateral positions

6. If a patient is in traction for a cervical spine injury, the radiograph should normally be accomplished: 1:350:I

A. After removing traction apparatus
B. After removing the weights
C. In acute flexion and extension
D. None of the above

7. The best position for the evaluation of lateral scoliosis is the: 1:394:I

A. Supine frontal
B. Erect frontal
C. Supine lateral
D. Erect lateral

8. The swimmer's (Twining) position is used for the evaluation of the: 1:352:I

A. Atlanto-occipital region
B. Atlanto-axial region
C. Cervicothoracic region
D. Lumbosacral region

9. In the oblique position for the lumbar spine, the nose of the "Scotty dog" corresponds to the: 1:373:I

A. Pedicle
B. Lamina
C. Body
D. Transverse process

10. The most commonly employed oblique positions for the cervical spine require the body to be rotated: 1:348:I

A. 20 degrees
B. 45 degrees
C. 60 degrees
D. 90 degrees

11. The demonstration of the sternum is best accomplished in the right posterioroanterior oblique (RAO) position which projects the image of the sternum: 1:408:I

A. Away from the heart shadow
B. Over the heart shadow
C. Over the thoracic vertebrae
D. None of the above

12. The articulations between the superior and inferior articulating processes of the cervical spine are best seen in the: 1:337:I

A. Anteroposterior (AP) projection
B. AP axial (Pillar) position
C. AP oblique position
D. Lateral position

13. To determine motion in the area of a spinal fusion of the lumbar spine, which of the following is indicated? 1:398:I
1. AP - legs flexed and extended 2. AP - right and left bending positions 3. Laterals - flexion and extension

A. 1 & 2 only
B. 1 & 3 only
C. 2 & 3 only
D. 1, 2 & 3

14. In the lateral position for the sacrum, the central ray is directed ____________ to the midaxillary line. 1:388:I

A. 13cm (5") anterior
B. 8cm (3") posterior
C. 45° cephalic
D. On

15. For the anteroposterior (AP) axial position of the cervical vertebral arch, the central ray is directed: 1:346:I

A. 25 cephalic
B. 25 caudad
C. Perpendicular
D. 15 cephalic

16. In the lateral position for the upper thoracic and lower cervical vertebrae, the central ray is directed 5° caudad and the dependent arm is: 1:354:I

A. Raised
B. Lowered
C. 90° from the body
D. None of the above

17. In order to improve contrast in the lateral position for the sternum, which of the following would be true? 1:410:I

A. 180cm (72") SID is used
B. Exposure is made on full inhalation
C. Exposure is made on full expiration
D. The shoulders are rolled back

18. The anteroposterior oblique positions for the lumbar spine are used in the evaluation of the: 1:373:I

A. Dependent zygapophyseal articulations
B. Independent zygapophyseal articulations
C. Dependent intervertebral foramina
D. Independent intervertebral foramina

19. The degree of obliquity required for proper radiographic evaluation of the sternoclavicular joints is: 1:417:I

A. 15 degrees
B. 30 degrees
C. 45 degrees
D. 60 degrees

20. In the lateral position for the thoracic spine, the ribs are moved from obscuring the intervertebral foramina by: 1:358:I

A. Directing the central ray 15° caudad
B. Adjusting the arms to right angles
C. Taking the exposure on full inhalation
D. Directing the central ray 5° caudad

21. A patient is positioned for an anteroposterior (AP) oblique position of the cervical spine with the central ray directed 15-20 degrees to the head. This will demonstrate the pedicles: 1:341:I

A. Closest to the film
B. Farthest from the film
C. Both of the above
D. Neither of the above

22. A patient comes from the emergency room with a possible cervical spine fracture. The preliminary evaluation radiograph for this purpose is the: 1:350:I

A. Cross-table lateral
B. Lateral in extension
C. Lateral in flexion
D. Erect true lateral

23. In the oblique positions for the lumbar spine, "Scotty dogs" are seen when the patient is properly positioned. The front feet of the "dog" corresponds to the: 1:373:I

A. Lamina
B. Pedicle
C. Superior articular process
D. Inferior articular process

24. The position most commonly employed to show the intervertebral disc space, of the thoracic spine is the: 1:359:I

A. Anteroposterior projection
B. 45° oblique position
C. 15° oblique position
D. Lateral position

25. With the patient supine, the technologist obliques the patient for proper demonstration of the right sacroiliac (SI) joint. In order to accomplish this, the patient is rotated: 1:380:I

A. Toward the left
B. Toward the right
C. Neither of the above

26. In order to obtain a uniform density when radiographing the thoracic spine, which of the following would be true? 1:356:I

1. The anode should be over the patient's head. *2. The anode should be directed caudally.*
3. The cathode should be directed caudally.

A. 1 only
B. 2 only
C. 1 & 2 only
D. 2 & 3 only

27. The presence or absence of cervical ribs would best be demonstrated in the: 1:335:I

A. Anteroposterior (AP) oblique position
B. Lateral position
C. Posteroanterior oblique position
D. Anteroposterior (AP) projection

28. A lateral cervical spine radiograph is done at 180cm (72") distance, in order to: 1:337:I

A. Reduce exposure factors
B. Increase magnification
C. Decrease magnification
D. Improve contrast

29. A frontal projection of the coccyx with the patient in the prone position requires that the central ray be angled: 1:387:I

A. 10° cephalic
B. 10° caudad
C. 25° cephalic
D. 25° caudad

30. A lordotic curve of the spinal column is best demonstrated in a/an: 1:313:I

A. Anteroposterior (AP) projection of the cervical spine
B. Lateral position of the thoracic spine
C. Anteroposterior (AP) projection of the lumbar spine
D. Lateral position of the cervical spine

31. The zygapophyseal articulations of the lumbar spine are best visualized using the: 1:372:I

A. Frontal projection
B. 30° AP oblique position
C. 45° oblique position
D. Lateral position

32. To help insure that the odontoid process will be free of superimposition in the open-mouth position, a line drawn from the lower edge of the upper incisor to the ____________ is placed perpendicular to the image receptor. 1:327:I

A. Mastoid tip
B. Mental process
C. External auditory meatus
D. Inion (EOP)

33. To compensate for the normal degree of kyphosis of the sacrum for the anteroposterior (AP) projection, the central ray is directed: 1:386:I

A. 5° cephalic
B. 10° caudad
C. 15° cephalic
D. 25° caudad

34. When performing a lateral for the thoracic spine, the patient's arms should be elevated no more than 90° from the body. Any more than this will cause: 1:359:I

1. Correction of kyphosis *2. The scapulae to obscure the upper vertebrae*
3. Discomfort to the patient

A. 1 only
B. 2 only
C. 3 only
D. 1, 2 & 3

35. In the posteroanterior (PA) oblique axial position of the cervical spine, the structures best demonstrated are the ____________ closest to the image receptor. 1:343:I

A. Vertebral foramina
B. Transverse foramina
C. Zygapophyseal articulations
D. None of the above

36. The zygapophyseal articulations of the thoracic spine can be demonstrated by adjusting the midcoronal plane to form an angle of ____________ with the plane of the image receptor. 1:361:I

A. 15°
B. 25°
C. 45°
D. 70°

37. An open mouth anteroposterior (AP) projection of the odontoid is taken. The resultant radiograph demonstrates the occipital bone obscuring the dens. In order to correct the positioning, the technologist should: 1:327:I

A. Tuck the patient's chin more
B. Raise the patient's chin more
C. Oblique the patient's head 10°
D. Nothing; the positioning is correct.

38. In the anteroposterior (AP) oblique position for the sacroiliac (SI) joints, the joint space is best visualized when the patient is rotated ____________ to the plane of the image receptor. 1:382:I

A. 15 degrees
B. 25 degrees
C. 45 degrees
D. 70 degrees

39. In the anterioroposterior (AP) projection of the thoracic spine, the central ray is directed to the: 1:356:I

A. Second thoracic vertebra
B. Fourth thoracic vertebra
C. Seventh thoracic vertebra
D. Tenth thoracic vertebra

40. In order to demonstrate the L5-S1 interspace in the anteroposterior (AP) projection, the central ray is directed: 1:378:I

A. Perpendicular to the ASIS
B. 5°-7° caudad
C. 30° cephalic
D. 10° cepalic

41. The best demonstration of the spinous processes of the cervical vertebrae is seen on: 1:337:I

A. Anteroposterior(AP) projection
B. Posteroanterior (PA) projection
C. 45° oblique position
D. Lateral position

42. A patient has injured his anterior 5th rib. The ideal method for the rib radiograph of this area would be: 1:404:I

1. Erect PA *2. Inspiration* *3. Prone*

A. 1 & 2 only
B. 1 & 3 only
C. 2 & 3 only
D. 1, 2 & 3

43. The anteroposterior (AP) oblique position for the cervical spine is employed in the evaluation of the: 1:341:I

A. Vertebral foramina
B. Transverse foramina
C. Intervertebral foramina
D. Zygapophyseal articulations

44. Lateral position for the lumbar spine is done to show the: 1:369:I

1. Transverse processes of the vertebrae *2. Spinous processes of the vertebrae*
3. Height of the vertebral disc space

A. 1 & 2 only
B. 1 & 3 only
C. 2 & 3 only
D. 1, 2 & 3

45. If the lower vertebral column is not elevated to the horizontal position for the lateral recumbent position of the thoracic spine, the central ray should be directed: 1:359:I

A. 10°-15° cephalic
B. Perpendicular to T7
C. 10°-15° caudad
D. None of the above

46. In the demonstration of the odontoid utilizing the anteroposterior (AP) (Fuchs) method, the line between the mastoid tip and ____________ must be perpendicular to the table. 1:328:I

A. Infra-orbital margin
B. Acanthion
C. Upper incisors
D. Tip of the chin

47. In the anteroposterior (AP) axial position of the costal joints for the demonstration of rheumatoid spondylitis, the central ray is directed ____________ through T7. 1:432:I

A. 5° caudad
B. 50° cephalic
C. 20° caudad
D. 20° cephalic

48. In the lateral position (Kurzbauer) for the demonstration of sternoclavicular joints, the patient is positioned according to the following: 1:418:I

A. Cleaves method
B. Pawlow method
C. Trendelenburg method
D. Simms method

49. The use of grids, in general, are not employed for radiographs of the lateral cervical spine because of the scatter reduction occuring from: 1:350:I

A. The anode heel effect
B. Tight collimation
C. An air gap
D. A high kVp technique

50. To better define the intervertebral spaces in the anteroposterior projection of the cervical spine, the central ray is angled: 1:334:I

A. 45° to the feet
B. 15° to the feet
C. 45° to the head
D. 15° to the head

51. The patient is placed in the right lateral recumbent position and rotated anteriorly 20°. The central ray is directed through T7. The structures best demonstrated would be: 1:361:I

A. Right zygapophyseal articulations
B. Left zygapophyseal articulations
C. Right intervertebral foramina
D. Left intervertebral foramina

52. The posteroanterior oblique axial position (Kovacs method) is utilized to demonstrate the: 1:376:I

A. Last lumbar zygapophyseal articulation
B. Last lumbar spinous process
C. L5-S1 interspace
D. Last lumbar intervertebral foramen

53. The Ottonello (wagging jaw) method is employed for the evaluation of: 1:344:I

A. Basal fractures
B. Atlanto-occipital articulations
C. Lateral cervical spine
D. AP of the entire cervical spine

54. Abnormal sacroiliac (SI) motion is best demonstrated, when the patient is positioned for: 1:384:I

A. Weight-bearing PA positions of the sacroiliac
B. Weight-bearing PA positions of the pubic bones
C. PA lateral bending positions
D. None of the above

55. In the anteroposterior (AP) projection of the lower cervical vertebrae, the central ray is directed 15° cephalic and enters at the: 1:334:I

A. Mental point
B. Gonion
C. Thyroid cartilage
D. Hyoid bone

56. The intervertebral disc spaces of the lumbar spine will more closely parallel the divergence of the beam of radiation in the: 1:364:I

A. Anteroposterior (AP) projection
B. Left lateral position
C. Right lateral position
D. Posteroanterior (PA) projection

57. When performing scoliosis images, which of the following will reduce the radiation dosage to the thyroid, gonads and breasts? 1:394:I

A. Anteroposterior (AP) projection
B. Anterior oblique position
C. Posteroanterior (PA) projection
D. None of the above

58. In the open mouth anterioroposterior (AP) projection of the odontoid, the instructions to the patient for the exposure are: 1:327:I

A. Slow inspiration
B. Phonation - ahhh
C. Slow expiration
D. Suspended respiration

59. During the lateral position for the lumbar spine on a female patient, if the spine is not adjusted to the horizontal, the central ray should be directed: 1:369:I

A. Vertically to the image receptor
B. 5° cephalic
C. 8° caudad
D. 12° caudad

60. The Axillary portion of the right ribs is best demonstrated with the patient in the: 1:428:I

A. AP oblique (LPO) position
B. PA oblique (LAO) position
C. RT Lateral position
D. Apical Lorbotic position

Pertaining to the radiographic image, answer questions 61 and 62.

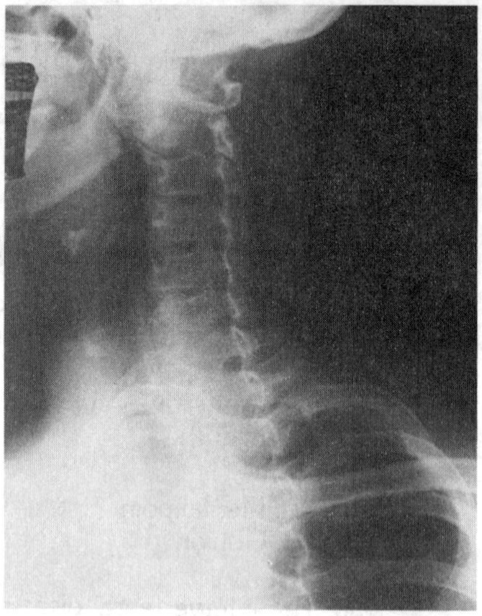

61. This image was obtained with the patient positioned for the: 1:343:I

A. AP projection
B. 20° AP oblique position
C. 45° AP oblique position
D. Lateral position

62. This position is primarily used in the visualization of the: 1:343:I
1. Intervetebral foramina 2. Odontoid process 3. Zygapophysial joints

A. 1 only
B. 2 only
C. 3 only
D. 1, 2 & 3

Pertaining to the radiographic image, answer questions 63 and 64.

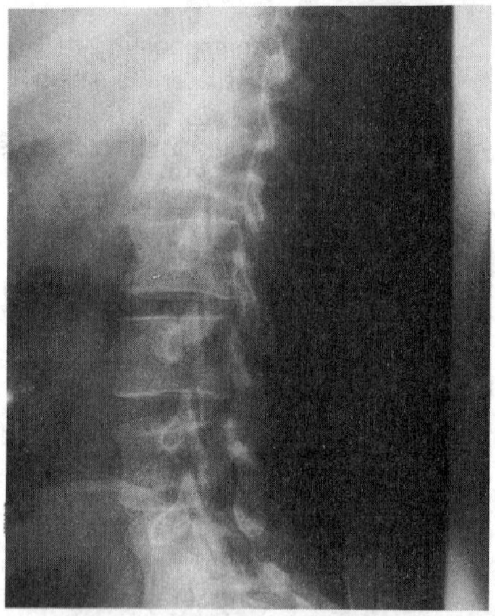

63. This image was obtained with the patient positioned for the: 1:373:I

A. AP projection
B. 20° PA oblique position
C. 45° AP oblique position
D. Lateral position

64. This position is primarily used in the evaluation of the: 1:373:I
1. Spinous processes 2. Zygapophysial joints 3. Intervetebral foramina

A. 1 only
B. 2 only
C. 3 only
D. 1, 2 & 3

Pertaining to the radiographic image, answer question 65.

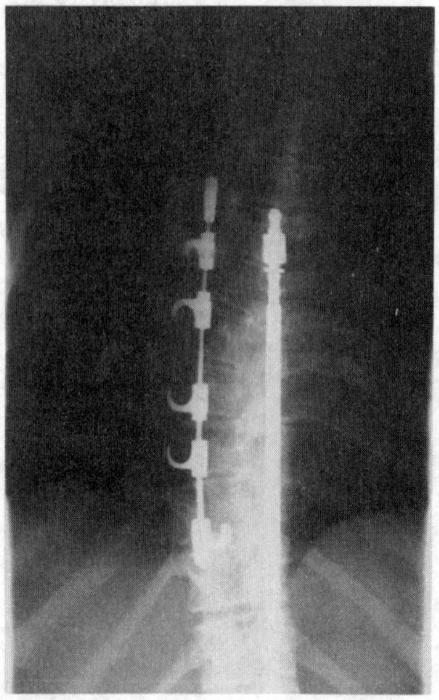

65. The low density (Harrington) rods seen on this PA projection of the spine are used in the treatment of:

A. Scoliosis
B. Whip lash
C. Disc slippage
D. None of the above

Skull Radiography

Chapter 49

1. The line extending from the smooth prominence between the superciliary arches to the external auditory meatus is the __________ line. 1:233:II

A. Acanthiomeatal
B. Orbitmeatal
C. Supraorbitmeatal
D. Glabellomeatal

2. The position that best demonstrates all of the paranasal sinuses on a single radiograph is the: 1:379:II

A. Submentovertical position
B. Parietoacanthial position
C. Lateral position
D. Posteroanterior projection

3. In a lateral position of the nasal bones, the central ray is directed perpendicular to: 1:314:II

A. 3/4" distal to the glabella
B. The anterior nasal spine
C. 3/4" distal to the nasion
D. The level of the supraorbital ridge

4. Which of the following would be most useful for the demonstration of a single zygomatic arch? 1:325:II

A. Parietoacanthial position
B. Tangential oblique submentovertical position
C. Posteroanterior axial position
D. Axial (intraoral) position

5. The cranial type in which the petrous ridges project anteriorly and medially from the midsagittal plane at an angle of 54° is described as: 1:234:II

A. Mesocephalic
B. Hypercephalic
C. Brachycephalic
D. Dolichocephalic

6. The most superior portion of the skull is referred to as: 1:217:II

A. Lambda
B. Vertex
C. Corona
D. Mentum

7. The parietoacanthial position (Waters) is useful in the evaluation of the foramen: 1:383:II

A. Rotundum
B. Spinosum
C. Magnum
D. All of the above

8. When performing an examination of the temporomandibular joints, it is important to include: 1:359:II
1. Open mouth views *2. Closed mouth views* *3. Bilateral views*

A. 1 only
B. 2 only
C. 3 only
D. 1, 2 & 3

9. The radiographic line from the anterior nasal spine to the external auditory meatus (EAM) is known as: 1:233:II

A. Glabellomeatal
B. Nasomeatal
C. Acanthiomeatal
D. Infraorbitomeatal

10. The mesocephalic cranium measures approximately __________ between the parietal eminences. 1:217:II

A. 15cm
B. 18cm
C. 25cm
D. 20cm

11. For a lateral position of the sella turcica, the central ray is directed to enter __________ to the external auditory meatus (EAM): 1:239:II

A. 2cm anterior and superior
B. 5cm posterior and inferior
C. 2cm anterior and inferior
D. 5cm anterior and superior

12. In the posterior profile position (Stenvers), the midsagittal plane is adjusted to form an angle of __________ to the plane of the image receptor. 1:408:II

A. 30° degrees
B. 37° degrees
C. 45° degrees
D. 53° degrees

13. In the verticosubmental position of the cranium, the central ray is directed to enter at the: 1:255:II

A. Bregma
B. Nasion
C. Throat
D. Foramen magnum

14.-20. Pertaining to the radiograph of the AP axial position of the skull, place the letter that corresponds to the following structures:

14. ________ Occipital bone

15. ________ Mastoid air cells

16. ________ Petrous ridge

17. ________ Dorsum sellae

18. ________ Parietal bone

19. ________ Condyloid process of mandible

20. ________ Foramen magnum

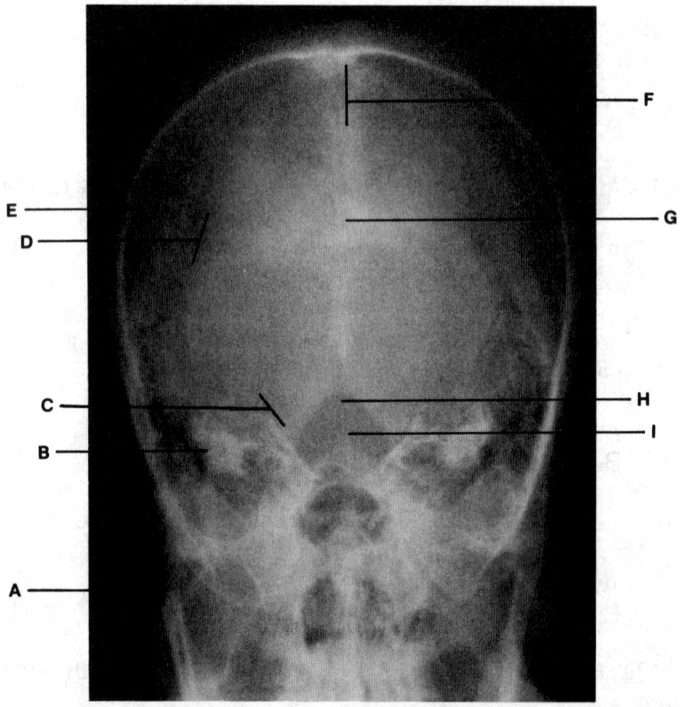

21. There is a __________ difference between the orbitomeatal line and the infraorbitomeatal line. 1:233:II

A. 3°
B. 7°
C. 12°
D. 17°

22. The orbitoparietal oblique position of the optic foramen requires that the midsagittal plane form an angle of __________ to the plane of the image receptor. 1:274:II

A. 23°
B. 37°
C. 45°
D. None of the above

23. In order to remove overlying shadows during mastoid radiography, it is recommended that: 1:398:II

1. The mouth be opened *2. A small focal spot be used* *3. The auricles are taped forward*

A. 1 & 2 only
B. 1 & 3 only
C. 2 & 3 only
D. 1, 2 & 3

24. The full basal projection (Schuller) is recommended for the demonstration of the: 1:253:I

1. Foramen ovale   2. Foramen spinosum   3. Atlas

A. 1 only
B. 2 only
C. 3 only
D. 1, 2 & 3

25. The radiographic line that passes from the outer canthus of the eye to the external auditory meatus (EAM) is described as the: 1:233:II

A. Orbitomeatal line
B. Canthiomeatal line
C. Acanthiomeatal line
D. None of the above

26. The submentovertical projection is often employed for the evaluation of: 1:253:II

1. Sphenoid sinuses   2. Zygomatic arches   3. Foramen magnum

A. 1 & 2 only
B. 1 & 3 only
C. 2 & 3 only
D. 1, 2 & 3

27. The average measurement of a skull from the vertex to the submental area is: 1:217:II

A. 15cm
B. 22cm
C. 12cm
D. 28cm

28. The radiographic line that bisects the pupils of the eye is termed: 1:233:II

A. Intercanthial
B. Interorbital
C. Interpupillary
D. Orbitomeatal

29. In evaluating the proper positioning of a submentovertical projection of the skull, the structure(s) which will be seen entirely anterior to the petrous ridges is/are: 1:253:II

1. Odontoid process   2. Mandibular condyles   3. The atlas

A. 1 & 2 only
B. 2 only
C. 3 only
D. 1, 2 & 3

30. The posteroanterior axial (Caldwell) method of demonstrating the paranasal sinuses is used primarily for the sinus(es): 1:281:II

1. Anterior ethmoid   2. Maxillary   3. Frontal

A. 1 & 2 only
B. 1 & 3 only
C. 2 & 3 only
D. 1, 2 & 3

31. There is a/an ____________ difference between the orbitomeatal line and the glabellomeatal line. 1:233:II

A. 30°
B. 5°
C. 8°
D. 12°

32. In the parieto-orbital oblique projection (Rhese), the head is positioned so that the ____________ are in contact with the table: 1:272:II

A. Forehead, nose and chin
B. Nose, zygoma and chin
C. Forehead, zygoma and nose
D. None of the above

33. In the anteroposterior (AP) axial position for the demonstration of the closed temporomandibular joints, the patient must occlude the ____________ to avoid protruding the condyles. 1:356:II

1. Incisors   2. Molars   3. Canines

A. 1 only
B. 2 only
C. 3 only
D. 1, 2 & 3

34. The patient is placed prone, and the head is adjusted so that the acanthiomeatal line is perpendicular and the midsagittal plane forms an angle of 37° to the perpendicular. The central ray is directed vertically to pass through the orbit. This will demonstrate the: 1:272:II

A. Maxillary sinus
B. Eustachian tube
C. Internal auditory canal
D. Optic foramen

35. The radiographic line that passes from the inferior border of the orbits to the auricular point is described as: 1:233:II

A. Cranial base line
B. Orbitomeatal line
C. Acanthiomeatal line
D. None of the above

36. The average orbit projects anteriorly from the midsagittal plane at an angle of: 1:232:II

A. 23°
B. 30°
C. 37°
D. 62°

37. The best demonstration of mediolateral displacement fractures of the mandibular ramus would be seen in the: 1:343:II
1. Axiolateral position *2. Anteroposterior axial position* *3. Posteroanterior projection*

A. 1 only
B. 3 only
C. 1 & 3 only
D. 2 & 3 only

38. In the posterior profile position (Stenvers), the central ray is directed ____________ through the petrous ridges ____________ to the image receptor. 1:408:II

A. 10° caudad, away from
B. 45° caudad, away from
C. 12° cephalic, closest to
D. 10° cephalic, closest to

39. In the lateral position for the cranium, the line that will be perpendicular to the image receptor is the: 1:238:II

A. Orbitomeatal
B. Interpupillary
C. Infraorbitomeatal
D. Glabelloalveolar

40. When performing a parietoacanthial position, failure to extend the chin sufficiently will result in the ____________ obscuring the maxillary sinuses. 1:383:II

A. Frontal sinuses
B. Zygomatic arches
C. Teeth
D. Petrous ridges

41.-48. Pertaining to the radiograph of the lateral skull place the letter that corresponds to the following structures:

41. ________ Parietal bone

42. ________ Frontal sinus

43. ________ Posterior clinoid process

44. ________ Anterior nasal spine

45. ________ Sphenoid sinus

46. ________ Coronal suture

47. ________ Maxillary sinus

48. ________ Occipital bone

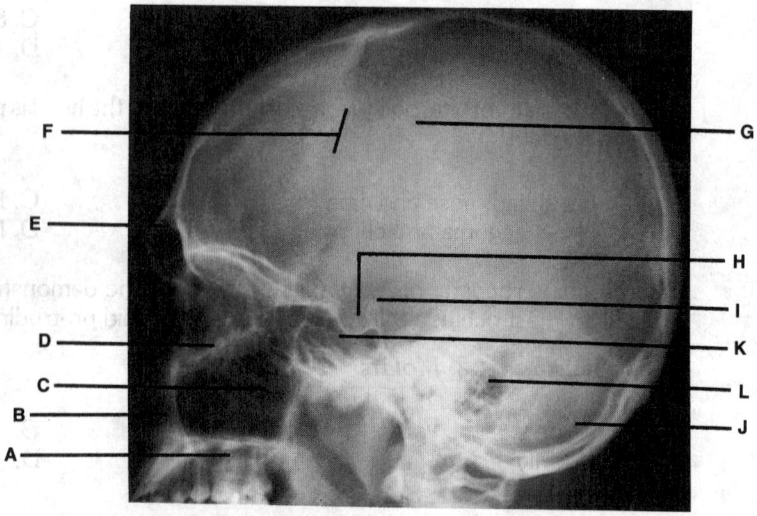

49. When radiographing the skull in the lateral position, the ____________ is adjusted so that it is parallel to the plane of the image receptor. *1. Midsagittal plane* *2. Interpupillary line* *3. Coronal plane*

A. 1 only
B. 2 only
C. 3 only
D. 1, 2 & 3

50. In the anterior profile position (Arcelins), the petrous ridges demonstrated are: 1:410:II

A. Perpendicular and farthest from the image receptor
B. Perpendicular and closest to the image receptor
C. Parallel and farthest from the image receptor
D. Parallel and closest to the image receptor

51. The anteroposterior (AP) axial (Grashey) position of the skull is useful for the demonstration of all of the following, except: 1:249:II

A. Zygomatic arches
B. Dorsum sella
C. Pars petrosa
D. Frontal sinuses

52. In the anterior profile position (Arcelin) of the mastoid process, the central ray is directed: 1:410:II

A. 12° cephalic
B. 37° caudad
C. 10° caudad
D. 45° caudad

53. The patient is supine with the infraorbitomeatal line parallel to the plane of the film. The central ray enters vertically at a coronal plane 2cm anterior to the EAM. This describes a/an ____________ projection. 1:253:II

A. True lateral
B. Submentovertical
C. Anteroposterior
D. Verticosubmental

54. Which of the following position would be most helpful in diagnosing a blow-out fracture of the orbit? 1:270:II

A. Parieto-orbital oblique
B. Orbitoparietal oblique
C. Parietoacanthial
D. Posteroanterior

55. The posteroanterior transorbital position with a perpendicular central ray uses a tightly collimated x-ray beam for the demonstration of the____________ within the shadow of the orbits. 1:428:II

A. Optic foramen
B. Internal auditory canals
C. Antra of Highmore
D. Jugular foramen

56. The posteroanterior (PA) axial position (Haas) is often employed to give a comparable projection to the: 1:251:II

A. Parietoacanthial
B. Orbitoparietal oblique
C. Anteroposterior axial
D. Submentovertical

57. The patient is prone. The infraorbitomeatal line is parallel to the image receptor and the midsagittal plane is perpendicular. In order to obtain an oblique tangential position (May) of the zygomatic arch, which of the following would be true? 1:330:II

A. The midsagittal plane is rotated 15° away from the affected side.
B. The midsagittal plane is rotated 15° toward the affected side.
C. The central ray is directed 15° caudad.
D. The central ray is directed 15° cephalic.

58. In the axiolateral oblique position (Mayer) of the auditory ossicles, the midsagittal plane is rotated 45° toward the affected side, the infraorbitomeatal line is parallel to the transverse axis of the image receptor, and the central ray is directed: 1:412:II

A. 12° cephalic
B. 23° caudad
C. 30° cephalic
D. 45° caudad

59. In order to obtain an anteroposterior axial (Grashey) position on an infant or young child, the central ray angulation is ____________ and directed to enter the ____________. 1:22:III

A. Less than 30°, nasion
B. More than 30°, superciliary
C. Less than 30°, anterior hairline
D. More than 30°, nasion ridge

60. In the parieto-orbital oblique position (Rhese), the radiographic line that must be perpendicular to the image receptor is the: 1:272:II

A. Acanthiomeatal line
B. Infraorbitomeatal line
C. Mentomeatal line
D. Interpupillary line

61. Which of the following is/are indicative of a properly rotated posterior profile (Stenvers) position? 1:409:II
1. Mastoid process in profile 2. Petrous filling 2/3 of the orbit
3. Mandibular ramus superimposed lateral aspect of C-spine

A. 1 only
B. 2 only
C. 3 only
D. 1, 2 & 3

62. In the posteroanterior projection of the cranium for a general survey examination, the orbitomeatal line is perpendicular and the central ray is directed: 1:242:II

A. 23° caudad to the glabella
B. 15° cephalic to the nasion
C. 15° caudad to the nasion
D. 23° cephalic to the glabella

63. The lateral position (Modified Law) is used for demonstration of the: 1:401:II
1. Mastoid air cells 2. Maxillary sinus 3. Nasal septum

A. 1 only
B. 2 only
C. 3 only
D. 1, 2 & 3

64. With the patient in the posteroanterior position and the glabellomeatal line and midsagittal plane perpendicular to the image receptor, in order to demonstrate the frontal sinus, the central ray would be directed: 1:380:II

A. 15° caudad
B. 23° caudad
C. 23° cephalic
D. 37° cephalic

65. In order to obtain a posteroanterior axial (Bertel) position of orbital floors and the inferior orbital fissures, the infraorbitomeatal line is perpendicular to the image receptor and the central ray is directed: 1:284:II

A. 35° caudad
B. 25° caudad
C. 25° cephalic
D. 5° cephalic

66. In the parieto-orbital oblique position, the optic foramen should be seen: 1:272:II
1. In the upper outer quadrant 2. In the lower outer quadrant
3. On end at the end of the sphenoid ridge

A. 1 & 2 only
B. 1 & 3 only
C. 2 & 3 only
D. 1, 2 & 3

67. When radiographing the paranasal sinuses, which of the following is indicative of proper exposure factors? 1:376:II
1. Penetration of the sella area 2. Even density of the upper and lower orbits
3. Visualization of maxillary sinuses

A. 1 only
B. 2 only
C. 3 only
D. 1, 2 & 3

68. Cahoon, Wigby-Taylor, and Fuchs all described axial or oblique positions for the demonstration of: 1:442:II

A. Temporal styloid
B. Sphenoid sinuses
C. Mastoid antrum
D. Hypoglossal canal

69. The patient is prone, and the orbitomeatal line and midsagittal plane are perpendicular to the plane of the image receptor. The central ray is directed 25° caudad to pass through the inferior border of the orbit. This position is used to demonstrate: 1:282:II

A. Supraorbital fissure
B. Foramen magnum
C. Petrous ridges
D. Mastoid air cells

70. In the oblique tangential position for the demonstration of a zygomatic arch, the patient is supine and the infraorbitomeatal line is parallel to the plane of the film. The central ray is perpendicular to the image receptor and the head is rotated: 1:320:II

A. 15° away from affected side
B. 45° toward the affected side
C. 15° toward the affected side
D. 45° away from the affected side

71. For a posteroanterior (PA) projection of the cranium in which the frontal bone is the primary area of interest, the central ray is directed: 1:242:II

A. 5° cephalic
B. 15° cephalic
C. 30° cephalic
D. Perpendicular to the image receptor

72. Which of the following positions will demonstrate the least angular distortion of the bilateral petrous ridges? 1:261:II

A. Posteroanterior axial (Valdini)
B. Posteroanterior axial (Haas)
C. Laterial (Laws)
D. Axiolateral (Lysholm)

73. In the anteroposterior (AP) axial position (Grashey) for the demonstration of the jugular foramen, with the orbitomeatal line perpendicular, the central ray is directed: 1:247:II

A. 25° cephalic
B. 30° caudad
C. 37° cephalic
D. 50° caudad

74. When radiographing a skull to determine the presence of sphenoid sinus effusion, the optimum method would be: 1:240:II

A. Cross-table lateral
B. Transoral posteroanterior (Pirie)
C. Posteroanterior 25° caudad
D. Submentovertical

75. The patient is supine and the infraorbitomeatal line is perpendicular. In order to obtain an acanthioparietal position of the facial bones, the central ray must be directed: 1:308:II

A. 7° caudad
B. 30° cephalic
C. 45° caudad
D. Perpendicular to the film

76. A radiograph of a parieto-orbital oblique is taken. The radiograph demonstrates the optic foramen in the upper outer quadrant of the orbit. In order to improve the positioning, the technologist should: 1:272:II

A. Rotate the affected orbit medially
B. Extend the patient's chin
C. Rotate the affected orbit laterally
D. Tuck the patient's chin

77. It is recommended that radiographs of the paranasal sinuses be taken erect in order to demonstrate the existence of: 1:376:II

A. Fracture fragments
B. Layering of stones
C. Fluid levels
D. Lacrimal duct obstructions

78. The lateral projection (Law) for the mastoid air cells requires a double tube angulation of: 1:399:II

A. 15° anteriorly, 15° caudad
B. 15° anteriorly, 15° cephalic
C. 15° anteriorly, 45° caudad
D. 25° anteriorly, 45° caudad

79. A posteroanterior (PA) axial position of the mandible requires a 30° cephalic angulation. This is used for the demonstration of: 1:344:II

A. Mandibular body
B. Superior alveolar border
C. Mandibular symphysis
D. Condylar processes

80. In the axial transoral position (Pirie), the central ray passes through a line drawn from the sella turcica and the open mouth. This projection is used to demonstrate the: 1:390:II

A. Styloid process
B. Sphenoid sinus
C. Hypoglossal canal
D. Frontal sinus

81. The patient is supine with the infraorbital meatal line parallel with the plane of the film. The central ray enters at a right angle with the image receptor through a coronal plane 2cm posterior to the outer canthus. This position is used for the demonstration of: 1:253:II

1. Mandibular symphysis *2. Nasal septum* *3. Mastoid processes*

A. 1 & 2 only
B. 1 & 3 only
C. 2 & 3 only
D. 1, 2 & 3

82. A patient is prone with the orbitomeatal line perpendicular to the plane of the image receptor. The central ray is directed 25° cephalic exiting 5cm above the nasion. This is termed the: 1:251:II

A. Axiolateral position
B. Full basal projection
C. Parietotemporal position
D. Posteroanterior axial position

83. In the posteroanterior (PA) axial position (Valdini), the orbitomeatal line is adjusted to form an angle of 50° to the plane of the film. The central ray is directed perpendicularly through the forehead. This projection is used for the demonstration of the: 1:259:II

A. External auditory meatus
B. Nasal septum
C. Optic foramen
D. Anterior nasal spine

84. When performing the parietoacanthial position (Waters), the orbitomeatal line in the average adult forms an angle of ____________ to the plane of the film. 1:382:II

A. 15°
B. 23°
C. 37°
D. 48°

85. When preoperative measurements are to be made of the sinuses in the lateral position, it has been recommended that the radiographs be taken at: 1:378:II

A. 28cm SID
B. 40cm SID
C. 100cm SID
D. 180cm SID

86. The patient is prone with the orbitomeatal line perpendicular to the film. The midsagittal plane is perpendicular. The central ray is directed vertically to pass through the nasion. In this position, the petrous ridges will be seen: 1:245:II

A. In the lower third of the orbits
B. Below the maxillary sinuses
C. Filling the orbits
D. Below the orbits

87. When a submentovertical (SMV) projection or its reverse cannot be otained on a patient for the demonstration of the cranial base, Lysholm recommends that two ____________ positions be taken. 1:257:II

A. Acanthioparietal
B. Posteroanterior oblique
C. Tangential oblique
D. Axiolateral

88. Intra-oral positions utilizing occlusal film can be used to demonstrate: 1:317:II

1. Mandibular symphysis *2. Nasal bones* *3. Hard palate*

A. 1 only
B. 2 only
C. 3 only
D. 1, 2 & 3

89. In order to obtain an anteroposterior (AP) projection of the skull on an infant or young child for the purposes of general survey, the central ray is directed ____________ through the canthomeatal line. 1:22:III

A. Cranially
B. Caudad
C. Perpendicular
D. Parallel

90. An axial transcranial position is taken on a patient for the demonstration of the left temporomandibular joint (TMJ). The interpupillary line is perpendicular to the image receptor, and the central ray is directed 25° caudad. Upon viewing the radiograph, both temporomandibular joints are seen. One is slightly above the center of the image receptor and the other below it. Which is the left temporomandibular joint (TMJ)? 1:359:II

A. The most inferior joint
B. The most superior joint
C. It is not possible to determine.

Pertaining to the radiographic image, answer questions 91 and 92.

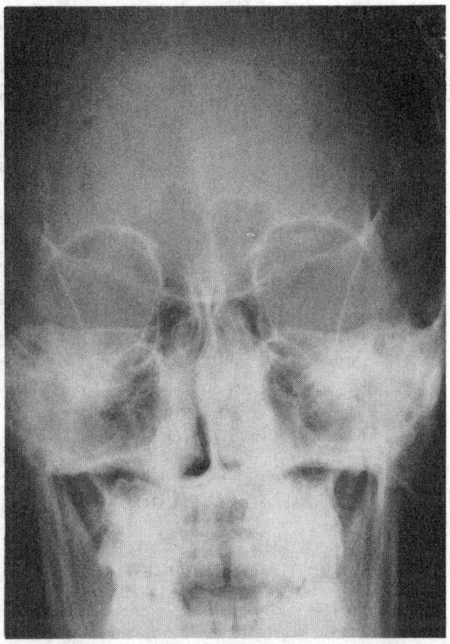

91. In this AP projection of the skull, the position of the petrous ridges within the orbits is most consistent with a __________ of the central ray to the orbitomeatal line: 1:245:II

A. No angulation
B. 15° cephalic angulation
C. 23° cephalic angulation
D. 23° caudad angulation

92. This projection is often employed for the visualization of the: 1:245:II

1. Mastoid air cells *2. Crista galli* *3. Frontal sinus*

A. 1 & 2 only
B. 1 & 3 only
C. 2 & 3 only
D. 1, 2 & 3

Pertaining to the radiographic image, answer questions 93 and 94.

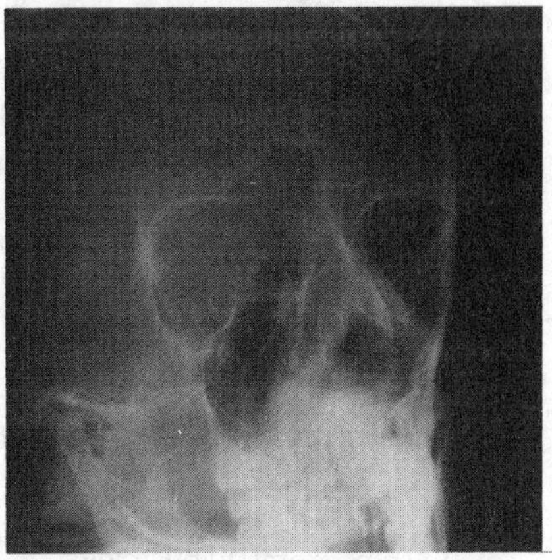

93. This position was taken with the patient in the: 1:275:II

A. PA projection
B. SMV projection
C. Oblique position
D. Lateral position

94. The principle structure of interest in this projection is the: 1:275:II

A. Sella turcica
B. Optic foramen
C. Nasal septum
D. Zygomatic arch

Pertaining to the radiographic image, answer questions 95 and 96.

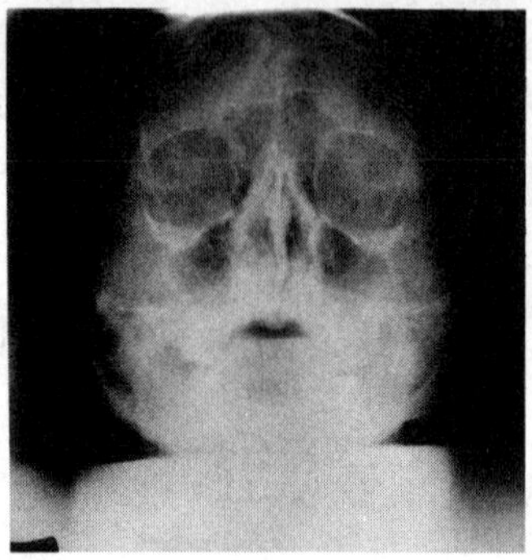

95. In the parietoacanthial (Waters) position, the bony process seen obstructing the maxillary sinuses are the: 1:382:II

A. Zygomatic arches
B. Infraorbital ridges
C. Alveolar process
D. Petrous ridges

96. The desired visualization of the entire maxillary sinuses could best be accomplished by: 1:382:II

A. Tilting of the head
B. Angling the tube cranially
C. Extending the head & chin
D. Flexing the head & chin

Pertaining to the radiographic image, answer questions 97 and 98.

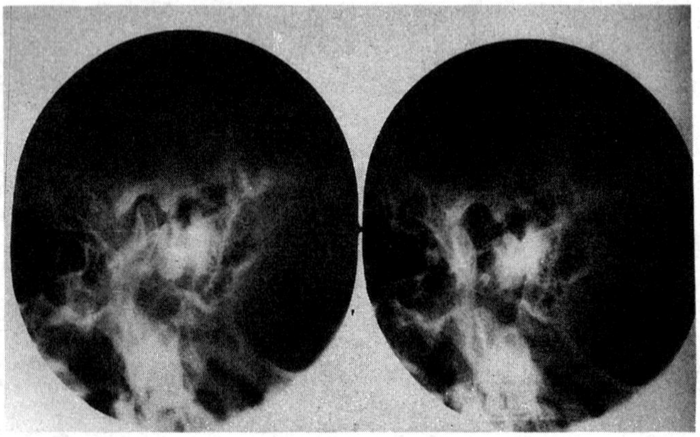

A B

97. The images shown are examples of lateral transcranial positions of the: 1:360:II

A. Tempomandibular joints
B. Sella turcica
C. Sphenoid sinuses
D. Mastoid processes

98. Which of the following radiographic images corresponds to the open mouth view? 1:360:II

A. Film A
B. Film B

Pertaining to the radiographic image, answer questions 99 and 100.

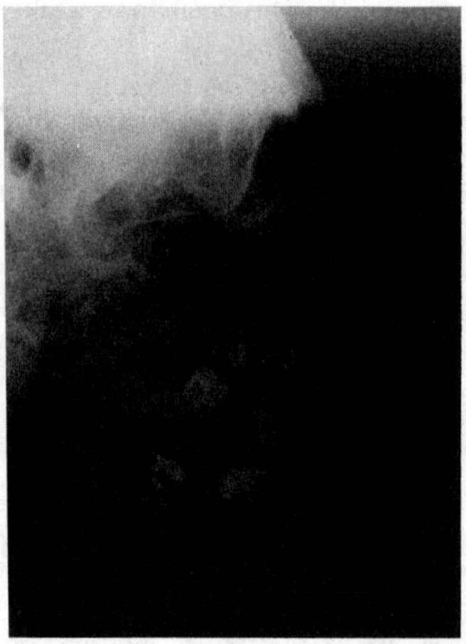

99. This image can be described as a/an: 1:350:II

A. True lateral projection
B. Axiolateral projection
C. PA axial projection
D. Infrosuperior projection

100. Which of the following structures is not well demonstrated in this view? 1:350:I

A. Temperomandibular joint
B. Body of the mandible
C. Alveolar process
D. Ramus of the mandible

Respiratory System

Chapter 50

1. In the lateral position for the chest (erect), it is important that the midsagittal plane is: 1:456:I

A. Vertical to the floor
B. Perpendicular to the image receptor
C. Horizontal to the floor
D. None of the above

2. During a posteroanterior (PA) projection of the chest, the shoulders are rolled forward to: 1:454:I

A. Move the diaphragm to its highest position
B. Move the diaphragm to its lowest position
C. Remove the clavicles from the lung fields
D. Remove the scapulae from the lung fields

3. In the posteroanterior (PA) projection of the chest, the central ray is directed to the level of the: 1:454:I

A. Second thoracic vertebra
B. Fourth thoracic vertebra
C. Seventh thoracic vertebra
D. Ninth thoracic vertebra

4. A patient is positioned for a (PA) posteroanterior projection of the chest. In order to demonstrate the apices free of superimposition of the clavicles, the central ray would be directed: 1:471:I

A. 15° cephalic
B. 15° caudad
C. 40° cephalic
D. 40° caudad

5. An 180cm source-to-image (SID) distance is used in chest radiography in order to: 1:444:I

A. Reduce respiratory motions
B. Remove the scapulae from the lung fields
C. Demonstrate true heart size
D. Project the clavicals above the lungs

6. A right lateral position for the chest is most commonly employed for the visualization of lesions involving the: 1:457:I

A. Mediastinum
B. Right lung
C. Left ventricle
D. Left hemidiaphragm

7. In an erect posteroanterior (PA) projection of the chest, depression of the shoulders directs the clavicles: 1:452:I

A. Above the scapulae
B. Below the hilum
C. Above the apices
D. Below the apices

8. Whenever possible, chest radiography is performed erect in order to: 1:452:I

A. Prevent engorgement of the great vessels
B. Reduce heart magnification
C. Demonstrate fluid levels
D. More than one but not all above

9. In order to evaluate free air in the left pleural cavity, a frontal image is produced when the patient is in the: 1:476:I

A. Prone position
B. Supine position
C. Right lateral decubitus position
D. Left lateral decubitus position

10. Two posteroanterior (PA) projections are taken of a chest; one on inspiration and one on expiration. This is commonly used to evaluate: 1:454:I

1. Foreign bodies *2. Peripheral pneumothoraces* *3. Diaphragm excursion*

A. 1 only
B. 2 only
C. 3 only
D. 1, 2 & 3

11. A __________ of obliquity from the posteroanterior (PA) projection is recommended for the demonstration of the heart and great vessels. 1:460:I

A. 15°
B. 30°
C. 45°
D. 55°

12. During radiography of the lungs, the diaphragm moves to its lowest position during: 1:452:I

A. Erect inspiration
B. Erect expiration
C. Supine inspiration
D. Supine expiration

13. Rotation on a posteroanterior (PA) radiograph of the chest can best be detected by evaluating the: 1:442:I

1. Scapulae *2. Costophrenic angles* *3. Sternoclavicular joints*

A. 1 only
B. 2 only
C. 3 only
D. 1, 2 & 3

14. When performing a posteroanterior (PA) projection of the chest when the heart is of primary interest, the breathing instructions to the patient is to stop breathing at the end of: 1:442:I

A. Full inspiration
B. Normal inspiration
C. Full expiration
D. Normal expiration

15.-24. Pertaining to the radiograph of the chest, place the letter that corresponds to the following structures.

15. __________ Axillary border of the ribs

16. __________ Soft tissue shadow of the breast

17. __________ Apical region of the right lung

18. __________ Right clavicle

19. __________ Hylar region

20. __________ Air-filled trachea

21. __________ Left hemidiaphragm

22. __________ Left ventricle of the heart

23. __________ Posterior portion of the eighth rib

24. __________ Right costophrenic angle

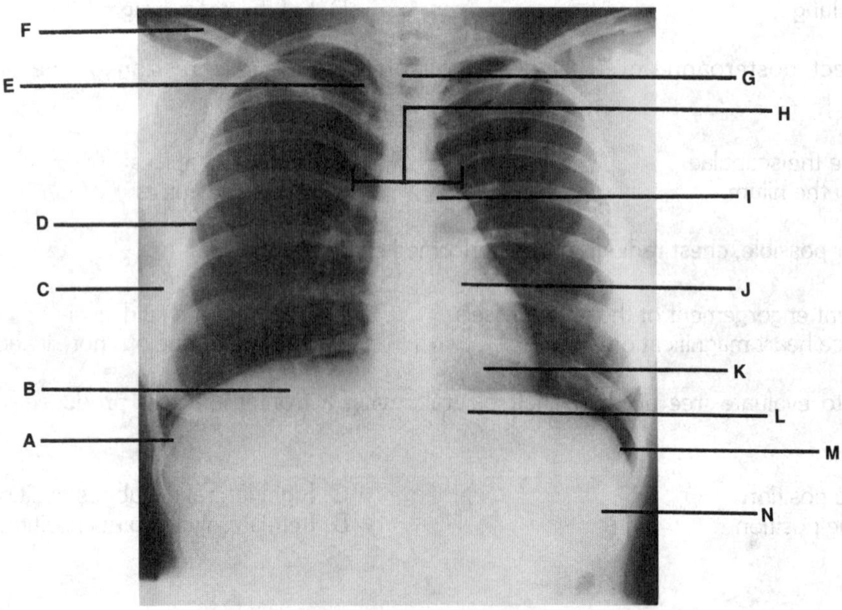

25. The best demonstration of the left atrium is obtained in the: 1:460:I

A. Right posteroanterior (PA) oblique position
B. Left posteroanterior (PA) oblique position
C. Left lateral position
D. Posteroanterior (PA) projection

26. For an anteroposterior (AP) projection of the trachea, the central ray is directed perpendicular to the: 1:446:I

A. Xiphoid
B. Sixth thoracic vertebra
C. Level of the acromioclaviclar joints
D. Manubrium

27. The position of the chest that is used for the demonstration of interlobar effusion and an unobstructed view of the pulmonary apices is the: 1:469:I

A. Left anteroposterior (AP) oblique
B. Apical lordotic
C. Lateral decubitus
D. Posteroanterior (PA)

28. The degree of obliquity from the posteroanterior projection that will best demonstrate pulmonary disease is: 1:461:I

A. 15°
B. 30°
C. 45°
D. 60°

29. The projection of the chest that is commonly employed for the demonstration of free fluid in the costophrenic recess is the: 1:477:I

A. Supine anteroposterior
B. Erect posteroanterior
C. Erect apical lordotic
D. Supine lateral

30. An adequate inspiratory effort of a patient for an erect posteroanterior projection of the chest would be indicated by the visualization of ____________ pairs of ribs above the diaphragms. 1:455:I

A. 10
B. 8
C. 6
D. 4

31. In the lateral position for the chest, the central ray is directed to pass through the coronal plane: 1:456:I

A. Of the midaxillary line
B. 5cm anterior to the midaxillary line
C. 5cm posterior to the midaxillary line
D. None of the above lines

32. When performing a frontal projection of the chest, the posteroanterior (PA) is preferred because: 1:442:I
1. It lowers the diaphragms 2. It demonstrates air/fluid levels 3. It places the heart closest to the film

A. 1 only
B. 2 only
C. 3 only
D. 1, 2 & 3

33. For the demonstration of the maximum amount of lung area on a hypersthenic patient, which of the following is recommended? 1:456:I

A. Erect positions
B. Decubitus positions
C. Second breath inspiration
D. Inhalation, exhalation images

34. In a left posteroanterior oblique (45°) position of the chest (LAO), the lung that will be demonstrated as foreshortened is: 1:460:I

A. The right
B. The left
C. Both will be foreshortened.

35. When radiographing a patient in the supine position for a frontal projection of the chest, movement of the scapulae may be accomplished by: 1:466:I

A. Flexing the elbows and pronating the hands
B. Raising the arms overhead
C. Supinating the hands
D. Abducting the upper arms 90°

Pertaining to the radiographic image, answer questions 36 and 37.

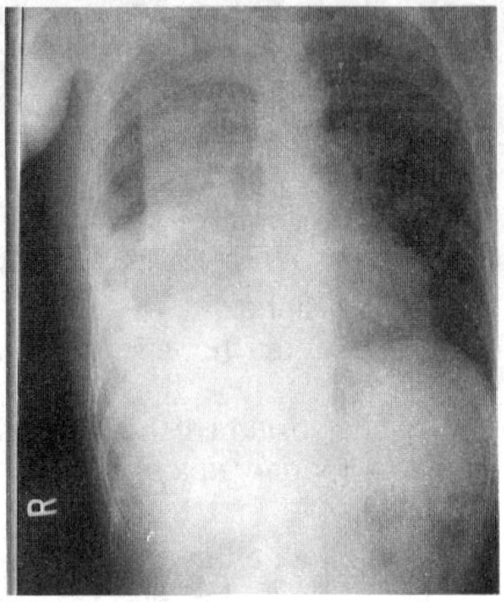

36. This radiographic image represents a __________ position of the chest. 1:476:I

A. Frontal
B. Oblique
C. Lateral
D. Axial

37. This radiographic image was obtained with the patient in the: 1:476:I

A. Right lateral decubitus position
B. Left lateral decubitus position
C. Ventral decubitus position
D. None of the above

Pertaining to the radiographic image, answer questions 38 and 39.

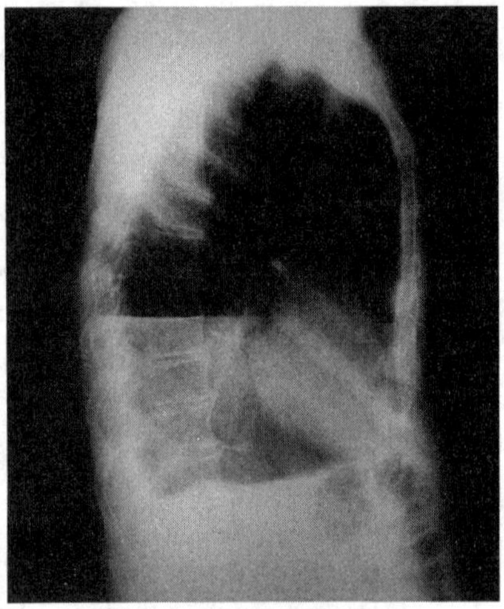

38. This radiographic image represents a __________ position of the chest.: 1:456:I

A. Frontal
B. Oblique
C. Axial
D. Lateral

39. The radiograph was obtained with the patient:

A. Lying on his side
B. Lying on his back
C. Lying on his stomach
D. None of the above

Pertaining to the radiographic image, answer question 40.

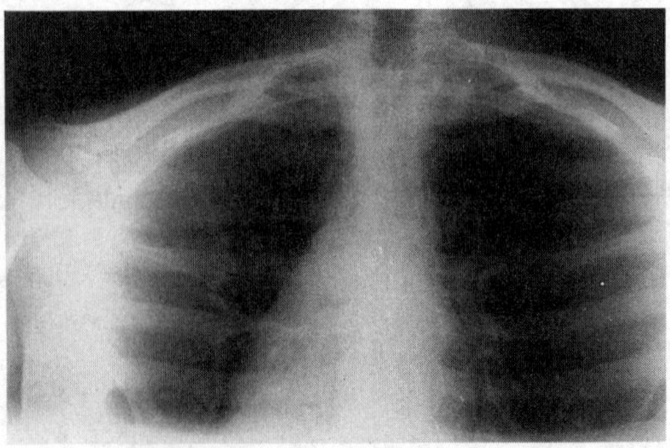

40. This image can best be described as a __________ of the chest. 1:426:I

A. AP axial position
B. Frontal projection
C. PA oblique position
D. Lateral position

Routine Contrast Studies

Chapter 51

1. On the day preceding the administration of the oral contrast media for a cholecystogram, the patient is instructed to eat lunch consisting of: 1:59:II

A. Clear fluids
B. High fat content foods
C. Low fat content foods
D. High fibrous foods

2. What is the maximum amount of air that should normally be used to inflate a retention tip (Bardex)? 1:123:II

A. 10cc
B. 30cc
C. 90cc
D. 250cc

3. In order to accomplish a double-contrast examination of the upper GI tract, barium is administered orally mixed with: 1:100:II

A. Air
B. Glucagon
C. Gas-producing tablets
D. Gastrographin

4. Diaphragmatic (hiatal) hernias are common gastrointestinal problems and can be best demonstrated with the patient in the: 1:112:II

A. Erect position
B. Trendelenburg position
C. Sims position
D. Billings position

5. The demonstration of the upper urinary system can be accomplished during a contrast study termed: 1:167:II

1. Intravenous urography *2. Infusion nephrotomography* *3. Retrograde urography*

A. 1 only
B. 2 only
C. 3 only
D. 1, 2 & 3

6. In a patient with intestinal fluid, or free air, additional anteroposterior (AP) radiographs can be made in the: 1:40:II

1. Erect position *2. Lateral decubitus position* *3. Supine position* *4. Prone position*

A. 1 & 2 only
B. 1 & 3 only
C. 2 & 3 only
D. 1, 2 & 3

7. During exposures of the esophagus, breathing instructions are not normally required because: 1:96:II

A. Rapid exposure times eliminate motion
B. The esophagus is not affected by respiration
C. Deglutition inhibits respiration
D. Motion aids visualization

8. In order to insure that the entire small intestines have been demonstrated in a small bowel series, delayed films are taken until the barium passes into the: 1:115:II

A. Distal jejunum
B. Proximal colon
C. Duodenal bulb
D. Cardiac orifice

9. Before a study of the large intestines is performed, it is important that the patient be prepared. This will generally involve: 1:122:II

1. Colonic lavage *2. Dietary restrictions* *3. Gastric lavage*

A. 1 & 2 only
B. 1 & 3 only
C. 2 & 3 only
D. 1, 2 & 3

10. Which diagnostic procedure requires dietary restrictions preceding the radiographic examination? 1:122:II

1. GI series 2. Barium enema 3. Gallbladder study

A. 1 only
B. 2 only
C. Gallbladder study
D. All of the above

11. The pelvicalyceal system will demonstrate its greatest concentration about ____________ after the intravenous administration of an aqueous iodinated contrast media. 1:168:II

A. 1 min.
B. 3 min.
C. 15 min.
D. 90 min.

12. When performing survey abdominal radiography, the patient is instructed to: 1:37:II

A. Suspend respiration at the end of exhalation.
B. Suspend respiration at the end of inhalation.
C. Breathe normally throughout the exposure.

13.-18. Pertaining to the radiograph of the barium-filled stomach, place the letter that corresponds to the following structures.

13. ________ Pylorus

14. ________ Cardiac orifice

15. ________ Sweep of the duodenum

16. ________ Fundus of the stomach

17. ________ Lesser curve of the stomach

18. ________ Duodenal bulb

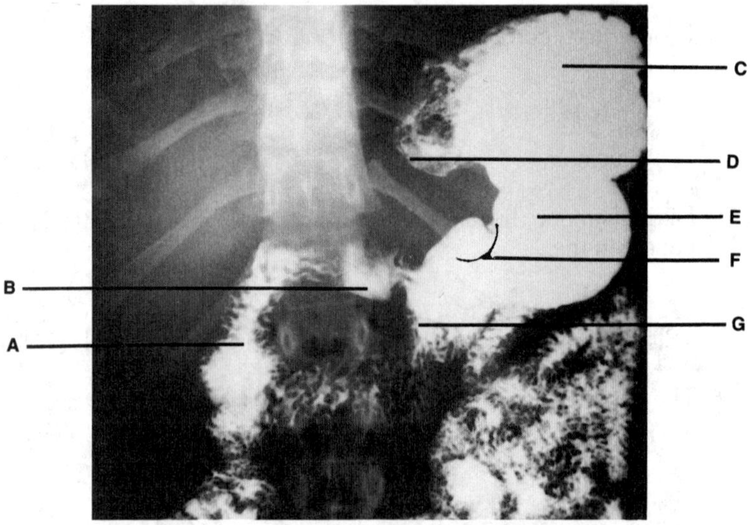

19. To insure proper preparation of the stomach prior to a gastrointestinal series, the patient is instructed to avoid foods and liquids for ____________ before the procedures. 1:98:II

A. 30 min.
B. 1 hr.
C. 8 hrs.
D. 24 hrs.

20. When radiolucent foreign bodies are suspected in the pharynx and upper esophagus, ____________ are sometimes employed to demonstrate these obstructions. 1:93:II

A. Nasogastric tubes
B. Mercury-filled bags
C. Barium saturated cotton balls
D. Iodinated tablets

21. In order to insure that all abdominal structures are included on a survey film, it is important to include the ____________ on the film. 1:37:II

A. Greater trochanters
B. Diaphragm
C. Cardiac shadow
D. Male gonads

22. During a barium enema, the lateral projection and Chassard-Lapine positions are useful in the evaluation of lesions in the: 1:137:II

A. Rectosigmoid junction
B. Ileocecal junction
C. Splenic flexure
D. Distal ascending colon

23. The intravenous injection of a contrast media to demonstrate the urinary system is called: 1:160:II

A. Excretory urography
B. Retrograde pyelography
C. Voiding urethrography
D. Intravenous cholangiography

24. The optimum technique employed for non-contrast abdominal radiography should demonstrate: 1:36:II

A. Long scale contrast
B. Short scale contrast
C. Contrast is not important.

25. The routine procedure for a gallbladder should include films in erect or decubitus positions for evaluation of:
1. Solitary calculi *2. Cholesterol stones* *3. Layered calculi* 1:72:II

A. 1 only
B. 2 only
C. 3 only
D. 1, 2 & 3

26. The best demonstration of the biliary ducts generally occurs in the: 1:72:II

A. Anteroposterior (AP) projection
B. Posteroanterior (PA) projection
C. Right posterior oblique (RPO) position
D. Left posterior oblique (LPO) position

27. During a GI series, the visualization of an airfilled fundus and duodenum is most readily accomplished with a/an ____________ position. 1:104:II

A. Supine
B. Prone
C. Oblique
D. Lateral

28. In intravenous urography, the Trendelenburg position is often employed for the optimum visualization of the: 1:171:II

A. Lower urinary bladder
B. Upper urinary bladder
C. Upper ureters
D. Lower ureters

29. During a barium enema, the visualization of an open hepatic flexure is best seen in the: 1:140:II

A. Posterior anterior position
B. Left AP oblique (LPO) position
C. Right AP oblique (RPO) position
D. Left lateral position

30.-38. Pertaining to the radiograph of the barium-filled colon, place the letter that corresponds to the following structures.

30. ________ Descending colon
31. ________ Hepatic flexure
32. ________ Ileocecal valve
33. ________ Rectum
34. ________ Sigmoid colon
35. ________ Transverse colon
36. ________ Cecum
37. ________ Ascending colon
38. ________ Splenic flexure

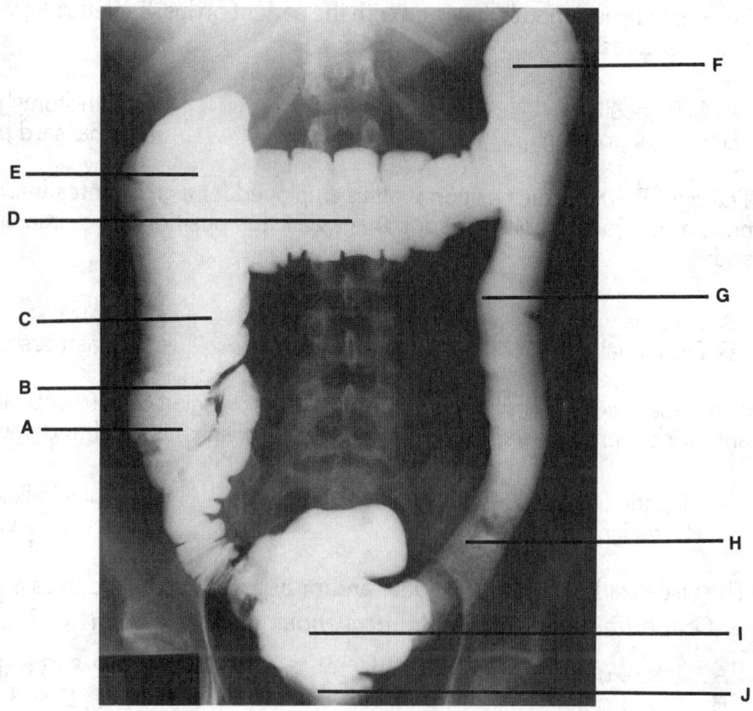

39. The procedure by which a long hollow tube is passed into the stomach and or intenstine is termed: 1:119:II

A. Enteroplasty
B. Gastrointestinal intubation
C. Gastrointestinal reflux
D. Retrograde gastrography

40. When possible, an abdomen survey image is performed after the elimination of: 1:36:II

1. Intestinal gas *2. Fecal material* *3. Radio-opaque medication*

A. 1 only
B. 2 only
C. 3 only
D. 1, 2 & 3

41. During a barium enema, which of the following structures is normally filled first? 1:125:II

A. Cecum
B. Hepatic flexure
C. Descending colon
D. Splenic flexure

42. The amount of urographic contrast media used for an intravenous urogram for a normal adult is about: 1:168:II

A. 5-10cc
B. 30-100cc
C. 100-200cc
D. 200-300cc

43. To insure a maximum concentration of oral cholecystographic contrast media after ingestion, it is important for the patient to avoid: 1:58:II

1. Laxatives *2. Liquid foods* *3. Solid foods*

A. 1 only
B. 2 only
C. 3 only
D. 1, 2 & 3

44. The preliminary radiographs that should be taken on infants who have swallowed or aspirated foreign objects include the area of the: 1:517:I

1. Neck *2. Chest* *3. Abdomen*

A. 1 & 2 only
B. 1 & 3 only
C. 2 & 3 only
D. 1, 2 & 3

45. The movement of an opacified gallbladder in relation to the phase of respiration can be quite dramatic (5-15cm), therefore, it is advisable to expose an image at the end of: 1:60:II

A. Inhalation
B. Exhalation

46. An enema tip is generally inserted with the patient on his/her left side, leaning forward, and the right knee flexed. This describes the: 1:125:II

A. Sims position
B. Billings position
C. Lithotomy position
D. Chassard position

47. The right P.A. oblique position is often employed for a gastrointestinal examination because it tends to increase peristalsis. Radiographs, or spot images, in this position are primarily intended to visualize the pyloric canal and: 1:105:II

A. Fundus
B. Duodenal bulb
C. Antra
D. Gastroesophageal junction

48. The patient is placed supine and the central ray angled 30°-40° cephalad and centered 4cm (2") inferior to the anterior superior iliac spines. This projection (Billings) is used to demonstrate the: 1:139:II

A. Hepatic flexure
B. Rectosigmoid area
C. Splenic flexure
D. Ileocecal valve

49. The visualization of the gallbladder and/or biliary ducts by a contrast media can be accomplished by: 1:53:II

1. Oral administration *2. Intravenous administration* *3. Direct (T-tube) administration*

A. 1 & 2 only
B. 1 & 3 only
C. 2 & 3 only
D. 1, 2 & 3

50. The Trendelenburg position is often employed in a G.I. series to show: 1:112:II

A. Gastroesophageal reflux
B. Duodenal ulcerations
C. Pyloric stenosis
D. Hypermotility

51. When radiographing for the presence of fluid in the semierect patient, the central ray is directed: 1:376:II

A. Perpendicular to the image receptor
B. Horizontal to the floor
C. Vertical to the floor
D. Perpendicular to the patient

52. The minor side effects that result from the use of oral cholecystographic agents are nearly all related to the action on the: 1:58:II

A. Liver
B. Gallbladder
C. Kidney
D. Intestinal mucosa

53. The centering for the RAO position for thè esophagus is at the vertebral level of: 1:95:II

A. C7-D1
B. T5-T6
C. T10-T11
D. D12-L1

54. When the normal urogram images are completed, the patient is instructed to go to the lavatory and void. Upon returning, post-voiding images are taken to show: 1:170:II

1. Residual urine *2. Calyceal retention* *3. Prostate glad enlargement*

A. 1 & 2 only
B. 1 & 3 only
C. 2 & 3 only
D. 1, 2 & 3

55. During intravenous cholegraphy, maximum opacification of biliary ducts occurs about ____________ minutes after the injection of contrast media. 1:72:II

A. Two
B. Five
C. Thirty
D. One hundred and twenty

56. Preliminary preparation for the non-acute abdomen include all of the following, except: 1:36:II

A. Administration of laxatives
B. Colonic lavage
C. Gastric lavage
D. Controlled food intake

57. In certain barium enema studies, a retention tip (Bardex) is used to help the patient retain the enema. Because of the danger of intestinal rupture, it is recommended that inflation of the tip be done in conjunction with (a): 1:123:II

A. Fluoroscopic control
B. Pressure injector
C. Compression paddle
D. None of the above

58. A patient has a suspected ruptured viscus. Which position for the abdomen is often used to demonstrate this condition? 1:40:II

A. Supine
B. Lateral
C. Erect AP
D. Chassard-Lapine

59. During a double contrast study, when gas is distending the colon, it is important to ____________ the exposure technique in order to avoid overpenetration. 1:142:II

A. Increase
B. Decrease
C. No change is necessary.

60. When tomography is performed on the average patient, the level of the kidney is found about ____________ above the table top. 1:61:III

A. 3cm
B. 9cm
C. 15cm
D. 20cm

61. In recent years the use of high density barium suspensions have gained a greater acceptance because of their: 1:122:II

A. Lower toxicity
B. Higher absorption rate
C. Superior retention
D. Uniform coating of the lumen

62. Which of the following examinations can be performed without any type of patient preparation? 1:93:II

1. Cholecystogram *2. Nephrotomography* *3. Esophagram*

A. 1 only
B. 2 only
C. 3 only
D. 1, 2 & 3

63. Radiographic examination of the small intestine with barium sulfate suspension can be accomplished by: 1:119:II

1. Intravenous injection *2. Reflux enema* *3. Oral administration*

A. 1 & 2 only
B. 1 & 3 only
C. 2 & 3 only
D. 1, 2 & 3

64. In order to prevent possible damage to the rectum, the tip should not be inserted more than: 1:124:II

A. 1 inch (2.5cm)
B. 2 inches (5cm)
C. 3 inches (7cm)
D. 4 inches (10cm)

65. Most radiologists recommend a posteroanterior (PA) projection in the visualization of the gallbladder. This serves to: 1:59:II

A. Reduce liver motility
B. Increase bladder opacification
C. Diffuse overlying rib shadows
D. Reduce part-film distance

66. To obtain optimum visualization of the distal ureters during an intravenous urogram, a/an ____________ is sometimes employed. 1:166:II

A. Compression band
B. Foley catheter
C. Ambu bag
D. Compression paddle

67. During a double contrast enema, a right lateral decubitus position is used with a horizontal central ray and a stationary grid. This projection is primarily intended to show lesions of the: 1:142:II

A. Medial ascending colon
B. Ileocecal valve
C. Transverse colon
D. Medial descending colon

68. The anteroposterior (AP) projection (supine) of the stomach best shows the barium-filled structures of the: 1:110:II

A. Cardiac orifice and esophagus
B. Fundus and duodenum
C. Greater and lesser curves
D. Pylorus and fundus

69. The esophagus is filled with a thick barium mixture, and right P.A. oblique position radiographs are taken of the chest. These images will often aid in the demonstration of the ____________ the heart. 1:460:I

A. Right atrium
B. Right ventricle
C. Left atrium
D. None of the above

70. The routine for a gallbladder study requires that a scout image is taken before the administration of the contrast agents. This is useful in the evaluation of: 1:59:II

1. Cholecystitis *2. Radiolucent stones* *3. Radiopaque stones*

A. 1 only
B. 2 only
C. 3 only
D. 1, 2 & 3

71. During intravenous urography, some radiologists recommend a 30 second image be taken immediately after the bolus of contrast is injected, which shows the outline of the kidney. This is called a: 1:175:II

A. Tomogram
B. Nephrogram
C. Cystogram
D. Pyelogram

72. A posteroanterior (PA) oblique position in the ____________ and ____________ position are often included in gallbladder filming routines to free it from adjacent structures: 1:68:II

A. LAO, right lateral
B. LPO, right lateral
C. RAO, left lateral
D. LPO, left lateral

73. During urography of infants and children, it has been recommended that the administration of a carbonated drink may improve visualization of the kidneys by: 1:62:II

A. Dispersing the fecal material
B. Compressing the ureters
C. Distending the stomach
D. Distending the intestines

74. The Valsalva maneuver in which the patient is asked to increase abdominal pressure by "bearing down" is often used to demonstrate or evaluate: 1:95:II

A. Polyps
B. Diverticuli
C. Esophageal varices
D. Peptic ulcers

75. The reduction of colonic spasm, as well as, better retention and mild anesthetic effects, are generally seen when ____________ enema preparations are employed. 1:124:II

A. Hot
B. Warm
C. Room temperature
D. Cold

Pertaining to the radiographic image, answer questions 76 and 77.

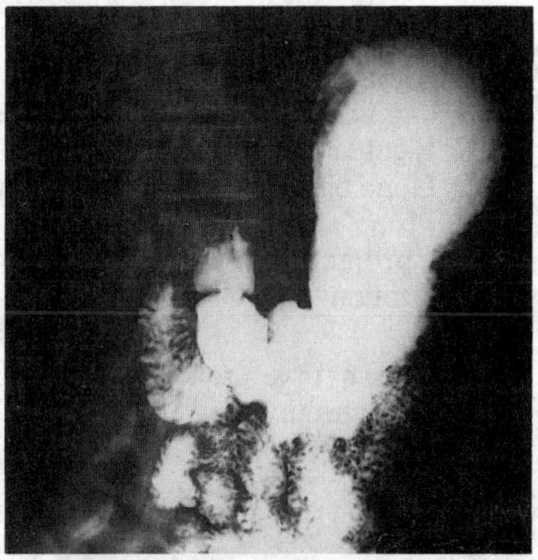

76. The radiographic image was obtained with the patient in the: 1:106:II

A. Supine position
B. Prone position
C. Right anterior oblique position
D. Lateral position

77. This position of the barium filled stomach is most often employed to provide a non-superimposed image of the: 1:106:II
1. Pylorous 2. Duodenal bulb 3. Retrogastric space

A. 1 only
B. 2 only
C. 3 only
D. 1 & 2 only

Pertaining to the radiographic image, answer questions 78 and 79.

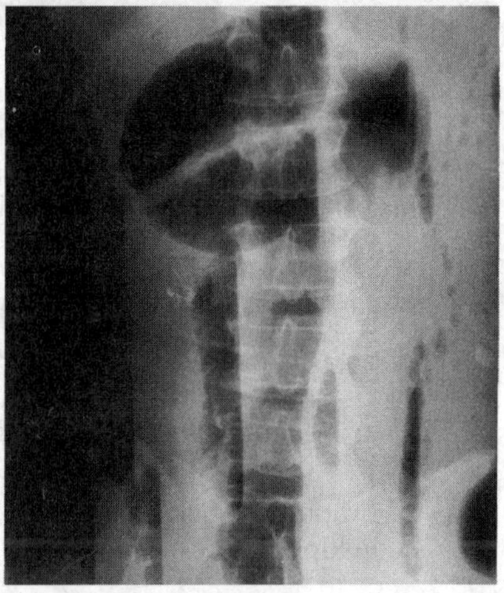

78. This position of the abdomen was obtained with the patient in the: 1:140:II

A. Right lateral decubitus position
B. Left lateral decubitus position
C. Ventral decubitus position
D. Dorsal decubitus position

79. This radiographic image represents a __________ of the abdomen. 1:40:II

A. Frontal projection
B. Oblique position
C. Lateral position
D. Axial position

Pertaining to the radiograph, answer questions 80 and 81.

80. The air contrast enema image shown was taken with the patient in the: 1:143:II

A. Erect position
B. Supine position
C. Lateral decubitus position
D. Dorsal decubitus position

81. The radiograph represents a ____________ of the large intestine. 1:143:II

A. Frontal projection
B. Oblique position
C. Lateral position
D. Axial position

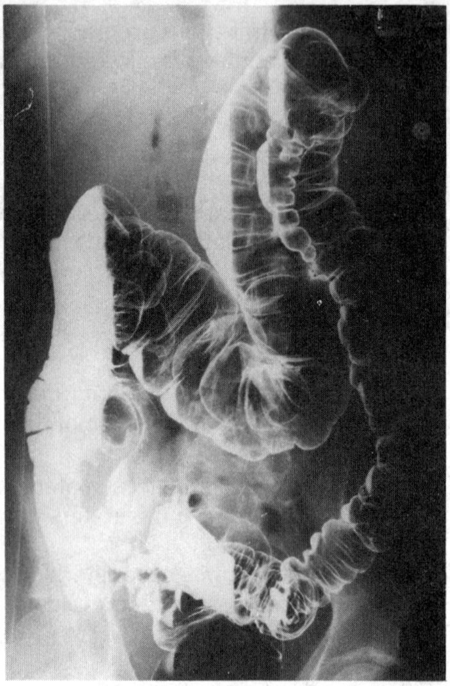

Pertaining to the radiographic image, answer questions 82 and 83.

82. This projection of the barium filled colon was obtained in the: 1:140:II

A. Right posterior oblique position
B. Left posterior oblique position
C. Right lateral position
D. None of the above

83. This projection is most commonly employed to improve the visualization of the: 1:140:II

A. Lesser curvature
B. Greater curvature
C. Hepatic flexure
D. Splenic flexure

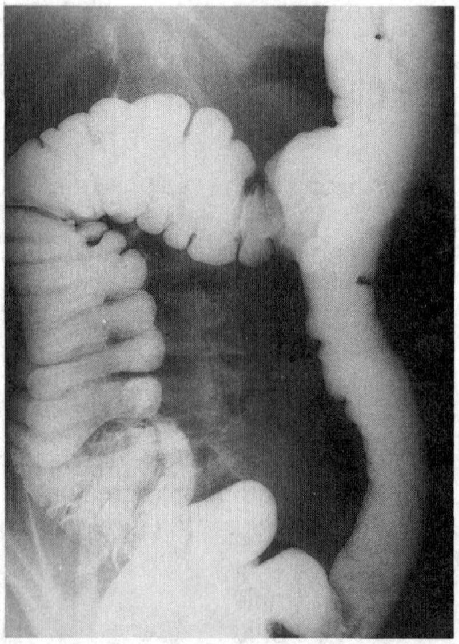

Pertaining to the radiographic image, answer questions 84 and 85.

84. This projection of the abdomen is an example of a patient in the: 1:42:II

A. Erect frontal projection
B. Erect lateral position
C. Supine oblique position
D. Supine frontal projection

85. This projection is most useful in the determination of: 1:42:II

A. Diaphragm hernias
B. Pyloric stenosis
C. Colonic spasms
D. Air/fluid levels

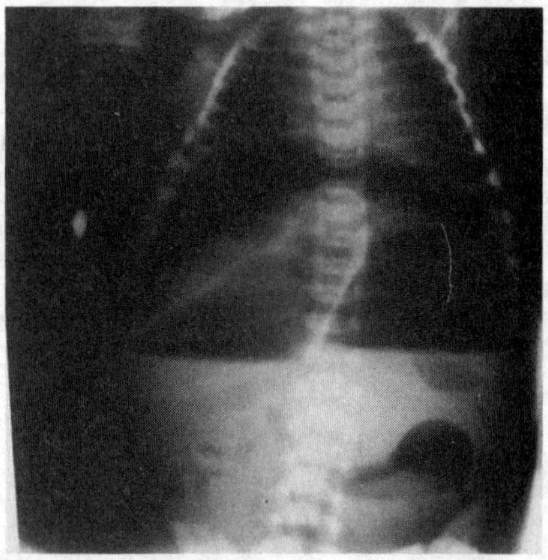

Special Procedures

Chapter 52

1. The radiographic evaluation of the central nervous system's structures located within the spinal canal is termed: 1:500:II

A. Meningography
B. Myography
C. Ventriculography
D. Myelography

2. The specific term applied to any radiographic opacification study of the veins is called: 1:535:II

A. Angiography
B. Venography
C. Digital angiography
D. Arteriography

3. The radiographic examination of the breast is termed: 1:456:II

A. Sialography
B. Mammography
C. Salpingography
D. Duodenography

4. Arthrography is employed for the radiographic visualization of the soft tissue structures of the: 1:487:II

A. Mediastinum
B. Blood vessels
C. Pelvic cavity
D. Joints

5. The placement of a T-tube into the biliary duct network for postoperative drainage is often used in conjunction with a contrast injection to evaluate: 1:54:II

1. Gallbladder function *2. Cystic duct patency* *3. Common bile duct patency*

A. 1 only
B. 2 only
C. 3 only
D. 1, 2 & 3

6. The delineation of the uterine cavity and lumen of the fallopian tubes can be accomplished by the introduction of a positive contrast media. This procedure is called: 1:199:II

A. Hysterosalpingography
B. Pelvic pneumography
C. Pelvimetry
D. Vaginography

7. Cephalopelvimetry is a radiographic procedure used for the evaluation and measurement of the: 1:200:II

1. Fetal head size *2. Pelvic inlet* *3. Pelvic outlet*

A. 1 only
B. 2 only
C. 3 only
D. 1, 2 & 3

8. The evaluation of the pelvicalyceal system and ureters by a direct catheterization of the ureter is termed: 1:179:II

A. Intravenous urography
B. Infusion tomography
C. Retrograde urography
D. Nephrotomography

9. In order to maximize the contrast of human breast tissue, the kilovoltage range recommended is: 1:456:II

A. 5-15 kVp
B. 15-35 kVp
C. 45-60 kVp
D. 70-80 kVp

10. The most common site for a spinal puncture is between the: 1:500:II

A. First and second lumbar vertebrae
B. Third and fourth lumbar vertebrae
C. Tenth and eleventh thoracic vertebrae
D. Sixth and seventh cervical vertebrae

11. During a retrograde urogram, the vesicoureteral orifices are located and catheterized using a: 1:179:II

A. Colonoscope
B. Gastroscope
C. Proctoscope
D. Cystoscope

12. During thoracic and cervical myelography, the ascent of contrast media into the ventricles of the brain can be limited by acute ________________ of the head. 1:506:II

A. Flexion
B. Extension
C. Lateral flexion
D. Lateral rotation

13. The normal position(s) employed for proper examination of the breast includes the: 1:470:II

1. Craniocaudal position *2. Mediolateral oblique* *3. Axillary position*

A. 1 only
B. 2 only
C. 3 only
D. 1, 2 & 3

14. Sialography is a radiographic procedure used for the evaluation of the: 1:4:II

A. Nasolacrimal ducts
B. Salivary glands
C. Mastoid air cells
D. Paranasal sinuses

15. During a dorsal spine myelogram, the most likely injection site will be the: 1:505:II

A. Cisterna magna
B. Upper dorsal region
C. Lower dorsal region
D. Lumbar region

16. The principal type of contrast media necessary to outline the renal parenchyma is a/an: 1:156:II

A. Aqueous barium suspension
B. Aqueous iodinated solution
C. Oil-based iodine
D. Gaseous media

17. In a myelogram, spinal fluid is removed before the introduction of a contrast media. This helps to prevent excessive: 1:503:II

A. Cerebral hemorrhaging
B. Bacterial contamination
C. Spinal reflux
D. Intracranial pressure

18. The term cystography is employed for the radiographic examination of the: 1:183:II

A. Gallbladder
B. Urinary bladder
C. Splenic vein
D. Pancreas

19. A specialized mammography system is likely to include: 1:465:II

1. Molybdenum target *2. Fractional focal spots* *3. Compression devices*

A. 1 only
B. 2 only
C. 3 only
D. 1, 2 & 3

20. After the removal of the gallbladder, a tube may be placed in the collecting system for evaluation of the ducts. This is termed a/an: 1:54:II

A. Cantor tube
B. NG tube
C. T-tube
D. Ostomy tube

21. The insertion of a spinal needle into a lumbar interspinous space is often facilitated by ____________ of the spine. 1:502:II

A. Flexion
B. Extension
C. Lateral Rotation
D. Medial Rotation

22. In postoperative (T-tube) cholangiography, the status of the sphincter of Oddi is determined by contrast media spillage into the: 1:54:II

A. Peritoneum
B. Splenic vein
C. Duodenum
D. Gallbladder

23. The compression of breast tissue during mammography serves to: 1:64:III

1. Equalize breast thickness *2. Reduce exposure time* *3. Increase image quality*

A. 1 only
B. 2 only
C. 3 only
D. 1, 2 & 3

24. During myelography, the principal advantage of using a water-soluble contrast media is: 1:502:II

A. Its greater viscosity
B. Its slower absorption rate
C. Post-procedural removal is unnecessary
D. The elimination of toxic reactions

25. Which of the following anatomical areas are examined by contrast arthrography? 1:488:I

1. Elbow joints *2. Hip joints* *3. Wrists*

A. 1 & 2 only
B. 1 & 3 only
C. 2 & 3 only
D. 1, 2 & 3

26. The position most commonly employed for male cystourethrography is the: 1:189:II

A. Anteroposterior (AP)
B. 35° RPO
C. Erect true lateral
D. Supine true lateral

27. Expiratory and inspiratory phonation are two common techniques employed during: 1:15:II

A. Dacryocystography
B. Arthrography
C. Laryngopharyngography
D. Thyroid angiography

28. During lower extremity venography, the x-ray table is tilted to the semierect position or tourniquets are placed on the extremities to help insure filling the: 1:542:II

A. Superficial leg veins
B. Deep leg veins
C. Inferior vena cava
D. Pelvic veins

29. During an ERCP a small cannula is passed through the endoscope and placed into the: 1:80:II

A. Uretal orfice
B. Common bile duct
C. Pyloric Antrum
D. Hepatic Vein

30. Which of the following examinations is normally performed on a gravid patient? 1:207:II

A. Cephalometry
B. Hysterosalpingography
C. Pelvic pneumography
D. Vaginography

31. During a hysterosalpingogram, the contrast media is introduced into the uterine cavity and should normally spill over into the: 1:200:II

A. Ovary
B. Cervix
C. Peritoneum
D. Vagina

32. Which mammographic imaging system will result in the lowest patient exposure? 1:465:II

A. Cardboard-film combination
B. Single-screen film combination
C. Electrophographic technique (xeroradiograph)
D. All give about the same dose.

33. In cases of lumbar region obstruction, a myelogram may be performed by using a ____________ injection site. 1:107:III

A. Ventricular
B. Subdural
C. Cisternal
D. Cranial sinus

34. The injection of contrast media during arthrography is made into the: 1:488:I

A. Capsular space
B. Bursae
C. Menisci
D. Ligaments

35. During cystography, when ureteral reflux is suspected, it is often necessary to include a: 1:182:II

A. Voiding view
B. Trendelenberg view
C. Nephrotomogram view
D. Breathing view

36. The most common secretory stimulant used in radiographic contrast examination of the salivary glands is (a): 1:5:II

A. Lemon wedge
B. Fatty meal
C. Chocolate bar
D. Esophotrast

37. A relatively new procedure called endoscopic retrograde cholangiopancreatography (ERCP) can be used to evaluate the: 1:61:II

A. Biliary duct network
B. Upper respiratory system
C. Cystourinary system
D. Lower GI tract

38. Operative cholangiography is often employed during biliary tract surgery to evaluate the patency of the:
1. Intrahepatic ducts *2. Extrahepatic ducts* *3. Sphincter of Oddi* 1:56:II

A. 1 only
B. 2 only
C. 3 only
D. 1, 2 & 3

39. A myelogram is being performed after needle placement, and an amount of spinal fluid equal to contrast media is removed. This is true for: 1:503:II
1. Water-based media *2. Oil-based media* *3. Negative media*

A. 1 only
B. 2 only
C. 3 only
D. 1, 2 & 3

40. The orthoroentgenographic measurements of the extremities (scanography) requires the use of: 1:482:I
1. Metallic ruler *2. Sussman ruler* *3. Monometer*

A. 1 only
B. 2 only
C. 3 only
D. 1, 2 & 3

41. In recent years, interventional angiography has expanded rapidly. A procedure that can be used for the occlusion of arteriovenous malformations is termed: 1:566:II

A. Transluminal angioplasty
B. Elective embolization
C. Peripheral angiography
D. Spontaneous thrombolization

42. Dilatation of stenotic vessels is often possible using a Gruntzig balloon tip catheter inserted into the desired vessel, with the balloon expanded. This describes a procedure termed: 1:561:II

A. Therapeutic anastomosis
B. Elective vascularization
C. Elective embolization
D. Transluminal angioplasty

43. At the present time the advantages of digital subtraction angiography over conventional angiography include: 1:173:II
1. Post processing capabilities *2. Improved spatial resolution* *3. Reduced motion sensitivity*

A. 1 only
B. 2 only
C. 3 only
D. 1, 2 & 3

44. During pelvimetry, it is often necessary to obtain precise measurements of the pelvis and fetal head. This is often performed using a: 1:207:II

A. Colcher-Sussman ruler
B. Step wedge
C. Wedge filter
D. Parallax motion technique

45. The injection site for a lower extremity venogram is normally the/a: 1:542:II

A. Femoral artery
B. Femoral vein
C. Abdominal aorta
D. Superficial foot vein

46. For the lateral projection in pelvimetry using the Colcher-Sussman method the legs are partially extended to prevent: 1:207:II

1. Pelvic rotation *2. Magnification* *3. Obscuring of the pubic bones*

A. 1 only
B. 2 only
C. 3 only
D. 1, 2 & 3

47. For routine projections of the breast, the structure that should most often be seen in profile is the: 1:470:II

A. Sternal angle
B. Axilla
C. Tail
D. Nipple

48. Cerebral pneumography was, until recently, used for the examination of the air-filled ventricles of the brain. This procedure has been replaced by the use of: 1:130:III

A. Computed tomography
B. Digital radiography
C. Ultrasonography
D. Polytomography

49. The general term employed for the visualization of all of the vascular structures of the body by a contrast media is: 1:520:II

A. Arteriography
B. Venography
C. Angiography
D. Lymphangiography

50. The percutaneous introduction of catheters in most angiographic studies involves a multiset process using compound needles and guidewire, which is referred to as the: 1:524:II

A. Valsalva technique
B. Haschek technique
C. Lindenthal technique
D. Seldinger technique

51. When performing a venogram of the lower extremities, the legs should be positioned: 1:542:II

A. True anteroposterior (AP)
B. Internally rotated 30°
C. Internally rotated 15°
D. Externally rotated 15°

52. The demonstration of the lymphatic vessels and nodes by a contrast media is termed: 1:574:II

A. Sialography
B. Lymphangiography
C. Splenography
D. Portography

53. During fetography, fetal movement can often be prevented by: 1:207:II

1. Hyperaeration of the mother *2. Use of a compression band* *3. Long exposure times*

A. 1 only
B. 2 only
C. 3 only
D. 1, 2 & 3

54. In the Colcher-Sussman method of pelvimetry, the pelvimeter is placed 10cm below the level of the symphysis. This corresponds to the level of the: 1:207:II

A. Tip of the coccyx
B. Anterior superior iliac spine
C. Greater trochanters
D. Ischial tuberosities

55. In all mammographic projections, the side marker is placed: 1:470:II

A. Toward the axilla
B. Toward the sternum
C. Anterior to the nipple
D. None of the above

56.-60. In the following, match the angiographic procedure to the principal structures demonstrated:

56. ____ Pulmonary arteriography
57. ____ Cerebral angiography
58. ____ Celiac arteriography
59. ____ Renal arteriography
60. ____ Superior mesenteric arteriography

A. Vessels of the stomach
B. Vessels of the lungs
C. Vessels of the kidneys
D. Vessels of the small intestines
E. Vessels of the brain

Pertaining to the radiographic image, answer questions 61 and 62.

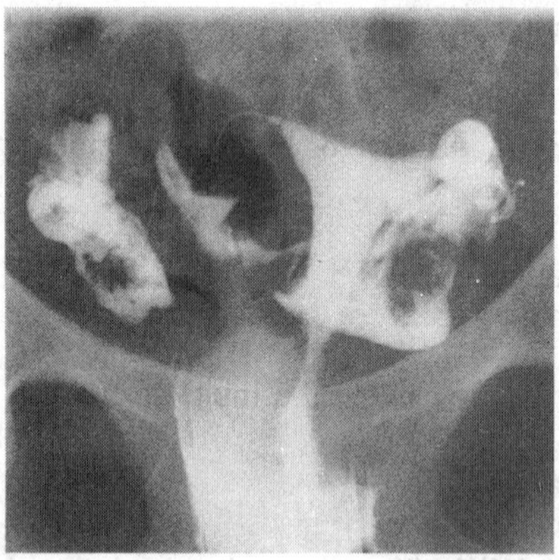

A

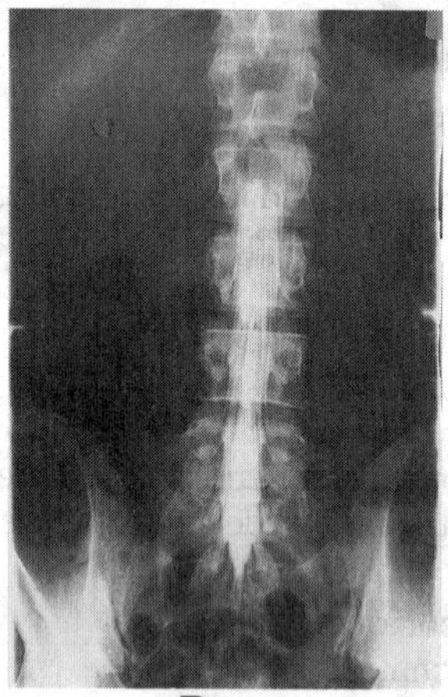

B

61. The radiographic image of Film A represents special contrast study used in the evaluation of the: 1:200:II

A. Urinary bladder
B. Rectum
C. Uterus and oviducts
D. Ovaries

62. The radiographic image of Film B represents a special contrast study used in the evaluation of the: 1:500:II

A. Intervertebral discs
B. Ventricular system
C. Spinal canal
D. Venous system

Pertaining to the radiographic image, answer question 63.

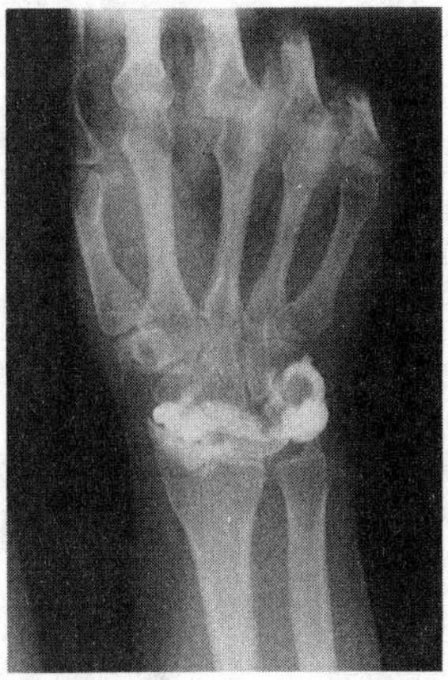

63. This radiographic image is an example of an examination used in the evaluation of (the): 1:494:I

1. Hairline fracture of a bone *2. Soft tissue structure of a joint*

3. Joint calcifications

A. 1 only
B. 2 only
C. 3 only
D. 1, 2 & 3

Pertaining to the radiographic image, answer questions 64 and 65.

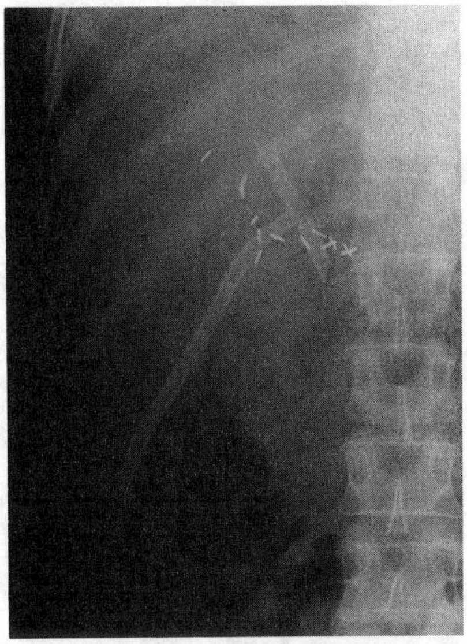

A

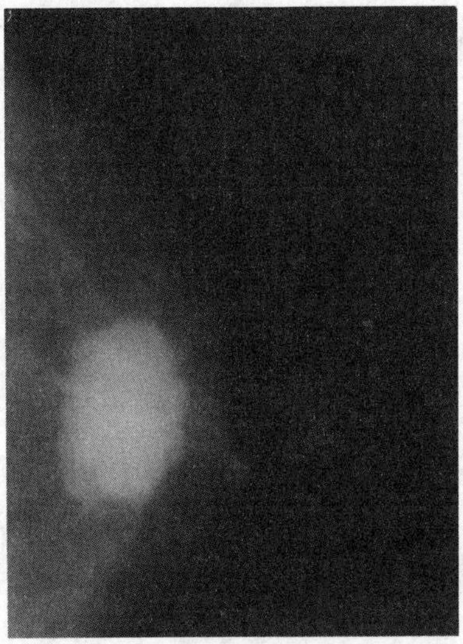

B

64. The radiographic image of Film A represents a scout film prior to a contrast study called: 1:56:II

A. Endoscopic retrograde cholangiopancreatography
B. Retrograde urography
C. T-Tube cholangiography
D. Abdominal angiography

65. The radiographic image of Film B represents a noncontrast study of the: 1:461:II

A. Salivary ducts
B. Ventricular system
C. Breast
D. None of the above

Section VI
Radiographic Procedures
Explanations

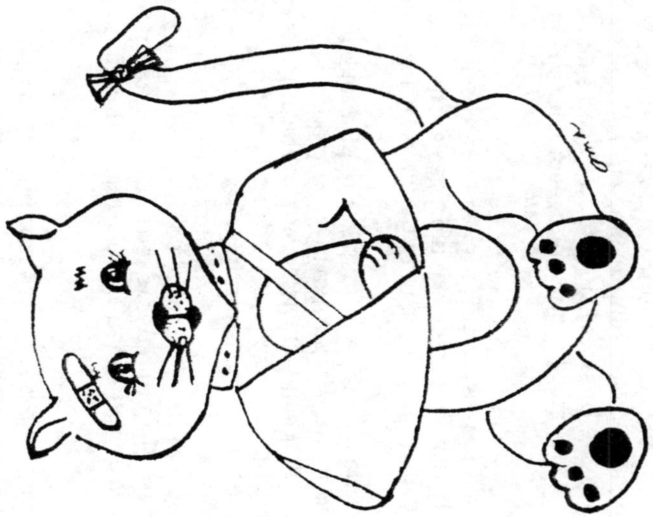

Positioning Terminology

Chapter 45

1. D

There are three fundamental planes used to help the technologist properly position an anatomic structure to obtain the best possible diagnostic radiograph. (1) The midsagittal, or median sagittal plane (MSP) passes vertically through the midline of the body dividing it into equal right and left halves. Any plane passing through the body that is parallel with the midsagittal plane is called a sagittal plane. (2) The midcoronal, or midfrontal plane is named for the coronal suture, and passes vertically through the midaxillary region of the body, at right angles to the midsagittal plane. It divides the body into anterior (ventral) and posterior (dorsal) portions. Any vertical plane passing parallel to the midcoronal plane is called a coronal plane. (3) The transverse, or axial, plane passes crosswise through the body at right angles to the midsagittal and coronal planes, dividing it into superior and inferior portions.

2. B

Bodily habitus refers to the general form of the body and can be used to help predict the size, shape and position of the internal organs. Individuals are classified in one of four types: (1) Sthenic type: the most common or average type consisting of 50% of the population; (2) Hypersthenic type: (5%) A massive build characterized by a high laterally placed gallbladder, high transverse stomach and large laterally positioned colon; (3) Hyposthenic type: (35%) Smaller than average individual found to have long vertical stomach, low vertical gallbladder and more centralized smaller colon; (4) Asthenic type (10%) : An extremely slender person characterized by organs which are smaller and lower than in a hyposthenic individual.

3. A

Cranial and cephalic refer to skull, and parts toward the head end of the body. Caudad or caudal refer to parts away from the head, or toward the feet; dorsal refers to the posterior or back side of the part, i.e., the dorsum of the hand is the back of the hand.

4. D

In order to provide a single reference for describing a patient position, the concept of anatomic position was developed. This refers to a person facing forward, standing erect, feet together with arms at the sides, palms facing forward. The radiograph placed on a view box or illuminated viewer should follow this convention.

5. A

The term axial or semiaxial normally refers to a central ray angulation following the long axis of the body or part i.e. AP axial (Towne) skull. The definition also includes extension and flexion movements on longitudinal planes which can also provide axial projections. In a tangential projection, the central ray skims a rounded body segment in an attempt to profile and eliminate superimposition with adjacent structures; i.e., Settegast (Sunrise) method for the patella.

6. A

When a patient is placed recumbent or lying down and the central ray is directed horizontal to the floor, a decubitus position has been obtained. These positions are especially useful in the diagnosis of air-fluid levels especially in the chest, abdomen and skull. A lateral decubitus position is defined as a patient lying on the affected side, with the anterior or posterior body surface closest to the image receptor. Because of the horizontal central ray, a frontal (PA/AP) image will be demonstrated. In a ventral decubitus position, the patient is lying on the ventral (anterior) surface of the body as in the prone position. In a dorsal decubitus position, the patient is lying on the dorsal (posterior) surface of the body facing up or supine. The resulting radiograph for the dorsal and ventral decubitus positions provide a lateral image of the anatomic region.

7. B See explanation number 2 of this section.

8. C See explanation number 1 of this section.

9. D The term distal designates a structure farthest from the center or midline of the body. Proximal refers to a structure or part closest to the center or midline of the body; i.e., the foot is distal to the ankle; the radial head is located at the proximal end of the radius.

10. A Inversion refers to turning of a body part inward. Eversion refers to turning of a body part outward; i.e., movement of the foot by turning it outward at the ankle joint.

11. D Anterior and ventral are both terms that describe the front surface of the body. Posterior and dorsal describe the back surface of the body.

12. D Extension, the opposite of flexion, increases the angle of the articulating bones by straightening the joint and stretching the anatomic structure. Flexion is a bending movement where the angle of the bones of the joint is diminished. In the vertebral column a backward bending movement is called extension, as opposed to forward bending movement known as flexion.

13. A The torso of the body is divided into two main sections, called the thoracic and abdominal cavities, which are separated by a dome shaped muscle called the diaphragm. The respiratory and cardiovascular structures associated with the heart occupy the thoracic cavity while most of the major organs of digestion and the urinary tract occupy the abdominal cavity. The pelvis is a subdivision of the lower portion of the abdominal cavity.

14. A See question number 5.

15. B Since most anatomical structures cannot be directly visualized, external indicators must be used to accurately position the patient. The gonion (angle of the mandible) is an external landmark that is located at the level of the 2nd and 3rd cervical vertebrae.

16. B Portions of the body located away from the midline or middle of an anatomic structure are referred to as lateral. The opposite, medial, refers to anatomic structures located toward the middle or median plane of the body; i.e., the pisiform is located on the medial side of the wrist.

17. B See question number 2.

18. B The tenth thoracic vertebra is located in the same axial plane as the xiphoid tip on the inferior aspect of the sternum.

19. C The ventral decubitus position places the patient prone. Due to gravity, free fluid will drain toward the patient's anterior surface. The lighter free air will rise and accumulate on the patient's posterior surface.

20. B Projection is a term applied to recording of an image resulting from the passage of the x rays through the patient. It is defined by the surfaces passed by the beam; for example, an AP projection indicates that the beam enters on the anterior surface and exits on the posterior surface of the patient.

21. B Adduction refers to the movement of an anatomic structure toward the central axis of the body. Abduction refers to the movement of an anatomic structure away from the central axis of the body. Example: Moving the extended arm from the anatomical position "down to the side" is called adduction, lifting the arm away from the body is called abduction.

22. C See question number 6.

23. A Dorsum usually refers to the posterior aspect or the back of a part or organ. When referring to the foot, the dorsum is the superior aspect, while the inferior surface is called the plantar aspect.

24. C The most superior aspect of the palpable iliac crest is at the same level as the body of the fourth lumbar vertebra. This external landmark is widely used when radiographing the abdomen, spine, kidneys (IVU), colon (BE), stomach (GI), and pelvis.

25. C Pronation refers to the rotation of the hand so that the palm faces the back of the body; supination, the opposite of pronation, refers to the rotation of the hand into true anatomical position.

26. A See question number 2.

27. D The term position most often describes position of the body or anatomic structure as seen from the imaging system when entry and exit points are not specified. In an oblique position, the patient is rotated between the frontal and lateral position; producing an oblique image of the anatomic structure. In the right anterior oblique (RAO) position, the patient's right anterior surface is closest to the imaging system, and is also described as a right PA oblique. By convention, it is always the side of the body closest to the film that defines position. A right posterior oblique (RPO) position has the patient with the right posterior surface in contact with the imaging system and is also described as a right AP oblique. Describing oblique positions is sometimes confusing. The majority of obliques describe the body surface that is closest to the image receptor (RAO/right anterior oblique) rather than the projection description (right PA oblique). The abbreviations are being used with the PA/AP oblique description parenthetically listed as follows: RAO/right anterior oblique (right PA oblique); RPO/right posterior oblique (right AP oblique); LAO/left anterior oblique (left PA oblique); and LPO/left posterior oblique (left AP oblique). Example: The patient is rotated 45 degrees from the PA so that the right anterior chest wall is placed closest to the image receptor (RAO). The central ray enters the posterior chest and exits the anterior chest wall with the right side closest to the image receptor (right PA oblique). This will produce an oblique image of the chest cavity.

28. B See question number 1.

29. C See question number 10.

30. C The greater trochanter is located on the proximal lateral aspect of the femur at the level of the symphysis pubis and the coccyx (see question number 1). The ischial tuberosity is located inferior to these structures. The greater trochanter is sometimes easier to palpate than the symphysis and can be used to find the lower portion of the abdomen or the hip joint.

31. C The anterior superior iliac spine (ASIS) is an external landmark located at the level of the first sacral segment and is commonly used to locate the sacroiliac and hip joint.

32. A The abdomen is divided into four quadrants by the transverse and midsagittal planes which intersect at the umbilicus. The four quadrants, located between the diaphragm and pelvic inlet, are named as follows: right upper quadrant (RUQ), left upper quadrant (LUQ), right lower quadrant (RLQ), and left lower quadrant (LLQ).

33. A See question number 6.

34. C See question number 13.

35. C — The gall bladder of a sthenic or "average" patient is normally located at the level of the ninth costal cartilage. Depending on the body habitus, the gallbladder can vary in position by as much as 8 inches both vertically and laterally from the level of the 8th rib down to well within the iliac fossa and anywhere from the midsagittal plane to the lateral wall of the abdomen. For example, for a hypersthenic patient, an initial radiograph may be centered 2" higher than the sthenic; for an asthenic patient, 2" lower. See question number 2.

36. D — See question number 27.

37. C — A process is an anatomic term used to describe a projection that extends or juts out from the main body of a structure. Malleolus is an example of a club shaped process found on the medial and lateral aspect of the distal end of the tibia and fibula (ankle).

38. B — See question number 13.

39. A — See question number 27.

40. D — A foramen is an anatomic term used to describe a hole in a bone for the transmission of blood vessels and nerves. For example, the foramen magnum allows the spinal cord (medulla oblongata) to pass into the cranial cavity; the jugular foramen allows passage of the jugular vein out of the cranium and into the neck.

Upper Extremity

Chapter 46

1. C — The 45 degree oblique is most commonly employed for the hand because it tends to open the metacarpal and interphalangeal joints. A properly positioned oblique will allow for a slight overlapping of the bases and heads of the metacarpals allowing separation of the shafts of the first three digits.

2. A — The clavicle is directed obliquely between the manubrium and the acromion process of the scapula. Due to its location on the anterior surface of the body, a shorter object-to-image distance improves definition in the PA projection. The erect position is often preferred to make the patient more comfortable. The recumbent PA projection should be avoided in cases of severe trauma to eliminate the possibility of secondary injury and pain.

3. C — With the body AP, the external rotation position of the arm places the humeral head in true anatomic position. This position shows a profile view of the greater tuberosity and is accomplished by placing a coronal plane passing through the humeral epicondyles parallel to the image receptor.

4. D — The lateral position of the elbow demonstrates the joint space, semilunar and olecranon processes. This is accomplished by placing the forearm, elbow and humerus in the same plane by adjusting the epicondyles perpendicular to the imaging system, bending the arm to 90 degrees, and superimposing the styloid processes of ulna and radius.

5. B — Magnification of the wrist bones can be reduced by flexing the fingers in the PA projection of the wrist, reducing its distance from the imaging system and improving the definition of the joint spaces.

6. C Bilateral views are discouraged because centering to the midpoint of the image receptor and not the anatomical structure may obscure the narrow joint spaces found in the hands and fingers.

7. C When acromioclavicular (AC) joint separation is suspected, an erect view with weights is employed to maximize dislocation and improve its visualization on a radiographic image.

8. C With the shoulder in anatomical position, and a plane between the humeral epicondyles is perpendicular to the imaging system, the lesser tuberosity is profiled in the lateral position of the humerus.

9. B The anteroposterior (AP) 45 degree oblique projection of the wrist will give an improved visualization of the bones on the medial side of the wrist including a non-superimposed pisiform bone and improved visualization of the lunate and triquetrum.

10. C The two bones of the forearm, the ulna and radius, are found in the uncrossed position when the hand has been supinated.

11. A The carpal interspaces, which extend obliquely, are most closely aligned with the divergent primary beam in the AP projection.

See accompanying diagram.

12. K	13. D	14. I	15. A
16. H	17. C	18. F	19. E
20. G			

21. C To include all of the phalanges and the distal portion of the metacarpal bones, the central ray enters at the level of the proximal interphalangeal joint.

22. A In the true lateral position of the scapula, the vertebral and axillary borders should be superimposed. This is best accomplished by placing the patient in the PA oblique position with the coronal plane forming about a 60 degree angle with the plane of the image receptor. In the PA projection, the affected scapula is placed closest to the image receptor.

23. B A true lateral projection of the humerus can be made without internal rotation of humerus by using the transthoracic (Lawrence) lateral position. In cases of severe trauma, this provides the only method by which right angle projections of a stabilized shoulder or humerus can be made. When possible, a plane passing through the humeral epicondyles should be perpendicular to the imaging system in the Lawrence method.

24. C A lateral position of the wrist with acute palmer flexion improves the visualization of carpe bossu, a bony lesion on the third metacarpocarpel joint first described by Fiolle.

25. D The bony tip located on the expanded proximal end of the ulna, the olecranon process is best visualized in the true lateral position of the elbow. The 90 degree flexion is preferred for bony anatomy while a 30 degree flexion is used when soft tissue structures are of primary concern.

26. C Palmar and ventral refer to the anterior surface of the hand. In order to perform a PA projection, the palmar surface is placed against the image receptor.

27. A In order to draw the scapula laterally away from its superimposition with the lungs and rib cage, the patient's arm is abducted to a 90 degree angle.

28. B — In the lateral position of the hand, the fingers are fanned when they are of primary interest. In cases of foreign body localizations and metacarpal fractures, the lateral in extension is the preferred position.

29. D — A quiet breathing technique for the AP projection of the scapula will tend to blur the overlying lung and ribs.

30. C — In order to place the humerus in the lateral position, the arm is internally rotated until the epicondyles are perpendicular to the plane of the imaging receptor. This projection will give a lateral image of the elbow and profile the lesser tuberosity.

31. A — The inferosuperior tangential position of the wrist (carpal canal) is used to improve the visualization of the anterior aspect of the carpal bones including the hook of the hamate bone and an unobstructed image of the pisiform.

See accompanying radiograph.

32. C 33. E 34. H 35. J

36. D 37. I

38. D — A partial flexion, 35 degrees, of the elbow is used instead of the right angle flexion to reduce stretching and compression of the soft tissues of the elbow.

39. D — In the majority of individuals, the counter rotation used to place the thumb in the AP position will place the hand perpendicular to the image receptor or in a lateral position.

40. C — The lower portion of the Y, which appears in the lateral position of the scapula, is formed by the superimposed body, vertebral and axillary borders. The portion of the upper Y closest to the ribs is formed by the coracoid process and the acromion forming the other lateral arm of the Y.

41. D — In order to place the humerus in an anatomic position (AP) and demonstrate the greater tuberosity in profile, a plane passing through the epicondyles should be parallel to the plane of the image receptor.

42. C — Radial flexion (ulnar deviation) and AP oblique positions of the wrist are used to improve the visualization of the carpal bones on the medial aspect of the wrist. Radial deviation (ulnar flexion) is used in the visualization of the scaphoid (navicular) bone.

43. C — A 40 degree external rotation oblique position of the elbow gives an unobstructed image of the radial head without superimposition of the ulna.

44. C — The approximate center of the hand is over the third metacarpophalangeal joint.

45. A — In order to project the clavicle above the apical region of the lungs and first rib, the central ray requires a 25 degree caudad angulation if the patient is prone and a 25 degree cephalic angulation directed to pass between the chest wall and the clavicle with the patient supine.

46. D — See explanation number 30 of this section.

47. D — With the hand flexed to form a 90 degree angle with the forearm and resting on the dorsal surface of the hand, the central ray is directed tangentially (45 degrees) to the dorsum of the wrist. This position is recommended for visualization of lunate dislocations, navicular fractures and foreign bodies or calcifications of the posterior aspect of the wrist.

48. A — A 45 degree medial (internal) oblique position of the elbow rotates the coronoid process medially until it is freed from self-superimposition.

49. B — A modified radial tilt of the carpal canal is used to demonstrate the first metacarpocarpal joint and the trapezium or greater multangular bone.

50. D — The scapular spine which projects posteriorly from the body of the scapula can be visualized by directing the central ray 45 degrees caudad which profiles the spine, free of nearly all bony superimposition.

See accompanying diagram.

51. H	52. K	53. C	54. E
55. B	56. A	57. F	58. D
59. I	60. G		

61. B — In order to demonstrate the coracoid process free of self superimposition, the patient is placed supine with the central ray directed 15-45 degrees cephalic depending on the roundness of the patient's shoulder.

62. D — See explanation number 31 of this section.

63. A — In order to cover the entire circumference of the radial head, four lateral positions are required with the hand supinated, in lateral extension, in pronation and in extreme internal rotation.

64. A — The ball catcher's (Norgaard) bilateral AP 45 degree oblique position of the hands is a position used to diagnose early rheumatoid arthritis. The physician will look for changes on borders of the distal phalanges and slight demineralization of the bones of the fingers.

65. B — The PA oblique positions, as described by Lorenz and Lilienfeld, require the patient to be placed in the lateral position with the affected scapula closest to the image receptor. The dependent arm is placed at a right angle to the body so that the scapula is projected free of the rib cage.

66. C — In order to demonstrate the joint space between the glenoid fossa and the humeral head, the patient is placed in the 35-45 degrees AP oblique position with the affected scapula closest to the image system. This position places the broad surface of the scapula perpendicular to the imaging plane.

67. D — The scaphoid (navicular) bone which is directed obliquely to the bones of the forearm can be placed at right angles to the central ray by placing the wrist on a 20 degree incline toward the elbow; or when the wrist is horizontal the central ray is angled 20 degrees toward the elbow (Stecher method). Other variations include radial deviations, ulnar flexion with a perpendicular central ray, and clenching of the fist to place the scaphoid parallel to the image receptor.

68. D — The profile view of the dorsum of the olecranon process can be obtained using an axial position of the elbow. The patient is placed with the elbow in acute flexion and a 20 degree angle of the central ray toward the elbow to best demonstrate the olecranon. A perpendicular central ray is used to show the distal humerus and olecranon.

69. D — The inferosuperior axial (Lawrence method) position for the glenohumeral joint and axillary soft tissue requires the arm to be externally rotated and abducted to 90 degrees.

70. C The forward tilting of the finger in the PA oblique of the hand results in a great deal of distortion of the phalanges. This distortion can be reduced by the use of the 45 degree stepped wedge sponge.

71. D The PA oblique Y scapula position places the scapula in the lateral position on the image because the body of scapula is directed at an angle of 30 degrees to the image receptor, a 60 degree oblique is necessary to obtain the right angle position of the scapula.

72. C Partial or complete flexion can elevate fat pad and distort appearance of the joint; the relaxation of this pad by the 90 degree flexion is therefore recommended.

73. B The bicipital (intertubular) groove which passes inferosuperiorly between the greater and lesser tubercles on the anterior surface of the humeral head, is best shown in the tangential position in which the central ray parallels the groove.

74. D When the hand is placed in the PA projection, the thumb is normally directed obliquely at about a 45 degree angle.

75. A The inferosuperior axial position of the shoulder which is used to show bony defects of the rim of the glenoid fossa, requires a double angulation of 25 degrees anteriorly and 23 degrees medially when the patient is laying prone with the arm abducted to a 90 degree angle.

76. A See explanation number 31 of this section.

77. B See explanation number 31 of this section.

78. B See explanation number 43 of this section.

79. D See explanation number 43 of this section.

80. A See explanation number 7 of this section.

81. D See explanation number 7 of this section.

82. C See explanation number 3 of this section.

83. D See explanation number 3 of this section.

84. D See explanation number 22 of this section.

85. C See explanation number 22 of this section.

Lower Extremity

Chapter 47

1. A The lateromedial lateral position of the foot (medial surface of the foot against image receptor) should be utilized to obtain a true lateral position. The more common mediolateral lateral position is usually used because it can be obtained with more ease.

2. A A 20-30 degree flexion of the knee is normally associated with a relaxation of the leg muscles and the improved visualization of the joint space.

3. C — In the relaxed AP position of the leg the femoral neck is tilted obliquely from the pelvis (anteversion). When the toes are inverted 15 degrees, the femoral necks will more closely parallel the plane of the image receptor thus improving the visualization of both the femoral neck and greater trochanter.

4. C — The relatively long object-to-image distance of both the ankle and knee joint combined with the use of the peripheral portion of the beam requires that the image receptor extend a few centimeters beyond the actual location of the joints.

5. B — The calcaneus (os calcis) sits just below the talus bone which forms the lower portion of the ankle joint with the medial and lateral malleoli.

6. C — In order to avoid the movement of a hip in case of suspected fracture, the Cleaves, Lauenstein, and Hickey (Frog) positions may put the patient at risk for further injury and should only be attempted after fractures have been ruled out.

7. C — The PA medial oblique position of the foot is used to demonstrate the cuboid bone and the articulation of the fourth and fifth metatarsal bones, as well as the articulations between the navicular and talus, navicular and cuneiform, and the sustentaculum tali and talus bone.

8. D — In a PA projection of the knee the elevated position of the foot will compensate for the inclination of the tibial plateaus so no central ray angulation is required. In the AP position, this inclination is more dependent on the patient's bodily habitus. In thinner hyposthenic patients, a 5 degree caudal angle is preferred to maximize the joint space. No angle is recommended for the sthenic patient and a 5 degree cephalic angulation is suggested on the hypersthenic individuals.

9. B — In order to place the femoral necks parallel to the image receptor in the AP axial (Cleaves) position, the thighs are abducted with the soles of the feet together. The thigh should form a 40 degree angle with the vertical. The central ray is directed perpendicular to the imaging system for the modified Cleaves and 40 degrees cephalad for the original Cleaves method for the femoral head, neck and trochanteric region.

10. C — In the lateral (mediolateral) position of the ankle, the toes are dorsiflexed to eliminate lateral rotation of the ankle. The broad medial surface of the foot provides a more consistent surface on which to obtain a true lateral position.

11. C — The tangential or sunrise projection as described by Settegast, provides a profile view of the patella for the evaluation of vertical fractures and the patellofemoral articulation. The central ray is directed inferosuperiorly at a right angle to the patellofemoral joint space with the knee in acute flexion.

12. D — A measurement of the transverse diameter of the pelvis; the relationship of the femoral heads to the acetabula, the evaluation of opacified rectosigmoid areas; and visualization of the junction between the ureter and the posterior portion of the urinary bladder are determined the axial superoinferior (Chassard-Lapine) position of the pelvis.

13. A — See accompanying radiograph.

14. F — See accompanying radiograph.

15. D — See accompanying radiograph.

16. A — Because of the natural curve of the anterior aspect of the foot, a lateral oblique position improves the visualization of the articulations between the medial and intermediate cuneiforms and the first and second metatarsal bones.

17. C — A slight 5 degree cephalic angulation of the central ray on a lateral position of the knee will reduce the magnification of the more distal femoral condyle and improve the visualization of the joint space.

18. A — When a grid is employed, its visualization is improved by placing the grid strips parallel with the plane of the femoral neck in the axiolateral position of the hip.

19. A — Because of the relationship of the malleoli in the lateral plane, a medial oblique of about 15-20 degrees is required to place them parallel to the image receptor improving the visualization of the ankle mortise.

20. C — In order to obtain a frontal projection of the calcaneus, the central ray is directed axially to a vertical foot. It is directed to enter the foot at the level of the base of the fifth metatarsal with a 40 degree cephalic angulation, exiting proximal to the posterior ankle joint.

21. C — See explanation number 9 of this section.

22. C — Not unlike the PA medial oblique (explanation number 7) the AP medial oblique in which the sole of the foot forms a 30 degree angulation with the plane of the image receptor will better demonstrate the articulating surfaces of the cuboid and navicular bones.

23. B — Since acute flexion is required for all tangential projections of the knee, transverse fractures of the patella which can be displaced by this motion must be ruled out prior to attempting these positions.

24. C — The head of the femur is located 5 cm (2") medial and 5 cm (2") inferior to the adjacent anterior superior iliac spine (ASIS). In order to center this structure, a sagittal plane passing through this region is used to approximate central ray location.

25. A — The flattened medial side of the foot in the lateromedial lateral position enables the ankle joint to be closer to the image receptor than in the mediolateral lateral position.

26. C — The central ray should be directed to the center of the third digit at the third metatarsophalangeal joint.

27. B — The ankle contains three malleoli; the lateral malleolus of the fibula, the medial malleolus and the lateromedial malleolus of the tibia.

28. B — In the AP dorsoplantar oblique position, the leg is rotated 45 degrees medially to form a 30 degree angle between the sole of the foot and the plane of the image receptor.

29. A — There are three methods for visualizing the intercondylar fossa of the knee: the PA axial (Camp Coventry) position in which the patient lays prone with the lower leg supported at an angle of about 40 degrees, the PA axial (Holmblad) in which the patient rests on the knee with the lower leg resting on the table, and the AP axial (Beclere) where the knee rests on a curved cassette. In all positions it is essential to direct the central ray parallel to the tibial plateau and perpendicular to the long axis of the tibia.

See accompanying radiograph.

30. I	31. G	32. D	33. A
34. F	35. B	36. J	

37. D A line drawn between the ASIS and the symphysis pubis, is intersected at a right angle by a line parallel to the long axis of the femoral neck which localizes the dome of the acetabulum.

38. A See explanation number 20 of this section.

39. B See explanation number 19 of this section.

40. B The PA or prone position for the anterior pelvic bones is preferred because of its close part-to-image receptor distance which reduces the amount of magnification of the image.

41. A The ischial and pubic rami are projected free of superimposition in an AP axial position employing a 25-40 degrees cephalic angulation depending on the shape of the pelvis. Taylor recommends a central ray entering at a point 2" distal to the symphysis pubis with a 25 degree angle in males and a 40 degree angle in female patients.

42. A The center of the foot is found at the base of the third metarsal bone. A perpendicular central ray is used for general survey and foreign body localization; a 10 degree posterior (cephalic) angle may be used to improve the visualization of the bony articulations in the middle region of the foot.

43. D The patella is generally the simplest palpable landmark used in the localization of the joint space of the knee. The articulation between the femur and tibia usually occurs about 1.5 cm (1/2") distal to the inferior (apex) portion of the patella.

44. D See explanation number 3 of this section.

45. B The appearance of a widened joint space, when stressed, is nearly always related to tears in the collateral ligaments. Inversion stress is used to indicate damage to the lateral collateral ligaments, eversion stress is used for medial tears. Stress studies are normally performed after the acute pain of the injury has subsided or with the use of local anesthesia.

46. A Since the joint spaces between the toes are directed at about a 15 degree angle with the horizontal, a 15 degrees cephalic central ray angulation, or a 15 degree wedge placed under the toes with a perpendicular central ray to best demonstrate these articulations.

47. A When the unaffected leg cannot be raised to enable a shoot-through position of the hip, a lateral may be obtained using the Clements Nakayama method. With the patient supine, a cassette is placed vertically alongside of the affected hip parallel to the femoral neck and perpendicular to the table. The central ray is directed 15 degrees posteriorally through the femoral neck.

48. A See explanation number 42 of this section.

49. D When a patient cannot be turned onto his/her back for a tangential position of the patella, Hughston recommends that the lower leg be flexed 50-60 degrees to enable the central ray to skim the patella by using a 45 degree cephalic angulation directed through the patellofemoral joint.

50. A In an axiolateral positon of the hip, the ischial tuberosity lies posterior to the dome of the acetabulum, femoral head and neck; the greater trochanter is seen just distal to the neck projecting posteroanteriorly with the lesser trochanter seen toward the posterior surface of the body.

51. A The metatarsophalangeal sesamoid bones which are located at the ball of the foot are best demonstrated by the inferosuperior or superoinferior tangential position(Lewis) or the lateromedial tangential positions (Causton).

52. D — The inferosuperior tangential position with the toes dorsiflexed with the foot perpendicular, uses a vertical central ray to project the metatarsophalangeal sesamoid bones away from the head of the first metatarsal bone.

53. B — Because the head of the fibula is located more laterally and distal to the knee joint, a medial rotation from the AP position will best demonstrate tibiofibular articulation.

54. B — The entire prosthesis must be shown radiographically to ensure its alignment and position within the shaft of the femur.

See accompanying radiograph.

55. K	56. D	57. A	58. F
59. H	60. I	61. C	62. G
63. J	64. E		

65. C — There is no angle required on the weight bearing projection of the knees. This projection is used in the evaluation of the degree of valgus (outward turning) varus (inward turning) deformities.

66. B — The Kite and Kandel methods describe the various positions that can be used to evaluate talipes equinovarus or club feet. Kite described the appropriate dorsoplantar (AP), and lateral position to best show this condition. Kandel described the dorsoplantar (superoplantar) axial position with the infant standing and the central ray angled 40 degrees caudad, entering the posterior lower leg just superior to the calcaneus and exiting the plantar portion of the foot.

67. D — Martz and Taylor have recommended a pair of AP pelvis images; the first taken with a perpendicular central ray and the second with a 45 degree cephalic angulation. The no-angle projection is used to show the lateral displacement of the hip while the axial position will project anterior displaced femoral heads above the acetabulum and posterior displaced heads below this socket.

68. B — Since the ankle joint is located in the same plane as the malleoli, a vertical central ray directed to midmalleoli will best demontrate the joint.

69. A — The subtalar (talocalcaneal) joint which may be compromised in comminuted fractures of foot, is demonstrated by any number of axiolateral or oblique axial positions as described by Broden, Isherwood, Feist-Mankin.

70. D — See explanation number 29 of this section.

71. C — The wing of the ilium is projected obliquely at an angle of about 40 degrees. An AP oblique position with the unaffected side elevated 40 degrees will place the affected (down side) ilium parallel to the image receptor, producing a frontal image.

72. D — Since all intercondylar projections (see explanation number 29) require that the central ray be directed perpendicular to the tibial plateaus and long axis of the lower leg, no tube angulation is required in this position.

73. A — The longitudinal arch which extends from the metatarsal bones along the plantar surface of the foot to the calcaneus is shown in the lateral projection of the foot. The flattening of the arch, pes planus, (flat feet) is most clearly seen when weight bearing lateral images are obtained.

74. D The use of two axial projections, one directed posteriorly at an angle of 15 degrees and one anteriorly at 25 degrees angulation, on a single film, provides a unique composite image of the weight bearing foot without the shadows of the bones of the lower leg.

75. C The length of the legs and the relative position of the joints can be obtained without magnification by using a metallic ruler and maintaining the central ray over the three joints of the lower extremity. Recent advances in CT has allowed accurate measurements of the extremities at reduced exposures by the use of computed tomographic scanography.

See accompanying radiograph.

76. D	77. E	78. K	79. G
80. B	81. H	82. L	83. F
84. A	85. J		

86. A See explanation number 29 of this section.

87. D See explanation number 29 of this section.

88. B See explanation number 11 of this section.

89. C See explanation number 11 of this section.

90. C See explanation number 9 of this section.

91. C See explanation number 9 of this section.

92. D This knee radiograph was taken on a child. The technical factors selected by the radiographer were insufficient to provide the appropriate radiographic density and should be corrected for by a doubling of the mAs or a 15% increase in the kVp.

93. B The radiolucent lines represent the cartilage at the growth plates called the epiphyses.

94. B Not unlike the stress studies of the ankle discussed in explanation number 45, inversion stress of the knee will demonstrate damage to the lateral collateral ligament by the widening of the lateral joint space. The overexposed radiograph could be improved by a reduction of the mAs by 1/2 or a 15% reduction in the kVp. The low density structure is a sandbag placed between the knees which helps create the internal stress.

95. A See explanation number 94 of this section.

Axial Skeleton

Chapter 48

1. D In the lateral aspect of the lumbar vertebrae the space between the vertebral bodies, the intervertebral foreman and spinous processes are readily visible radiographically.

2. A After cervical fractures have been ruled out, lateral positions in hyperflexion and extension may be ordered to determine the degree of anterior/posterior motion of the spine. The appropriate extension images are obtained with the chin raised so the body of the mandible is parallel to the floor and fully flexed when the body of the mandible is perpendicular to the floor.

3. A In the lateral position, a large percentage of scatter radiation is produced which reduces the radiographic contrast of the image. The use of a leaded shield behind the patient helps absorb some of the scattered radiation that would otherwise reach the image receptor.

4. A The normal kyphosis of the dorsal spine can be reduced by flexing the knee and hips to place the back in contact with the table.

5. A The transverse processes of the entire spinal column are best shown in the frontal (AP, PA) projections of the spine.

6. D Cervical collars or other types of support devices placed on the patient by a physician should remain in place during imaging to reduce possible injuries due to subluxation or complications due to movement of bone fragments.

7. B The abnormal lateral curvature due to the spinal scoliosis is best demonstrated in the upright frontal (AP, PA) with the foot elevated on the concave side. This helps distinguish compensatory and deforming curves. Other commonly requested positions for scoliosis include upright laterals and PA projections with right and left flexion.

8. C The overlaying structures of the shoulder joints can make radiography of the lower cervical and upper dorsal spines difficult. In the Twinning (swimmers) position, the patient is placed in the lateral position with the dependent arm raised above the head and the independent shoulder depressed. This will project the C7-T2 vertebrae between the shoulders, minimizing the degree of bony superimposition.

9. D The AP (45 degree) oblique position of the lumbar spine shows the dependent zygapophyseal joint between the superior and inferior articulating processes of the vertebra. Radiographically, this joint forms the ears and front foot of the scotty dog. The nose is formed by the transverse process, the eye from the pedicle and the body from the lamina of the vertebra.

10. B In the cervical region, the intervertebral foramina are directed obliquely from the spinal canal at about a 45 degree angle laterally a 15-20 degrees inferiorly. The AP (45 degree) axial oblique (RPO, LPO) positions utilize these relationships in demonstrating the foramina and pedicles furthest from the image receptor in the 45 degree oblique position and a 15-20 degree cephalic angulation.

11. B Though it is, at best difficult to contrast the sternum from overlaying ribs, vertebral and lung shadows, the right PA (15-20 degrees) oblique (RAO) position which superimposes the heart and sternum will often provide the best visualization of the sternum. The use of a short souce-to-image distance of 15" (38 cm) will magnify and blur the posterior ribs. Breathing techniques to blur the lung tissue have been recommended by some authors.

12. D The zygapophyseal joints form between the superior and inferior articulating processes of the 6 lower cervical vertebrae a 90 degree angle with the midsagittal plane and are therefore most clearly visualized in the lateral projection. This position also demonstrates the cervical bodies, cervical interspaces and the spinous processes.

13. C — Spinal motion prior to or following a lumbar laminectomy or other lower spinal surgery is determined by the degree of lateral bending motion on a frontal projections and/or forward (flexion) and backward (extension) bending on a lateral projections.

14. B — The sacrum, which extends from the top to the middle of the pelvis lies near the posterior surface of the body about 3" (8 cm) behind the midaxillary plane. The central ray is directed perpendicular to this plane. The centering point for the coccyx is nearly 5 inches behind the midaxillary line.

15. B — The posterior aspect of the vertebra (vertebral arch) which includes the laminae, zygapophyseal elements and spinous processes are often best demonstrated in the axial projections. From the AP position, the central ray is angled 25-30 degrees to place it parallel with the plane of the articular facets.

16. A — See explanation number 8 of this section.

17. B — Deep inspiration, which fills the lung with more air, will provide a higher contrast between the lung and sternum.

18. A — See explanation number 9 of this section.

19. A — The junction of the clavicle and sternum near the midline of the body makes its visualization difficult, without being at least in part superimposed with the spine. A shallow PA 15 degree oblique (RAO, LAO) position is used to demonstrate the dependent sternoclavicular (SC) articulation.

20. B — The movement of the arm to a right angle location during the lateral position of the thoracic spine will help to elevate ribs and clear the intervertebral foramina. If the spine is not adjusted to the horizontal, a 10-15 degrees cephalic angulation is employed for the lateral position.

21. B — See explanation number 10 of this section.

22. A — The single best position for the evaluation of cervical spine fractures is a lateral projection. A shoot-through lateral can be performed without moving the patient thus minimizing the risk of additional spinal cord injury.

23. D — See explanation number 9 of this section.

24. D — The intervertebral disk spaces of the cervical, thoracic and lumbar spine are best visualized in the lateral projection.

25. A — The sacrum joins with the two innominate bones at their medial surface. This articulation which is called the sacroiliac (SI) joint, projectes obliquely at an angle of 25-30 degrees from the midsagittal plane. A 25-30 degree AP oblique (RPO, LPO) places the independent (upside) SI joint perpendicular to the image receptor.

26. C — In order to utilize the anode heel effect for thoracic spine radiography, a greater amount of exposure is desired towards the foot end of the table. This can be accomplished by placing the cathode over the thicker or lower parts of the body and the anode toward the head of the table.

27. D — Cervical ribs which project anteriorly from the lateral surface of C7, are well visualized in the AP projection of cervical spine.

28. C — The degree of magnification due to long object-to-image distance occurring in the lateral position of the cervical spine, is to a large extent compensated for by the use of a long 72" (180 cm) SID.

29. A Due to the kyphotic curve of the coccyx, a 10 degree cephalic angle with the patient PA, will tend to reduce the foreshortening that would occur without this angulation. This angulation is reversed or caudal in the AP projection.

30. D All of the normal anterior (kyphotic) or posterior lordotic curves of the spinal column are well visualized in the lateral projection. Scoliosis or the lateral curvature of the spine is best seen in the frontal projections.

31. C See explanation number 9 of this section.

32. A When the occlusal plane (the line along the upper incisors), and the tip of the mastoid process is perpendicular to the image receptor, shadows from the back of the head and teeth should be eliminated from the odontoid process in an open mouth AP projection. The phonation of "ah" during the exposure helps to fix the tongue to the bottom of the mouth.

33. C The foreshortening of the sacrum and the anteroposterior position can be compensated for by the use of a 15 degree cephalic angulation or by the use of a 15 degree caudal angulation in the PA position.

34. B An increase of the 90 degree arm position in the lateral position of the thoracic spine will result in the scapula being projected over the upper thoracic vertebrae.

35. D (See explanation number 10 of this section). In the PA obliques (RAO, LAO), with a 45 degree obliquity angle of the neck the central ray is directed 15-20 degrees caudad to demonstrate the intervertebral foramina and pedicles closest to the film.

36. D The articulation between the inferior and superior articulating processes (zygapophyseal joints) in the thoracic spine forms an angle of 70 degrees with the image receptor. In the AP 70 degrees obliques (RPO,LPO) these articulations are closest to the image receptor. In the PA 70 degrees oblique (LAO, RAO) the joints furthest from the image receptor are shown.

37. A If the odontoid process is superimposed on the teeth in an open mouth position the head must be extended or directed away from the chest. The base of the skull will appear over the odontoid when the head is not tucked close enough to the chest wall.

38. B See explanation number 25 of this section.

39. C Because the upper thoracic vertebrae are smaller than those of the lower thoracic spine, the appropriate centering point is at the level T7 instead of T6.

40. C The fifth lumbar and first sacral vertebrae articulate at an angle to the plane of the table of about 30 degrees in males and 35 degrees in females. In order to open this articulation a 30-35 degree cephalic angulation is used in the AP axial position. This angle is reversed in the PA position.

41. D See explanation number 2 of this section.

42. A An anterior rib injury would be best examined in the PA upright position which eliminates potential complications due to lung involvement while maintaining the best definition and true image size. Images taken at the end of inspiration are used to place the upper 10 pairs of ribs above the diaphragm.

43. C See explanation number 10 of this section.

44. C See explanation number 1 of this section.

45. A — During lateral positions of the spine the vertebral column should be adjusted to be parallel with the image receptor. When recumbent the spine must be supported through the use of radiolucent sponges placed under the lateral thorax or lumbar region to achieve this goal. When unsupported a 10 degree cephalic angle for females and a 15 degree angle for males will compensate for this curve in the thoracic region. In the lumbar region a 5 degree caudal angle for males and an 8 degree caudal angle for females is required on the unsupported spine.

46. D — Fuchs describes a method for demonstrating an AP projection of the odontoid process lying within the foramen magnum. This is accomplished by having the patient extend the chin (closed mouth) and directing the central ray vertically to the table and parallel to the body of the mandible exiting at the foramen magnum.

47. D — The normal kyphosis of the thoracic spine causes the costovertebral and costotransverse joints to be best visualized in the AP axial position employing a 20-30 degree cephalic angulation. A 5-10 degree increase in this angle is recommended for patients with noticeable kyphosis.

48. B — The sternoclavicular (SC) joints in the lateral position are usually superimposed on each other. Kurzbauer recommends placing the patient in a modified swimmers position and directing the central ray 15 degrees caudad to the lowest SC joint.

49. C — The long object-to-image distance between the cervical spine and the image receptor creates an air gap enabling most of the divergent scattered radiation produced in the patient to miss the image receptor. This eliminates the need for a grid and improves the contrast of the image.

50. D — See explanation number 10 of this section.

51. A — See explanation number 36 of this section.

52. D — The intervertebral foramina formed by L5 and SI is not well visualized in the lateral position and requires an additional forward rotation of 30 degrees to improve its visualization.

53. D — In the AP cervical spine position, the mandible will obscure C1 and C2 when the mouth is closed and C3, C4, and C5 when open. The Ottonello method (wagging jaw) uses a long exposure time and the movement of the jaw to blur its image on the receptor, thus providing an image of the entire cervical spine on a single radiograph.

54. B — Chamberlain described the two PA weight bearing projections for the determination of movements of the SI joints. A slipped SI joint is most readily seen by the relationship of the pubic bones at the symphysis pubis where the images are centered.

55. C — In an AP projection of the cervical spine, the central ray is angled about 15-20 degrees to enter at the thyroid cartilage. This position will demonstrate the lower cervical vertebra intervertebral disk spaces, and transverse processes.

56. D — A lateral radiograph of the lumbar region clearly indicates that the intervertebral spaces will be more aligned with the divergence of the x-ray beam in the PA position. The main disadvantage of this position is an increased magnification of the vertebrae due to a longer object-to-image distance.

57. C — During the examination of the and thorax a PA position is preferred to reduce the exposure to the thyroid, ovaries and breast found near the anterior surface of the body.

58. B — See explanation number 32 of this section.

59. C See explanation number 45 of this section.

60. B The sides or axillary borders of the ribs are best seen in the 45 degree oblique position with the patient in the AP oblique (LPO, RPO) positions. The effected side is placed closest to the image receptor. In the PA obliques (LAO, RAO) positions the effected side is rotated away from image receptor.

61. C See explanation number 10 of this section.

62. A See explanation number 10 of this section.

63. C See explanation number 9 of this section.

64. B See explanation number 9 of this section.

65. A See accompanying radiograph.

Skull

Chapter 49

1. D The glabellomeatal line (GML) is the most superior of the four common radiographic lines used in positioning of the skull. The orbitomeatal line (OML) extending from the outer canthus to the external auditory meatus (EAM) forms an angle of 8 degrees with the GML. The infraorbitomeatal line (IOML) which extends from the lower orbital margin to the EAM is separated from the OML by 7 degrees. The most inferior acanthiomeatal line (ACML) extends from the point just below the nose (acanthion) to the EAM and is about 15 degrees below the IOML.

2. C Because of the close proximity of the paranasal sinuses with each other, they tend to be superimposed in the frontal and axial projections. The lateral projection demonstrates all of the four paranasal sinus groups separately, and is used to compare their relationship to each other.

3. C To include the anterior nasal spine, the frontal suture, and the nasal bones in the lateral projection of the nasal bones, the central ray is directed to 3/4" distal to the nasion (the center of the frontonasal suture).

4. B The tangential oblique submentovertical (SMV) projection will often provide the best demonstration of a single zygomatic arch, by projecting it free of superimposition of the adjacent facial and skull bones.

5. C The "cranial type" influences the location of the petrous ridges and their relationship to the midsagittal plane (MSP). In the more broad brachycephalic cranium an angle of 54 degrees is found. The mesocephalic or most common skull is associated with an angle of 47 degrees, the thin long skull of the dolichocephalic averages about 40 degrees. Variation of patient positioning and/or compensation of the central angulation may be required when radiographing these different cranial types.

6. B The word vertex literally means "summit", and refers to the superior aspect of the cranium.

7. A The foramina rotundum are horizontally located in the anteromedial portion of the sphenoid bone, projecting through its greater wings at an angle of 37 degrees to the vertical. In the parietoacanthial Water's method, the orbitomeatal line adjusted 37 degrees to the plane of the image receptor, to open these round foramina.

8. D The mandibular condyle at the superioposterior border of the mandible and the mandibular fossa on the inferior lateral surface of the temporal bone, articulate to allow opening and closing of the jaw. During the examination of this joint, it is important to include both open and closed mouth views to compare the movement of the condyles into and out of the fossa.

9. C See explanation number 1 of this section.

10. A The side to side measurements of the cranium, are made between the prominences (eminences) found on the external surface of the parietal bones. For the average (mesocephalic) skull a value of 15 cm is considered normal.

11. A The sella turcica is a saddle like bony depression located on the upper surface of the body of the sphenoid bone. In a lateral position, the sella is located at the level of 2 cm anterior and 2 cm superior to the external auditory meatus (EAM).

12. C In order to demonstrate the pars petrosa in profile, it must be placed parallel to the image receptor. The 45 degree rotation, employed in the posterior profile projection (Stenvers) best accomplishes this alignment of the petrous ridges for the mesocephalic skull type.

13. A The bregma is the point on the vertex of the skull where the sagittal and coronal sutures meet. If the central ray is directed to the bregma for the verticosubmental (full basal) projection, the structures of the anterior cranial base and the auditory structures are well demonstrated.

See accompanying radiograph.

14. G	15. B	16. C	17. I
18. E	19. A	20. H	

21. B See explanation number 1 of this section.

22. D Because the optic foramen opens at an angle of 53 degrees to the midsagittal plane and 37 degrees to the vertical, an AP oblique position with the head rotated to form a 53 degree angle with the table, is used to demonstrate the optic foramen furthest from the image receptor.

23. C The mastoid air cells are thin and may be difficult to visualize because of their superimposition with other skull anatomy. Taping the cartilaginous auricles of the ear forward helps prevent their obscuring of the air cells in the axiolateral and tangential positions.

24. D The full basal projection (SMV) is used for the demonstration of the foramina ovale and spinosum in the base of the skull, the maxillary, sphenoid, and anterior ethmoid sinuses, the odontoid process of the axis and the entire atlas.

25. A See explanation number 1 of this section.

26. D The inferosuperior path of the central ray in the full basal (SMV) projection is used to provide a tangential view of the zygomatic arches as well as visualization of the foramen magnum, sphenoid sinus, auditory canals and mandibular condyles. (see explanation number 24).

27. B — The distance between the vertex and submentum in an "average" skull is 22 cm. Any variation of this external measurement will result in deviation of the internal structures and must be compensated by central ray or part angulations.

28. C — The interpupillary line (IPL) which bisects the pupils of the eyes should be placed perpendicular to the image receptor and the midsagittal plane in lateral skull radiography.

29. B — In the full basal or SMV projection, the axis, atlas, clivus, and foramen magnum are found in the mid skull region posterior to the petrous ridges. The paranasal sinuses, and mandibular structures are located anterior to the ridges.

30. B — In a PA axial position with a 15 degree caudal angle of the central ray to the OML the petrous ridges are projected into the lower 1/3rd of the orbital shadows, freeing the frontal and anterior ethmoids from their superimposition with the petrous ridges.

31. C — See explanation number 1 of this section.

32. B — When the nose, zygoma and chin are in contact with the table, the midsagittal plane forms a 53 degree angle to the image receptor. This will enable a vertical central ray to exit through the dependent optic foramen and provide an oblique image of the facial bones and the dependent optic foramen.

33. B — Because of the normal overbite occluding the incisors would likely dislocate the mandibular condyles. In the normal bite, the molars are occluded, avoiding the unnatural displacement of the mandible.

34. D — The perpendicular placement of the acanthiomeatal line (ACML) to the plane of image receptor helps to project the optic foramen into the outer lower quadrant of the orbit. This, along with the 37 degree angle of the MSP to the perpendicular (53 degrees to the image receptor), opens the optic foramen for its optimum visualization.

35. D — See explanation number 1.

36. C — The long axis of the orbit is directed obliquely at an anterior angle of 37 degrees from the midsagittal plane.

37. D — Either of the frontal projections: the AP axial or the PA will demonstrate side to side (mediolateral) fracture displacement of the mandibular ramus. The axiolateral projection will demonstrate anteroposterior displacement of the mandible.

38. C — In the posterior profile (Stenvers) projection, the head is rotated 45 degrees to the midsagittal plane to place the dependent petrous ridge parallel to the image receptor. In order to direct the central ray through the orbit the central ray is directed 12 degrees cephalad.

39. B — In a true lateral projection of the skull, the midsagittal plane is parallel to the image receptor, while the interpupillary line and coronal planes are perpendicular to the image receptor.

40. D — In the parietoacanthial (Waters) projection, the extension of the head and chin places the petrous ridges below the maxillary sinuses. In order to determine the appropriate amount of extension, the orbitomeatal line (OML) forms an angle of 37 degrees to the plane of the image receptor leaving the nose about .5- 1.5 cm from the table.

See accompanying diagram.

41. G 42. E 43. I 44. B

45. K 46. F 47. C 48. J

49. A See explanation number 39 of this section.

50. C The anterior profile (Arcelin) position of the petrous portion of the temporal bone is parallel to the image receptor to improve its visualization.

51. D A 30-40 degree angulation of the central ray in an AP axial projection of the skull is used to give a symmetrical image of the petrous pyramids and zygomatic arches as well demonstrating the dorsum sella, foramen magnum, and occipital bone.

52. C In the anterior profile (Arcelin) position with the head adjusted 45 degrees to the midsagittal plane, and the infraorbitomeatal line is perpendicular, the central ray should be directed 10 degrees caudad to best visualize the petrous portion of the temporal bone.

53. B The infraorbitomeatal (IOML) line is parallel to the image receptor in both the full basal submentovertical (SMV) and the verticosubmental (VSM) projections. The central ray is directed vertically, to the the base skull in the SMV projection.

54. C The floor of the orbit and roof of the maxillary sinus are formed by the maxillary bones. In cases of facial trauma, pressure on the orbit can fracture this bony plate filling the sinus with fluid. A parietoacanthial (Waters) projection is commonly included in all facial bones and sinus routines.

55. B The internal auditory canals are located within the medial portion of the petrous ridges of the temporal bones. The transorbital projection produces a bilateral symmetrical image of these canals projected within the orbital shadows. The prone position is preferred because it may result in a dramatic reduction in the exposure to the eye.

56. C The PA axial (Haas) position may be used on a hypersthenic patient who, because of a large torso and short neck, is unable to drop the chin enough to place the orbitomeatal (OML) line perpendicular to the image system. By turning the patient to a prone position and angling the central ray cephalad, the problem is corrected.

57. A In the tangential ("May Away") projection of the zygomatic arch, the midsagittal plane is rotated 15 degrees away from the affected side, producing a clearly delineated arch, free of superimposition with the adjacent facial bones.

58. D The 45-45 degree double angulation in the axiolateral oblique (Mayer) position of the skull of the midsagittal plane and the caudad central ray projects the petrosa inferior to the mastoid air cells, allowing the visualization of the auditory ossicles.

59. C Compared to the overall body size of a child or infant, the skull is disproportionally large. When laying supine, the child's neck is more flexed than in an adult and the orbitomeatal line cannot be placed perpendicular to the image receptor. To compensate, the central ray angle must be less than the usual 30 degrees for an adult.

60. A In order to project the optic foramen into the inferolateral orbital quadrant, the acanthiomeatal line must be perpendicular to the image receptor.

61. D A 45 degree rotation of the midsagittal plane will place the dependent petrosa parallel to the image receptor. The 12 degree cephalad angle projects the petrous ridges into the lower 2/3rds of the orbits and superimposes the mandibular ramus on the lateral aspect of the cervical spine.

62. C In a PA projection of the skull with the orbitomeatal line perpendicular to the image receptor, no angulation is desirable when the frontal bone is of primary interest; a 15 degree caudal angle is used for general survey; 25 degree caudad angle for the superior orbital fissure, and a 30 degree caudal angle to show the foramen rotundum.

63. A The axiolateral (modified Law) position, using double angulation of 15 degrees to the face and 15 degrees to the feet, provides an image of the mastoid air cells free of superimposition of adjacent cranial structures.

64. B In order to demonstrate the frontal sinuses, when the orbitomeatal line is perpendicular to the image receptor, the central ray is directed 15 degrees caudad. Since there is an 8 degree difference between the GML and OML, a central ray angulation of 23 degrees caudad is used to produce a comparable image when the GML is perpendicular.

65. C The inferior orbital fissure is a narrow cleft extending anterolaterally to the lateral wall of the orbit, at an angle of 25 degrees. With the patient prone and the infraorbitomeatal (IOML) line is perpendicular to the image receptor. A 25 degree cephalad angle of the central ray opens the fissure and enable its demonstration within shadow of the orbit.

66. C The optic foramen located at the medial end of the sphenoid bone can be demonstrated by rotating the head at an angle of 53 degrees with the image receptor and adjusting the acanthiomeatal line perpendicular to the table. In the parieto-orbital oblique, the optic foramen is projected into the lower outer portion of the orbit.

67. B When demonstrating the paranasal sinuses an even radiographic density between the upper and lower orbits is an indication that proper exposure factors have been employed.

68. A Cahoon describes a PA axial projection demonstrating a symmetrical image of the styloid processes free of superimposition of adjacent bony structures and the coronoid process of the mandible. Wigby-Taylor describes an AP oblique projection with the mouth opened to demonstrate an oblique image of the styloid free of superimposition of the coronoid process. Fuchs describes a lateral projection with the mouth opened producing a lateral image of the styloid process free of superimposition of the coronoid process.

69. A The superior orbital fissure is a cleft in the roof of the orbit between the greater and lesser wings of the sphenoid bone. When the patient is prone, a caudal angle of 20-25 degrees of the central ray projects the petrosa just below the inferior margin of the orbit, allowing shadows of the superior orbital fissures to be seen on the medial side of the orbit. (See explanation number 62 of this section.)

70. C When the infraorbitomeatal line (IOML) is parallel to the image receptor, the head is in good position to allow an inferosuperior projection of the zygomatic arch. The 15 degree rotation of the midsagittal plane, along with the 15 degree tilt, ensures that the central ray will be tangent to the lateral surface of the skull, projecting the arch closest to the image receptor free of superimposed shadows. This position is particularly useful with patients who have depressed fractures or "flat" cheekbones.

71. D The PA projection places the frontal bone parallel and closest to the image receptor. If the orbitomeatal line (OML) and the central ray are perpendicular, a true frontal image is provided.

72. A Less angular distortion of any part is accomplished by the correct positioning of that anatomic structure, rather than using a tube (central ray) angulation. The Valdini PA axial method which places the forehead so that the OML is 50 degrees to the image receptor rather than angling the central ray (as in the Grashey or Haas position) demonstrates the petrous ridges with the least amount of distortion.

73. D The jugular foramina are located lateral to and at the level of the foramen magnum. When the OML is perpendicular, a 50-60 degree angle is required to open these foramina enough to demonstrate them in their entirety.

74. A To determine the presence of sphenoid sinus effusion due to basal fracture, a cross table lateral is recommended. This position will best demonstrate a fluid filled sphenoid sinus, which may be the only clue to the presence of a basal skull fracture.

75. B This AP axial position is often called the reverse Waters (acanthioparietal) because it gives a similar image to the parietoacanthial projection of the facial bones. In this position, the patient is placed supine with the infraorbitomeatal line perpendicular to the image receptor and a central ray angle of 30 degrees cephalic to the acanthion.

76. B By extending the chin more in the Rhese method, the acanthiomeatal line becomes perpendicular providing optimum visualization of the optic foramen and improving the geometrically recorded detail of the image.

77. C Because fluid levels can only be detected by having the central ray horizontal to the floor, only upright or decubitus projections are acceptable for its visualization.

78. A The double 15 degree tube angulation of the axiolateral (Law) separates the mastoid cells from surrounding bony structures. This is accomplished by placing the skull in a true lateral position and angling the central ray 15 degrees caudad and 15 degrees anteriorly, demonstrating the mastoid air cells closest to the image receptor.

79. D The PA axial projection requiring a 30 degree cephalic angle of the central ray is used to project the mastoid process and the base of the skull above the condylar processes of the mandible.

80. B To help remove the superimposition of the teeth that would otherwise obstruct visualization of the sphenoid sinuses, the Pirie method projects the sinuses through the open mouth. The phonation "ah" during the exposure will immobilize the tongue in the floor of the mouth producing better demonstration of the sphenoid sinuses.

81. C The full basal submentovertical (SMV) projection is often used in the evaluation of the mastoid processes, auditory canals, nasal septum, zygomatic arches and numerous foramina on the base of the skull. The mandibular symphysis is superimposed on the frontal bone of the skull so is therefore poorly visualized.

82. D The PA axial position (reverse Towne) was developed by Haas principally to overcome the difficulty in radiographing hypersthenic patients, for the visualization of the petrous pyramids and occipital regions of the cranium.

83. A The PA axial (Valdini) method provides a means for obtaining a projection similar to a Towne or Haas without requiring a tube angulation. The patient is positioned so that the infraorbital meatal line forms a 50 degree angle with the plane of the image receptor. The projection shows the sella turcica and petrous ridges including the auditory canals and mastoid air cells.

84. C The orbitomeatal line (OML) forms a 37 degree angle to the plane of the image receptor when performing a parietoacanthial (Waters) projection. This projects the dense shadows of the petrous ridges below the antral floors allowing visualization of the maxillary sinuses, foramen rotundum and axial images of the facial bones. It is important to note that when the neck is extended too much (less than 37 degrees) the antral shadows will be foreshortened; when the neck is extended too little, the shadows of the petrous ridges are projected into the antra and obscure underlying pathology.

85. D When preoperative measurements of the sinuses are required a 72" (180 cm) source-to-image distance (SID) is employed tol minimize image magnification and distortion allowing for more precise measurements.

86. C See explanation number 62 of this section.

87. D Lysholm recommends an axiolateral position to demonstrate the cranial base for those patients who are unable to extend their head enough for a satisfactory submentovertical or verticosubmental projection. With the patient lying prone, the head placed in a true lateral position with the central ray directed 30-35 degrees caudad through a point 1" inferior to the lower external auditory meatus. Both sides are done for comparison.

88. D An inferosuperior projection of the mandibular symphysis is demonstrated using an occlusal film that is placed in the mouth with the pebble (exposure) side facing downward. When the pebbled surface of the film is directed upward a tangential projection of the nasal bones or an axial projection of the hard palate can be obtained depending upon the angulation of the central ray.

89. A Compared to the size of the trunk, a young child or infant's skull is relatively large. When lying supine, the neck is somewhat flexed. To compensate, the tube must be angled cranially so the central ray is directed along the orbitomeatal line to produce a true anteroposterior projection. This also holds true for an AP axial (Towne) projection where the central ray is angled less than the usual 30 degrees.

90. B To demonstrate the left temporomandibular joint (TMJ) in an axial transcranial position, the left TMJ should be placed closest to the image receptor with the skull in a true lateral position. The central ray is directed through the upper parietal region and lower TMJ at an angle of 25-30 degrees caudad. The TMJ furthest from the image receptor will be projected more inferiorly and will also appear somewhat larger because of magnification.

91. B In an AP projection of the skull a 15 degree caudad angle to the orbitomeatal line will project the petrous ridges into the lower 1/3 of the orbits, as demonstrated in the illustration. A perpendicular central ray will demonstrate the petrous ridges filling the orbits. To completely free the orbits of the shadows of the petrous ridges, an angle of 23 degrees caudad should be employed.

92. C This AP projection is used to demonstrate the crista galli and frontal sinuses free of bony super imposition.

93. C The size differences seen between the orbits indicates that this is an example of an oblique position.

94. B See explanation number 66 of this section.

95. D See explanation number 40 of this section.

96. C See explanation number 40 of this section.

97. A The lateral transcranial positions are employed to demonstrate the temporomandibular joints closest to the image receptor.

98. B The closed (A) and open mouth (B) are easily recognized by the relationship of the mandibular condyle to the mandibular fossa. The condyle lies in the fossa when the mouth is closed and inferior to the articular tubercle when the mouth is open. The closed mouth is employed to demonstrate fractures of the neck and condyle of the ramus, and the open mouth best demonstrates anterior and inferior excursion of the condyle from the mandibular fossa.

99. B The radiographic image shows an axiolateral projection of the mandibular body, ramus and alveolar process. The patient's neck must be extended and head rotated so the mandibular body of the affected side is placed parallel to the plane of the image receptor. A 20 degree cephalad angulation directs the central ray to a point 2" distal to the angle of the mandible furthest from the image receptor, separating the two mandibular bodies enough to visualize each side. By increasing the angle of the central ray by 5 degrees (25 degrees) more of the ramus is visualized, but the condyle and mandibular fossa of the temporomandibular joint are not usually visible.

100. A See explanation number 99 of this section.

Respiratory System

Chapter 50

1. A In order to insure that the chest is in the true lateral position, it is important to have the midsagittal and midaxillary planes vertical to the floor.

2. D During the upright posteroanterior (PA) projection of the chest the anterior chest is placed in contact with the image receptor. The patient is instructed to place his/her hands on the hips, palms facing upward and to roll the shoulders forward to remove the shadows of the scapulae from the lung fields.

3. C When a chest is to be imaged on a 35 x 43 cm image receptor, the entire thoracic cavity will be included if the central ray is directed to T7 and the chest is centered to the image receptor.

4. A Because the clavicles are normally superimposed on the apical regions of the lungs, an unobstructed view of this region is difficult. The AP aalso employing a 15-20 degree cephalic angulation, or the apical lordotic position with a perpendicular central ray all demonstrate the apical region.

5. C The use of a long 72" or 180 cm source-to-image distance (SID) reduces image unsharpness and decreases magnification of the entire chest and heart. The PA projection is preferred to further reduce the size of the heart shadow by placing the heart closest to the image receptor.

6. B When the principal structures of interest are located in the right lung, their visualization can be improved by the use of the right lateral position. Because of the importance of the heart, the left lateral position is normally employed for routine chest radiography.

7. D In the upright position, the depression of the shoulders is encouraged to direct the clavicles below the apical regions of the lungs. Special positions used for the demonstration of the apices are discussed in explanation number 4 of this section.

8. D — In order to demonstrate the presence of free fluid in the lung or free air in the abdominal cavity, and reduce the engorgement of the pulmonary vessels, the upright or erect position is preferred. Images are taken at the end of inspiration to maximize the size of the lung tissue by lowering the diaphragm.

9. C — When a patient cannot be placed upright, air-fluid levels can be identified in the decubitus positions. The right lateral decubitus position is preferred for the visualization of air in the axillary region of the left lung and fluid levels in the right lung.

10. D — Inspiration and expiration radiographs are most frequently taken to show the differences in position of the diaphragm (excursion) and small pneumothoraces that may only be visible on the expiration image. Foreign bodies may also be better localized on these images.

11. D — The area between the lungs, the mediastinum, contains the heart, great vessels, and the esophagus. To avoid superimposition of these and other bony structures, the chest is obliqued 45 degrees from the PA projection. Other degrees of obliquity and the principal structures they demonstrate are summarized 10-20 degrees for pulmonary disease, 35-40 degrees to demonstrate the esophagus and 5-60 degrees for the heart and great vessels.

12. A — The diaphragm moves to its lowest position during erect inspiration; forced inhalation should be avoided because it may cause an abnormal elongation of the heart distorting its normal appearance.

13. C — Rotation of the chest is most easily seen by comparing the spacing of the sternoclavicular (SC) joints located at the sides of the third thoracic vertebra.

14. B — The proper breathing instruction for chest radiography of the heart, lungs and mediastinum is to stop breathing at the end of full inspiration. When the heart is of primary interest, the patient is instructed to stop breathing at the end of normal inspiration.

See accompanying diagram.

15. D	16. C	17. E	18. F
19. H	20. G	21. L	22. K
23. J	24. A		

25. A — The right anterior oblique (RAO) and left posterior oblique (LPO) positions are used to provide the optimum visualization of the left atrium, when the patient is rotated 45 degrees.

26. D — The trachea extends from the cricoid cartilage at a level near the first thoracic vertebra to it's bifurcation near the level of the 5th thoracic vertebra. In order to center this structure on the image, the central ray is directed through the body of the manubrium.

27. B — The positions of the pulmonary apices (see explanation number 4) will project the clavicles above the apices and may help in the evaluation of interlobar effusions.

28. A — See explanation number 11 of this section.

29. B — The upright AP or PA projection is the preferred projection for the demonstration of free fluid levels in the costophrenic angles.

30. A — During full inspiration, the diaphragm is lowered to a position at or near the eleventh rib enabling the visualization of the first ten pairs of ribs.

31. B When the midaxillary line is centered to the image receptor, the curve of the thoracic spine results in a greater volume of the anterior aspect of the chest cavity. The center of the lung field where the central ray is directed, is about 5 cm (2") anterior to the midaxillary line.

32. C See explanation number 5 of this section.

33. C Normal inhalation on the second breath will most often increase the lung volume for chest radiography, especially in those patients with a heavy or hypersthenic body habitus.

34. B In the LAO position, the left side of the chest is foreshortened, the right side is somewhat elongated showing maximum area of the right lung.

35. A Even during anteroposterior (AP) radiography of the chest in either the upright or supine position, the shoulders should be rolled forward and hand positioned on the hips as they would be in the PA projection to remove the scapula from the lung fields.

36. A The air-fluid level appearing on the right side of the chest cavity indicates that this frontal projection was obtained with the patient in the left lateral decubitus position.

37. B See explanation number 36 of this section.

38. D This lateral projection and air-fluid level seen at the ninth thoracic vertebra, is visualized because of the use of a horizontal central ray to this upright patient.

39. D See explanation number 38 of this section.

40. A This is an AP axial projection of the chest for the apical region. The 20 degree cephalic tube angulation has projected the clavicles above the top of the lungs.

Routine Contrast Studies

Chapter 51

1. B The main function of the gall bladder is to temporarily store bile until its release into the duodenum to aid in the digestion of fats. To improve the visualization of the biliary ducts, a fatty meal; i.e., milkshake, can be given to the patient to empty the gall bladder and encourage the flow of the contrast agent into the ducts.

2. C In many single or double contrast barium enema examinations, the use of an inflatable retention (Bardex) type enema tube is employed. The inflatable collar which sits against the anal sphincter lessens the chance of premature evacuation of the contrast media. Because of the potential danger of intestinal rupture from over inflation, many come with a 90 cc reusable squeeze-type inflation as a safety precaution. The insufflation of air should be performed under fluoroscopic control.

3. C The use of a double (air) contrast technique during the examination of the gastrointestinal tract has become increasingly popular for the demonstration of the mucosal linings. A combination of a thin coating of barium followed by the introduction of air or gas provides improved visualization of small lesions not normally seen with barium alone.

4. B Small sliding gastroesophageal (hiatal) hernias are found in the thoracicoabdominal region passing through the esophageal hiatus of the diaphragm. The placement of the head lower than the feet as in the Trendelenburg position, increases the intra-abdominal pressure, and helps to fill the hernia with barium.

5. D The main differences in the 3 common procedures which demonstrate the upper collecting system are based on the means by which the contrast media are introduced. During an intravenous urogram, an intravenous (IV) injection of iodinated media is rapidly filtered by the kidneys with subsequent filling of the rest of the urinary tract. In nephrotomography, improved visualization is possible through the infusion of contrast agents and the use of tomographic slices. Retrograde pyelography involves the placement of a catheter transurethrally into the proximal ureter and the direct injection of the contrast media into the proximal end of the ureter.

6. A The routine AP (supine) projection of the abdomen is unable to free air-fluid levels. Upright and/or decubitus positions, employing a horizontal central ray will be included to show free air and/or fluid that may have escaped from the viscera into the peritoneum. When the right lateral decubitus position is used, the patient should be placed in this position for at least 15-20 minutes prior to the exposure to give the gases time to rise.

7. C The act of swallowing, deglutition, will inhibit respiration for about two seconds, enabling subsequent exposures without the appearance of breathing motion.

8. B The 22 feet of small intestine extends from the pyloric sphincter of the stomach to the ileocecal valve at the beginning of the colon. Digested food or barium takes about 12 hours to travel the 3 portions of the small intestine; the duodenum, jejunum and ileum.

9. A Colonic lavage (cleansing enemas), dietary restrictions of solid food, and the use of laxatives, are all employed to remove fecal material, intestinal gas or previous contrast media from the colon. Any remaining fecal material may mimic the appearance of small tumor masses or polypoid lesions of the large bowel.

10. D Dietary restrictions are necessary prior to all contrast studies of the gastrointestinal tract; i.e., GI, SI, BE, GB. Food or other liquids can dilute or simulate the appearance of small masses or mucosal defects. During gall bladder examinations, food can trigger the release of bile from the gall bladder.

11. C Following the intravenous injection or infusion of about 30-100 ml of iodinated contrast media, the pelvicalyceal system begins to visualize within 2-8 minutes. Depending on a number of physiologic variables, the maximum concentration will occur in about 15-20 minutes.

12. A The exhalation phase of breathing relaxes the intra-abdominal organs and maintans the abdominal organs at a consistent location. The exposure should be delayed 1-2 seconds after the patient stops breathing to allow more complete cessation of abdominal organ movement.

See accompanying diagram.

13. G 14. D 15. A 16. C

17. F 18. B

19. C Food and water should be withheld prior to an upper gastrointestinal (GI) series for a period of at least 8 hours to insure that the stomach is free of all solids and liquids.

20. C Barium flows rapidly through the gastrointestinal tract due to its high mass and the force of gravity. In the upper esophagus, the use of barium soaked cotton balls or thick barium paste will slow the media's descent and improve the visualization of obstructed regions.

21. B The abdomen is bounded superiorly by the diaphragm and inferiorly by the area at the level of the symphysis pubis. When possible, both the diaphragm and symphysis should be included on the radiograph. In larger patients, two crosswise 34 x 43" cm films may be necessary to accomplish this.

22. A The sigmoid colon and rectosigmoid junctions are well visualized in the axial, lateral, ventral and dorsal decubitus and Chassard-Lapine positions which may be used to supplement routine projections used in a barium enema study.

23. A See explanation number 5 of this section.

24. A In order to produce a radiograph with a greater number of gray scales or long scale contrast, a kilovoltage setting in the 80-90 range is desired for abdominal radiography.

25. C Due to the superimposition of the ribs, the vertebrae or overlying fecal material, small calculi in the gall bladder are often poorly visualized. For this reason, upright and/or decubitus positions are often included in gall bladder studies. Low density cholesterol stones may float on the bile producing the stratified appearance on these radiographs.

26. C The biliary ducts extend medially and posteriorly from the gall bladder and are better visualized in the right posterior oblique (RPO) position which places them parallel to the image receptor.

27. B Since the fundus and duodenum are more posteriorthan the body of the stomach, the prone position will show these structures filled with air or gas.

28. D In the Trendelenburg position, the upper torso is lowered allowing retrograde spillage of the contrast media from a filled bladder into the distal ureter.

29. B The right colic or hepatic flexure formed at the junctions of the ascending and transverse colons is best visualized by rotating the patient 35-45 degrees away from the affected side. This is more commonly referred to as the left AP oblique or left posterior oblique position.

See accompanying diagram.

30. G.	31. E	32. B	33. I
34. H	35. D	36. A	37. C
38. F			

39. B Gastrointestinal intubation which can remove gas or fluids from the GI tract is accomplished through the placement of a thin rubber tube into the stomach. The tube is most often inserted through the nose and passed into the oropharynx, esophagus, stomach and small intestine.

40. D See explanation number 9 of this section.

41. C During an enema, the barium flows in a backward direction, through the colon, passing from the rectum to the sigmoid colon, descending colon, splenic flexure, transverse colon, hepatic flexure, ascending colon and into the cecum.

42. B The normal amount of iodinated contrast media used to opacify the urinary system is between 30-100 ml (cc) given by intravenous injection. Amounts may vary depending on the age and size (weight) of the patient.

43. D After the ingestion of the cholecystographic contrast media, it is important to avoid foods, liquids and laxatives which can all stimulate the emptying of the gall bladder.

44. D Localization of foreign bodies of the respiratory or GI tracts will normally require films of the neck, chest and abdomen. In cases where nonopaque objects are suspected, the use of contrast agents may also be indicated.

45. B See explanation number 12 of this section.

46. A The Sims position, with the patient lying on his/her left side leaning forward, relaxes the abdominal muscles and decreases the intra-abdominal pressure, making the insertion of the enema tip easier.

47. B The right anterior oblique position (RAO, right PA oblique) increases the peristaltic action of the stomach speeding its emptying into the duodenum.

48. B The axial positions of the rectosigmoid area described by Billings, utilize a 35-40 degree angulation to project the S-shaped sigmoid colon free of superimposition of itself.

49. D Because of its double blood supply (arterial and venous), oral agents are absorbed by the GI tract as they pass into the portal circulation, while intravenous agents take the arterial pathways, to show the function of the liver and biliary tract. A number of other direct methods involving a direct needle or catheter placement into the biliary ducts called T-tube cholangiography and endoscopic retrograde cholangiopancreatography (ERCP) may also be performed.

50. A Gastroesophageal reflux is most frequently seen radiographically by increasing pressure on the gastroesophageal junction through the use of the Trendelenburg position or the Valsalva maneuver.

51. B Regardless of the position of the patient, air-fluid levels can only be demonstrated if the central ray is directed horizontal to the floor.

52. D Though oral cholecystographic contrast agents may cause a myriad of systemic symptoms, the vast majority affect the GI tract, causing nausea, vomiting and diarrhea.

53. B The 30-40 degree right anterior oblique (RAO) position for the esophagus helps free it from its superimposition with the heart and vertebral column. The central ray is directed vertically through T5-T6.

54. B Post voiding films which serve to detect residual urine in the bladder are often related to small tumor masses or enlargements of the prostate gland.

55. C In a noncholecystectomized patient, following an intravenous injection of an iodinated contrast media, the biliary ducts will become visible in about 15 minutes, reaching their maximum concentration at about 40 minutes. The gall bladder will reach its maximum concentration at about 2 hours.

56. C Similar to a barium enema preparation, the visualization of a non-acute abdomen can be dramatically improved by colonic lavage, laxatives and limiting the intake of foods.

57. A See explanation number 2 of this section.

58. C A ruptured viscus (a perforated stomach or small intestine) is often associated with a spillage of gastric contents and gases into the peritoneum. These will be seen most readily as free air-fluid levels on upright or decubitus images.

59. B The lower tissue density of the gas-filled structures of a double contrast enema require a reduction of above 10 kVp to avoid over penetration.

60. B In a small adult patient, the kidneys are about 7 cm above the table top; 9 cm above the table in the average patient; and 11 cm in large patients, when the patient is supine.

61. D A uniform coating of the intestinal walls is most easily accomplished with a high density barium suspension such as polybar.

62. C See explanation numbers 9 and 10 of this section.

63. C The small bowel can be examined in 3 ways: by mouth, complete reflux or by enteroclysis. The most common and comfortable is by mouth through the ingestion of 36-60 ounces of liquid barium sulfate. The complete reflux method requires the filling of the colon and the retrograde filling of the small intestine. This uncomfortable procedure is rarely done in the modern department. Enteroclysis involves introduction of barium through a nasogastric tube placed in the duodenum.

64. D The maximum distance for the insertion of an enema tip is four inches.

65. D In order to minimize magnification and improve definition, the gall bladder should be placed as close to the image receptor as possible. This is accomplished in the prone position.

66. A In order to improve the filling of the ureters and renal pelves, compression should be applied over the distal ureters by placing a band above the symphysis pubis. This retards the filling of the bladder and encourages the retention of urine within the drainage apparatus.

67. A In the right lateral decubitus position air will rise, demonstrating the medial border of the ascending colon and lateral border of the descending colon.

68. D The somewhat C-shaped stomach is directed anteriomedially from the fundus in the upper posterior abdomen to the more anterior body before extending posteriorly to the pylorus. In the supine position, air will fill the body while the barium settles in the more posterior fundus and pylorus.

69. C In the 45 degree right anterior oblique (RAO) position of the chest, a barium filled esophagus is in contact with the left atrium of the heart.

70. C Prior to most iodinated contrast studies, a scout film is taken to visualize radiopaque stones whose appearance is masked by positive contrast media. Radiolucent (cholesterol) stones are not seen on the scout film but may be demonstrated during the contrast enhanced images.

71. B The nephrogram is the first image taken of the kidney within 30 seconds of the bolus injection and shows the blush phase associated with the uptake of contrast by the parenchyma of the kidney.

72. A In order to avoid superimposition of the gall bladder with the lateral aspect of the spine, the left anterior oblique (LAO), right posterior oblique (RPO) positions and right lateral positions are generally employed.

73. C In order to displace the overlying shadows caused by intestinal gas or fecal material, the stomach may be filled with carbonated liquids which push the intestine inferior to the renal pelvis and kidneys.

74. C — The increased venous pressure resulting from the Valsalva (bearing down) maneuver places the esophagus under stress causing the varices to enlarge, thus improving their appearance.

75. D — The mild anesthetic effects and the stimulation on the anal sphincter are the two main reasons for the use of cold (41 degrees F., 5 degrees C) barium enema suspensions. Since the colon is unable to detect temperature this can help to make the procedure more comfortable and improve the retention of the barium.

76. C — See the following explanation.

77. D — In the right anterior oblique (RAO) position free gas in the stomach will rise into the fundus while the pylorus, duodenum and body of the stomach remain filled with barium.

78. B — See the following explanation.

79. A — This frontal radiograph clearly shows air-fluid levels along the long axis of the body indicating that this frontal projection was obtained in the left lateral decubitus position.

80. C — See the following explanation.

81. A — The presence of barium on the lateral surface of the ascending colon and medial surface of the descending colon, is characteristic of the right lateral decubitus position of a double contrast enema series.

82. B — See the following explanation.

83. C — The hepatic flexure is best demonstrated in a 35-45 degree left posterior oblique (LPO) position seen in this radiograph. The more superior splenic flexure is superimposed on itself in this projection.

84. A — See the following explanation.

85. D — The air-fluid levels which are seen crossing the abdomen from side to side indicates that this child was upright and PA/AP during the filming of this image.

Special Procedures

Chapter 52

1. D — The fluid filled space around the spinal cord (subarachnoid space) can be accessed by way of a spinal needle. A contrast media introduced into this space will outline the spinal cord and spinal nerves in a procedure called a myelogram.

2. B — The three opacification studies of the vessels are: arteriography involving the opacification of the arteries; venography involving the opacification of the veins; and angiography involving the opacification of any vessels including capillaries and lymphatic vessels.

3. B — A mammogram is the non-contrast radiographic examination of the breast.

4. D — Arthrography is a procedure involving the injection of an iodinated contrast media into the fluid filled joint cavity for demonstration of the major support structures, the ligaments, and cartilages of the joint.

5. C — A T-shaped tube is sometimes placed into the common hepatic and common bile duct after the removal of the gall bladder and cystic duct. Contrast agents are injected into the bottom of the T-Tube to demonstrate undetected stones in the common bile duct. This procedure is called operative cholangiography.

6. A — In cases where a blockage of the fallopian tubes is suspected, a cannula can be introduced through the uterus into the opening of the tubes. An iodinated media is passed into the oviducts to demonstrate their lumen. When the duct is patent (open) the contrast media will spill over into the peritoneum.

7. D — Rarely performed in a modern department due to the advances in OB ultrasound, a cephalopelvimetry is a study in which the fetal head and pelvis are measured to determine if a vaginal birth is feasible.

8. C — The demonstration of the ureter and kidney by direct means, requires the passage of a catheter through the urethra and bladder into the proximal ureter. An iodinated contrast media is introduced against the normal urine flow; hence the name retrograde urogram for examination of the pelvicalyceal system and ureters.

9. B — The soft and fatty tissue of the human breast are best demonstrated by the use of low kVp in the 15-35 kVp range.

10. B — A spinal tap or a myelographic injection is nearly always done at the L3-4 interspace to avoid damage to the spinal cord, which ends at the level of L1 or L2.

11. D — Prior to the placement of a catheter into the ureter, the opening is located on the posterolateral surface of the urinary bladder, by way of a cystoscope.

12. B — The greater density of the myelographic contrast media allows it to be moved by gravity. As the head of a table is tilted down, the contrast flows toward the brain. Acute extension of the head and neck closes the opening to the fourth ventricle, and prevents the spillage of the media into the ventricles.

13. D — In a two view mammogram, the craniocaudal (cc) and mediolateral oblique (MLO) are two of the most common projections. The axilla position is often used in a three view routine or as an adjunct view to show the axillary region.

14. B — The demonstration of the salivary ducts by the canalization of one of the salivary ducts and the introduction of an iodinated contrast media is called sialography.

15. D — Because the spinal cord doesn't end until about L2, the injection sight for nearly all myelograms is at L3-4.

16. B — The kidney or renal parenchyma will be demonstrated by newly formed urine which contains the water-based iodinated contrast media.

17. D — After the puncture of the subarachnoid space around the spinal cord, an amount of spinal fluid equal to the amount of contrast to be injected, is removed, to prevent an increase in intracranial pressure.

18. B — The root "cysto" refers to the urinary bladder; the suffix "graph" is from the word picture.

19. D — A dedicated mammographic system will often possess molybdenum targets for the low energy x rays, micro or fractional focal spots to provide for high recorded detail, and compression devices to improve contrast and control motion.

20. C — See question number 5 of this section.

21. A — During a spinal tap or myelogram, acute flexion of the spine will help open the L3-4 interspace and facilitate access to the meningeal coverings.

22. C — During a T-tube injection, contrast media flows through the common bile duct and passes through the sphincter of Oddi on its way to the duodenum.

23. D — The use of compression during a mammogram serves three main functions; namely to reduce breast tissue thickness differences, reducing the amount of scattered radiation, and providing a higher contrast.

24. C — The use of water-soluble compounds such as amnipaque or metrizamide for myelography have the advantage of a lower viscosity, improved nerve root visualization, and eliminate the need for post-procedural contrast removal.

25. D — Arthrography can currently be performed on all joints of the extremities and TMJ's, the less invasive magnetic resonance imaging (MRI) has become the method of choice for the examination of most soft tissue joint injuries.

26. B — The oblique position during male cystourethrography is used to reduce the bony superimposition of the urethra, the neck of the bladder, and the bony pelvis.

27. C — The opening of the vocal cords during inspiration, or the closing of the vocal cords with the "eee" phonation is helpful in the evaluation of the normal function of the larynx and pharynx.

28. B — The use of tourniquets and/or a 45 degree semierect projection of the contrast-filled leg, will aid in the filling of the deep veins during lower leg venography.

29. B — Endoscopic retrograde cholangiopancreatography (ERCP) involves the scoping of the duodenum and the passage of a cannula into the hepatopancreatic ampulla and common bile ducts system.

30. A — A gravid or pregnant patient may undergo one of four radiographic procedures: fetography, placentography, or cephalometry or pelvimetry.

31. C — Because the ovary is not directly connected to the fallopian tube, any contrast within these tubes will spill over into the peritoneum.

32. B — With a new mammographic single screen imaging system, the dose to the breast has been reduced to about 1/100th of a non-screen and 1/5th of a xeroradiographic system.

33. C — When a blockage above the normal lumbar injection sight is present, the contrast media may be introduced into an area near the back of the head called the cisterna magna.

34. A — In order to demonstrate the soft tissue structures of the joints (arthrogram), a needle is passed into the joint capsule for introduction of an iodinated contrast media.

35. A — The back pressure exerted during voiding is often sufficient to cause ureteral reflux at the junction between distal end of the ureters and urinary bladder.

36. A — The emptying of the salivary gland prior to and after a contrast injection to the gland is readily accomplished by sucking on a wedge of fresh lemon.

37. A — An ERCP is a procedure for the introduction of a cannula into the biliary duct network. (See explanation number 29).

38. D During an operative cholangiogram, a direct injection into the biliary ducts will enable the evaluation of intra and extrahepatic ducts as well as the sphincter of Oddi.

39. D In all myelographic procedures, it is important to remove spinal fluid prior to the contrast injection to avoid the complication of raising the intracranial pressure.

40. A The measurement of the lower leg length is accomplished most readily through the placement of a specially designed metallic ruler which is radiographed between the desired extremities.

41. B With the advances in catheterization techniques, many disorders that previously required surgery such as arteriovenous malformations (AVM's) and some vascular bleeds, can now be repaired with an injection of embolic materials.

42. D At the present time dilatation of constricted vessels by the manipulation of a balloon tip catheter and/or the more recent addition of a permanent stint to keep the vessel open are two of the types of transluminal angioplasty.

43. A The digital angiography has all but replaced many types of conventional angiographic techniques. The main reasons are the simplicity of the procedure and the ability to post process the images to improve the contrast resolution available.

44. A In order to overcome the degree of magnification which may negate precise measurement of the fetus or pelvis, a specially designed perforated metallic Colcher Sussman ruler can be utilized to enable a more accurate calculation of pelvic dimensions.

45. D Because the blood flow in a vein is toward the heart, the injection site for most venograms is a superficial vein of the foot or hand.

46. C In order to provide accurate pelvic measurements, it is critical that the pubic bones are not obscured by the femora in the lateral position.

47. D In nearly every mammography position, the nipple will be profiled. More recently, less emphasis is placed on this and more on the maximization of the quantity of breast tissue visualized.

48. A The advent of computed tomography (CT), and more recently, magnetic resonance imaging (MRI), both enable the improved visualization of the soft structures of the central nervous system and many joint cavities.

49. C See explanation number 2 in this section.

50. D The placement of a blunt cannula or flexible catheter guide wire device into an artery will often require the use of a multiple step process called the Seldinger technique.

51. B To improve the visualization of the deep veins of the legs, the position most often selected is the AP with a 30 degree internal rotation.

52. B The demonstration of the lymphatic vessels and nodes may be essential in determining the type and spread of many types of lymphatic cancers and is termed lymphangiography. The visualization of these small vessels requires special injection techniques, and slow introduction of an oil based iodinated contrast agent into the system.

53. A During fetography, as well as obstetrical ultrasonography, the use of hyperaeration is often beneficial in reduction of fetal movement.

54. D — To provide the appropriate magnification adjustment factor, the Colcher Sussman ruler needs to be placed at the middle of the pelvis in the AP position, at the same level as the ischial tuberosities.

55. A — By convention, most physicians desire that the marker on a mammogram be placed near the axilla, whenever possible.

See corresponding information.

56. B	57. E	58. A	59. C
60. D	61. C	62. C	63. B
64. C	65. C		

Section VII
Patient Care and Management

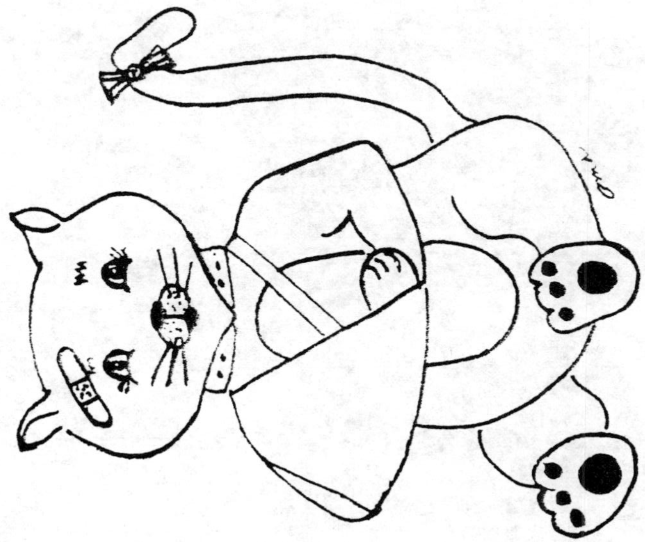

Vital Signs And Equipment Monitoring

Chapter 53

1. When I.V. administration is taking place, the solution should always remain __________ above the level of the vein. 3:149

A. 6"-8"
B. 8"-12"
C. 18"-20"
D. 36"-40"

2. A sphygomomanometer is a device that is employed in the evaluation of: 3:94

A. Respiration
B. Cardiac sounds
C. Pulse rate
D. Blood pressure

3. Chest drainage apparatus should always be placed: 9:157

A. 12'-20' above the chest
B. At the level of the chest
C. Below the level of the chest
D. None of the above

4. To enlarge the vein to make penetration easier, a tourniquet is often placed on the patient's arm: 3:151

A. Proximal to the injection site
B. Distal to the injection site
C. Over the injection site
D. None of the above

5. The pulse rate for individuals is dependent upon many factors. Among these is: 3:90
1. Age of the patient *2. Sex of the patient* *3. Patient's anxiety*

A. 1 only
B. 2 only
C. 3 only
D. 1, 2 & 3

6. A persistently high arterial blood pressure is called: 9:80

A. Hyperthermia
B. Hyperkinesia
C. Hypotension
D. None of the above

7. A rapid heart rate is one of the common indicators of cardiac insufficiency or heart damage and is termed: 9:80

A. Bradycardia
B. Hypertension
C. Tachycardia
D. Hypoxia

8. You are asked by a physician to draw up 20 milligrams of epinephrine. The vial reads 5 milligrams per cc. How many cc's should be drawn for the proper dosage? 3:142

A. 100cc
B. 20cc
C. 4cc
D. 2 1/2cc

9. When IV solutions are employed, to insure the proper flow of the solution, the fill chamber below the IV bottle should be: 3:150

A. Completely empty
B. About half full
C. Completely full
D. Three-quarters full

10. A tube that is placed into the upper respiratory tract to improve access to the lower respiratory tract is called a: 9:155

A. Nasogastric tube
B. Chest tube
C. Tracheostomy tube
D. T-tube

11. A patient has a blood pressure reading 140mm Hg systolic and 70mm diastolic. The reading for this person indicates he/she has: 9:81

A. Above normal systolic pressure
B. Below normal systolic pressure
C. Below normal diastolic pressure
D. Normal range of pressures

12. During an intravenous infusion, a swelling develops around the injection site from a dislodged needle or catheter. The immediate action to be taken by the radiographer is: 9:223

A. Remove the needle
B. Clamp off the tubing
C. Lower the IV bottle
D. Apply warm compresses

13. The most common route by which medications that cannot be absorbed by the intestinal mucosa are administered is called: 3:142

A. Topical administration
B. Oral administration
C. Intrathecal administration
D. Parenteral administration

14. Which of the following can be classified as a physical sign? 3:85
1. Cyanosis 2. Diaphoresis 3. Skin Pallor

A. 1 & 2 only
B. 1 & 3 only
C. 3 & 3 only
D. 1, 2 & 3

15. Which of the following types of medical apparatus should normally remain at a level above the patient at all times? 9:223

A. Chest drainage tubes
B. Suction apparatus
C. Intravenous infusion equipment
D. Urinary drainage equipment

16. Gastrointestinal intubation can be used for: 1:119:II
1. Visual inspection 2. Aspiration of contents 3. Direct feeding

A. 1 & 2 only
B. 1 & 3 only
C. 2 & 3 only
D. 1, 2 & 3

17. When a patient is connected to the chest tube drainage equipment, it is important to ____________ when moving the patient. 9:157
1. Clamp off the chest tube 2. Raise the bottle above the chest
3. Avoid kinks in the drainage tube

A. 1 only
B. 2 only
C. 3 only
D. 1, 2 & 3

18. Which of the following equipment is considered a part of the high-flow oxygen system? 9:85
1. Respirator 2. Oxygen tent 3. O^2 mask

A. 1 only
B. 2 only
C. 3 only
D. 1, 2 & 3

19. The normal range for diastolic blood pressure is about: 9:81

A. 30-50 mmHg
B. 60-90 mmHg
C. 90-110 mmHg
D. 110-140 mmHg

20. The common sites for the injection of subcutaneous medications are the: 3:143

A. Elbow and wrist
B. Arm and thigh
C. 3rd & 4th fingers
D. Webs of fingers and toes

21. In children under five years of age, to avoid damage to the sciatic nerve the preferred site for intramuscular injections is the: 3:143

A. Deltoid muscle
B. Gluteus muscle
C. Vastus lateralis muscle
D. Quadraceps femorus

22. The lumen of a hypodermic needle is measured by the unit called a gauge. Which of the following possesses the largest diameter opening? 3:143

A. 10 gauge
B. 15 gauge
C. 20 gauge
D. 25 gauge

23. A patient has a pulse rate of 35. This may be an indication of: 9:78

A. Shock
B. Fever
C. Hypothermia
D. Hemorrhage

24. Chest tube drainage systems are introduced to ____________ to/in the pleural cavity. 9:157

1. Create a negative pressure *2. Remove air* *3. Create a positive pressure*

A. 1 & 2 only
B. 1 & 3 only
C. 2 & 3 only
D. 1, 2 & 3

25. An increase in body metabolism, usually in responce to an infectious process is called (a): 3:86

A. Fever
B. Shock
C. Defibrillation
D. Sepsis

26. The escape of fluids into the surrounding tissues, often seen when intravenous needles infiltrate, is termed: 3:156

A. Extravasation
B. Emasculation
C. Embolization
D. Exacerbation

27. The normal adult resting respiration is about: 9:78

A. 3-5 times/min.
B. 7-10 times/min.
C. 16-20 times/min.
D. 25-30 times/min.

28. If pressure is not maintained on a vein immediately after the removal of the needle, a collection of blood may result called a: 3:156

A. Hemolysis
B. Hemorrhage
C. Hematoma
D. Diaphoresis

29. Which of the following special precautions must be made in a patient's room, who is receiving oxygen? 3:164

1. Avoid open flames *2. Keep doors closed* *3. Wear masks*

A. 1 only
B. 2 only
C. 3 only
D. 1, 2 & 3

30. The flow rate of a oxygen delivery system is normally measured in units of: 3:164

A. Cubic feet per hour
B. Liters per minute
C. Cubic centimeter per hour
D. Respirations per minute

31. In general, oral temperature readings possess a high degree of accuracy. A common factor that may cause erroneous readings is: 3:86

A. Oxygen therapy
B. Tachycardia
C. Dyspnea
D. Chest pain

32. A normal body temperature reading of 38° C (99.6° F) can be expected for: 9:72

A. Oral temperatures
B. Axillary temperatures
C. Rectal temperatures
D. None of the above

33. Cardiac and respiratory sounds and apical pulse determinations are made using a device called a: 9:77

A. Sphygmomanometer
B. Stethoscope
C. Cardiac monitor
D. Spirometer

34. In order to reduce the need for oxygen during surgery and to reduce high fever, a technique called ____________ is sometimes employed. 8:699

A. Hypoalimentation
B. Hypothermia
C. Hyperpnea
D. Hypochondria

35. The determination of the pulse can be made at various sites. Among these is/are: 9:78
1. Radial artery 2. Carotid artery 3. Abdominal aorta

A. 1 & 2 only
B. 1 & 3 only
C. 2 & 3 only
D. 1, 2 & 3

36. During the evaluation of a patient's pulse rate it is important to: 9:78
1. Determine the pulse using the thumb 2. Record both strength and regularity
3. Compare readings on opposite extremities

A. 1 only
B. 2 only
C. 3 only
D. 1, 2 & 3

37. Blood or other secretions, which may be accidentally aspirated, are often removed by using: 9:153

A. A chest drainage tube
B. T-tube drainage
C. Suction equipment
D. Respirator

38. The unlawful touching of a person without their consent 3:20

A. Assault
B. Battery
C. Invasion of privacy
D. Negligence

39. In order to obtain the appropriate vital signs for a patient will require the use of a: 9:72
1. Watch 2. Stethoscope 3. Sphygmomanometer 4. Thermometer

A. 1 & 2 only
B. 1 & 3 only
C. 2 & 3 only
D. 1, 2 & 3

40. The normal retention period for which radiographs must legally be kept on file for an adult patient is: 3:51

A. 6 months
B. 1-2 years
C. 3-4 years
D. 5-7 years

41. The five phases of the grieving process were identified by: 3:27

A. Elizabeth Kubler Ross
B. Sigmund Freud
C. Abraham Maslow
D. Miller Abbott

42. In many emergency situations, a rapid systemic response to the medication is necessary. The route that will in most cases provide this is: 3:147

A. Subcutaneous
B. Intravenous
C. Oral
D. Intramuscular

43. In order to properly identify a radiograph, the following items should be on the film: 1:10:1
1. Patient's Name 2. Institutional identity 3. The date of the exam

A. 1 only
B. 2 only
C. 3 only
D. 1, 2 & 3

44. Which of the following is a common type of nasogastric or nasoenteric tube? 3:261
1. Miller-Abbott 2. Levin 3. Cantor

A. 1 & 2 only
B. 1 & 3 only
C. 2 & 3 only
D. 1, 2 & 3

45. Which of the following will be noted if an injury has effected an individual's motor control functions? 8:219
1. Muscle coordination 2. Pain sensitivity 3. Visual reflex

A. 1 only
B. 2 only
C. 3 only
D. 1, 2 & 3

The Control Of Infection And Injuries

Chapter 54

1. The practice that helps reduce the spread of micro-organisms is termed: 9:32

A. Lethargy
B. Contracture
C. Asepsis
D. Isolation

2. In general, the best method for a radiographer to maintain medical asepsis and avoid the spread of infection is by frequent and proper: 9:32

A. Mask usage
B. Hand washing
C. Isolation usage
D. Personal hygiene

3. The most common means by which infections are spread is by: 9:29

A. Airborne particulates
B. Indirect (fomite) contact
C. Direct (touching) contact
D. Endogenous contact

4. Nosocomial infections are those that are contracted by an individual: 9:23

A. From contaminated water
B. From occupational environment
C. Within a hospital
D. With poor hygiene

5. Any medical procedure that involves the penetration of the skin requires the use of: 9:164

A. Medical asepsis
B. Surgical asepsis
C. Isolation techniques
D. Protective isolation techniques

6. A sterile pack is considered to be unsterile or contaminated if it has been exposed to: 9:166
1. Excessive heat 2. Excessive cold 3. Water or moisture

A. 1 only
B. 2 only
C. 3 only
D. 1, 2 & 3

7. A surgical procedure involving the repair of a fractured bone is termed a/an: 3:180

A. Fram reduction
B. Compound reduction
C. Open reduction
D. Closed reduction

8. Which is a common method of the direct spread of infectious agents: 9:29
1. Sexual transmission 2. Fomite transmission 3. Vector transmission

A. 1 only
B. 2 only
C. 3 only
D. 1, 2 & 3

9. Surgical aseptic technique should be employed during which of the following procedures? 9:164
1. Myelography 2. Intravenous pyelography 3. Cystography

A. 1 only
B. 2 only
C. 3 only
D. 1, 2 & 3

10. Which is a common form of indirect contact method of spreading contagious diseases? 9:29
1. Droplet transmission 2. Touching 3. Sexual contact

A. 1 only
B. 2 only
C. 3 only
D. 1, 2 & 3

11. A person is bitten by a tick or mosquito and develops an infection. This is classified as ____________ transmission. 9:29

A. Vector
B. Fomite
C. Endogenous
D. Nosocomial

12. The center of gravity for a person standing in the erect position is at the level of the: 9:49

A. Mid-abdomen
B. Mid-thorax
C. Mid-pelvis
D. Mid-thigh

13. A burn patient is generally very susceptible to infection and is often protected by the use of a/an: 3:122

A. Enteric isolation unit
B. Respiratory isolation unit
C. Strict isolation unit
D. Protective isolation unit

14. To preserve sterile integrity when a person's path is crossed, contact should be made: 9:168

A. Face to face
B. Side to side
C. Back to back
D. None of the above

15. A barium enema is a type of procedure that requires the use of: 9:32

A. Surgical asepsis
B. Medical asepsis
C. No asepsis

16. A radiographer is accidentally punctured by a used syringe needle. This is classified as a/an ____________ transmission. 9:29

A. Endogenous
B. Nosocomial
C. Fomite
D. Vector

17. The spread of infection may occur from: 9:29
1. Personal contact *2. Aerial transmission* *3. Fomites*

A. 1 only
B. 2 only
C. 3 only
D. 1, 2 & 3

18. The smallest disease-producing organism yet discovered is (the): 9:26

A. Yeast
B. Virus
C. Fungus
D. Bacteria

19. The two most common and effective methods of sterilization for use in a hospital are: 9:165

A. Pressurized steam and dry heat
B. Radiation and dry heat
C. Gas and pressurized steam
D. Chemical and dry heat

20. When a sterile gown is worn during an operative procedure, the portion of the gown considered to be sterile is the: 9:168
1. Front above the waist *2. Back above the waist* *3. Front below the waist*

A. 1 only
B. 2 only
C. 3 only
D. 1, 2 & 3

21. The muscle groups that should be employed for lifting heavy objects are located in the: 9:49
1. Arms *2. Legs* *3. Back* *4. Neck*

A. 1 & 2 only
B. 1 & 3 only
C. 2 & 3 only
D. 1, 2 & 3

22. A patient has a serious complication during a myelographic procedure, and a malpractice suit is lodged. The claimant is able to sue the: 3:19

1. Physician 2. Hospital 3. Radiographer

A. 1 only
B. 2 only
C. 3 only
D. 1, 2 & 3

23. Which of the following diseases is likely to require some type of strict isolation procedure? 9:39

1. Small pox 2. Diphtheria 3. Staphylococcus auricus

A. 1 only
B. 2 only
C. 3 only
D. 1, 2 & 3

24. The tissue of the spinal cord is protected by the canal through the vertebrae. Due to the mobility of the ____________ and ____________ spines, injuries in these areas are most common. 9:122

A. Cervical/dorsal
B. Dorsal/lumbar
C. Cervical/lumbar
D. Dorsal/lumbar

25. The term that best describes the complete removal of all micro-organisms is called: 9:164

A. Medical asepsis
B. Disinfection
C. Antisepsis
D. Sterilization

26. The two most common methods for the control of the spreading of a contagious disease are: 9:38

A. Vaccination and inoculation
B. Decompression and contamination
C. Disinfection and isolation
D. Inoculation and asepsis

27. The most susceptible to infection "at risk" patients are usually: 3:122

1. The very young 2. The very old 3. Those with poor hygiene

A. 1 only
B. 2 only
C. 3 only
D. 1, 2 & 3

28. When opening a sterile pack, the first fold of cloth or paper should be folded: 9:168

A. Away from the individual
B. To the right of the individual
C. To the left of the individual
D. Towards the individual

29. With patients getting in and out of wheelchairs and on and off x-ray tables, the radiographer should give: 9:54

A. Verbal assistance only
B. Help only when necessary
C. Help only when asked
D. Help at all times

30. A patient with a communicable disease should be separated from other patients and placed in a/an: 9:38

A. Protective isolation unit
B. Strict isolation unit
C. Intensive care unit
D. Burn unit

31. The proper use of body mechanics is primarily intended to prevent damage to the muscles and structures of the: 9:48

A. Upper extremities
B. Lower extremities
C. Synarthritic joints
D. Spinal column

32. After a portable machine has been used in a strict isolation unit, care must be taken to disinfect the ____________ after leaving the room. 9:38

1. Cassette 2. Mobile unit 3. Radiographer

A. 1 only
B. 2 only
C. 3 only
D. 1, 2 & 3

33. Which of the following items in the O.R. are usually considered free of all micro-organisms? 9:164

1. Surgical instruments *2. O.R. table* *3. Suction device*

A. 1 only
B. 2 only
C. 3 only
D. 1, 2 & 3

34. The best method to prevent the spread of droplet infections is by: 9:44

1. Hand washing *2. Wearing a mask* *3. Using gloves*

A. 1 only
B. 2 only
C. 3 only
D. 1, 2 & 3

35. According to the Center for Disease Control, the Acquired Immunodeficiency Syndrome (AIDS) requires that ____________ isolation technique be employed. 9:38

A. Strict
B. Respiratory
C. Enteric
D. Blood/body fluid

36. In a system of protective precautions (reverse isolation), two persons are required. The "clean" member is responsible for: 9:38

1. Patient positioning *2. Positioning of the tube head* *3. Making the exposure*

A. 1 only
B. 2 only
C. 3 only
D. 1, 2 & 3

37. The contaminated linens from an isolation patient should be placed in a plastic bag and: 9:40

A. Disposed of in the usual manner
B. Marked and sent to the laundry
C. Maintained in the isolation unit for disposal
D. Any of the above.

38. Because of the availability of anti-infective agents and good medical asepsis, hospital infections are on the decline. 9:23

A. True
B. False

39. A common unit used for high pressure steam sterilization is called the: 3:125

A. Thermal ventilator
B. Autoclave
C. Vector chamber
D. Inhalator

40. The cycle of infection can be broken by the: 3:101

1. Destruction of the infectious organism *2. Removal of the source of infection* *3. Preventing the means of transmission*

A. 1 & 2 only
B. 1 & 3 only
C. 2 & 3 only
D. 1, 2 & 3

41. Which of the following is normally classified as a type of vehicle spread infection? 9:29

1. Mosquito bite *2. Contaminated blood transfusion* *3. Coughing*

A. 1 only
B. 2 only
C. 1 & 3 only
D. 1, 2 & 3

42. A common disease caused by a vector is: 3:105

1. Rabies *2. Bubonic plague* *3. Lymes disease*

A. 1 & 2 only
B. 1 & 3 only
C. 2 & 3 only
D. 1, 2 & 3

43. Micro-organisms that cause infections can be classified as: 9:23

1. Fungi *2. Bacteria* *3. Vectors* *4. Isogens*

A. 1 & 2 only
B. 1 & 3 only
C. 2 &3 only
D. 1, 2 & 3

44. Pertusis, influenza and tuberculosis are three common disorders that require a patient to be placed in: 9:39

A. Enteric isolation
B. Protective isolation
C. Respiratory isolation
D. No isolation is required

45. Most strict isolation techniques are designed to: 9:39

A. Control the spread of infection
B. Speed the healing process
C. Protect the patient from disease
D. Control secondary infection

46. A chemical substance that is used to kill pathogenic bacteria on instruments and equipment is called a/an: *1. Germicide (disinfectant)* *2. Antiseptic solution* *3. Antibiotic* 1:6:1

A. 1 only
B. 2 only
C. 3 only
D. 1, 2 & 3

47. The viral agent that is responsible for acquired immunity deficiency syndrome (AIDS) is: 3:106

A. Human immune virus
B. Escherichia coli
C. Candida albicans
D. Herpes simplex 2

48. In order to prevent needle stick injuries to hospital staff, used needles should be: 3:113
1. Recapped after use *2. Bent or broken after use* *3. Placed in puncture-resistant containers*

A. 1 only
B. 2 only
C. 3 only
D. 1, 2 & 3

49. A patient with salmonella is placed in an enteric isolation room. All linens used should be: 9:39

A. Disposed of in the usual manner
B. Incinerated immediately
C. Double bagged and marked contaminated
D. Sterilized in an autoclave

50. The log-roll method of patient transfer should be used to turn patients with suspected or known injuries to the: 9:54
1. Spinal column *2. Pelvis* *3. Upper extremities*

A. 1 only
B. 2 only
C. 3 only
D. 1, 2 & 3

51. Which of the following areas will generally require the use of protective isolation procedures? 3:122

A. Neonatal nursery
B. Tuberculosis ward
C. Recovery room
D. Pediatric ward

52. The flu, measles, herpes and hepatitis are some of the commen types of infections caused by ___________ agents: 9:26

A. Bacterial
B. Yeast
C. Viral
D. Fungal

53. During an intravenous infusion, a swelling develops near the injection site from a dislodged needle. The first action that should be taken is: 9:223

A. The application of a warm compress
B. Increase the rate of the infusion
C. Clamp off the tubing
D. Squeeze the IV bag

54. Providone-iodine and isopropyl alcohol are two common agents used to: 9:178

A. Sterilize surgical instruments
B. Lubricate enema tips
C. Disinfect the skin
D. Reduce skin temperature

55. A disease that is usually transmitted by accidental blood or needle stick injury is: 3:106

1. Acquired immunodeficiency syndrome *2. Hepatitis B* *3. Creutzfeldt-Jokob disease*

A. 1 only
B. 2 only
C. 3 only
D. 1, 2 & 3

The Medical Emergency

Chapter 55

1. The process of identifying and assigning priorities to a patient for emergency care is termed: 3:189

A. Sequencing
B. Scheduling
C. Triage
D. Prophylaxis

2. A patient's request contains contradictory information, i.e. right hand is injured but request asks for left hand. The radiographer should:

A. Do the ordered exam without question.
B. Do the ordered exam and opposite hand, also
C. Call the ordering physician or station and get proper information
D. Perform the exam you feel is appropriate

3. The patient has the right to: 9:6

A. Refuse medical treatment
B. Confidential treatment of his records
C. Know the qualifications of the medical personnel
D. All of the above

4. Prognosis is defined as: 8:1165

A. The science of pathology
B. Forecast of the probable result of disease
C. Method of examination
D. Detailed program of treatment

5. A simple technique used to dislodge foreign bodies from the respiratory system is called the: 3:174

A. Heimlich maneuver
B. Syncope technique
C. Cardiopulmonary resuscitation
D. Cardiac arrest

6. A type of shock occurring in response to a massive buildup of bacteria and their toxic byproducts is termed: 3:176

A. Septic shock
B. Neurogenic shock
C. Hypovolemic shock
D. Allergic shock

7. The major sign or symptom associated with cardiac failure is: 9:98

1. Apnea *2. Unconsciousness* *3. Weak or absence of a pulse*

A. 1 only
B. 2 only
C. 3 only
D. 1, 2 & 3

8. The common indicators of shock generally include: 9:91

1. Hypotension *2. Tachycardia* *3. Convulsive seizures*

A. 1 & 2 only
B. 1 & 3 only
C. 2 & 3 only
D. 1, 2 & 3

9. A comminuted fracture is best described as a: 3:180

A. Displaced fracture
B. Compound fracture
C. Fragmented fracture
D. Simple fracture

10. The back, chest and abdominal thrust maneuvers can be of assistance in the relief of: 9:101

A. Convulsive seizures
B. Anaphylactic shock
C. Airway obstructions
D. Cardiac tamponade

11. During cardiopulmonary resuscitation, the cardiac compression rate should be about: 9:100

A. 10-20 compressions/min.
B. 20-30 compressions/min.
C. 30-40 compressions/min.
D. 40-60 compressions/min.

12. When a patient is vomiting, the head is lifted or turned to the side to prevent: 3:183

A. Abdominal cramps
B. Dehydration syndrome
C. Drowsiness
D. Aspiration

13. A strong indication of a low oxygen content in the tissues of the body is: 3:85

A. High body temperature
B. Cyanosis
C. Eclampsia
D. Diaphoresis

14. The failure to metabolize glucose may result in a condition termed: 9:96

A. Antisepsis
B. Septicemia
C. Ketoacidosis
D. Anaphylaxis

15. The control of bleeding in the extremities is often helped by: 3:182

1. Direct pressure over the wound *2. Raising the extremity* *3. Cold compresses*

A. 1 only
B. 2 only
C. 3 only
D. 1, 2 & 3

16. A young patient is radiographed and shows an incomplete fracture of the radius. This is often classified as a/an: 3:180

A. Epiphyseal fracture
B. Greenstick fracture
C. Compound fracture
D. Comminuted fracture

17. Cardiopulmonary resuscitation should be started immediately on patients with: 3:173

A. Hypovolemic shock
B. Grand mal seizures
C. Cardiac failure
D. Neurogenic shock

18. During a convulsive seizure, the most important action to be taken by a radiographer is: 9:103

1. Prevention of patient injury from falling *2. Inserting the fingers into the mouth*
3. Administer some type of sugar immediately

A. 1 only
B. 2 only
C. 3 only
D. 1, 2 & 3

19. Fainting (syncope) is caused by an insufficient blood supply to the brain. A patient experiencing this sensation should be instructed to: 9:103

A. Stand quietly
B. Stand and walk
C. Sit in a chair
D. Lay supine

20. During cardiopulmonary resuscitation, the rate of cardiac compressions to assisted respirations should normally be about: 9:100

A. 60:1
B. 30:1
C. 315:1
D. 4:1

21. A medication that increases urine output is called a/an: 3:136

A. Narcotic
B. Diuretic
C. Vaso dilator
D. Antipyretic

22. A fracture that causes an external wound because of protruding fragments is classified as a/an: 3:180

A. Simple fracture
B. Compound fracture
C. Incomplete fracture
D. Comminuted fracture

23. Anaphylactic shock is usually preceded by: 9:93

A. Profuse bleeding
B. Systemic infection
C. Drug administration
D. Cardiac failure

24. During cardiopulmonary resuscitation, the external cardiac massage is started by placing the heel of the hand 1 1/2" above the: 9:99

A. Manubrium
B. Gladiolus
C. Xiphoid
D. Solar plexus

25. Rigid muscles and jerky body movements are often seen in a patient experiencing (a/an): 9:103

A. Heart attack
B. Shock
C. Convulsive seizure
D. Airway obstruction

26. The human brain can survive without oxygen for about: 9:97

A. 2-4 minutes
B. 7-9 minutes
C. 11-14 minutes
D. 15-20 minutes

27. When a patient with a fractured extremity is moved, special care must be taken to support the site ____________ to the injury. 3:180

A. Distal
B. Proximal
C. Both of the above
D. Neither of the above

28. Constriction of the blood vessels going to the heart often leads to a painful sensation of the chest called angina pectoris. This is often relieved by administration of: 8:97

A. Heprin
B. Spirits of ammonia
C. Benedril
D. Nitroglycerin

29. In many emergency situations, a rapid systemic response to the medication is necessary. The route that will in most cases provide this is: 3:143

A. Parenteral
B. Intravenous
C. Oral
D. Intramuscular

30. Uremia is caused by excessive buildup of the byproducts of protein metabolism that is often associated with: 8:1943

A. Gastric disorders
B. Kidney disorders
C. Respiratory disorders
D. Circulatory disorders

31. A patient is brought into the radiographic room and during the course of the examination, he/she begins to bleed profusely from a surface wound of the arm. Before help arrives, the following assistance should be given: 8:647

A. Lower the arm to a level below the heart
B. Compress an artery distal to the injury
C. Apply direct pressure to area of the bleeding
D. Place the patient in a sitting or upright position

32. Convulsive seizures may be associated with: 9:103

1. Hyperthermia *2. Epilepsy* *3. High intercranial pressure*

A. 1 & 2 only
B. 1 & 3 only
C. 2 & 3 only
D. 1, 2 & 3

33. A diabetic has taken an insulin injection but is NPO for a G.I. series examination. You note excessive sweating and dizziness. This person may be experiencing: 9:95

A. A hypoglycemic reaction
B. A hyperglycemic reaction
C. Ketoacidosis
D. Cardiac failure

34. A common allergic reaction to iodine contrast media is bronchial constriction. This is often treated with a/an: 9:204

A. Vaso dilator
B. Vaso constrictor
C. Antihistamine
D. Antidiuretic

35. A patient comes into the X-ray Department with a fractured rib which has punctured the lung. This is an example of a: 3:133

A. Compound fracture
B. Complicated fracture
C. Comminuted fracture
D. Greenstick fracture

36. The dizziness associated with sitting or standing after prolonged bed rest is called: 3:185

A. Vertigo
B. Decubitus seizure
C. Postural hypotension
D. Dehiscence

37. A patient has been in a car accident and is being radiographed. Without warning, he begins to projectile vomit. This should be reported to a nurse or physician because it is often associated with: 3:178

A. High intercranial pressure
B. Internal bleeding
C. Respiratory distress
D. Cardiac tamponade

38. Severe respiratory dysfunction often results from: 9:97

1. Grand mal seizure *2. Airway obstruction* *3. Ketoacidosis*

A. 1 only
B. 2 only
C. 3 only
D. 1, 2 & 3

39. A patient is brought to the Radiography Department from a car accident with a weak, rapid pulse, low blood pressure, and feels cold and lacks color. The patient is showing the signs of: 9:91

A. Tetanus
B. Shock
C. Cardiovascular disease
D. Epilepsy

40. A patient with a pre-existing disease, such as osteomalacia, develops a non-trauma fracture. The term for this is a/an: 8:557

A. Greenstick fracture
B. Incomplete fracture
C. Pathologic fracture
D. Colles fracture

41. A patient that is said to be experiencing dyspnea has: 3:173

A. Breathing difficulties
B. High blood pressure
C. A fever
D. A urinary problem

42. A car accident victim has lost a large amount of blood, and while performing the radiographs, you see his condition worsen. The patient is most likely lapsing into: 9:91

A. Septic shock
B. Hypovolemic shock
C. Grand mal seizure
D. Anaphylactic shock

43. Angina pectoris is a common indication of __________ disease. 3:142

A. Heart
B. Lung
C. Thyroid
D. Kidney

44. The most important consideration in the care of unconscious patients is to insure that: 3:178

1. Fluids are maintained *2. Airways remain clear* *3. No medications are given*

A. 1 only
B. 2 only
C. 3 only
D. 1, 2 & 3

45. A sharp turning of the ankle often results in a fracture of the distal fibula called a/an: 8:557

A. Evulsion fracture
B. Pott's fracture
C. Slipped epiphysis
D. Depressed fracture

46.-50. In the following, match the medications with their principal action. 9:204

46. ____ Penicillin, tetracycline A. Analgesics

47. ____ Aminophylline, epinephrine B. Sedatives

48. ____ Aspirin, codeine C. Bronchodilators

49. ____ Phenobarbital, ethanol D. Antianxiety

50. ____ Vallium, thorazine E. Antibiotic

Contrast Agents

Chapter 56

1. Before a contrast media is injected, the patient should be questioned about his/her: 3:160

1. History of allergies *2. Alcohol consumption* *3. Smoking habits*

A. 1 only
B. 2 only
C. 3 only
D. 1, 2 & 3

2. One of the major danger signs after administration of iodine contrast media is: 3:154

A. Nausea
B. Urticaria
C. Dysphagia
D. Sneezing

3. When iodinated contrast media is employed, the maximum kilovoltage that should be used is: 6:428

A. 60 kVp
B. 80 kVp
C. 95 kVp
D. 110 kVp

4. The principal route of elimination of most aqueous iodine contrast agents is the: 3:215

A. G.I. tract
B. Respiratory system
C. Biliary tract
D. Urinary tract

5. The major property that determines the x-ray attenuation properties of the contrast media is its: 7:66

A. Atomic number
B. Valence
C. Viscosity
D. Concentration

6. The resistance of a fluid to change form or position due to the molecular cohesion is termed: 8:1558

A. Viscosity
B. Miscibility
C. Toxicity
D. Osmolality

7. A severe allergic reaction to a chemical substance (contrast media) may result in: 7:67

A. Homeostatic shock
B. Anaphylactic shock
C. Histotoxemia
D. Toxonosis

8. A serious drug reaction when water-based iodine contrast is administered intravenously occurs in: 3:158

A. Less than 1% of the patients
B. About 2% of the patients
C. About 30% of the patients
D. About 50% of the patients

9. Which of the following contrast media is classified as a type of colloidal suspension? 3:198

A. Sodium diatrizoate
B. Iopanoic acid
C. Nitrous oxide
D. Barium sulfate

10. When a low atomic number material is used to demonstrate the density difference between two structures, it is termed a ____________ contrast media. 7:66

A. Negative
B. Positive
C. Saline
D. Hypotonic

11. The type of contrast media most commonly employed for intravenous administration is: 9:210

A. Water-based barium solutions
B. Water-based iodine solutions
C. Oil-based iodine solutions
D. Gaseous solutions

12. The type and severity of complications associated with contrast media is related to: 9:212

1. Route of administration *2. Amount used* *3. Patient's sensitivity*

A. 1 only
B. 2 only
C. 3 only
D. 1, 2 & 3

13. The extent to which a material is found to be poisonous is termed: 3:218

A. Toxicity
B. Miscibility
C. Contraindication
D. Persistence

14. A serious complication often associated with intravenous administration of iodinated contrast media is: 9:211

1. Renal failure *2. Hypotension* *3. Impactions* *4. Dizziness*

A. 1 & 2 only
B. 1 & 3 only
C. 2 & 3 only
D. 1, 2 & 3

15. Iodized oil, such as Pantopaque, is most often employed in structures that are extremely sensitive to: 1:100:III

A. Blood clotting
B. Dehydration
C. Irritation
D. Infection

16. The use of sodium diatrizoate (Hypaque) is not considered safe for ________ injections.

A. Intravenous
B. Intramuscular
C. Intrathecal
D. Intra-arterial

17. A contrast media that absorbs more radiation than the organ in which it is placed is termed a: 9:210

A. Negative agent
B. Positive agent
C. Suspension agent
D. Miscible agent

18. An allergic reaction to iodinated contrast media is more frequent in patients known to have: 7:69

1. Diabetes mellitus *2. Multiple myeloma* *3. Asthma*

A. 1 & 2 only
B. 1 & 3 only
C. 2 & 3 only
D. 1, 2 & 3

19. The signs of flushing, sneezing and urticaria from an iodine contast media injection are some of the warning signs of: 7:69

A. Prophylaxis
B. Iodine coma
C. Extravasation
D. Anaphylaxis

20. An intravenous urogram patient has an anaphylactic shock reaction. You pull the emergency tray out. The drug that is most likely to be administered first is: 3:135

A. Heparin
B. Epinephrine
C. Antibiotics
D. Atropine

21. The introduction of a small dose of contrast media (sensitivity testing) before full dosage is given has proven to be ____________ in predicting reactions.

A. Totally reliable
B. Very reliable
C. Somewhat reliable
D. Very unreliable

22. The most commonly employed contrast media for intravenous cholangiography is: 3:713

A. Iopanoic acid
B. Iodipamide meglumine
C. Iophendylate
D. Diatrizoate meglumine

23. Which of the following contrast media is classified as a negative contrast media? 3:212

A. Room air
B. Iopanoic acid
C. Sodium diatrizoate
D. Iodipamide meglumine

24. Barium sulfate suspensions are nearly totally safe if they remain within the G.I. tract. Their toxicity increases dramatically if spillage occurs into the: 9:131

1. Duodenum *2. Peritoneum* *3. Circulatory system*

A. 1 & 2 only
B. 1 & 3 only
C. 2 & 3 only
D. 1, 2 & 3

25. In infants with swallowing disorders and adults with suspected bronchoesophageal fistulas, it is often prudent to select a/an ____________ contrast media. 3:200

A. Water-soluble iodine
B. Water-based barium
C. Oil-based iodine
D. None of the above

26. Iopanoic acid, sodium thyropanoate, and sodium ipodate are among the most common agents used for: 3:213

A. Oral cholecystography
B. Intravenous urography
C. Angiography
D. Myelography

27.-35. Match the radiographic procedure with the class of contrast media most often selected to improve visualization. 3:213

27. _____ Arthrography
28. _____ Sialography
29. _____ Esophography
30. _____ Cholangiography
31. _____ Urography
32. _____ Mammography
33. _____ Hysterosalpingography
34. _____ Cholecystography
35. _____ Venography

A. Ionic water-based iodine
B. Barium sulfate
C. Oil-based iodine
D. No contrast media

36. The principle advantage of a low osmolality non-ionic contrast agent as its decreased: 3:218

A. Opacity
B. Toxicity
C. Miscibility
D. Viscosity

37. A patient is to be scheduled for a barium enema, intravenous urogram and a gastrointestinal series. The proper sequence for these exams is: 3:191

A. IVP, BE, GI
B. BE, IVP, GI
C. GI, BE, IVP
D. IVP, GI, BE

38. As the concentration of an iodine contrast media is increased, and increase in __________ should be anticipated. 3:218
1. Viscosity 2. Toxicity 3. Opacity

A. 1 & 2 only
B. 1 & 3 only
C. 2 & 3 only
D. 1, 2 & 3

39. A patient is to be scheduled for the following series of radiographic examinations. The proper sequence for these exams from first to last is: 3:191
1. GI series 2. Gallbladder 3. Lumbar spine

A. 3, 2, 1
B. 1, 2, 3
C. 2, 3, 1
D. 1, 3, 2

40. Because of the possibility of residual contrast media which of the following studies is normally performed last in a sequence of exams. 3:191

A. Esophagram
B. Barium enema
C. Intravenous urography
D. Gallbladder series

41. A 4 year old is brought down to the Radiology Department for an intravenous urogram. The Radiologist orders 40% of the normal 50cc adult dosage be used. The amount to be prepared is:

A. 45cc
B. 38cc
C. 30cc
D. 20cc

42. A barium enema is to be performed on a patient suspected of having a perforated bowel. The contrast media preferred for this procedure would be a/an: 3:200

A. High density barium sulfate preparation
B. Low density barium sulfate preparation
C. Water-based iodine preparation
D. Oil-based iodine preparation

43. A low osmolality contrast agent is able to deliver a relatively ____________ concentration of iodine with ____________ particles in solution than conventional ionic contrast agents. 3:218

A. High, fewer
B. Low, more
C. High, more
D. low, fewer

44. During radiographic exams requiring long intravenous infusions, a/an ____________ is often preferred over typical intravenous needles. 9:218

A. Angiographic catheter
B. Butterfly set
C. Seldinger technique
D. Transdermal injection

45. Most of the new low osmolality contrast agents which are classified as multipurpose can be employed for: 3:218
1. Arthrography 2. Angiography 3. Urography

A. 1 only
B. 2 only
C. 3 only
D. 1, 2 & 3

46. A commonly employed category of drugs that are used to promote defecation in patients are termed: 8:305

A. Emetics
B. Cathartics
C. Diuretics
D. Stimulants

47. The principle reason that iopamidol and iohexal are now beginning to replace melglamine salts as the primary agents for intravenous urography is their: 3:218

1. Lowered iodide content *2. Lower toxicity* *3. Lower cost*

A. 1 only
B. 2 only
C. 3 only
D. 1, 2 & 3

48. In order to prevent possible damage to the rectum, the enema tip should never be inserted more than ________ into the rectum: 1:125:II

A. 1" (2.5cm)
B. 2" (5cm)
C. 4" (10cm)
D. None of the above

49. Which of the following is a major compound used in the formulation of a low osmolality (non-ionic) contrast agent? 2:218

1. Iohexol *2. Iopamidol* *3. Sodium meglumin*

A. 1 & 2 only
B. 1 & 3 only
C. 2 & 3 only
D. 1, 2 & 3

50. A serious complication following the administration of an iodine contrast agent is development of a moving blood clot that obstructs the blood supply to the lung. This is called a ________. 8:575

A. Phlebolith
B. Pulmonary edema
C. Pulmonary embolus
D. Pericardial effusion

51. The contrast agent that does not dissociate into charged particles when placed in a solution is designated as a/an ________ contrast agent. 3:218

A. High osmolar
B. Ionic
C. Anaphylactic
D. Non-ionic

52. The cleansing of the bowel after a barium enema to avoid barium impactions can be accomplished by the use of: 9:208

1. Milk of Magnesia *2. Benadryl* *3. Castor oil*

A. 1 & 2 only
B. 1 & 3 only
C. 2 & 3 only
D. 1, 2 & 3

53. A common medication given for relief of bronchial spasm which occasionally accompanies iodine contrast injections is: 9:206

A. Nitroglycerin
B. Heprin
C. Benadryl
D. Spirits of ammonia

54. A male infant is brought to the department with a diagnosis of pyloric stenosis. The examination that will most likely to confirm this condition is a/an: 8:1533

A. Intravenous urogram
B. Cholangiography
C. Gastrointestinal series
D. Barium enema

55. The types of contrast media that can be employed for use in the subarachnoid space of the spinal canal are: 3:233

1. Oil based iodides *2. Ionic water-based iodides* *3. Non-ionic water-based iodides*

A. 1 & 2 only
B. 1 & 3 only
C. 2 & 3 only
D. 1, 2 & 3

BIBLIOGRAPHY Sections V, VI & VII

Referencing Questions To the right of each question is a multidigit number, eg. 2:342. The first number indicates the reference text, after the colon is the page number for the information. An additional Roman numeral follows the number indicating a volume number when applicable.

1. Ballinger P.W., Merrill's Atlas of Radiographic Positions and Radiologic procedures, 8th Edition, C.V. Mosby Co., 1995
2. Chafee, E.E., Lytle, I.M., Basic Physiology and Anatomy, 4th Edition, J.B. Lippincott Co., 1980
3. Ehrich, R.A., Givens, E.M., Patient Care in Radiography, 4th Edition, C.V. Mosby Co., 1993
4. Hole, J.W., Human Anatomy and Physiology, 3rd Edition, W.C. Brown, 1984
5. Mallett, M., Anatomy and Physiology for Students of Medical Radiation Technology, 3rd Edition, Burnell Co., 1981
6. McInnes, J., Clark's Positioning in Radiography, 9th Edition, II Ford Limited, 1973
7. Snopek, A.M., Fundamentals of Special Radiographic Procedures, 3rd Edition, W.B. Saunders Co., 1992
8. Thomas, C., Taber's Cyuclopedic Medical Dictionary, 16 Edition, F.A. Davis Co., 1989
9. Torres, L.S., Morrill, C., Basic Medical Techniques and Patient Care for Radiologic Technologists, 4th Edition, J.B. Lippincott Co. 1993
10. Tortora, G.J., Principles of Anatomy and Physiology, 6th Edition, Harper and Row Publishers, 1990

Section VII
Patient Care & Management Explanations

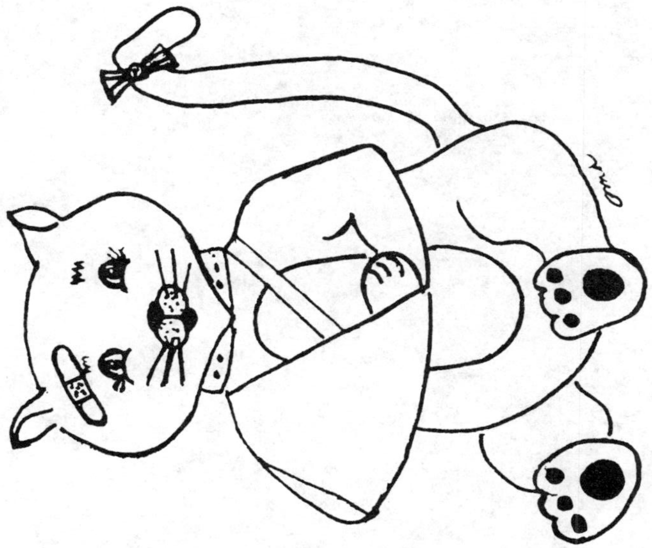

Vital Signs and Equipment Monitoring

Chapter 53

1. C — During the intravenous administration of fluids, a bag or bottle which is too low will slow the delivery rate. Raising the bag too much can increase the hydrostatic pressure and cause infiltrations into the tissues near the injection site.

2. D — A blood pressure cuff or sphygmomanometer consists of an inflatable cuff, a mercury or aneroid manometer and a pressure bulb for the inflation of the cuff. A stethoscope is used to listen for pulsations in the brachial artery.

3. C — In order to maintain the negative pressure required with a chest tube, it must remain below the level of the chest. Raising the unit will cause a positive pressure cause fluids to flow back into the chest cavity.

4. A — Because venous blood flows toward the heart a tourniquet is placed proximal to the injection site to increase the venous pressure and make the vein more accessible. The tourniquet is released after the penetration of the vein but prior to the injection of the medication to avoid the formation of a hematoma at the injection sight.

5. D — The number of heart beats (pulse rate) is most easily determined by detecting the pulsations felt in a superficial artery. The average adult rate of between 60-90 is dependent upon age, sex, anxiety, physical exertion, and many other physiological conditions.

6. D — Blood pressure is a measurement of the pressures associated with the contraction (systole) and relaxation (diastole) phases of the heart. Blood pressures above the normal range are referred to as hypertensive, while low blood pressures are called hypotensive.

7. C — A rapid heart rate called tachycardia may be indicative of low blood volume, anxiety, cardiac problems or any number of other physiological states.

8. C — The proper dosage is determined by dividing the unit dose per cc into the total dose desired.

9. B — Failure to partially fill the drip chamber below the IV bottle may result in an insufficient flow and the development of air bubbles in the tubing.

10. C — Severe injuries to the nose, mouth, pharynx, or larynx which obstruct the respiratory passages may require an artificial opening into the trachea just below the cricoid cartilage of the larynx called a tracheostomy.

11. D — These pressures fall within the normal adult range of 110-140 mm Hg systolic and 60-80 mm Hg diastolic pressures.

12. B — The failure of a needle to remain within a vein will cause the release or infiltration of the medication into the soft tissues. If this occurs, the infusion should be stopped immediately and a physician notified to administer additional treatment.

13. D — Though most patients prefer oral medications, not all drugs can be absorbed by the intestinal organs. Parental administrations, including subcutaneous, intramuscular, intravenous, and intradermal injections are used to introduce substances directly into the body.

14. D Physical signs such as skin color, sweating, and tremors can all be determined by careful observation of the patient.

15. C All devices which operate by maintaining a negative pressure (suction) or gravity, such as chest tubes and urinary catheters, should be placed below the level of the patient.

16. C Many tubes are available to give direct access to the stomach. Among the more common are the Levin, Nutriflex, Moss and sump tubes. These are used to aid the healing process by removing fluids and gas or for the introduction of foods, medication, or contrast agents. Visual inspection of the gastric mucosa can be accomplished through the use of an endoscope.

17. C In order to maintain the suction of a chest drainage tube as a patient is moved, the tubing must be free of kinks and tightly sealed in addition to keeping the unit below chest level.

18. A A respirator or ambu bag which is involved with the forcible introduction of respiratory gases into the body is classified as a part high flow oxygen system. Passive or low flow systems include masks, cannulas and oxygen tents.

19. B See explanation number 11 of this section.

20. B The injection sites for most subcutaneous medications are selected for their relatively low pain sensitivity and easy access. Among the more common sites are the upper arm, posterior and lateral thigh.

21. C In adults with fully developed muscles, intermuscular injections are performed into the deltoid and gluteus muscles with little complication. In children, a more lateral sight over the vastus lateralis muscle is preferred to avoid possible damage to the sciatic nerve.

22. A The normal range for a hypodermic needle is from about 14 gauge for the largest and 28 gauge for the smallest needle diameters.

23. C A pulse of 35 which is well below the normal 60-90 beats/min range is called bradycardia and may be associated with hypothermia shock. A fever or hemorrhaging will normally cause tachycardia or an increased heart rate.

24. A A chest drainage device is used to extract fluids or air from the pleural cavity.

25. D The natural response to an invading organism is frequently the development of a fever. The increased blood flow and higher body temperature creates a better environment for the elimination of the organism.

26. A During an injection into a vessel, leakage at the injection site is termed extravasation. This frequently painful condition is treated with warm compresses which helps to speed the dissipation of the medication into the surrounding tissues.

27. C A number of conditions can effect the normal respiration rate of 12-16 breaths/minute. Blood loss and shock will normally increase this rate while the administration of oxygen will often slow this rate.

28. C A collection of blood in the soft tissue (Hematoma) following the removal of an intravenous needle can usually be avoided by the removal of the tourniquet prior to the drug administration and 1 minute of direct pressure over the injection site.

29. A Because oxygen supports combustion, it can increase the potential for a fire. When oxygen is in use, smoking, open flames, and the use of ungrounded electrical equipment must be avoided.

30. B The normal flow rate of oxygen delivery is about 4-5 liters per minute. When a patient is transferred to a new oxygen source a radiographer should always maintain the rate that was previously set. Patients with emphysema normally receive about half of the normal rate.

31. A Inaccurate oral temperatures may result from a patient who has recently ingested hot or cold beverages and those on oxygen therapies.

32. C Normal body temperature varies depending upon the area from which it is taken. Rectal readings are highest at 99.6 F. (37.5 C.), oral readings of 98.6 F. (37 C.) are considered normal with a 97.6 F. (36.4 C.) being average for an axillary temperature.

33. B The distal end of a stethoscope usually consists of a dual side bladder or bell which transmits sound vibrations through flexible tubing to the bilateral ear pieces.

34. B A technique by which the body temperature is lowered to between 80-90 degrees F. called hyperthermia, is used to reduce persistent fever, or reduce the body's need for oxygen.

35. A The common sites for the determination of the pulse include the radial, carotid, femoral, popliteal, temporal, dorsalis pedis and brachial arteries.

36. B During the recording of the pulse, it is important to note both the rate and quality. The thumb, which has its own pulse, is not used because it may give erroneous readings. A comparison of the opposite extremity is only required in cases of injury or an interventional procedure which might compromise the blood flow to that area.

37. C The collection of fluids or mucus in the upper respiratory passages can be removed by negative pressure equipment. For most adult patients the usual pressure setting is 120 mm Hg.

38. B Personal injury law suits may involve 1) assault - the threat of touching a person in an injurious way, 2) battery - the unlawful touching of a person, 3) false imprisonment - detaining a patient against their will, 4) invasion of privacy - failure to maintain the patient's right to privacy, 5) negligence - failure to use reasonable care or caution.

39. D Obtaining the vital signs of a patient will require a sphygmomanometer and stethoscope for blood pressure reading, a watch for taking a pulse, and a thermometer for determination of the body temperature.

40. D Most radiology departments will retain films for 5-7 years based on the statute of limitations for most law suits. Radiographs of children are normally maintained until the child is 21 years of age. Mammographic films are also retained indefinitely so with comparison of previous images are possible.

41. A Dr. Elizabeth Kubler-Ross identified the five stages that a terminally ill patient will undergo. They are denial, anger, bargaining, depression and acceptance.

42. B When an immediate response to medication is desired, the intravenous route is most commonly employed.

43. D In order to ensure the proper identification and access to a patient's radiographs the following data is required: patient name and/or ID number, the date of the exam, appropriate side marker, and the institution's identity.

44. D The three most common nasoenteric tubes are the Cantor, Harris and Miller Abbott. Nasogastric tubes include the Levin, sump, and Sengstaken-Blakemore tubes. The Swan-Ganz catheter is a balloon tipped catheter used in pressure measurements of the pulmonary artery.

45. A — The motor control functions are related to the neuromuscular system i.e. movement and reflexes. Pain sensitivity, sight, sound, taste, and touch are classified as sensory in nature.

Control of Infection

Chapter 54

1. C — Reduction in the probability of infectious organism transmission to susceptible individuals is accomplished through proper medical asepsis which involves keeping the room clean and washing of the hands between patients.

2. B — Microbes are most commonly spread from one individual to another by the direct physical contact of two persons.

3. C — Touching, a form of direct contact, is the most common method by which infections are transmitted from one person to another.

4. C — Patients entering the hospital for one condition sometimes acquire secondary infections due to hospital environment which are called nosocomial infections.

5. B — The increased potential for the introduction of microorganisms when the skin is broken requires the use of the more strict surgical asepsis.

6. C — The sterility of a pack may be compromised by its exposure to moisture which can penetrate through the outer layers and bringing bacteria or viruses toward the interior of the pack.

7. C — A serious fracture which requires surgical intervention to accomplish proper alignment is termed an open reduction. The more usual non-surgical repair is called a simple or closed reduction.

8. A — Direct contact with infectious agents may occur from touching or sexual contact. Indirect agents for the spread of infection include fomites, vectors, air borne, food, and water contaminates.

9. D — In addition to any breakage of the skin, surgical asepsis should be practiced with any procedures which penetrate the urinary tract; i.e., catheterization and tracheostomy tubes.

10. A — Droplet or airborne contaminates associated with respiratory infections are substantially reduced by the use of surgical type masks and negative air flow systems.

11. A — When an animal serves as an intermediary in an infectious process, it is termed a vector. Mosquitoes that transmit malaria, fleas that carry bubonic plague, ticks that spread yellow fever or Lyme disease and many types of animal bites are types of vector spread infections.

12. C — The point around which the weight of the body is balanced, the center of gravity, is found in the mid pelvis region. The weight distribution in females results in a lower center of gravity compared to the males.

13. D — An isolation procedure that is designed to prevent a compromised patient's exposure to infectious agents is a form of strict isolation often called protective or reverse isolation.

14. C — In an operating room or other sterile area, the accidental contamination of the gown can be avoided by passing back to back to keep the sterile surfaces separated.

15. B — In all routine radiographic and patient care procedures the main component for medical asepsis is proper hand cleanliness.

16. C — Any inanimate object that can spread a pathogenic organism is termed a fomite. X-ray tables, cassettes, sponges, calipers, needles, and even pillow cases are potential carriers of infection.

17. D — The spread of infection may be caused by direct contact by touching or sex, or indirectly by air, fomites, vectors, droplets, or food and water.

18. B — Diseases are caused by a number of organisms. The smallest of these is a virus which injects its genetic instructions (DNA) into a host cell turning it into a virus factory.

19. C — The destruction of all micro-organisms can be accomplished by exposing pathogens to pressurized steam (autoclave) or gasses such as ethylene oxide.

20. A — Only a small portion above the waist in the front of the gown is considered sterile.

21. A — Due to their strength and lower potential for serious injury, the muscles of the extremities are preferred for lifting heavy objects.

22. D — Law suits may be brought against any caregiver or organization responsible for the patient. The most serious is medical malpractice which can include both inappropriate action or lack of action by the medical personnel.

23. D — Serious infections which can be easily transmitted by air or indirect contact will require the strict isolation of the patient in which masks, gowns, and gloves are worn. These diseases include diphtheria, pneumonic plague, chicken pox, smallpox, and staph infections.

24. C — The greater flexibility of the cervical and lumbar spines make them the more common sites for spinal injuries.

25. D — The destruction of all microorganisms and their spores, sterilization, can be accomplished by steam, dry heat, ionizing radiation, gasses, or chemical agents.

26. C — The most effective means for the control of a contagious disease is by a combination of an isolation technique and disinfection.

27. D — Three classes of individuals most likely to contract infections; the very young, the very old, and those with poor hygiene.

28. A — The first fold of a sterile pack is always unfolded away from the body so that all other folds can be made without leaning or reaching over the open sterile pack.

29. D — Wheel chairs and stretchers are inherently dangerous if used improperly. To avoid problems, direct physical contact and assistance with the patient is required at all times.

30. B — Strict isolation units are category related. Specific isolation categories include respiratory, enteric, blood and body fluids and are all employed to prevent specific types of infections.

31. D — During lifting or moving patients, the structures at greatest risk to injury are those of the spinal column. Good body mechanics are essential to help minimize potential injury to the medical workers.

32. D After completing a study in a strict isolation unit, all equipment and members of the care team must be disinfected prior to leaving the unit to avoid the spread of infection.

33. A Sterilization is only possible on relatively small items which can tolerate heat and chemical agents. Suction devices, OR tables, mobile x-ray units, and the patients themselves can only be disinfected, not sterilized.

34. B The spread of droplets or air borne contaminates is most easily prevented by the use of a mask. As with all patient contact, hand washing is also essential for all types of asepsis.

35. D Blood and body fluid precautions should be observed for AIDS, tick born infections, hepatitis B, hepatitis non A, non B, and malaria.

36. A During strict isolation techniques for a compromised patient, the clean member of the team is responsible for all contact with the patient.

37. C The disposal of contaminated linens or materials requires that a double wrapped clearly marked (red) bag be used for all materials.

38. B Hospital acquired (nosocomial) infections are unfortunately on the rise. This is due to a combination of a more aged population, the existence of more resistant bacterial strains, and an increase in the use of immunosuppressive drugs in organ transplant patients.

39. B The most convenient type of device for the sterilization of most types of reusable instruments is the pressurized steam device called an autoclave.

40. D The four factors included in the cycle of infection are the pathogenic organism, reservoir of infection, means of transmission and susceptible host. The infection is controllable when any link in the cycle is eliminated or controlled.

41. B Vehicle routes of transmission spread include water, food, medications, and contaminated blood.

42. D Animals or insects which carry and transmit infections are called vectors. Malaria, bubonic plague, Lyme disease, and rabies are all spread in this manner.

43. A The four major classes of microorganisms are viruses, bacteria, fungi, and protozoa.

44. C Because pertussis, influenza and TB are all spread by airborne means, the use of a respiratory isolation technique is required.

45. A Most isolation procedures are designed to protect the caregiver and reduce the spread of infection.

46. A The destruction of all pathogenic bacteria can be accomplished by the use of a disinfectant or germicide.

47. A AIDS is caused by the human immune virus, which unlike most infections, does not attack a specific organ, but instead infects the T-cells which are an essential component of the immune system.

48. C The recapping and breakage of needles, practiced up until the mid 80's, has been replaced with the less dangerous policy requiring their disposal into a puncture resistant container.

49. C All linens from a patient in enteric isolation should be double bagged using the appropriate tag or color coded bag.

50. A The log roll, requiring a five person team, is used to prevent further injury to a patient with neck or back trauma.

51. A Special techniques to prevent infecting compromised patients are routinely used for burn patients and for neonates.

52. C The smallest microorganism. the virus, is responsible for a number of disorders, including AIDS, hepatitis A & B, herpes, mumps, measles, the flu, and the common cold.

53. C The first action that should be taken when a needle infiltrates is the clamping of the IV tubing. A doctor or nurse should then be notified immediately.

54. C The disinfection of the skin can be accomplished by an antiseptic solution such as alcohol or povidone-iodine (Betadine).

55. D Accidental needle-sharp (needle stick) injuries are related to the transmission of AIDS, Hepatitis A & B, and Creutzildt-Jakob disease.

Medical Emergency

Chapter 55

1. C When medical services must be provided to a large number of individuals, patients are assigned according to the severity of their illness in a process called triage.

2. C Prior to performing a requested examination, a radiographer must review and confirm any questionable information contained on the requisition.

3. D The American Hospital Association affirms the following patient rights:

1) considerate care 2) complete information concerning their diagnosis and prognosis 3) informed consent 4) refusal to be treated 5) privacy 6) confidential treatment of their records 7) receive appropriate services 8) know the relationship which may exist between the caregiver and the institution 9) when and if experimental treatment is being employed 10) have any bills explained 11) expect continuity of care 12) know appropriate conduct expected of a patient.

4. B Diagnosis is the use of scientific methods to establish the cause of an illness. Prognosis is the forecast of the likely outcome of an illness.

5. A Improper ventilation caused by a foreign blockage of the throat can often be alleviated by the abdominal thrust or Heimlich maneuver in which arms are placed around the patient and rapidly squeezed to force expelled air from the compression of the diaphragm.

6. A Some bacteria produce harmful compounds called endotoxins which may be released into the blood stream. This causes symptoms which may include fever, chills, and in severe cases, septic shock.

7. D Cardiac arrest or heart failure is associated with a complete loss of blood circulation in the body resulting in unconsciousness and the absence of the pulse and respiratory functions.

8. A The physiological response to a trauma or injury is often a condition called shock. The symptoms of this serious disturbance which effects the blood flow to the vital organs ,include a weak rapid pulse, shallow breathing, low blood pressure, and a decreased perfusion to the vital organs and skin.

9. C A comminuted fracture, commonly associated with a severe blunt trauma, appears radiographically as a splintered or crushed area of a bone.

10. C Thrust maneuvers, such as the Heimlich procedure, are used in the relief of blocked respiratory passages.

11. D During cardiopulmonary resuscitation, the cardiac compression rates should be about one compression per second. The breathing rate is about five times slower to maintain 12-15 respirations per minute.

12. D A serious complication associated with an unconscious patient is the aspiration of vomitus which may result in an airway obstruction. This can be prevented by laying the patient on his/her side or stomach.

13. B A low oxygen content blood obtains a dark red tint when seen through the skin of the extremities, this results in a characteristic blue color (cyanosis) of the skin and in the nail beds of the fingers and toes.

14. C When insufficient insulin is present in the blood stream, glucose metabolism is reduced causing the mobilization of fatty acids resulting in an acidic state termed ketoacidosis.

15. D The control of bleeding in an extremity is assisted by lowering its blood pressure. This is most easily accomplished by raising the effected area above the level of the heart. The use of direct pressure over the bleeding sight and application of cold compresses to constrict the blood vessels are also useful.

16. B An incomplete fracture of a bone most often involving the more flexible bones of a developing child is called a greenstick fracture, due to its similarity to the appearance of a bent live tree branch.

17. C Cardiopulmonary resuscitation is essential in cases of cardiac arrest and respiratory failure with or without airway obstructions.

18. A During a seizure, little medical intervention is necessary or helpful, the radiographer's main responsibility is that of preventing secondary injury from falling. Trying to restrain the patient or taking other actions may cause more harm than letting the seizure run its few minute course.

19. D Fainting has many possible causes, such as heart irregularities, hunger, poor ventilation, fatigue, and emotional shock. Assisting the patient to a recumbent position or lowering the head to a level below that of the heart increases the blood flow to the brain and may prevent fainting in some cases. After fainting, if the patient does not rouse immediately, spirits of ammonia held under the nose will bring a rapid return to consciousness.

20. D The ratio between the normal heart rate (72) and normal respiration rate (18), the vital sign ratio, is useful in determining the correct compression to breathing rate to employ in cardiopulmonary resuscitation.

21. B The abnormal build up of fluids in the body's tissues can be reduced by the use of a diuretic such as furosemide (lasix).

22. B When external bone fragments are present, the fracture is described as compound. This type of injury is more prone to secondary infections such as osteomyelitis.

23. C The severe allergic reaction to a chemical agent; e.g., contrast media is classified as a type of anaphylactic reaction.

24. C External cardiac compression during CPR is started by placing the heel of the hand at a level of the midsternum thus avoiding possible injury to the xiphoid process located at the lower end of the sternum.

25. C Seizure disorders and their associated involuntary muscle contractions can be a result of epilepsy, internal cranial pressure, poisoning, uremia, and high fevers.

26. A During respiratory or cardiac failure it is critical to supply the brain with adequate oxygen in order to avoid brain damage which can occur within 2-5 minutes.

27. C All non-splinted fractures require support on both ends of the injured segment.

28. D A severe pain or constriction in the area of the heart usually radiating down the left arm, angina pectoris, is caused by a low cardiac blood supply and can often be relieved by the administration of a vasodilator such as nitroglycerin or amylnitrite.

29. B A direct injection into the circulatory system provides for the most rapid systemic response. The intravenous route is the simplest and least complicated procedure for entering the circulatory system.

30. B A severe toxic condition associated with renal failure or insufficiency results in a dangerous build up of nitrogenous by-products. Renal or hemodialysis is normally required to clean these toxins from the blood.

31. C Surface bleeding is most effectively treated with direct pressure over the wound site. When the veins are effected 1-2 minutes of pressure is sufficient, whereas for arterial bleeding 15-20 minutes of pressure may be required.

32. D Though epilepsy is by far the most common cause of seizures, high intracranial pressure, toxins, infection, fever, tetanus, uremia, and low oxygen levels can also trigger seizures.

33. A An inadequate diet following an insulin injection will result in a low blood sugar or hypoglycemic reaction. This serious condition requires medical treatment which should include the immediate administration of sugar to prevent the lapse of the patient into a coma.

34. C Bronchial constriction due to an allergic reaction is triggered by histamine production. The treatment most often prescribed is an antihistamine such as diphenhydramine hydrochloride (Benadryl).

35. B Any fracture which is associated with an injury to another organ or structure is termed a complicated fracture.

36. C Sometimes confused with fainting, postural hypotension results from blood pooling in the extremities while recumbent. Upon sitting, a transient cerebral anoxia may occur.

37. A Head trauma, depending on its severity, may result in minor symptoms to a full blown coma. Careful observation of a head injury patient is important. Symptoms of lethargy, irritability, lowered vital signs, and projectile vomiting may indicate the potential for high intracranial pressure or brain injury.

38. B Among the most common causes of respiratory dysfunctions are mechanical airway obstructions, chronic lung disease, and severe cardiovascular disorders.

39. B After any type of mental or physical trauma, a thready rapid pulse, low blood pressure, shallow respirations, and a reduced peripheral circulation are the classic signs associated with shock.

40. C Undiagnosed bone tumors, osteoporosis, and any number of other conditions which weaken the bone and make a fracture more likely are called pathological predisposers.

41. A The word root, pnea, for breathing is often combined with the common prefixes: dys - difficult, brady - slow, tachy - fast, and u - normal.

42. B The type of shock most frequently observed in individuals experiencing a severe blood or fluid loss (burns) is termed low volume or hypovolemic shock.

43. A The constriction of blood vessels around the heart can lead to a sharp pain over the chest radiating down the left arm called angina pectoris.

44. B All caregivers must be aware of the importance of breathing and the special care that must be taken to maintain clear airways on unconscious patients.

45. B A fracture of the lower end of the fibula and medial malleolus of the tibia with outward dislocation of the foot is termed a Pott's fracture.

See accompanying information.

46. E 47. C 48. A 49. B

50. D

Contrast Media

Chapter 56

1. A Contrast sensitivity is dependent upon a number of patient factors including general health, age, weight, a previous history of allergies, and prior contrast reactions which may predispose an individual to an allergic reaction.

2. C Among the more serious warning signs preceding a severe contrast reaction are dysphasia, laryngeal edema, and breathing difficulties.

3. B The dilution of an iodide contrast media by the blood and the relatively low concentrations achieved in the target organs will require the use of a 60-75 kVp range to avoid over penetration.

4. D Nearly all of the currently employed aqueous ionic and non-ionic contrast agents are eliminated by way of the kidneys.

5. A The degree of absorption or attenuation of a contrast agent is related to its atomic number. As the atomic number of a substance is increased its ability to absorb radiation also increases.

6. A — The rate of an injection is greatly effected by the ability of a liquid to flow, which is a measure of its viscosity. Highly concentrated contrast agents, as is often indicated by a higher number on the label of the media, are more viscous or thicker and less easily injected.

7. B — The most common type of allergic reactions seen in the radiology department are in response to iodinated contrast agents. Recently a number of cases of Latex reactions from enema retention tips have been reported.

8. A — Contrast reactions are reported in about 5% of all patients in which these agents are used. Serious or life threatening side effects occur in less than 1% of patients, with one death occurring in about every 40,000 patients.

9. D — The type of positive (radiopaque) contrast media used most frequently in the gastrointestinal tract is a white insoluble powdered compound called barium sulfate. When mixed with water it forms a type of colloidal suspension.

10. A — A number of gasses such as air, carbon dioxide, and nitrous oxide are employed as a negative or radiolucent contrast agents.

11. B — The only type of contrast media suitable for intravenous uses are the water based (aqueous) iodinated contrast agents. Barium sulfate and oil based iodinated compounds are not readily absorbed by the tissues and will cause embolic formations in the lungs if introduced into the blood stream.

12. D — The safety of a contrast media is related to the patient's physical condition, type and amount of the agent used, as well as the route by which the agent is introduced into the body.

13. A — The toxicity or extent to which media is poisonous is principally related to the concentration it achieves in the body.

14. A — The administration of iodinated contrast agents may cause a number of serious side effects including shock, coma, renal failure, respiratory failure, and cardiac irregularities.

15. C — Before the development of the less reactive non-ionic contrast agents, a major drawback of ionic media was the high degree of irritation they caused. When imaging of sensitive organs or cavities was required, such as the lungs or subarachnoid space, oil-based mixtures were employed to overcome this difficulty.

16. C — One of the first water-based ionic contrast agents called sodium diatrizoate (Hypague) was found to be unsuitable for intrathecal or subarachnoid space injections.

17. B — Iodine and barium sulfate, which both have a high atomic number, attenuate large amounts of x-radiation and are the most commonly used materials in positive contrast agents.

18. C — A number of pre-existing conditions may predispose an individual to contrast reactions. Among the most notable are multiple myeloma, sickle cell disease, asthma, and thyroid disorders.

19. D — An allergic reaction to contrast agents is called anaphylaxis and is classified as mild or severe;the most common mild reactions are nausea, vomiting, hives, and sneezing.

20. B — The most important medication for the relief of the symptoms associated with a severe contrast reaction is epinephrine. Also known as adrenalin, this drug relieves bronchospasms, raises the blood pressure and increases the heart rate.

21. D The use of a sensitivity test prior to contrast administration has proven to be very unreliable and is rarely used in a modern radiology department. Tests included 1/2 cc subdermal injections (skin pop), 1/2 cc IV, or a single drop placed in the eye. All tests were subject to a high false positive response rate.

22. B In order to demonstrate the biliary tract, the contrast media must be eliminated by the liver. Iodipamide (Cholografin) is carried to the liver in the blood stream and is rapidly secreted by the liver. Melglumine serves as the carrier salt.

23. A The most commonly used negative contrast agent is room air.

24. C Though barium sulfate is extremely safe, when it is introduced into the gastrointestinal tract it becomes highly toxic if a leakage occurs into the peritoneum (intestinal perforations) or if injected into the blood stream.

25. C An abnormal connection between the upper GI tract and the respiratory system, such as a fistula, may allow swallowed barium to collect in the lungs. In these situations, an oil based iodinated agent may be the contrast media of choice.

26. A Oral cholecystography requires a type of water based iodide which is absorbed by the GI tract and eliminated by the liver. Two most common compounds for this purpose are iopanoic acid (Telepague) or sodium iodate (Oragrafin).

See accompanying information.

27. A	28. C	29. B	30. A
31. A	32. D	33. A	34. A
35 A			

36. B Osmolality is the degree to which fluid will be moved in and out of a cell. Contrast agents possessing a low osmolality tend to cause fewer contrast reactions and therefore have a lower toxicity.

37. A Proper exam sequencing is beneficial in performing a series of tests in the least amount of time and without compromising subsequent exams due to retained media. General guidelines recommend that non-contrast exams be performed first, followed by iodide studies, then lower GI tract barium studies, and lastly upper GI tract barium studies.

38. D As the concentration or weight per unit volume of a contrast media is increased, the viscosity and opacity of the media also increases.

39. A See explanation number 37 of this section.

40. A The esophogram is performed last in a series of exams because it may take up to 2 days for the barium to be completely eliminated from the GI tract.

41. D The new dose is obtained by multiplying the adult dose by 40% (See explanation number 4).

42. C Because of the high toxicity of barium entering the peritoneum, a water based iodinated media such as a diatrizoate meglumine solution (Gastrografin) is strongly recommended.

43. A Since there is little movement of fluids when a low osmolar agent is used, a higher concentration with more particles in the solution can be introduced.

44. B In order to make a patient more comfortable during long intravenous infusions, a needle with flat wings (butterfly) or a flexible angio-catheter may be placed into the vein.

45. D A multi-purpose contrast agent, as its name implies, can be used for numerous exams including angiography, urography, arthrography, and myelography.

46. B One of the common ways to eliminate unwanted substances from the lower gastrointestinal tract is by the use of cathartics or laxatives.

47. B The anal canal, in which the enema tip is placed, extends into the rectum. To avoid injury, the maximum safe distance for the tube insertion is about 4 inches.

48. C The new general purpose non-ionic contrast agents, though about 10 times more expensive than the ionics they replace, have much lower toxicity.

49. A The vast majority of the water based non-ionic contrast media employ iohexal or iopamidol as their active components.

50. C One of the most serious complications associated with a contrast media reaction is the formation of a pulmonary embolus blood clot in the lung which can result in death in just a few minutes.

51. D The principle difference between ionic and non-ionic contrast media is related to the ability to form charged particles or ions. Non-ionics are chemically less reactive and therefore cause fewer harmful effects.

52. B Geriatric patients may be less active and are often given laxatives such as castor oil or milk of magnesia to help eliminate barium following examination of the gastrointestinal tract.

53. C Diphenhydramine (Benadryl) is commonly administered to block the action of histamine and relieve the upper respiratory constriction it may cause.

54. C The constriction of the end of the stomach, pyloric stenosis, is best evaluated during an upper gastrointestinal series.

55. B Until the advent of non-ionic contrast agents, subarachnoid or intrathecal injections were made with oil based iodinated contrast agents such as Pantopaque. Ionic water based iodides should never be employed in this region because they may cause meningits.

Order Form

				Quantity	X Amount	=Total
PAS Series Radiography	Full Series (Four Exams	Minimum order 4-10 sets	$36.00 per set			
		11-200 sets	$34.00 per set			
	Limited Series (Three Exams)	10-20 sets	$30.00 per set			
		21-200 sets	$28.00 per set			
	Limited Series (Two Exams	10-20 sets	$21.00 per set			
		21-200 sets	$19.00 per set			
	Limited Series (One Exam)	10-20 sets	$13.00 each			
		21-200 sets	$12.00 each			
	Grading System Per Exam		$7.50 each			
	Individual Student Orders		$16.95 each			
Mammography Simulated Exam			$16.95 each			
Cardiovascular Simulated Exams			$21.95 each			
Computed Tomography Simulated Exams with Explanations			$24.95 each			
Magnetic resonance Imaging Exam with Explanations			$16.95 each			
NEW The Integrated Radiography Workbook			$44.95 each			
Computer Simulated Registry Exams	Exam A	31/2 or 5 1/4	$79.99 each			
	Exam B	31/2 or 5 1/4	$79.99 each			
	Additional DIskettes	31/2 or 5 1/4	$3.00 each			
NEW	Easy Six Pack		$18.95 each			

Name ____________________ Total __________

Address ____________________ Shipping +$4.00

City __________ State __________ Zip __________ Total Order __________

Make checks payable to H.E.A.L.S. Orders totaling over $100.00 will be billed separately for shipping

HEALTH AND ALLIED SCIENCE PUBLISHERS

PO Box 6036 • Rutland, VT 05702 • (802) 773-4444